Frontiers in numerical relativity

Frontiers in numerical relativity

Edited by

CHARLES R. EVANS

Department of Theoretical Astrophysics, California Institute of Technology

LEE S. FINN

Center for Radiophysics and Space Research, Cornell University

DAVID W. HOBILL

National Center for Supercomputing Applications, University of Illinois

CAMBRIDGE UNIVERSITY PRESS

Cambridge

New York New Rochelle

Melbourne Sydney

CAMBRIDGE UNIVERSITY PRESS
Cambridge, New York, Melbourne, Madrid, Cape Town,
Singapore, São Paulo, Delhi, Tokyo, Mexico City

Cambridge University Press
The Edinburgh Building, Cambridge CB2 8RU, UK

Published in the United States of America by Cambridge University Press, New York

www.cambridge.org
Information on this title: www.cambridge.org/9780521115957

First published 1989
First paperback edition 2011

A catalogue record for this publication is available from the British Library

ISBN 978-0-521-36666-3 Hardback
ISBN 978-0-521-11595-7 Paperback

CONTENTS

Preface

During the week of 8 May 1988, an international workshop devoted to research in numerical relativity convened at the University of Illinois in Urbana-Champaign. For the five days of the meeting, representatives of the numerical relativity community from four continents met to discuss the most recent advances in, and the future directions of, computational general relativity. The name of the workshop, and the title of this volume, *Frontiers in Numerical Relativity,* reflects the discipline's march toward the farthermost limits of knowledge.

As organizers of the workshop, we wish first and foremost to thank the 53 participants whose enthusiastic participation made our efforts worthwhile. As editors, we owe a special thanks to the invited speakers, whose written contributions comprise this volume and whose careful preparation of their articles in camera-ready form allowed for the timely publication of these proceedings.

Frontiers in Numerical Relativity was funded by the National Center for Supercomputing Applications at the University of Illinois, and we are grateful for their generous and reliable support. The NCSA staff, and particularly the staff of the Interdisciplinary Research Center, provided technical support above and beyond the call of duty for both the organizers and the participants of the conference. Most importantly, however, we are indebted to the Director of NCSA, Larry Smarr, for his enthusiastic endorsement of *Frontiers,* and his encouragement and moral support during the year of organizational effort leading up to the meeting.

We also extend our thanks to the Office of Conferences and Institutes at the University of Illinois, which acted as liaison between the organizers, the University, and the staff at Jumers' Castle Lodge (where the conference sessions were held). In particular, we are grateful for the assistance of Roy Roper, Joan Cornell and Keri Blackwelder of C&I: their efforts before, during, and after the workshop helped contribute to its success.

Finally we thank our home institutions — the California Institute of Technology, Cornell University, and the University of Illinois — for their support during the organization of the meeting, and the Institute for Theoretical Physics at the University of California, Santa Barbara, for hospitality during the final preparation of these proceedings.

Charles R. Evans
L. Samuel Finn
David W. Hobill

Santa Barbara, California
September, 1988

Participants

Abrahams, Mr. Andrew
University of Illinois

Anderson, Professor James L.
Stevens Institute of Technology

Anninos, Mr. Peter
Drexel University

Bernstein, Mr. David
National Center for Supercomputing Applications

Bishop, Professor Nigel T.
University of Witwatersand

Bonazzola, Dr. Silvano
Observatoire de Paris, Meudon

Bowen, Professor Jeffrey M.
Bucknell University

Centrella, Professor Joan M.
Drexel University

Choptuik, Dr. Matthew
Cornell University

Clancy, Dr. Sean
Los Alamos National Laboratory

Cook, Mr. Gregory
University of North Carolina

Detweiler, Professor Steven
University of Florida

Duncan, Dr. Comer
Bowling Green State University

Eardley, Professor Doug
University of California, Santa Barbara

Evans, Dr. Charles R.
California Institute of Technology

Ferrell, Mr. Robert
University of California, Santa Barbara

Fette, Dr. William
Penn State University, McKeesport

Finn, Dr. L. Samuel
Cornell University

Goldwirth, Ms. Dalia
Hebrew University

Gomez, Mr. Roberto
University of Pittsburgh

Hawley, Professor John F.
University of Virginia

Hobill, Dr. David W.
National Center for Supercomputing Applications

Kochanek, Mr. Christopher S.
California Institute of Technology

Kojima, Dr. Yasufumi
Kyoto University

Kurki-Suonio, Dr. Hannu
Drexel University

Laguna-Castillo, Dr. Pablo
University of Texas at Austin

LeBlanc, Dr. James
Lawrence Livermore National Laboratory

Mathews, Dr. Grant J.
Lawrence Livermore National Laboratory

Mann, Dr. Patrick
University of Western Ontario

Marck, Dr. Jean-Alain
Observatoire de Paris, Meudon

Matzner, Professor Richard A.
University of Texas at Austin

Mezzacappa, Mr. Anthony
University of Texas at Austin

Moore, Dr. Thomas
Pomona College

Nakamura, Professor Takashi
Kyoto University

Oohara, Dr. Ken-ichi
National Lab. for High Energy Physics (KEK)

Ove, Dr. Roger
National Center for Supercomputing Applications

Piran, Dr. Tsvi
Hebrew University

Reilly, Mr. Paul
University of Pittsburg

Rauber, Professor Joel D.
South Dakota State University

Schinder, Dr. Paul
University of Pennsylvania

Schutz, Professor Bernard F.
University College, Cardiff

Seidel, Dr. Edward
Washington University

Shapiro, Professor Stuart
Cornell University

Smarr, Professor Larry
National Center for Supercomputing Applications

Stark, Dr. Richard F.
University of California, Santa Barbara

Sun, Mr. Shang Wen
Stevens Institute of Technology

Teukolsky, Professor Saul
Cornell University

Thornburg, Mr. Jonathan
University of British Columbia

Tuckey, Mr. Philip
Cambridge University

Wilson, Dr. James R.
Lawrence Livermore National Laboratory

Winicour, Professor Jeffrey
University of Pittsburgh

York, Professor James
University of North Carolina

INTRODUCTION

This book presents a review of the latest developments in numerical relativity and highlights the probable future directions of progress in the field. Numerical relativity involves the application of advanced computer technology and modern computational algorithms to study outstanding problems in gravitation. Over the last twenty years it has become apparent that many of the important problems in classical relativity theory cannot be addressed exclusively by analytic means. As an example, consider the gravitational radiation emission from the coalescence of a binary star system. Those who wish to solve important problems such as these are necessarily driven to numerical methods.

This volume can be considered the third in an informal series dealing with numerical relativity published by Cambridge University Press. The first, **Sources of Gravitational Radiation** (1979), edited by Larry Smarr, appeared a decade ago and chronicled the proceedings of a workshop held at the Battelle Seattle Research Center in 1978. That meeting was pivotal in establishing the viability of computational techniques to attack problems in general relativity that lie beyond the scope of analytic means. The second volume in the series, **Numerical Relativity and Dynamical Spacetimes** (1986), edited by Joan Centrella and held at Drexel University, reflected the state of the field in the fall of 1985. At that time, a major emphasis was placed upon increasingly sophisticated test-bed calculations, the burgeoning synergism between numerical and analytic methods, highly refined one-dimensional simulations, and results from two-dimensional collapse calculations. The present volume bears witness to the rapid development of numerical relativity in the two and a half years since the Drexel workshop. In the remainder of this Introduction, we provide a brief guide to the major themes connecting the thirty articles contained in this volume.

From its beginnings, numerical relativity has been concerned with astrophysical sources gravitational radiation. While most astrophysicists are convinced that the orbital decay of the binary pulsar PSR 1913+16 results from the emission of gravitational radiation, the *direct* detection of gravitational waves will provide an important observational test of general relativity theory. Significant computational efforts are necessary to guide those who are now constructing gravitational wave detectors, and to interpret their observations. The strongest bursts of gravitational radiation will probably come from the coalescence of binary black hole or neutron star systems. Another significant source could be the non-axisymmetric gravitational collapse of rapidly rotating stellar cores. Three-dimensional computations of this kind are analytically, and until recently computationally, intractable. Advances in supercomputing technology and its increased availability to the scientific researcher have made it possible to develop three-dimensional numerical simulations. In his contribution, Stark reports on the development of a non-axisymmetric relativistic collapse simulation. One important issue that may be addressed by such a code is the the development of non-axisymmetric instabilities. Should the collapse become non-axisymmetric it will significantly enhance the amount of gravitational radiation emitted.

Modeling binary neutron star coalescence provides additional challenges to the numerical relativist: not only is it an inherently three-dimensional phenomenon, but one with multiple and highly disparate timescales. Nakamura & Oohara and Wilson & Mathews have begun to attack the problem of binary coalescence, and describe their progress here. In addition, Oohara & Nakamura report on numerical studies of the initial value problem in three dimensions. Their results will form the basis for future three-dimensional simulations.

Owing to the large computational demands made by such simulations, it is worthwhile to find useful simplifications. It is astonishing, as Ferrell & Eardley show, that configurations of extremal Reissner-Nordstrøm black holes can be calculated *analytically* (in Jim Wilson's sense of regarding ordinary differential equations as "analytic"). Seidel & Moore use perturbative techniques to reduce the calculation of gravitational radiation from axisymmetric stellar core collapse to a dynamical one-dimensional problem involving additional radiative fields.

Detweiler has been investigating the problem of binary *black hole* coalescence, exploiting the weakness of the gravitational radiation emitted by the system until its final moments. He has made considerable progress in developing analytic approximations to the problem and these techniques may lead to simplifications in numerical computations of coalescence. Beyond these approximations, however, looms the problem of simulating a generic system of (two or more) black holes. Before we can attempt such simulations, we must be able to solve the initial value problem for black holes in full generality. York outlines one such procedure for constructing initial data for an arbitrary number of black holes, each with its own linear and intrinsic angular momentum.

While the projects described above are aimed at an all out attack on fully general relativistic problems, a number of researchers are writing codes to model self-gravitating Newtonian hydrodynamics in three dimensions. Even for this apparently simple task it will be difficult to obtain accurate results with current methods and technology. Bonazzola & Marck are attempting to calculate three-dimensional hydrodynamic collapse using pseudo-spectral methods. Kochanek & Evans test a three dimensional smooth particle hydrodynamics code against an axisymmetric finite difference code by calculating the gravitational radiation from a head-on stellar collision. They go on to show preliminary results from the non-axisymmetric collisions of equal mass stars.

Some researchers are exploring new numerical techniques to improve the accuracy and stability of their simulations (*cf.* review by Hobill & Smarr). Contributions in this volume include investigations into multigrid methods (Cook), predictor-corrector methods (Ove), automatic code generation via a "PDE compiler" (Thornburg), constraint preserving transport in MHD (Hawley & Evans), fully constrained evolution of the Einstein equations (Evans), and a method for ultra- (special) relativistic hydrodynamic flows (Wilson & Mathews).

Spherical symmetry offers a simple laboratory for testing new numerical techniques. Mann has championed the use of finite element methods in the 3+1 formalism for hydrodynamic systems. This represents a radical departure from the traditional dependence upon finite difference methods in numerical relativity. Choptuik's study of the collapse of minimally coupled scalar fields in general relativity provides an excellent testing ground for the study of "pop-up" grids and mesh refinement algorithms, all aimed at a quantitative determination of the accuracy and *precision* of numerical simulations. Finally, renewed attention to modeling the pure vacuum Schwarzschild spacetime has provided stringent tests of the accuracy and stability of various finite difference methods used to evolve the nonlinear hyperbolic equations (Bernstein *et al.*).

Spherically symmetric systems do not radiate gravitational waves. Nonetheless, simulations of spherical systems are a remarkably fertile ground for research. For example, detailed modeling of the microphysics known to be important in stellar core collapse is more easily refined and studied in spherical simulations. This includes the use of more detailed equations of state as well as the work by Mezzacappa & Matzner, and Schinder on radiation (neutrino) transport. Dynamical degrees of freedom may also be studied in spherically symmetric systems, if an additional scalar field is coupled to the Einstein field equations. In this context, Choptuik has been investigating the formation of black holes from imploding scalar waves.

An alternative to the traditional use of spacelike slices in numerical relativity calculations involves using a null cone foliation. With an improved understanding of how to pose useful initial data on such slicings, Gómez & Winicour are studying new finite difference methods needed to maintain a stable evolution.

Shapiro & Teukolsky report on their efforts to develop a numerical general relativistic treatment of stellar dynamical systems. In one newly begun project, they hope to use their code to probe the validity of the Cosmic Censorship conjecture through a combination of analytic and numerical means. Goldwirth *et al.* discuss two different approaches to the study of singularity formation and Cosmic Censorship. One examines the self-similar collapse of a fluid system, while the second emphasizes the use of imploding scalar fields.

Cosmological models that involve inhomogeneities in one direction (planar symmetry configurations) admit one gravitational wave polarization and can be used to study the interactions between very nonlinear gravitational waves (Anninos *et al.*). In cases where the vacuum fields are strong, there is interest in the nonlinear gravitational wave behavior: do they form black holes, or naked singularities? These studies can be extended to other topologies and higher dimensions, as Ove has done for $T^3 \otimes R$ vacuum cosmologies. He finds that the behavior of interacting gravitational waves has a strong dependence on global symmetries and topology. In an astrophysical application of planar cosmologies, Kurki-Suonio is studying the effects of inhomogeneities on primordial nucleosynthesis.

Researchers have long strived to understand how the perturbations required for galaxy formation arise. One recently proffered scenario suggests that cosmic strings collide and reconnect, forming loops that become these seed perturbations. In his contribution, Matzner presents detailed three-dimensional calculations of the probability of reconnection of colliding cosmic strings.

There were strong arguments at the Drexel meeting that research efforts in numerical relativity should be more closely coupled to analytic work. Participants at the *Frontiers* meeting saw evidence of how fruitful such a collaboration could be. A fine example of this was Kojima's calculation of a new class of rapidly damped quasi-normal stellar pulsation modes. These modes were first suggested by a mathematical model. In addition to their predictive power, analytic methods can provide powerful diagnostic tools. Abrahams has pioneered the use of perturbative expressions for the asymptotic gravitational field to produce a template for the extraction of radiation waveforms from fully relativistic calculations. Finn reports on a similar approach for determining gravitational radiation from *Newtonian* calculations, and discusses how this technique provides consistency checks on the implementation of the equations of motion.

We hope that this introduction has given the reader a clear impression of the many fronts along which numerical relativity research is advancing. With so many new avenues of research opening up, progress in numerical relativity is largely a function of available manpower. It is thus particularly heartening to see both new graduate students as well as established researchers embarking on new explorations on the frontiers of numerical relativity.

SUPERCOMPUTING AND NUMERICAL RELATIVITY: A LOOK AT THE PAST, PRESENT AND FUTURE

David W. Hobill and Larry L. Smarr
National Center for Supercomputing Applications
605, East Springfield Avenue
Champaign, IL 61820

Abstract. Reviewing the issues in supercomputing applications that are relevant to numerical relativity, we discuss the present state of the art of computing solutions of the Einstein equations. In particular we look at how recent developments in hardware and software have influenced the maturation of numerical relativity. Special emphasis is placed upon the contributions that are being made by symbolic manipulation and scientific visualization to the theory of general relativity.

I. INTRODUCTION

In the past few years, numerical relativity has witnessed a rapid growth not only in the number of researchers in the field, but also in the scale and difficulty of the problems being attacked. As analytic methods for studying the Einstein equations become exhausted, numerical methods will likely increase in importance, particularly as observational general relativity reaches its maturity and measurements of the relativistic gravitational field become possible.

Numerical relativity, and computational physics in general, can be considered as new mode of scientific inquiry, augmenting the already well established modes of theory and observation/experimentation. It is computation that takes a theory, replaces the spacetime continuum with a finite lattice, calculates derivatives with finite differences and, using numerical methods, simulates a phenomenon that represents some aspect of physical reality. As in any scientific endeavor, measurements and observations are still needed to verify the underlying theory.

The new synergism that exists between theory, computation and experimentation, is the basis of a number of important collaborations. In general relativity, perhaps the best known collaboration exists in the analysis of gravitational radiation. Theorists have long been interested in understanding the generation, propagation, and detection of gravitational waves. However their work has generally been based upon simple idealized models of sources and detectors. Numerical models of realistic gravitational wave sources and detection devices will provide the experimentalists with more precise information concerning the energy and signatures of the gravitational waves that reach an earth-based observer from violent astrophysical events. Similar collaborations also exist among cosmologists who studying the physical processes that occurred during the first moments in the life of our universe try to determine how they may be able to explain the observations that are made today.

Phase 1. In spite of the fact that such collaborations amongst relativists have been formed, they have not reached the same point of maturity occurring in some other realms of physics such as fluid dynamics or condensed matter physics. Numerical relativity is still at an early stage in its development compared to these fields. Long before electronic computers could realistically attack problems in numerical relativity, a few far-seeing individuals realized that numerical methods would be useful for solving the Einstein equations. This prophetic age, which began with the work of Lichnerowicz in 1944 and continued with further expositions by DeWitt (1957) and Misner and Wheeler (1957), established that the structure of the Einstein equations was amenable to computational methods.

Phase 2. Somewhat later a second stage of development began with pioneering efforts dedicated to attacking problems that at one time were considered intractable. Often problems are attempted long before the technology and methodologies are capable of providing the solutions to those problems and numerical relativity found itself in this situation at its birth. In general relativity, this heroic period began in the 1960's with the attempts by Hahn and Lindquist (1964), Cadez (1971) (the two black hole collision problem), May and White (1966) (spherical collapse) and Pachner (1973) (rotating collapse). Only the May and White code succeeded in obtaining the goals it had set out to reach. Later, successes were met in two dimensional vacuum solutions by Smarr (1979) and Eppley (1979) who completed the evolution of two colliding black holes and calculated the dynamics of Brill waves. At about the same time, one dimensional problems were extended to cylindrical [Piran (1979)] and planar [Centrella (1979)] symmetry, and 2-D hydrodynamic solutions were obtained by Wilson (1979). All of these codes were either highly idealized (1-D) or run on crudely zoned grids (2-D) since the computational power available at that time was insufficient to perform large-scale computations. Subsequently, other researchers entered the field and efforts to push the frontiers of numerical relativity further have continued to the point that now three-dimensional problems are being considered (see for example the articles by Nakamura and Stark in this volume).

Phase 3. While all of these ambitious explorations of unknown territory were being carried out, the third phase of numerical relativity began. Its purpose is to construct more accurate and stable codes capable of re-attacking those problems that now seem less intimidating. In some cases we have seen a return to one dimensional problems which can be performed with such an increased accuracy that these codes nearly rival the accuracy of codes designed to solve ordinary differential equations. In addition, analytic methods play an important role in this third phase. Not only are they being used to test and verify the results of a numerical calculation, but they are also being directly used in conjunction with the numerical calculations to produce self-consistent results. This third phase of computation is more than just a development of algorithms aimed at refining the results of previous work. It also is an exploration of some fundamental physical concepts and theories. For example in this volume we see one dimensional codes leading to new a understanding of the behavior of inhomogeneous cosmologies (Anninos et. al. and Kurki-Suonio), cosmic censorship (Shapiro & Teukolsky and Choptuik), and gravitational collapse to form star clusters and black holes (Shapiro & Teukolsky). Since each of these subjects requires high accuracy in order to precisely measure subtle effects that often escaped earlier investigators, a more careful treatment of these problems with better numerical methods is necessary.

As more sophisticated methods become commonplace the emphasis of the third phase will be focused on the solution of realistic problems with a complexity so great that computer methods are the only means of linking the theory to experiment. The field of hydrodynamics provides a good example of how complex physical reality can be, and no less will be expected of numerical relativity. General relativity has yet to reach the stage where it can simulate a large variety of complex phenomena and then correlate its conclusions either with measurements made in the laboratory or with observation.

Experimental relativity is presently undergoing its own parallel evolution. By the 1990's it is expected that these two modes should reach their maturity almost simultaneously, thereby ushering forth a new era of increased interest in general relativity.

In this paper we will outline some of the developments in computational relativity that have occurred as a result of the introduction of supercomputers. The term supercomputer in this article will always refer to those machines with the largest memories and fastest processors available. These machines are capable of solving problems that were once considered intractable on any previously available computer. We shall also discuss certain issues that are now being addressed by numerical relativists in order to ensure the continued growth of the field. While some examples of progress that have been made in our discipline will be presented, these proceedings speak for themselves concerning the present and future frontiers of numerical relativity.

II. SUPERCOMPUTERS AND THEIR IMPACT ON NUMERICAL RELATIVITY

1960's. In 1966 May and White performed the first successful 1-D general relativistic evolution on a computer (mostly on a CDC 3600). This was a spherically symmetric Lagrangian code that computed the gravitational collapse of a fluid sphere. The choice of time gauge was such that it could not avoid the central singularity that formed during a collapse to a black hole and therefore the code was limited to the hydrodynamics involved in neutron star formation.

1970's. Two dimensional numerical relativity met with its first successes on the CDC 6600 [Smarr (1975)] and eventually CDC 7600 [Smarr and Eppley (1979) and Wilson (1979)] machines. These were the first computers with speeds and memories capable of carrying out elementary finite differenced calculations on grids fine enough to resolve the relevant length scales in a 2-D general relativity problem. The core memories (on the order of 10's of KWords) of these machines were still small compared to what was needed in numerical relativity so ingenious operator splittings often had to be performed to take advantage of what memory was available. In addition the need to transfer data back and forth to rotating disk or magnetic tape introduced considerable overhead in the coding. Numerical relativity required an inordinate amount of patience and perseverance and only a few diehard computer specialists were willing to put up with the idiosyncrasies of the supercomputer of that era. Lack of access proved to be even a greater problem. There were few machines capable of performing numerical relativity codes and they were not readily available to academic researchers. As a result the number of practitioners of numerical relativity remained small.

1980's. Both the increased speed and memory of the Cray supercomputers, have relieved many of the frustrations associated with having to deal with time and memory constraints. For those codes that still require more memory and must use external storage devices, the new solid state disks can transfer data at a rate of 125 times faster than a rotating disk and can be treated almost as an extension of main memory. Computers with memories of hundreds of millions of 64-bit words are now available making older problems more tractable and solutions to new and more difficult problems possible. Table 1 presents the core memory and speed of those machines that have contributed to the development of numerical relativity in the past two decades.

Supercomputer Capabilities. In general the evolutionary pattern has been to simultaneously increase both speed and memory. The scalar speed up of the fast processors in the CDC 6600 meant that the results could be produced in a reasonable amount of time. The vector processing of the CDC 7600 with even faster CPU's made the speed up factor even greater. However vectorization required some changes in the way codes were written. The Cray-1 continued this trend and in addition made more memory available. Multiple processors became available on the Cray-XMP and the YMP will extend this number even further. Finally the Cray-2, while not significantly increasing the speed and number of processors compared to the Cray-XMP, will by virtue of its large memory open the door to more detailed calculations in numerical relativity. Along with the availability of powerful workstations that act as an intelligent interface between the user and the supercomputer, many of the earlier limitations to large-scale computing have been eliminated and new researchers are now entering the field.

date first used	computer	core memory (Words)	Peak Mflop rating	estimated time to complete a 2-D calculation
1964	CDC 6600	40KW	7M	1 week
1969	CDC 7600	65KW	30M	4 days
1978	Cray-1	2MW	160M	10 hours
1984	Cray-XMP	8MW	200M*	7.5 hours
1987	Cray-2	128MW	2000M**	1 hour

Table 1. A comparison of different supercomputers used for numerical relativity codes. The peak megaflop ratings are obtained from the literature supplied by different vendors and may vary depending upon the codes used. * Most XMP's are used in single processor mode. **Assuming code uses four processors.

The computationalist must still deal with the peculiar characteristics of particular hardware and software implementations, especially if they are dependent upon the latest developments in technology and methodologies. Like an experimentalist, the computationalist has an apparatus with which he/she must interact and continually coax to perform in the manner intended. It is unlikely that the forefront of numerical relativity will ever be 'user friendly', but advances are always being made to ensure that it will not be entirely user abusive.

Increased Complexity. Supercomputers present the researcher with the opportunity to increase the complexity of the problems that are now being attacked. In general relativity this complexity can occur in many ways. The most obvious among these are the use of locally finer grids and spatially larger domains, the extension of a particular model to higher dimensions and the addition of more realistic physics in the set of equations or initial data. The increased demands on computer memory and execution time increase geometrically with an increase in any one of these realms.

Grid Refinement. Augmentation of both the local spatial and temporal resolution of an existing code can lead to a better correspondence with analytic behavior. If for example a particular second order accurate finite difference method is employed to approximate the Einstein equations, then decreasing the grid size by a factor of two increases the accuracy by a factor of four. This will often (but not always) be enough to reproduce either the behavior of an analytic solution or the results from a simpler numerical test-bed calculation. An example of how a simple increase in the fineness of the numerical grid proceeds to correctly reproduce the known analytic results is presented in these proceedings by Bernstein, et. al. Unfortunately, for a given N-dimensional time-explicit evolution, a grid refinement by a factor of two requires a 2^{N+1} increase in computational time.

Domain Enlargement. The spatial extent and length of time over which the evolution is calculated is also an important factor in increasing the accuracy of a calculation. Outer boundary conditions for asymptotically flat spacetimes will always be more accurate if they can be posed at large distances (ideally, infinitely far) from compact sources with relativistic motions and/or regions of strong gravitational fields. With an increase in the spatial extent of the problem, longer time evolutions are necessary. Since one is not refining the Courant limit zone, the increase in computational time is less than that discussed above.

Higher Dimensions. The complexity that arises in the passage to a higher spatial dimension from a lower one has even a more profound effect. First, there is the obvious increase in the number of grid points that are the consequence of going to higher dimensions. Second, the increase in dimensions is accompanied by a larger number of non-zero metric coefficients which translates into a need to calculate and store more variables at each grid point. The equations and expressions appearing in them are more complex simply due to the increase in non-zero variables. While in principle the Einstein equations for a completely general metric can written on a finite number of pages, the computer used to construct a numerical solution may be unable to store all of the variables at enough grid points to make such a calculation meaningful. Finally, physics in three dimensions is significantly more complex than it is in fewer spatial dimensions and finer grids may be necessary in order to resolve some of this complexity. Some problems still need to be delayed until technological gains (primarily associated with available memory) can be made.

Variable Increase. An increase in variables is a result not only of increasing the number of spatial dimensions, but is also a consequence of the decision to employ auxiliary variables and more complicated gauge choices. Until recently, most researchers have followed the Wilson (1979) suggestion of keeping the number of variables in the three metric to a minimum through choices in the shift vector. However more geometric and natural choices of shift vectors do exist [Smarr and York (1978)], but they increase the number of elliptic equations that need to be solved at each time step and do not eliminate metric variables. The freedom to use alternative gauges has yet to be explored and it is quite likely that in the future, gauge choices will be made not for computational ease but for more fundamental theoretical reasons. There is still much work to be done in this relatively untouched area of research.

Algorithm Development. Algorithm development often leads to increased complexity in a numerical calculation. Numerical hydrodynamicists in their search for greater accuracy and stability, have been led to the construction of more complex and more elaborate numerical methods that yield more accurate solutions of the partial differential equations that govern their simulations [Woodward (1988)]. Often new and more complex algorithms for calculating a solution at a fixed order of accuracy will result in more stable methods that allow for longer time evolutions to be calculated. Rather than provide specific examples of algorithm development here, we shall consider this issue within the context of a general discussion on software.

Figure 1 provides an example of the progress that is a consequence of using finer grids, larger spatial domains and more accurate and stable algorithms. In this graph the value of the lapse function at the inner most grid point is measured for maximally sliced spherically symmetric black holes. Analytically, the lapse function at the throat of such a black hole has a time dependent behavior that falls off exponentially ("collapse of the lapse") with an e-folding time of 1.82M where M is the mass of the black hole [Smarr &

York (1978)]. The values of the lapse shown in Figure 1 represent the largest absolute value of the logarithm of the lapse that still obeys the analytic behavior. Without an increase in computing power some of these developments could not have occurred.

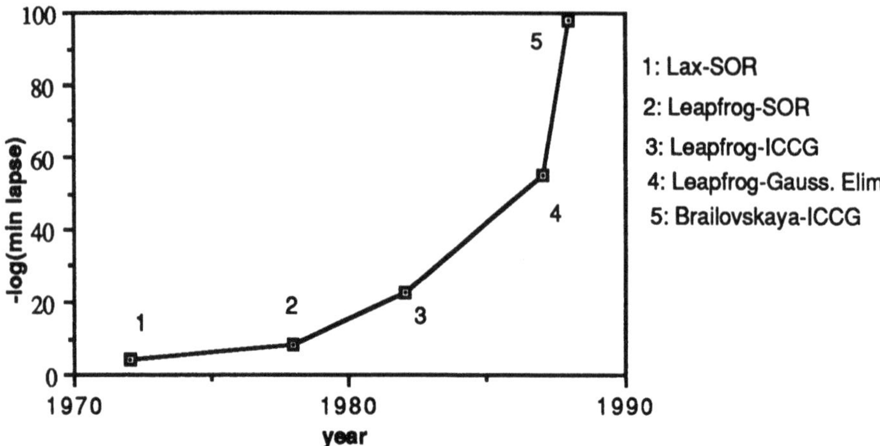

Figure 1. A graph of the increase in accuracy obtained with increased computing power and algorithmic development. The data points were obtained from the following papers: 1; Estabrook, et. al. (1973), 2; Smarr & York (1978), 3; Evans (1985), 4,5; Bernstein, et al. (this volume). The first two calculations were performed on a CDC 7600 and all subsequent results were obtained on the Cray X-MP.

More Physics. Nature itself presents us with an increase in complexity that is far more interesting but also more difficult to understand. Thus far we have discussed the more mathematically quantifiable or predictable increases in complexity (increases in memory requirements) but the introduction of a realistic source to the right hand side of the Einstein equations will produce consequences that should not be underestimated. Already we have seen from computations made in the simpler newtonian or special relativistic regimes, that hydrodynamics and many-body problems are subject to complicated and often unexpected behavior. If these systems act as sources and are coupled to nonlinear gravitational fields, it can be expected that the general relativistic systems will be no less complex. New methods of studying the behavior of such a complex physical system with many degrees of freedom will have to be developed in the future.

II. HARDWARE ISSUES

Supercomputing Demands. From the point of view of computational science as a whole,the hardware requirements of numerical relativity has yet to make a major impact in the field. Its computational demands are relatively small compared to computational fluid dynamics, computational chemistry and lattice gauge theory to name a few other disciplines that utilize the same resources. The largest two dimensional relativity code (written by Roger Ove) requires 4 million (64 bit) words of Cray-XMP memory and an additional 17 million words on the SSD (solid state disk). This is a fine resolution (200x500) code and requires between 2 and 3 CPU hours to run about 2000 time steps. Nakamura's three-dimensional code with 80^3 grid points requires 25 (64 bit) Million words of core memory. These examples are to be compared to some stellar evolution codes that need hundreds of millions of words of memory.

In a list of all of the 37 accounts on the NCSA Cray that have used 300 hours or more of computer time, numerical relativity ranks 25th and 29th in overall usage. Comparing the account with the largest usage (lattice gauge theory with 11,433 hrs) to the largest numerical relativity account (388 hrs), provides yet another demonstration of numerical relativity's relative rank as a consumer of supercomputer cycles. Numerical relativity is is still a young field, but it is expected that the 3-D codes will eventually place heavy demands upon computer power, especially when one considers the large number of variables (50 to 100) that need to be stored on a numerical grid fine enough to resolve dynamics at widely differing length scales.

An example of how important supercomputers are to numerical relativity in terms of productivity is presented in Table 2 where the execution times for different machines is compared. The code used to provide the benchmark was a one-dimensional predictor-corrector code for a maximally sliced spherically symmetric black hole (see Bernstein et. al.) The radial grid contains 1000 nodes and 20000 time steps were computed during the entire evolution. The code was written to be entirely portable so that system dependent instructions were eliminated from the very start. This allows coarse grid runs and code editing to be performed on local workstations leaving only the fine grid production runs to be performed on supercomputers. From the timing results, it is clear that if computational science is to be considered as 'experimental theory' then the benefits of fast machines (and turn-around) are obvious.

Computer	Processors	Mode	CPU Time (sec)	Ratio
Sun 3	1*	Scalar	586800	3738
Vax11-785	1	Scalar	317580	2023
Alliant FX/8	8	Vector/Parallel	5152.9	32.8
Cray-XMP	1	Vector	483.3	3.1
Cray-2	1	Vector	520.5	3.3
Cray-2	4	Vector/Parallel	157.6	1.0

Table 2. A comparison of CPU times to calculate the maximal time slices to 100M for the Schwarzschild solution (20000 time steps) with 1000 spatial nodes using a Brailovskaya finite differencing scheme to compute the evolution equations and shooting to solve the two-point boundary condition problem for the lapse function. Memory requirement: 1.1MWords. *Without floating point accelerator.

Supercomputer Upgrades. At the present time, the eight million word Cray XMP-48 (and SSD with its 128Mword memory) provides enough computing power and memory to perform most 2-D calculations. The Cray-2 will have a maximum memory of 512 million words (the NCSA Cray-2 will be smaller with 128Mwords) and this increase by a factor of sixty-four will make three-dimensional relativity a real possibility. One Giga-word machines will be available by 1990 (the Cray-YMP will have a 1Gword SSD by late 1989). If one extrapolates the 1-D results of Bernstein, et. al., then in order to obtain 3-D results with an accuracy of less than one per cent, a machine with approximately 20Gwords of memory will be needed. A computer with a memory of this order of magnitude may not be available until about 1995. It will be needed, however, if experimental and numerical relativity are to simultaneously open the door to the new field of gravity wave astronomy.

Parallel Supercomputing. Another critical hardware issue is that of parallelism. Almost all of the numerical relativity codes to date have run on single processor computers. However during the next five years a wide variety of parallel machines will become available and their use will be essential to achieve the speed-ups needed for frontier numerical relativity codes. This will require a whole new generation of codes that take advantage of parallelism. At the present time most parallel supercomputers use a small number of processors (<16) with a shared memory. Programming these machines is still an inherently serial process using loops to sequentially calculate portions of a system that have been split off from the whole. This is contrary to the idea that in a 3+1 decomposition of spacetime an entire spacelike hypersurface is updated simultaneously from local information at each grid point. Most of the equations of theoretical physics are partial differential equations that lead to explicit finite differenced equations with nearest neighbor interactions. Consequently, few if any data dependencies need appear in the algorithms that exploit the locality inherent in physical systems.

With a small number of processors, the operations occurring at each grid point cannot be calculated simultaneously. The operations must be distributed among the available processors until the entire loop is completed. If, on the other hand, there are as many processors as nodes in a finite differenced version of a PDE (each with their own local memory and capable of communicating with its nearest neighbors), then the loop structure and operator splitting is superfluous since each node performs its operations using information obtained from its neighboring nodes and itself.

The Connection Machine with its 65,536 processors pursues this philosophy with its 'news' (North-East-West-South) grid, and the get_from_news operation that in assembly language (Paris: an acronym for parallel instruction set) passes information from one node to another. Unfortunately in the connection machine the processors at each node only have 8Kbytes of memory, but it does not take a great deal of imagination to consider the possibilities that can arise from the use of processors with a larger memory at each node. One step in this direction is the MYRIAS machine that will use 1024 Motorola 68020 chips each with 4Mbytes of memory. However this still only allows a 30x30 2-D grid if each processor acts as a grid point. It is likely that this interim solution will be superceded when the price of high-speed large-memory processors is low enough that tens of thousands of such chips will be available in one machine.

Networking. Access to remote computers has been, and will remain in the near future one of the biggest problems in supercomputing. High speed T-1 (1.5Mbytes/sec transfer rates) networks are currently becoming available to researchers. Figure 2 shows a map of the upgraded NSFnet which consists of a T-1 backbone connecting the national supercomputer centers and other major universities in the United States. The network through-put to an individual is hampered by the fact that many regional networks which feed the T-1 backbone have lower bandwidths.

Networks not only provide the link to the supercomputer upon which the computationalist depends, they also act as a means of communication between researchers. In some cases it may necessary to exchange data and images quickly and computer networks are the only means of providing that service on time scales of hours or even minutes. New software is needed to allow the user to take advantage of multiple computer environments. NCSA's popular Telnet software uses TCP/IP protocols to link a researcher's workstation and multiple remote hosts. Its ability to transfer files, provide interactive computing capabilities on different machines and emulate a number of different terminal types (VT100, Tektronix, Raster, etc.) makes it the basis for a distributed computing environment and enhances the productivity of the individual scientist.

Local Workstations. Regardless of the capabilities of the network, it is always in the interest of the individual to perform as many tasks as possible in a local environment. Intelligent workstations are necessary in order to execute various chores such file editing, collection and analysis of data, routing of electronic mail, development and testing of prototypical codes, compilation of codes, symbolic calculation, and manipulation of visual images. The individual researcher is able to customize his/her workstation to the particular needs of the science being done through choices in hardware and software. As workstations increase in power to the point that the entire code development phase can be carried out in the office of the individual scientist, the supercomputers will need only be used for the purpose of 'number crunching'. The goal of network software development is to make the entire process of code development, file transfer, simulation and analysis of the results completely transparent to the scientist at his or her workstation.

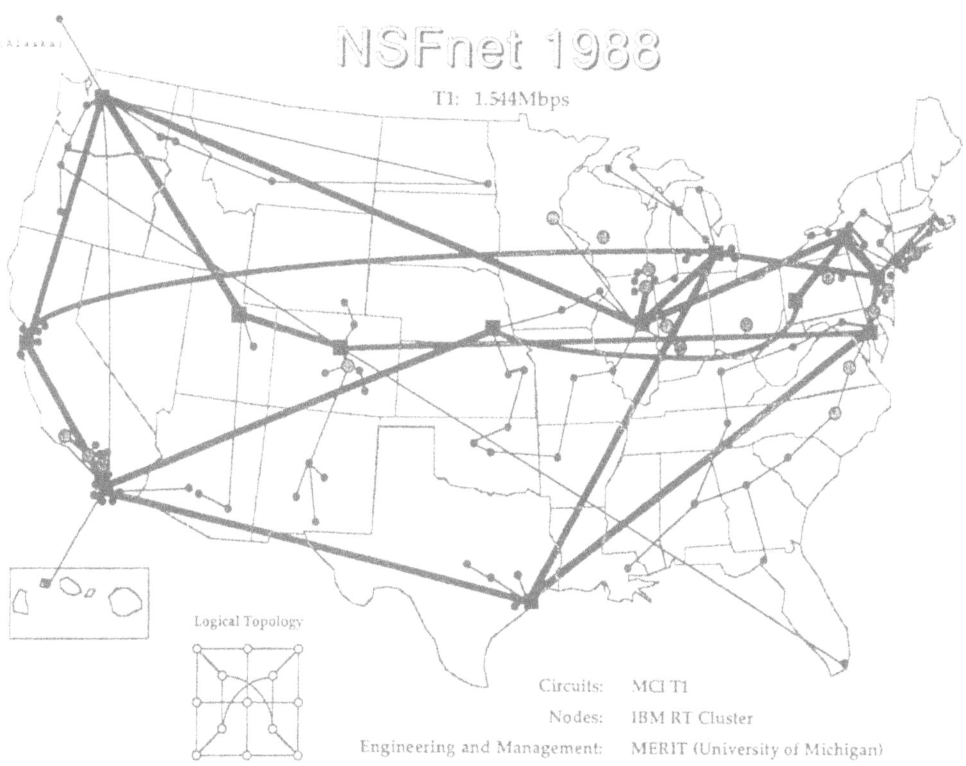

Figure 2. A map of the NSFNet used to connect the national supercomputer centers with smaller regional networks. The backbone that connects the major centers (thick lines) has recently been upgraded to T-1 status.

III. ALGORITHMS AND SOFTWARE

Operating Systems. The emergence of a single consistent operating system running on all of the links of a network from the researcher's workstation to the supercomputer will be of great help to the numerical relativist. Computing on different machines (with different operating systems) that form a distributed computing system has always been a programming nightmare. The passage from one system to another calls for different commands, protocols, and utilities, many of which are not compatible with each other. Added to the difficulties of having to learn the fundamentals of different operating systems in order to effectively perform interactive tasks, codes which have built in system dependencies often require a large expenditure of effort during the process of porting them from one machine to another.

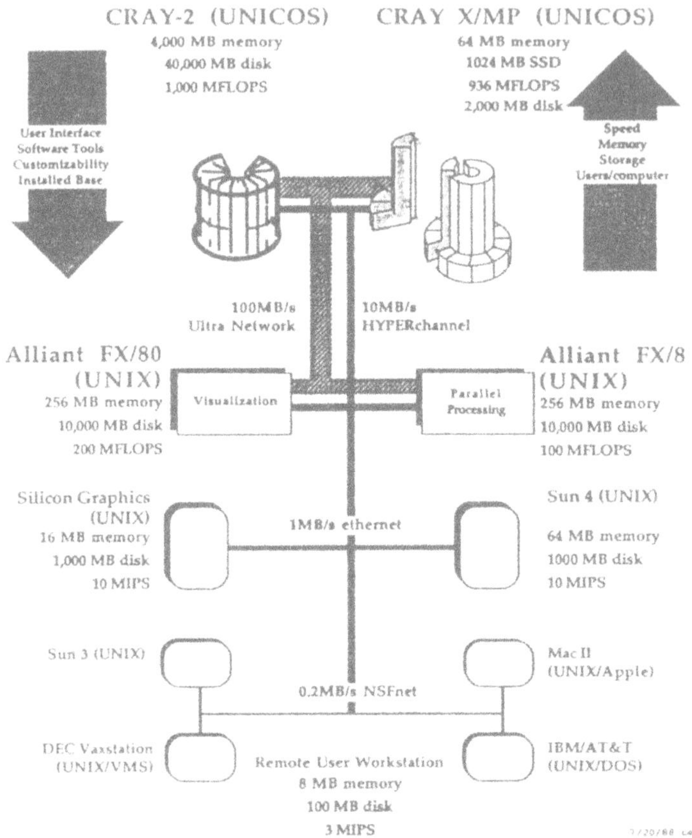

Figure 3. The NCSA Unix pathway showing the different computers and their connecting networks that use Unix or Unix-like operating systems. The user of such a system need only learn one set of commands, utilities, protocols, etc., thereby placing him/her in an efficient computing environment.

At the present time it appears that Unix in some guise or another will, at least for the near future, become the operating system of choice among computational scientists. This has been heavily influenced by the importance of scientific workstations (most of which run Unix or Unix-like operating systems) in computational research. The fact that this operating system may now be found on the most powerful supercomputers as well as widely available micro's is restructuring the way one works with computers. The step toward a complete Unix environment will undoubtedly accelerate the development of distributed computing. Figure 3 provides an example of a distributed Unix environment linking the supercomputers at NCSA to the workstations and personal computers used by researchers.

Programming Languages. Along with the shift to Unix, there has been a growing popularity of the C programming language, even among scientific programmers. Indeed we have seen the rapid publication of a C version of the popular *Numerical Recipes* book of Press et. al. (1988). At the University of Illinois, one computational physics course (taught by Stephen Wolfram) has as a prerequisite, the ability to program in C. It is not yet clear what the impact of C will be in scientific computing. While C gives the programmer a more intimate relationship with memory allocations and registers, numerical calculations are often carried out using a Fortran like syntax. Proposals for Fortran-8X include extensions that are aimed specifically at vector and parallel computing. Rather than be restricted to standard serial computations, the array extensions allow data to be operated on all at once. Many manufacturers of newer parallel and vector machines have already implemented these extensions. Fortran is not dead as far as the computational scientist is concerned and reports of its demise like that of Mark Twain's have been greatly exaggerated.

Compilers. Luckily compilers are becoming more intelligent with increasing version numbers. This is particularly important in vector and parallel processing. Scientists do not want to, nor should they have to, worry about adding compiler directives to every portion of code that can be vectorized or run concurrently. The bulk of this work should be recognized by the compiler and the listing files should provide information on where optimization, vectorization and concurrency are hindered or difficult to recognize. Expert compilers are being developed that are able to recognize vectorizable and parallelizable codes. Some of these even enter into a dialog with the user to ensure that the compiler and scientist mutually agree on the procedures to be executed. This important development allows the researcher to concentrate more on the physics of a particular simulation leaving it to the compiler to keep track of the various tasks being performed on the computer.

Applications Software. The issues surrounding software tools depend heavily upon the type of calculations that ones needs to perform. In computational engineering and chemistry, a number of general applications packages have been developed for use by a large majority of researchers in their respective fields. Nothing like this has occurred in general relativity (except perhaps for algebraic computing). Most numerical relativists still write their own code. This is largely due to the fact that many algorithms are still being explored. Relativity is not alone in these circumstances, hydrodynamicists who already have a great deal more experience in computation are still developing better algorithms to calculate the complex nonlinear behavior of fluids.

At the present time, numerical relativity lacks a nucleus of community software capable of performing the common tasks that each researcher encounters. The development of code generators, elliptic equation solvers, partial derivative to finite difference translators, and visualization software specific to the needs of numerical relativists will require many man-years of effort. Applications software for numerical relativity will only be produced from a group made up of a critical mass of computational relativists capable of collaborating with specialists in computer graphics, numerical analysis, and symbolic manipulation. The potential for such a situation exists currently at the national supercomputing centers, but requires a concerted effort by the numerical relativity community to pool its resources and expertise.

Elliptic Equations. For numerical relativists, perhaps the closest that one comes to having applications packages is in the area of elliptic equation solvers. The simultaneous over-relaxation (SOR) methods that were used in the original pioneering efforts of numerical relativity have been replaced with preconditioned conjugate gradient (CG) methods, alternating direction implicit (ADI), multigrid (see the Choptuik and Cook articles) and other methods. These algorithms, while being more complex, are more accurate, stable and often converge a good deal faster that the older SOR method. They generally require more storage and therefore have only recently begun to fall into favor. Evans (1986) has provided a good example of how much faster and accurate the incomplete Cholesky conjugate gradient (ICCG) method is and Fig. 4 reproduces his results. Similar results, reported in this volume, have been obtained by Nakamura and his colleagues.

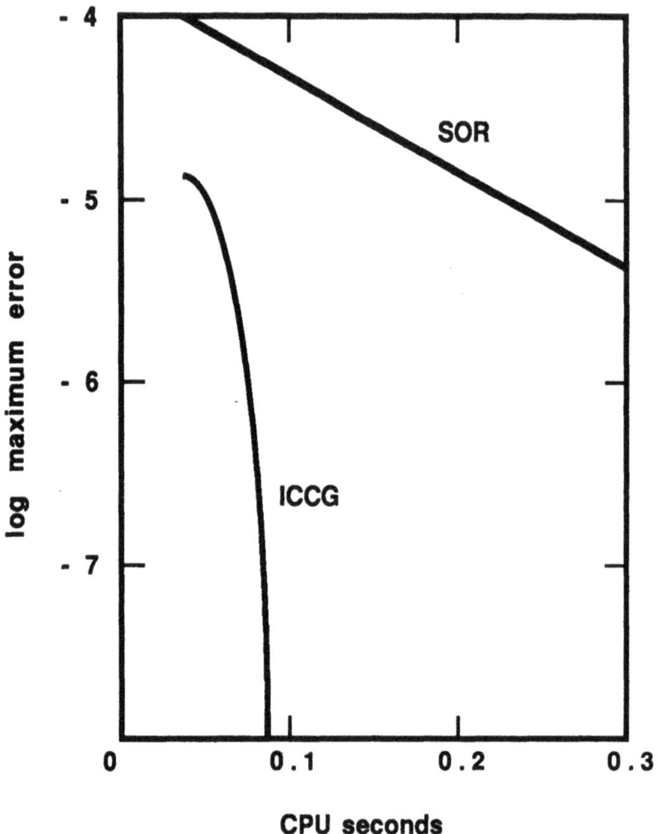

Figure 4. The rate of convergence for SOR and ICCG as a function of CPU time. These tests were run on a 900 node mesh (12x75) and solve the maximal slicing equation. [After Evans (1986)].

Outside of general relativity, there are efforts to create general purpose elliptic equation solvers. The ELLPACK project is a good example of a collaboration among computer scientists, numerical analysts and mathematicians designed to produce a software package useful for a large number of users. This package has its own programming language that sets up the equation, specifies a finite difference scheme and produces a graphical display of the solution. Unfortunately, ELLPACK has not yet reached a level of generality where it can be employed in relativity. Given the way it is set up, an entire numerical relativity code would have to be treated as subroutine called by ELLPACK. If the researcher performs the finite differencing of the elliptic equations, there is still an inadequate supply of methods capable of solving the large sparse, nonsymmetric matrix equations that are generated. Fewer still are optimized for parallel and vector processing. This is situation is beginning to change with extensions that are now being added to the Yale Sparse Matrix routines and there is a similar project now under way at Los Alamos. Until some of these projects produce software that is truly 'all purpose', numerical relativists have little alternative than to write their own special purpose elliptic solvers.

Hyperbolic Equations. The techniques for constructing solutions to the hyperbolic evolution equations are just now being explored and there is an incredible diversity of methods available to the computational scientist. Initial gains in the field were achieved in going from first order (e.g. Lax methods) to second order methods. The standard leapfrog scheme has provided accuracy to within five or ten percent for the finite differenced version of the Einstein equations. However there are limitations to these methods and greater accuracy (to less than one percent) will require the use of more modern predictor-corrector methods (see Bernstein et. al.). These techniques can add increased stability to the system and this has the advantage of allowing longer time evolutions to be performed. It is not yet clear what the 'best' algorithm is for GR hyperbolic evolution equations, but it is likely that the current standard methods will be enhanced or superceded as numerical relativity matures.

While multigrid schemes are now being used in numerical relativity to solve the elliptic constraint equations, adaptive grid methods for the hyperbolic equations have not generally been used. The exception to this has been the work of Evans and Choptuik (this volume). Adaptive methods have proven to be successful in fluid dynamics where they are used to follow the flow of fluids. Using adaptive methods with appropriate gauge choices to follow the geometry should make significant contributions to numerical relativity. This is a subject that will require more analysis and numerical experimentation in the future.

Higher order (beyond second order) finite differencing schemes are also being used to construct numerical solutions to the Einstein equations. These will probably be necessary as long as the memory available with supercomputers is such that increased accuracy cannot be gained through a decrease in step size. The price that one pays for this in the evolution is that variables need to be stored on more time steps. In this case one trades off the need for higher grid resolution for the storage of more variables at each node.

Finally, for the hydrodynamic equations, the modern monotonicity requirements implemented in piecewise parabolic methods have contributed to important advances in this field. These methods can treat the advective part of the hyperbolic equations more accurately than straight forward finite differencing schemes, but they are inherently more complicated and require more storage [Woodward (1988)]. Smooth particle hydrodynamics is another new method that is now being applied to relativistic problems (see Kochanek and Evans in these proceedings) with some success. The algorithm depends upon a binary tree structures and this makes vectorization and parallelization difficult nor can one determine apriori how many particles are needed to reproduce the accuracy obtained with finite difference methods requiring less computational effort. In addition these proceedings demonstrate that spectral methods (Marck and Bonazzola) and finite element techniques (Mann) which have produced a number of successes in newtonian hydrodynamics are now being considered in the relativistic theory.

From all of the efforts described above, it is clear that the third phase of numerical relativity is well underway. It is expected that the search for new algorithms will provide greater stability and accuracy to the finite differenced versions of the Einstein equations and not too far in the distant future the Holy Grail of numerical relativity mentioned in the Drexel meeting by Shapiro and Teukolsky (1986) will be attained.

IV. SYMBOLIC MANPULATION

Symbolic algebra and calculus on the computer is now being used by a large number of relativists. In fact, it might be said that more relativists use computers for symbolic manipulation than for numerical work. More and more, one is seeing that many of the calculations one needs to perform in general relativity are being executed by general purpose symbolic manipulation software packages running on stand-alone workstations. Having a dedicated machine for one's own use not only gives the user greater flexibility, but also prevents those others who have to share a computer of limited resources with that user from exhibiting antisocial behavior.

The fastest of the symbolic manipulation programs for general relativity can still be found in SHEEP since this language was designed specifically for the calculations one must perform in general relativity. In order to make it a more general purpose application, its output is readable by REDUCE which has the capability of performing the polynomial divisions, integrations and other procedures not available in SHEEP.

All of the general purpose algebraic programs now have or will have tensor manipulation packages written on top of them. REDUCE is able to use REDTEN, developed at the University of Toronto by Charles Dyer and John Harper. MAPLE also has a number of gravity packages written for it (including a version of REDTEN called MuTENSOR) and a recent development is a symbolic Newman-Penrose package written by R. McLenaghan at the University of Waterloo. MACSYMA is probably the most well known of the general purpose symbolic manipulation software and has a standard tensor manipulation packages written for component (CTENSOR) and indicial (ITENSOR) forms. The latter, equivalent to the STENSOR package in SHEEP, has not been implemented in all versions of MACSYMA and is known to have a number of bugs. Reportedly these have been eliminated in the forthcoming Common Lisp version of MACSYMA. This new version also has the capability of calculating the various quantities from tetrad frames as well. Finally, Mathematica, Stephen Wolfram's new algebraic computing software that is capable of running on almost everything from personal computers to the Cray-2, is the basis for a new tensor manipulation package being written by Steve Christensen (NCSA) and Leonard Parker at the University of Wisconsin (Milwaukee).

While these packages perform the calculations that any relativist using paper and pencil must do, i.e. calculate the Einstein equations and various quantities associated with a particular metric or manipulate indices in a symbolic fashion, they are not set up well to calculate the Einstein equations in a form useful for 3+1 numerical relativity. It should be a trivial problem to remedy this oversight. One can simply take the Einstein equations and substitute the appropriate definitions for the extrinsic curvature, lapse, shift, 3-D Ricci tensor, etc. and obtain the 3+1 decomposition. More manipulation needs to be done in order to take advantage of such procedures as numerical regularization, operator splitting, maintenance of operators in self-adjoint form, etc. At some point, the derivation of the 3+1 form of the Einstein equations will be so complicated that we should expect to see 3+1 packages being written either from scratch or on top of a pre-existing tensor package.

A number of numerical relativists are already using symbolic manipulation to derive and generate Fortran expressions for their codes. At the present time most Fortran (and in some cases C) code generation routines available with symbolic manipulation software is quite rudimentary, and a filter must be written that turns the analytic expressions into usable code. These proceedings are a witness to the time and effort that can be saved using such a process. For example Nakamura and his co-workers use REDUCE, Ove uses MAPLE , and Stark and Hobill use MACSYMA to aid in the process of code generation. At NCSA efforts are being made to incorporate symbolic manipulation as a 'front end' for more general numerical simulations. The present implementation uses GENTRAN, a code generation software package that presently runs on top of REDUCE and MACSYMA. While improvements are needed, it is clear that as computational physicists study more complex systems, a better interface between symbolic and numerical computations will have to be developed.

Finally, symbolic manipulation will be useful for translating PDE's into finite difference equations. Future implementations of symbolic manipulation packages will likely have libraries containing different finite difference methods that one can use to translate derivatives into finite differences on standard meshes. This feature should assist algorithm development in the future.

V. SCIENTIFIC VISUALIZATION

We are all familiar with graphs of analytic functions and how much information they can convey in a single glance. The graphical representation of three dimensional functions as geometric objects has a history that goes back well over a century. As computational science matures, graphical displays of large data sets will become increasingly important and methods of directly analyzing these images in a manner that parallels the methods of standard mathematical analysis will be developed.

Vector Graphics. Computational scientists already understand the importance of even the most rudimentary visual images for deciphering the results of a numerical calculation. Simple vector (line drawing) graphics still play an important role in isolating problems that arise during the early stages of code development. Using NCAR (National Center for Atmospheric Research) or DI-3000 (Precision Visuals Inc.) software running on a Cray or even a Sun workstation, the researcher can view 1-D (x-y plots), 2-D (contour and surface plots) or 3-D ('wireframe' plots) vector graphics. These proceedings present many examples of the use of vector graphics in 1- 2- and 3-D representations. The simplicity of these graphics routines allows them to be viewed remotely with graphics terminals (Tektronix or emulators), given a reliable network connection. If the numerical data exists in tabular form, simpler methods of producing graphical results exist in some spreadsheet software products (e.g. Apple Cricketgraph) or in symbolic manipulation packages such as Mathematica.

With NCSA Telnet, tektronix emulation includes zoom, cut and paste facilities that can be used to produce (laser printer or pen plotter) hardcopy versions for a more careful analysis or presentation. In the future, color vector graphics (with tektronix 4025 capabilities) will make this medium more useful to the scientist who must use color as another variable. Only a decade ago relativists were startled to see color movies of 'jiggling grids and contours' produced from the data created by the early numerical relativity codes. The production of time dependent vector graphic images that once required an incredible amount of human effort and computer time can, today, be re-created on an Apple Macintosh (using a combination of Apple Hypercard, NCSA Telnet, and NCAR graphics) in a matter of seconds. What once was considered the state of the art in presentation graphics is now commonly used for the day to day analysis of numerical results.

Raster Graphics. Given a data array on a 2-D grid, the floating point values are transformed into 8-bit integer format (a value between 0 to 255). This array is turned into a stream of bytes (raster file) that provides an integer value to each pixel in a raster scan. The color values of each integer are assigned by a palette that choses 256 colors (not necessarily different) out of a possible 16 million colors. A look-up table then translates the integer values into images whose colors are governed by the palette file. In many cases subtile changes in the palette can produce dramatic effects in the display of the images.

NCSA Imagetool, a raster graphics package conceived by astrophysicist, Michael Norman and computer scientist, Carol Song and developed by the NCSA Software Development Group, can display, animate and analyze such images on Sun workstations and Macintosh II's. The image processing can be controlled easily through the use of a mouse and multiple palettes can be used in the analysis of the images. One-dimensional cuts can be made through any two points on the image and can be displayed as linear or logarithmic x-y plots. The color images can also be transformed into vector graphic contour or surface plots from which hard copy images can be made. The recording of the color images onto photographic slides or videotape however requires the specialized equipment and expertise of the Scientific Media Services (SMS) at NCSA. Raster graphics are becoming increasingly useful in numerical relativity and recent examples of their ability to pick out detailed information from large data files are found in the work of Evans (neutron star collisions) and Ove (T^3xR universes).

At the present time raster files must reside locally on the researchers workstation. In the future, Telnet will be able to display images from files resident on the supercomputer itself thereby saving a great deal of time and effort. Multi-windowed raster images are now being used (Ove Windows and NCSA Datascope) to represent simultaneously all ten metric coefficients (or one function at ten different times) on a workstation terminal. Besides being able to correlate data through the images, numerical integrations and differentiations can also be performed to parallel analytic methods of studying functions of two variables.

Rendered Images. A fully rendered 3-D surface that can be viewed from any position in three-dimensional space represents the next level of complexity in scientific visualization. A finely resolved wireframe surface can be thought of as made up of a large number of very small, flat 2-D polygons (simplexes would be a name closer to the spirit of general relativity) defined by the vertices formed from the wireframe. If the polygons are assigned values of color, texture, shading, etc. the original wireframe image is turned into a colored solid surface with precise reflection and/or transmission properties determined by ray tracing (geometric optics) algorithms. Transparent surfaces provide the opportunity to visualize multiple level contours or self intersecting surfaces. Viewed on a 'fast' Silicon Graphics or Raster Technologies, Inc.workstation, one can actually 'fly around' the image in real time under the control of a mouse. For complex geometrical objects, this ability can provide a great deal of information vary rapidly.

The generation of rendered images is computationally intensive and using Wavefront Technologies, Inc. software, most multiple image animations must be produced on an Alliant FX/8 minisupercomputer (and a few smaller jobs may be performed on a Silicon Graphics workstation). In this case both the production and animation of the colored images at NCSA is now left to the expertise of the the Scientific Visualization Group and SMS. An example of a rendered animation relevant to numerical relativity would be that involving a dynamic isometric embedding diagram. One can imagine a color map representing the Newman-Penrose quantity, ψ_4, laid down on a time changing surface that is capable of studying gravitational wave generation and propagation on a dynamic spacetime background.

Volume Imaging. The imaging of functions dependent upon all four spacetime dimensions will provide the greatest challenges to scientific visualization. The problems of three-dimensional hydrodynamics being attacked by Nakamura and Stark will require the ability to see not only the surfaces of stars, but their interiors as well. These types of problems are similar to those now being attacked in geophysical, astronomical, meteorological and medical imaging. Algorithms are still being developed that will be able to visualize volumetric data. The most notable research is being done at Stellar Computer Corp. and PIXAR where methods that range from the use of multiple transparent rendered surfaces to ray-tracing through volume elements ('voxels' or extensions of polygons). Clearly the frontiers of numerical relativity have merged with the frontiers of computer graphics.

Referring to the famous quote by Richard Hamming on the purpose of computing being insight and not numbers, the report on *Visualization in Scientific Computing* [McCormick (1988)] states that "the goal of visualization is to leverage existing scientific methods by providing new scientific insight through visual methods". While it may be difficult to foretell the future, we should be prepared to meet profound and unexpected insights into general relativity through computational methods and scientific visualization.

REFERENCES

Cadez, A., (1971), Ph. D. Thesis, University of North Carolina.

Centrella, J., (1979), Ph. D. Thesis, Cambridge University.

DeWitt, B., (1957), Unpublished notes of a lecture given at GR1, Chapel Hill, North Carolina.

Eppley, K. (1979), *In Sources of Gravitational Radiation*, edited by Smarr, L., Cambridge University Press, Cambridge.

Estabrook, F., Wahlquist, H., Christensen, S., DeWitt, B., Smarr, L., and Tsiang, E., (1973) *Phys. Rev. D.*, **7**, 2814, 1973.

Evans, C., (1985), In *Numerical Astrophysics*, edited by Centrella, J., LeBlanc, J., and Bowers, R., Jones and Bartlett, Boston.

Evans, C., (1986), In *Dynamical Spacetimes and Numerical Relativity*, edited by Centrella, J., Cambridge University Press, Cambridge.

Hahn, S. G. and Lindquist, R. W., (1964), *Ann. Phys.* **29**, 304.

Lichnerowicz, A., (1944), *J. Math. Pures et Appl.*, **23**,37.

May, M., and White, R., (1966), *Phys. Rev.*, **141**, 1232.

Misner, C.W., and Wheeler, J.A.,(1957), *Ann. Phys.*, **2**, 525.

McCormick, B., DeFanti, T., and Brown, M., (1988), *Computer Graphics*, **21**, 1.

Pachner, J., (1973), *Can. J. Phys.*, **15**, 447.

Piran, T. (1979), *In Sources of Gravitational Radiation*, edited by Smarr, L., Cambridge University Press, Cambridge.

Press, W., Flannery, B., Teukolsky, S., and Vetterling, W., (1988), *Numerical Recipes in C*, Cambridge University Press, Cambridge.

Shapiro, S. and Teukolsky, S., (1986), In *Dynamical Spacetimes and Numerical Relativity*, edited by Centrella, J., Cambridge University Press, Cambridge.

Smarr, L., (1975), Ph.D. Thesis, University of Texas at Austin.

Smarr, L. (1979), *In Sources of Gravitational Radiation*, edited by Smarr, L., Cambridge University Press, Cambridge.

Smarr, L., and York, J., (1978), *Phys. Rev. D.*, **17**, 2529.

Wilson, J. (1979), *In Sources of Gravitational Radiation*, edited by Smarr, L., Cambridge University Press, Cambridge.

Woodward, P.R. (1988), In *High Speed Computing*, edited by R. Wilhelmson, Univ. of Illinois Press, Urbana.

Computational Relativity in Two and Three Dimensions

STUART L. SHAPIRO AND SAUL A. TEUKOLSKY
Center for Radiophysics and Space Research, and
Departments of Astronomy and Physics,
Cornell University, Ithaca NY 14853.

I. Introduction

We are pursuing a long-term program involving the numerical solution of Einstein's equations for the dynamical evolution of a *collisionless* gas of particles in general relativity. This is an important problem for two reasons. Astrophysically, the collapse of relativistic star clusters may be responsible for the formation of supermassive black holes. Computationally, collisionless systems are ideal for developing methods for solving Einstein's equations on the computer. Unlike fluid matter, collisionless particles move along geodesics, which are trivial to integrate. Accordingly, all attention can focus on solving the gravitational field equations without the complications associated with hydrodynamics.

Here we shall briefly report on three recent projects that are all loosely connected with this large-scale computational effort. The first project involves the construction of an axisymmetric, *Newtonian*, mean-field particle simulation code which follows the evolution of axisymmetric stellar dynamical systems. The scheme is a logical extension to 2+1 dimensions of our previous 1+1 dimensional relativisitic code for spherical systems in the Newtonian limit. It is an essential step toward the building of a fully relativistic code for collisionless systems in 2+1 dimensions. Details of this work may be found in Shapiro and Teukolsky (1987).

The second project involves the study of naked singularities and the Hoop Conjecture in general relativity. These are issues of long-standing interest in relativity theory. In the absence of global theorems, the formation of naked singularities and the validity of the Hoop Conjecture may require numerical simulations for definitive understanding. As a prelude to such simulations we have constructed analytic, time-symmetric dust configurations to explore singularities and the appearance of horizons in 2+1 dimensions. Details of this study are presented in Nakamura, Shapiro and Teukolsky (1988).

Finally, our third investigation concerns accretion onto a moving black hole in general relativity theory. We have been pursuing the classic Bondi-Hoyle problem of gas accretion onto a moving mass, which we have taken to be a black hole. Our focus has been numerical, and we have performed 2+1 hydrodynamic simulations in general relativity to find the steady-state flow solutions (Petrich, Shapiro, Stark and Teukolsky 1989). Interestingly, we have also discovered an *analytic* solution for the flow when the matter obeys the special equation of state $P = \rho$. The solution applies to accretion onto a Schwarzschild or a Kerr black hole and in the latter case is fully three dimensional. Complete details are given in Petrich, Shapiro and Teukolsky (1988).

II. Simulations of Axisymmetric Newtonian Star Clusters: Prelude to 2+1 General Relativistic Computations

We have previously constructed a numerical code that solves Einstein's equations for the dynamical evolution of a *collisionless* gas of particles in general relativity (Shapiro and Teukolsky 1985a,b,c and 1986). That investigation was restricted to spherically symmetric systems, but the gravitational field could be arbitrarily strong and particle velocities could be arbitrarily close to the speed of light. Our computational scheme combined the tools of numerical relativity with those of N-body particle simulation. We solved the Vlasov equation in general relativity by particle simulation and determined the gravitational field using the ADM $3 + 1$ formalism. Physical applications included the stability of relativistic star clusters, the binding energy criterion for stability, the collapse of star clusters to black holes, and relativistic violent relaxation. Astrophysical applications included the possible origin of quasars and active galactic nuclei via the collapse of relativistic star clusters to supermassive black holes. We found that our method is extremely accurate, even in the case of black hole formation. It provided a unique proving ground for testing different computational algorithms and gauge choices for the construction of numerical spacetimes.

Our present goal is to generalize our computational scheme to handle nonspherical clusters and nonspherical collapse. Our effort will focus on *axisymmetric* configurations initially. This restriction still permits two new dynamical features to arise which are entirely absent in spherical symmetry: *rotation* and *gravitational radiation*. Ours will be the first attempt to deal with these features numerically in the context of collisionless matter.

As a first step in extending our code to axisymmetric spacetimes, we recently completed the axisymmetric, *Newtonian* version of our numerical method (Shapiro and Teukolsky 1987). We again constructed a mean-field, particle simulation scheme to follow the evolution of axisymmetric dynamical systems. The scheme has been designed with the intent of subsequent generalization to fully relativistic spacetimes. The "mean-field" approach is ideally suited for treating collisionless gases. In the Newtonian domain, this approach is the most reliable way to solve the Vlasov equation with minimum contamination from spurious collisional effects (in contrast to a direct N-body integration scheme employing only a few thousand particles). Also, in the ADM 3+1 formalism that we adopt to treat the fully relativistic problem, the metric equations are naturally regarded as "mean-field" equations: they determine the background gravitational field arising from a smoothed-out distribution of matter sources.

We recognize that the scheme that we construct may not be the most efficient for purely Newtonian problems. The Newtonian gravitational field equation is Poisson's equation, which is strictly linear and separable. Elegant methods exist to handle elliptic equations with these special properties, such as those based on Fast Fourier Transforms (see e.g., Hockney and Eastwood 1981 and references therein). We choose instead to integrate Poisson's equation by finite-differencing. This approach is most easily adaptable to the strong field, fully relativistic domain, where the analogous field equations are highly nonlinear.

We assessed the reliability of our code by performing a number of nontrivial test-bed calculations. Some of these calculations possess known "analytic" solutions for the 2+1 axisymmetric evolution of both equilibrium and nonequilibrium configurations. Some of these "analytic" solutions require the integration of coupled *ordinary* differential equations, but this can be done to essentially unlimited accuracy. We solved these equations simultaneously with our numerical simulations in order to make detailed comparisons.

We have been particularly interested in the ability of our code to handle cases in which the collisionless configuration collapses to a *singularity*. Specifically, we followed the evolution of axisymmetric, homogeneous configurations that collapse to flat oblate pancakes and also to thin prolate spindles. Spindles—or string-like singularities—are especially significant because not only the density but also the Newtonian gravitational potential becomes infinite as the spindle thickness collapses to zero. Such outcomes are possible only in *collisionless* collapse, since gas pressure prevents line singularities and pancakes from forming in *hydrodynamic* collapse. We are interested in such singularities because we hope to explore their properties later in full general relativity. There it will be relevant to determine whether or not such singularities form during relativistic collapse. By exploring this issue, we can test the "Cosmic Censorship Hypothesis" (Penrose 1969). This hypothesis states that "naked" singularities do not form in general relativity from the evolution of nonsingular initial data in asymptotically flat spacetimes. Instead, all singularities must be "clothed" by event horizons, that is, be inside black holes. We can also examine the "Hoop Conjecture" (Thorne 1972), which requires that a configuration be sufficiently compact in all three spatial dimensions in order to form a black hole. Performing Newtonian test-bed calculations that lead to singularities and have "exact" solutions thus paves the way for these future relativistic investigations. Further discussion of these issues in general relativity is given in Section III.

As another application of our axisymmetric Newtonian code, we explored the dynamical stability of equilibrium Freeman spheroids against axisymmetric perturbations. These homogeneous, oblate, equilibrium spheroids are the collisionless analogues of rotating Maclaurin spheroids in fluid dynamics. We determined numerically where along an equilibrium Freeman sequence parametrized by eccentricity an axisymmetric (ring) instability sets in. Previously, this issue had only been addressed in linear perturbation theory using very approximate techniques (see Fridman and Polyachenko 1984 for a review and references). For comparison with our mean-field computations, we also performed direct N-body integrations using 1000 point masses to reassess the stability of Freeman spheroids. An N-body code automatically allows for the growth of nonaxisymmetric perturbations as well as axisymmetric perturbations. Hence we used the N-body code to determine under what circumstances the nonaxisymmetric modes, which are suppressed in our mean-field, particle simulation scheme, actually set in before the axisymmetric modes.

With our axisymmetric code, we found that sufficiently flat "cold" spheroids are unstable to the formation of rings, while "hot" spheroids are stable. To assess the importance of nonaxisymmetric modes, we evolved the same spheroids with an N-body code. We found that a Freeman spheroid is unstable to the formation of a rotating bar if its net angular momentum is sufficiently large.

We compared the gravitational collapse of cold, nonrotating, homogeneous spheroids with the collapse of inhomogeneous spheroids. An oblate homogeneous spheroid collapses to a pancake, overshoots, and ends up as a prolate string-like singularity. An inhomogeneous spheroid undergoes violent relaxation, which drives it to virial equilibrium. Our code can reliably distinguish between these outcomes.

III. Naked Singularities

It is well-known that general relativity admits solutions with singularities, and that such solutions can be produced by the gravitational collapse of nonsingular, asymptotically flat initial data. If Cosmic Censorship holds, then there is no problem with predicting the future evolution outside the event horizon. Alternatively, if it does not hold, then the formation of a naked singularity during such collapse would be a disaster for general relativity theory. In this situation, one cannot say anything precise about the future evolution of any region of space containing the singularity since new information could emerge from it in a completely arbitrary way.

But what guarantees are there that an event horizon will always hide a naked singularity? There are no completely definitive theorems as yet. In the case of spherical collapse, the solution outside the collapsing matter is the Schwarzschild metric. In all numerical and analytic studies, the singularity occurs inside[1] the event horizon at $r_s = 2M$. Results for nonspherical collapse are less complete. For this situation, Thorne (1972) has proposed the *Hoop Conjecture*: Black holes with horizons form when and only when a mass M gets compacted into a region whose circumference in *every* direction is $C \lesssim 4\pi M$.

If the Hoop Conjecture is indeed correct, aspherical collapse with one or two dimensions appreciably larger than the others might then lead to naked singularities. For example, consider the Lin-Mestel-Shu instability for the collapse of a nonrotating, homogeneous spheroid of dust in Newtonian gravity (Lin, Mestel and Shu 1965). If the spheroid is slightly oblate, the configuration collapses to a pancake, while if the spheroid is slightly prolate, it collapses to a spindle. While in both cases the density becomes infinite, the formation of a spindle during prolate collapse is particularly worrisome. The gravitational potential, gravitational force, tidal force, potential and kinetic energies all blow up. This behavior is far more serious than mere shell-crossing, where the density alone becomes momentarily infinite. In the case of collisionless matter, prolate evolution is forced to terminate at the singular spindle state, while for oblate evolution the matter simply passes through the pancake state. In fact, having passed through the pancake, the oblate configuration then evolves to a spindle singularity.

Does this example have any relevance to general relativity? We already know that *infinite* cylinders do collapse to singularities in general relativity, and, in accord with the Hoop Conjecture, are not hidden by event horizons (Thorne 1972; Misner, Thorne and Wheeler 1973). It has been argued that the ultimate singular state will be avoided by the presence of pressure, as long as the adiabatic index $\Gamma > 1$ (Thorne 1972; Misner, Thorne and Wheeler 1973; Piran 1979). However, we feel that this does not address the fundamental problem. Naked singularities could then still form in the case of *perfectly* collisionless matter. Does the possibility of forming naked singularities then depend on the details of the interactions affecting matter at high densities? Does the degree to which collapse becomes singular depend

---

[1] However, for a discussion of naked singularities related to shell crossing in spherical collapse, see Yodzis, Seifert and Muller zum Hagen (1973). Recently, Ori and Piran (1987) have presented a self-similar solution for the spherical collapse of a perfect fluid. Their solution exhibits a naked shell-focussing singularity if the equation of state is sufficiently soft. Lake (1988) has subsequently shown that the singularity is strong, in the technical sense used in general relativity. It is not known whether these solutions generalize to nonspherical collapse. See also the article by Piran in this volume.

on the complete zoo of coupling constants, gauge fields, supersymmetric partners, and so on? We would be far more comfortable knowing that Einstein's equations automatically prevent naked singularities with only very weak conditions imposed on the matter stress-energy tensor. The key question thus is not about pressure but is whether singularites form during the prolate collapse of a finite object in asymptotically flat spacetime.

Interestingly, there exists a class of static axisymmetric solutions of the vacuum Einstein equations that correspond to the exterior fields of prolate and oblate spheroidal configurations (Voorhees 1970). These configurations are finite in extent and the spacetimes are asymptotically flat. Yet, they have the *same* singularity structure that characterizes the corresponding Newtonian spheroids. Because these solutions do not contain matter, however, their relevance to the formation of naked singularities during gravitational collapse is not clear.

More troublesome are the simulations of axisymmetric fluid collapse by Nakamura and his collaborators (Nakamura and Sato 1982; Nakamura, Oohara and Kojima 1987). They simulated the general relativistic collapse of deformed stars with internal pressure. They found that if the initial internal energy was appreciable, then apparent horizons always formed, no matter how large the initial deformation was. However, if the initial internal energy was small and the initial deformation large, the results were different. Specifically, they describe an example of prolate collapse that appears to be evolving to a singular state, despite the presence of pressure obeying a very stiff equation of state (asymptotically $\Gamma = 2$). By the time the simulation was terminated, no apparent horizon had appeared. Of course it is always possible that an event horizon had already formed, but this was not studied because it is tedious in two dimensions.

Given all this, it is clearly desirable to perform detailed numerical simulations of the prolate collapse of realistic initial configurations with matter and to probe the resulting spacetimes for the growth of singularities and the development of event horizons. As a first step, one must construct a suitable family of initial configurations for such an evolution. Accordingly, we have solved the initial-value problem in general relativity for a class of axisymmetric prolate and oblate spheroids containing matter (Nakamura, Shapiro and Teukolsky 1988). The configurations are finite-size, inhomogeneous dust spheroids. The matter is instantaneously at rest at $t = 0$ (moment of time symmetry) and the dynamical components of the gravitational field are set equal to zero. For the cases we considered, the Hamiltonian constraint equation reduces to Poisson's equation. For the adopted density profile, the solutions are analytic and are determined from the solutions for homogeneous Newtonian spheroids.

We analyzed these solutions for apparent horizons to assess the validity of the

Hoop Conjecture. We were particularly interested in extreme configurations with eccentricities approaching unity as candidates for singularities. Should these configurations show singularities that are not hidden by horizons, this would suggest that naked singularities could actually arise in dynamical collapse. We explored this possibility by considering a sequence of momentarily static configurations of fixed rest mass but increasing eccentricity. We found that highly eccentric prolate and oblate spheroids are indeed singular, for example as measured by a curvature invariant. In agreement with the Hoop Conjecture, extended configurations have no apparent horizons. Hence, the validity of the unqualified Cosmic Censorship Hypothesis is somewhat suspect. Further dynamical calculations are urgently needed.

IV. Accretion onto a Moving Black Hole

Accretion of gas onto astronomical objects is an important phenomenon of long-standing interest to astrophysicists. There are many environments where such accretion provides the underlying source of energy for the emitted radiation. Examples include accretion onto compact objects in binary star systems, accretion onto compact objects moving through the interstellar medium, and accretion onto supermassive black holes in the cores of active galactic nuclei and quasars.

Consider a black hole moving at constant velocity through a gaseous, adiabatic medium at rest and with uniform density at infinity. Determining the steady-state flow poses a classic problem in accretion theory. The Newtonian version of this problem—accretion onto a Newtonian point mass moving nonrelativistically through a nonrelativistic gas—was first discussed by Bondi and Hoyle (1944), but only in qualitative terms. Only in the limit of spherical accretion, appropriate for a stationary black hole, have exact solutions been found (see Bondi 1952 for the Newtonian solution or Shapiro and Teukolsky 1983 and Michel 1972 for the solution in general relativity).

In general, numerical approaches are required to handle nonspherical accretion for either Newtonian or relativistic flow. Accordingly, we have performed numerical calculations of accretion onto a Scwarzschild black hole moving through a uniform gaseous medium at a constant velocity. These calculations have been carried out over a wide range of model input parameters (e.g., gas adiabatic index and sound speed, and black hole Mach number). Details of these calculations are given in Petrich, Shapiro, Stark and Teukolsky 1989.

Amazingly, there is one exact, fully relativistic, nonspherical solution which provides valuable physical insight into the more general cases and serves as a benchmark for testing numerical codes (Petrich, Shapiro and Teukolsky 1988). Our solution is for a black hole moving through a medium obeying a stiff $P = \rho$ equation of state. The black hole may be either Schwarzschild or Kerr. As the sound speed is equal to the speed of light, the flow is everywhere subsonic and the solution has no Newtonian analogue. Surprisingly, the angle between the angular momentum vector of the black hole and the direction of the incident flow can be arbitrary. Consequently, the solution can serve as a unique diagnostic not only of spherical and axisymmetric, but also of fully *three-dimensional*, hydrodynamic codes in general relativity.

ACKNOWLEDGEMENTS

This work has been supported in part by National Science Foundation grants AST 87-14475 and PHY 86-03284 at Cornell University. Computations were performed on the Cornell National Supercomputer Facility, which receives major funding from the National Science Foundation, IBM corporation, New York State, and members of the Cornell Research Institute.

References

Bondi, H., 1952, *Mon. Not. Roy. Astr. Soc.*, **112**, 195.
Bondi, H. and Hoyle, F., 1944, *Mon. Not. Roy. Astr. Soc.*, **104**, 272.
Fridman, A. M., and Polyachenko, V. L. 1984, *Physics of Gravitating Systems* (New York: Springer-Verlag), Ch. 4.
Hockney, R. W., and Eastwood, J. W. 1981, *Computer Simulation Using Particles* (New York: McGraw Hill).
Lake, K. 1988, *Phys. Rev. Lett.*, **60**, 241.
Lin, C. C., Mestel, L. and Shu, F. H. 1965, *Astrophys. J.*, **142**, 1431.
Michel, F.C., 1972, *Astrophys. Space Sci.*, **15**, 153.
Misner, C. W., Thorne, K. S., and Wheeler, J. A. 1973, *Gravitation* (San Francisco: Freeman).
Nakamura, T. and Sato, H. 1982, *Prog. Theor. Phys.*, **68**, 1396.
Nakamura, T., Oohara, K. and Kojima, Y. 1987, *Prog. Theor. Phys. Suppl.* No. 90, p. 57.
Nakamura, T., Shapiro, S. L. and Teukolsky, S. A. 1988, *Phys. Rev.*, **D**, in press.
Ori, A. and Piran, T. 1987, *Phys. Rev. Lett.* **59**, 2137.
Penrose, R. 1969, *Rivista del Nuovo Cimento* **1** (Numero Special), 252.
Petrich, L., Shapiro, S. L. and Teukolsky, S. A. 1988, *Phys. Rev. Lett.*, **60**, 1781.
Petrich, L., Shapiro, S. L., Stark, R. F. and Teukolsky, S. A. 1989, *Astrophys. J.*, in press.
Piran, T. 1979, in *Sources of Gravitational Radiation*, edited by L. L. Smarr (Cambridge University Press, Cambridge), p. 409.
Shapiro, S.L. and Teukolsky, S.A. 1983, *Black Holes, White Dwarfs, and Neutron Stars: the Physics of Compact Objects*, (New York: John Wiley and

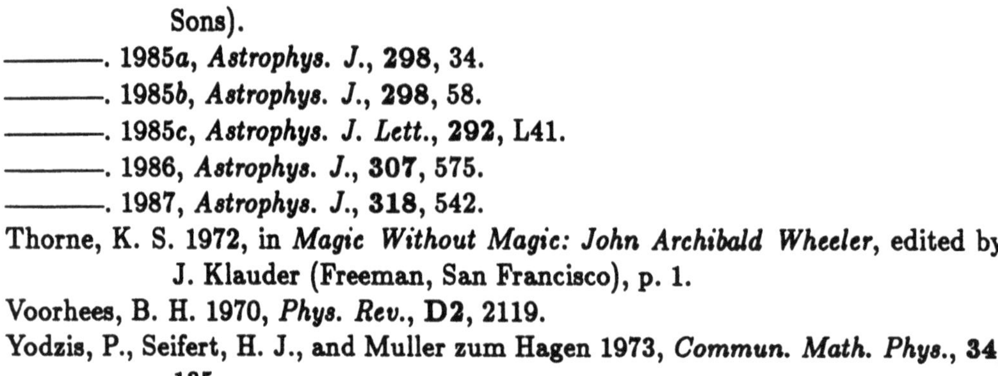

Sons).

———. 1985*a*, *Astrophys. J.*, **298**, 34.

———. 1985*b*, *Astrophys. J.*, **298**, 58.

———. 1985*c*, *Astrophys. J. Lett.*, **292**, L41.

———. 1986, *Astrophys. J.*, **307**, 575.

———. 1987, *Astrophys. J.*, **318**, 542.

Thorne, K. S. 1972, in *Magic Without Magic: John Archibald Wheeler*, edited by J. Klauder (Freeman, San Francisco), p. 1.

Voorhees, B. H. 1970, *Phys. Rev.*, **D2**, 2119.

Yodzis, P., Seifert, H. J., and Muller zum Hagen 1973, *Commun. Math. Phys.*, **34**, 135.

SLOWLY MOVING MAXIMALLY CHARGED BLACK HOLES

Robert C. Ferrell
Department of Physics
University of California Santa Barbara, CA 93106 USA

Douglas M. Eardley
Institute for Theoretical Physics
University of California Santa Barbara, CA 93106 USA

Abstract We study interactions of slowly moving, non-rotating, maximally charged ("Reissner-Nordstrøm") black holes. The slow motion approximation allows us to neglect both gravitational and electromagnetic radiation in the spacetime. Maximally charged black holes induce only small, velocity dependent, forces on one another so our analysis is valid even when the black holes approach each other closely. In the approximating spacetime we use, the field degrees of freedom, which would be associated with radiation, are not present. Instead, we foliate the spacetime with a sequence of spacelike 3-dimensional slices, and assume that on any given slice the gravitational and electromagnetic fields are determined entirely in terms of the positions and velocities of the black holes. The relations between the fields and the black holes are slow motion approximations to the initial value equations of general relativity and of electromagnetism. The motion of the black holes is found from an effective action which is the usual action for the coupled gravitational-electromagnetic system, but with the fields "integrated out". We find that there are both scattering orbits and orbits which result in coalescence of the two black holes. There are no stable periodic orbits, but there is one unstable circular orbit.

1. INTRODUCTION

In this lecture we will discuss an approximate solution to a special type of two body problem in general relativity — the motion of two slowly moving maximally charged but non-rotating black holes. Our approximate solution is of a different nature from most of the previous analytical work on the two body problem, and for this reason it is interesting. However, because the black holes are maximally charged it is probably not of direct astrophysical interest.

Much of the analytical work on the two body problem has been on either the "planetary" problem — a test mass moving in the background field of a massive star or black hole — or two massive bodies moving in the asymptotic — weak field — region of each other. This work generally falls into the realm of post-Newtonian and higher order corrections to Newtonian gravity. An extensive review of this work, and more references, can be found in Damour (1983). Eardley (1986) has studied the spacetime of two infalling Schwarzshild black holes, but the trajectories of the black holes had to be put in by hand. We have found that we can derive (and solve) equations of motion for two maximally charged black holes in the strong field region of each other, but with the restriction that they must be moving slowly.

2. PERTURBING A STATIC SOLUTION

2.1 The Majumdar-Papapetrou static spacetimes

It has been known for a long time that for a charged pressureless dust whose mass density = charge density (in units where $G = c = 1$) there are static solutions to the coupled Einstein-Maxwell-matter equations of the form:

$$ds^2 = -\psi^{-2}dt^2 + \psi^2 d\underset{\sim}{x} \cdot d\underset{\sim}{x} \tag{1}$$

$$A = -(1 - \psi^{-1})dt \tag{2}$$

$$\nabla^2\psi = -4\pi\rho\psi^3 \equiv -4\pi\tilde{\rho} \tag{3}$$

where ∇^2 is the flat Laplacian for the flat background 3-space and ρ is the mass or charge density. Roughly speaking, this can be a static solution since the electric force on each dust particle is exactly canceled by the gravitational force. Of course, classically this is obvious, since both forces go like $1/r^2$, but with opposite signs. It was discovered independently by Majumdar (1947) and Papapetrou (1947) that this is true also in general relativity.

Remarkably, there is no obstacle to letting the matter distribution become singular so that it actually describes black holes rather than dust. (This was also known to Majumdar and Papapetrou, and was studied further by Arnowitt *et al.* (1960).) As the dust density increases, a dimple in spacetime forms—the beginning of a throat. For uncharged matter we know that as the density increases this dimple gets deeper and then pinches off into a "bag of gold". For maximally charged matter, the dimple never pinches off, but rather the throat gets longer and longer. When the distribution finally becomes a singular delta function, representing a maximally charged black hole, the throat is infinitely long.

The solution to (3) when the sources are black holes,

$$\tilde{\rho} = \sum_a m_a \delta^{(3)}(\underset{\sim}{x} - \underset{\sim}{x}_a), \tag{4}$$

is

$$\psi = 1 + \sum_a \frac{m_a}{r_a} \tag{5}$$

where the black holes are at $\underset{\sim}{x} = \underset{\sim}{x}_a$, with masses m_a and $r_a = |\underset{\sim}{x} - \underset{\sim}{x}_a|$; the lengths are calculated in the flat metric on $\mathbf{R}^3$. In the coordinate system of (1) the event horizons are represented by the points $\underset{\sim}{x} = \underset{\sim}{x}_a$. One can check that each event horizon has surface area $4\pi m_a$. These are *maximally* charged black holes because a singularity with charge q_a greater than mass m_a does not have an event horizon, and is therefore a naked singularity.

2.2 The slow motion approximation

The question we address in this lecture is what happens when maximally charged black holes in a static configuration are perturbed slightly, so that they move slowly. The only forces they will feel will be the magnetic and gravito-magnetic ("frame-dragging") forces because the electrostatic and gravitostatic forces cancel exactly. As long as the relative velocities are small these forces will be small, even if the black holes are close to one another, *i.e.*, separated by distances of order m.

The small velocity approximation provides a great simplification because in this limit we can neglect radiation, both gravitational and electromagnetic. As discussed *e.g.*, in Landau and Lifshitz (1975) radiation is a consequence of the fact that field disturbances propagate with finite speed, so the fields can be independent of the sources. If there is no radiation, however, then the fields are completely specified by the source positions and velocities. On the one hand the fields have an infinite number of degrees of freedom. On the other hand, the black holes each have three degrees of freedom. The small velocity/no radiation approximation, then, is just a truncation of the infinite number of field degrees of freedom to a finite number. This is a great simplification. A familiar example is the field of a slowly moving point particle with charge q. In flat space, the electrostatic potential is $\phi(\underset{\sim}{r}) = q/|\underset{\sim}{r} - \underset{\sim}{r}_0|$ for a particle at position $\underset{\sim}{r}_0$. If the charge is moving slowly, so that $\underset{\sim}{r}_0 = \underset{\sim}{r}_0(t)$ then the scalar potential is approximately $\phi(\underset{\sim}{r}, t) = q/|\underset{\sim}{r} - \underset{\sim}{r}_0(t)|$. There is also a vector potential, which is approximately $\underset{\sim}{A}(\underset{\sim}{r}, t) = q\underset{\sim}{v}(t)/|\underset{\sim}{r} - \underset{\sim}{r}_0(t)|$, where $\underset{\sim}{v} = d\underset{\sim}{r}_0/dt$. These fields give an approximate solution to the Maxwell field equations.

3. DERIVING THE EQUATIONS OF MOTION

3.1 The action in the slow motion limit

To study the motion of maximally charged black holes we will derive an effective Lagrangian which is a function only of the black hole positions and velocities. Instead of the gravitational and electromagnetic fields being independent degrees of freedom in the field Lagrangian density, we will assume that they are solutions to the initial value equations. (We can do this because there is no radiation in the slow motion approximation.) The result will be an effective Lagrangian for maximally charged black holes. The problem is then reduced to classical mechanics of point particles, with the dynamics specified by the effective Lagrangian.

We shall assume a spacetime which is a simple generalization of the Majumdar-Papapetrou static spacetime. Namely, we introduce a shift vector and a vector potential, so that

$$ds^2 = -\psi^{-2}dt^2 + 2\underset{\sim}{\mathbf{N}} \cdot d\underset{\sim}{\mathbf{x}}dt + \psi^2 d\underset{\sim}{\mathbf{x}} \cdot d\underset{\sim}{\mathbf{x}} \tag{6}$$

$$A = -(1 - \psi^{-1})dt + \underset{\sim}{\mathbf{A}} \cdot d\underset{\sim}{\mathbf{x}}. \tag{7}$$

In this expression, ψ is given by (3) and will be a function of time because the source term is time dependent. We will do most of the calculations to derive the effective Lagrangian using a general matter density $\tilde{\rho}$. This is a kind of regularization that avoids some infinities in the calculations. In the end, we will choose $\tilde{\rho}$ as in (4) to describe a system of black holes. We will be interested in the lowest order dependence of $\underset{\sim}{\mathbf{N}}$ and $\underset{\sim}{\mathbf{A}}$ on $\tilde{\rho}$ and $\underset{\sim}{\mathbf{v}}$.

One way to derive equations relating $\underset{\sim}{\mathbf{N}}$ and $\underset{\sim}{\mathbf{A}}$ to $\tilde{\rho}$ and $\underset{\sim}{\mathbf{v}}$ would be to simply write out the initial value equations (also called the constraint equations in the Arnowitt-Deser-Misner (ADM) (Arnowitt *et al.* 1962; Misner *et al.* 1972) Hamiltonian formulation of general relativity) for the gravity-electromagnetic system and keep only first order terms in $\underset{\sim}{\mathbf{N}}$, $\underset{\sim}{\mathbf{A}}$ and $\underset{\sim}{\mathbf{v}}$, and also keep only single time derivatives. In particular, for the gravitational fields, these are the super-Hamiltonian constraint

$$G^{00} = 8\pi T^{00} \tag{8}$$

and the super-momentum constraint

$$G^{0i} = 8\pi T^{0i}. \tag{9}$$

To the order we need, (8) is satisfied by (3). For the electro-magnetic fields, the initial value equations are

$$D_i F^{i0} = 4\pi J^0 \tag{10}$$

$$D_i F^{ij} = 4\pi J^j. \tag{11}$$

These are just Gauss's law and Ampere's law in the coordinate system of (6). Gauss's law, (10), is satisfied to the order we need by (3) and $\phi = 1 - 1/\psi$ as in (7). We are, therefore, left with two first order vector equations, (9) and (11), relating $\underset{\sim}{N}$ and $\underset{\sim}{A}$ to $\tilde{\rho}$ and $\underset{\sim}{v}$. These are the initial value equations for $\underset{\sim}{N}$ and $\underset{\sim}{A}$. It is a simple exercise to derive these equations for the particular spacetime (6) and (7).

Rather than derive the first order initial value equations in this way, we derived them directly from the Lagrangian density for the Einstein-Maxwell-matter system. We did this because eventually we sought an effective point particle Lagrangian from which we could derive equations of motion for a system of maximally charged black holes.

In our approach, we wrote out the Lagrangian to second order in $\underset{\sim}{N}$, $\underset{\sim}{A}$ and $\underset{\sim}{v}$ (since we wanted first order field equations). We used the 3+1 formalism of ADM (Arnowitt *et al.* 1962) and the spacetime of (6) and (7). The time slicing is implied by (6), *i.e.*, the spacelike 3-surfaces are $t =$ constant hypersurfaces. The fields $\underset{\sim}{N}$ and $\underset{\sim}{A}$ are 3-tensors on this surface. The field equations will be the initial value equations on the initial surface. That is, we varied the action with respect to $\underset{\sim}{N}$ and $\underset{\sim}{A}$ to derive the initial value equations. This is, of course, exactly how one derives (9) and (11). However, since we had already truncated the Lagrangian to second order the equations we find are the approximate, first order initial value equations which we wanted. We sketch the derivation below. More details will be published in a subsequent paper.

The action for the Einstein-Maxwell-matter system can be written as a sum of four pieces:

$$S = S_{\text{gravity}} + S_{\text{Maxwell fields}} + S_{\text{current}} + S_{\text{matter}}.$$

For the fields, these are,

$$S_{\text{gravity}} = \frac{1}{16\pi} \int_M {}^{(4)}R\sqrt{-{}^{(4)}g}\, d^4x + \text{ Boundary Term} \tag{12}$$

and

$$S_{\text{Maxwell}} = -\frac{1}{8\pi} \int_M F \wedge {}^*F \tag{13}$$

where $F = dA$ is to be considered a derived field. For the matter, if we used point singularities then the action would be

$$S_{\text{matter}} = -\sum_a \int m_a d\tau_a$$

where $d\tau_a$ is the proper length of the world line, and a labels black holes. In our regularization scheme, we replace point masses by a dust distribution $\tilde{\rho}$, and we need the action for this dust. Each particle of dust has a world line $x^i(t)$. The tangent vector to this world line is $v^\mu = (1, v^i)$ where $v^i = \frac{dx^i}{dt}$. The action is

$$S_{\text{matter}} = -\int dt d^3x\, \tilde{\rho} \left(\psi^{-2} - \psi^2 \delta_{ij}(v^i + N^i)(v^j + N^j)\right)^{1/2}.$$

Similarly, for a charged dust the action is

$$S_{\text{current}} = \int dt d^3x\, \tilde{\rho}\, A_\mu \frac{dx^\mu}{dt}.$$

There is now a lot of straight forward algebra and index gymnastics involved in writing ${}^{(4)}R$ and $F_{\mu\nu}$ in terms of $\underset{\sim}{N}$ and $\underset{\sim}{A}$. Since we are interested in first order field equations, we need to keep only terms of second order in $\underset{\sim}{N}$, $\underset{\sim}{A}$ or $\underset{\sim}{v}$. Once the dust settles we arrive at the following approximate action:

$$\begin{aligned} S_{\text{approx}} = \int_M \Bigg[& -\frac{3}{8\pi}\dot{\psi}^2\psi^2 - \tilde{\rho} + \tilde{\rho}\frac{v^2\psi^3}{2} \\ & + (\underset{\sim}{A} + \psi\underset{\sim}{N}) \cdot \left(\tilde{\rho}\underset{\sim}{v} - \frac{1}{4\pi}\underset{\sim}{\nabla}\dot{\psi}\right) \\ & - \frac{1}{8\pi}\frac{|\underset{\sim}{\nabla} \times (\underset{\sim}{A} + \psi\underset{\sim}{N})|^2}{\psi^2} + \frac{1}{4\pi}\frac{(\underset{\sim}{\nabla} \times (\underset{\sim}{A} + \psi\underset{\sim}{N})) \cdot (\underset{\sim}{\nabla} \times (\psi^2\underset{\sim}{N}))}{\psi^3} \\ & - \frac{3}{32\pi}\frac{|\underset{\sim}{\nabla} \times (\psi^2\underset{\sim}{N})|^2}{\psi^4} \Bigg]. \end{aligned} \tag{14}$$

In this expression, $\dot{\psi} = \partial\psi/\partial t$. Also, to arrive at this form , we had to integrate by parts many times. Since our spacetime is asymptotically flat, total divergences vanish.

3.2 The effective Lagrangian

To derive the initial value equations, we vary (14) with respect to $\underset{\sim}{N}$ and $\underset{\sim}{A}$. One finds

$$\underset{\sim}{\nabla} \times \left[\frac{\underset{\sim}{\nabla} \times (\underset{\sim}{A} + \psi\underset{\sim}{N})}{\psi^2} - \frac{\underset{\sim}{\nabla} \times (\psi^2\underset{\sim}{N})}{\psi^3}\right] = \underset{\sim}{\nabla} \times (\underset{\sim}{\nabla} \times \underset{\sim}{K}) \tag{15a}$$

$$\underset{\sim}{\nabla} \times \left[\frac{\underset{\sim}{\nabla} \times (\underset{\sim}{A} + \psi\underset{\sim}{N})}{\psi^3} - \frac{3}{4}\frac{\underset{\sim}{\nabla} \times (\psi^2\underset{\sim}{N})}{\psi^4}\right] = 0 \tag{15b}$$

where

$$\underset{\sim}{K} \equiv -4\pi\nabla^{-2}(\tilde{\rho}\underset{\sim}{v}). \tag{16}$$

Here we had to use the fact that current is conserved, *i.e.*,

$$\underset{\sim}{\nabla}\cdot(\tilde{\rho}\underset{\sim}{v}) + \partial_0\tilde{\rho} = 0. \tag{17}$$

and also (3) to write $4\pi\tilde{\rho}\underset{\sim}{v} - \underset{\sim}{\nabla}\dot{\psi} = \underset{\sim}{\nabla}\times\underset{\sim}{K}$. Equations (15) are linear combinations of the first order approximations to (9) and (11). The zeroth order Hamiltonian constraint, (8), and the Gauss's law constraint, (10), are solved by (2,3).

Tantamount to expressing the fields in terms of only the matter degrees of freedom is evolving the matter in such a fashion that the constraints are preserved. To do this, we derived equations of motion for the black holes from the approximate action of above as follows. In (14), we let $\underset{\sim}{N}$ and $\underset{\sim}{A}$ be names for the solutions to (15). The only free variables left are the positions and velocities of the black holes. We find the equations of motion from the Euler-Lagrange equations. It is a bit of a miracle that the system (15) can be solved in terms of $\tilde{\rho}$ and $\underset{\sim}{v}$ so that $\underset{\sim}{N}$ and $\underset{\sim}{A}$ can be eliminated from (14). There are some subtleties due to functions of integration that arise which we will discuss in detail in another paper, but which are not important for our discussion here. The effective action for a system of black holes is found by using a sum of delta function sources in (3) and in (16). These equations are then easily solved:

$$\psi = 1 + \sum_a \frac{m_a}{r_a}$$

$$\underset{\sim}{K} = \sum_a \frac{m_a}{r_a}\underset{\sim}{v}_a$$

where

$$\underset{\sim}{v}_a = \text{velocity of } a^{\text{th}} \text{ source} = \frac{d\underset{\sim}{x}_a}{dt}.$$

Using this, the solution to (15) and $\dot{\psi} = \partial_0\nabla^{-2}(-4\pi\tilde{\rho}) = +4\pi\nabla^{-2}(\underset{\sim}{\nabla}\cdot(\tilde{\rho}\underset{\sim}{v})) = -\underset{\sim}{\nabla}\cdot\underset{\sim}{K}$ (from (3) and (17)) in (14) we found:

$$S_{\text{eff}} = \int dt\ L_{\text{eff}} = \int dt\ (L_{\text{free}} + L_{\text{int}})$$

where

$$L_{\text{free}} = \frac{1}{2}\sum_a m_a v_a^2 - \sum_a m_a \tag{18a}$$

$$L_{\text{int}} = \int d^3r\left(-\frac{3}{8\pi}\right)\left(1 + \sum_a \frac{m_a}{r_a}\right)^2 \times$$

$$\sum_{b\neq c}\frac{m_b m_c}{(r_b r_c)^3}\left\{(\underset{\sim}{v}_b\times\underset{\sim}{v}_c)\cdot(\underset{\sim}{r}_b\times\underset{\sim}{r}_c) - \frac{1}{2}|\underset{\sim}{v}_b - \underset{\sim}{v}_c|^2(\underset{\sim}{r}_b\cdot\underset{\sim}{r}_c)\right\}. \tag{18b}$$

The problem has been reduced to the study of the motion point particles, *i.e.*, classical mechanics.

3.3 The effective Lagrangian for two black holes

For an arbitrary number of black holes we have not been able to do the integral in (18b) explicitly. However, for two black holes we can, and we can also integrate the equations of motion. Therefore, for the remainder of this lecture we will discuss the case of two black holes.

The effective action for two maximally charged black holes is

$$S_{\text{eff}} = \int dt \left(\frac{1}{2} M \underset{\sim}{\mathbf{V}} \cdot \underset{\sim}{\mathbf{V}} - M + \frac{1}{2} \mu \underset{\sim}{\mathbf{v}} \cdot \underset{\sim}{\mathbf{v}} + \frac{3\mu M \underset{\sim}{\mathbf{v}} \cdot \underset{\sim}{\mathbf{v}}}{2} \left(\frac{1}{r} + \frac{M}{r^2} + \frac{(M^2 - 2\mu M)}{3r^3} \right) \right). \tag{19}$$

We have introduced reduced coordinates,

$$\begin{aligned} M &= m_1 + m_2 \\ \mu &= \frac{m_1 m_2}{m_1 + m_2} \\ \underset{\sim}{\mathbf{r}} &= \underset{\sim}{\mathbf{x}}_1 - \underset{\sim}{\mathbf{x}}_2 \\ \underset{\sim}{\mathbf{v}} &= \frac{d\underset{\sim}{\mathbf{r}}}{dt} = \underset{\sim}{\mathbf{v}}_1 - \underset{\sim}{\mathbf{v}}_2 \\ \underset{\sim}{\mathbf{V}} &= \frac{m_1 \underset{\sim}{\mathbf{v}}_1 + m_2 \underset{\sim}{\mathbf{v}}_2}{M}. \end{aligned} \tag{20}$$

If we use center of mass coordinates, so that $\underset{\sim}{\mathbf{V}} = 0$, and if we introduce the notation

$$\gamma(r) \equiv \frac{(1 + r/M)^3 - 2\mu/M}{(r/M)^3}$$

then the effective Lagrangian for two black holes takes the simple form

$$L_{\text{twobody}} = \frac{1}{2} \mu \gamma(r) \underset{\sim}{\mathbf{v}} \cdot \underset{\sim}{\mathbf{v}}. \tag{21}$$

This Lagrangian specifies the dynamics for the two maximally charged black hole system, *e.g.*, the force of interaction is $\underset{\sim}{\mathbf{F}} = \partial L / \partial \underset{\sim}{\mathbf{r}}$ and the equations of motion are the Euler-Lagrange equations.

4. SCATTERING OF TWO BLACK HOLES

4.1 Integrating the equations of motion

This turns out to be a remarkably simple problem in classical mechanics. In particular, we can exploit the spherical symmetry to reduce the equations of motion to quadratures, *i.e.*, the system is integrable. Since this is a central force problem, the motion is confined to a plane. In the standard (r, θ, ϕ) spherical coordinates, we choose that plane to be $\theta = \pi/2$. The two conserved quantities we need to integrate the equations of motion are the angular momentum and the energy. We will discuss the orbits in the context of scattering, so we introduce an impact parameter, b, and the initial velocity at infinity, v_∞. In terms of these parameters, we have:

$$p_\phi = \frac{\partial L}{\partial \dot{\phi}} = \mu\gamma r^2 \dot{\phi} = \mu b v_\infty \tag{22}$$

$$E = p_r \dot{r} + p_\phi \dot{\phi} - L = \frac{1}{2}\mu v_\infty^2. \tag{23}$$

The equation of motion for r reduces to a one dimensional, ordinary differential equation,

$$\frac{\dot{r}^2}{v_\infty^2} + \frac{1}{\gamma}\left[\frac{b^2}{\gamma r^2} - 1\right] = 0. \tag{24}$$

The second term is like an "effective potential",

$$U_{\text{eff}}(r) = \frac{1}{\gamma}\left[\frac{b^2}{\gamma r^2} - 1\right]. \tag{25}$$

Note that this is a bit different from the usual scattering problems of classical mechanics since the right hand side of (24) always vanishes. (This is similar to the problem of a massless particle in the field of an uncharged black hole (Misner *et al.* 1972 p674ff).) The only classically accessible regions are regions where $U_{\text{eff}}(r) \le 0$. In Figure 1 we have plotted $U_{\text{eff}}(r)$ for two different values of impact parameter b. It is apparent that the force is repulsive, and there are no stable circular orbits.

4.2 Scattering and coalescence

Both scattering orbits and orbits which result in coalescence of the two black holes are possible. In the diagram the $b = 0$ case corresponds to a head on collision which would result in coalescence of the two black holes. For the case $b = 6M$, however, the angular momentum barrier is too large, and the two black holes approach each other to some minimum distance, given by $U_{\text{eff}}(r_{min}) = 0$,

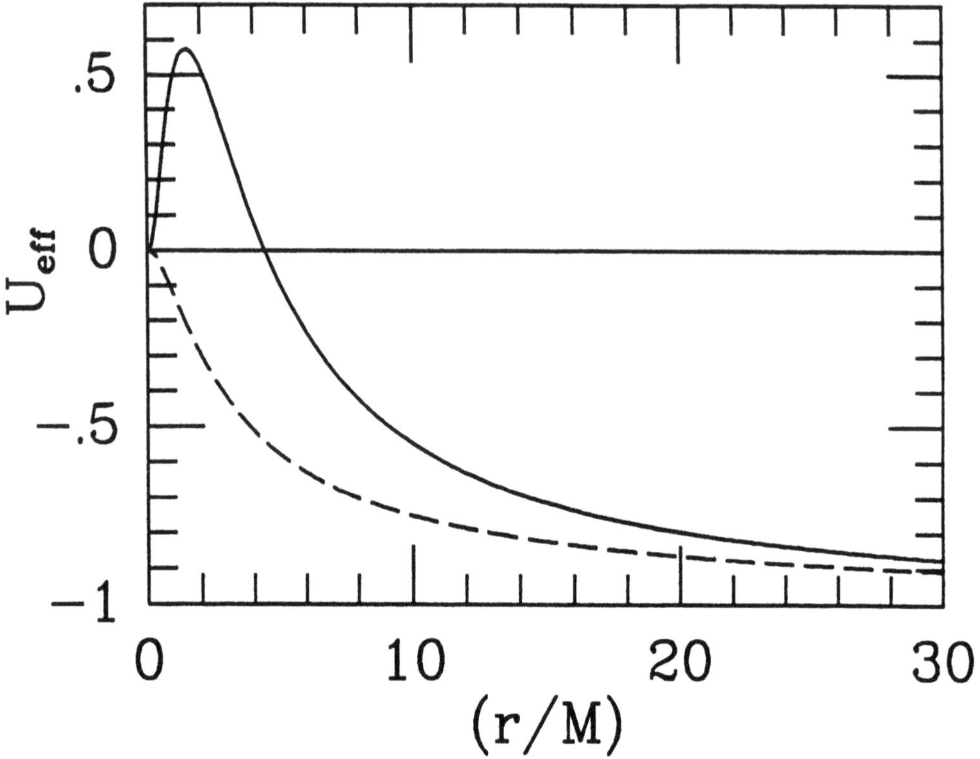

Figure 1. The effective potential which determines the dynamics for the two equal mass $(m1 = m2 = M/2)$ black hole motion. The solid line shows U_{eff} for an impact parameter $b = 6M$. The large angular momentum barrier shows that this corresponds to scattering of two black holes. The dashed line is for radial motion, $b = 0$. In this case, the two black holes coalesce into a nearly extremal Reissner-Nordstrøm black hole.

and then scatter back out to infinite separation. There is a critical b which divides the two classes of orbits. This satisfies the cubic equation

$$\left(\frac{(b_{crit}/M)^2}{3}\right)^3 = \left(\frac{\mu}{M} - \frac{(b_{crit}/M)^2}{2}\right)^2. \tag{26}$$

When $b = b_{crit}$, there is one circular orbit at $r = r_{circ}$ where

$$r_{circ}/M = \left(\frac{(b_{crit}/M)^2}{2} - \frac{\mu}{M}\right)^{\frac{1}{3}} - 1 = \left(\frac{(b_{crit}/M)^2}{3}\right)^{\frac{1}{2}} - 1. \tag{27}$$

In particular, for equally massive black holes, $m_1 = m_2$,

$$b_{crit} = \left(3 + (3/2)\sqrt{3}\right)^{1/2} M \approx 2.3660M$$

$$r_{circ} = \left(\left(1+\sqrt{3}/2\right)^{\frac{1}{2}} - 1\right) M \approx 0.36603M$$

to be compared with the known value for black hole and charged ($q = m$) test particle,

$$b_{crit} = (27/4)^{1/2} M \approx 2.5981M$$
$$r_{circ} = 1/2\ M.$$

In Figure 2 we have plotted both a coalescence and a scattering orbit. For both the scattering and the coalescence orbits, the two black holes may wind around each other many times. Even for the coalescence orbits, however, there are only a finite number of windings.

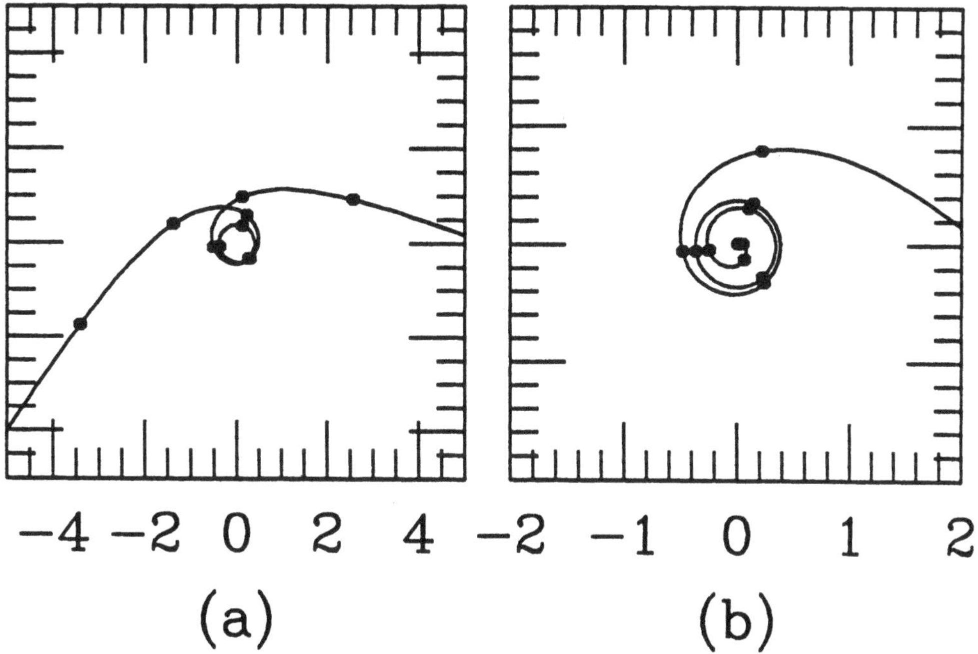

Figure 2. a) An example of a scattering orbit. This is for the equal mass case, $m_1 = m_2 = M/2$. Distances are in units of M and time is measured in units of (M/v_∞). The solid dots are separated by $\Delta t = 5(M/v_\infty)$. Here $b = 2.367M$, and we see that after winding around each other twice, the black holes return to infinite separation. b) An example of a coalescence orbit, again with $m_1 = m_2 = M/2$ and $\Delta t = 5(M/v_\infty)$. In this case $b = 2.3658M$. Since $b_{crit} \approx 2.3660M$, this orbit just barely results in coalescence. The two black holes orbit each other a few times with r near $r_{circ} \approx .366M$ but finally spiral inwards.

4.3 The DeWitt metric

An alternative and illustrative way to study the orbits is in the language of a DeWitt metric on mini-superspace (DeWitt 1967); similar ideas apply to soliton solutions in gauge theories (Manton 1982), and their applicability to maximally charged black holes was first suggested by Gibbons and Ruback (1986). As we mentioned above, the slow motion/no radiation approximation is a truncation of an infinite number of degrees of freedom down to a finite number. If we think of the general solution to the field equations as a geodesic in an infinite dimensional space of three metrics, superspace, then we are seeking an approximate solution in a finite dimensional subspace of superspace, often called mini-superspace. For the case we studied, the mini-superspace is the 6 dimensional configuration space for the two black holes — 3 dimensions each for the positions of the black holes. An approximate solution is a geodesic in mini-superspace. The $\mathbf{R}^6$ of mini-superspace splits into a flat $\mathbf{R}^3$ in the center of mass coordinates, $\{\underset{\sim}{\mathbf{X}}\}$, and a curved $\mathbf{R}^3$ in the relative coordinates, $\{\underset{\sim}{\mathbf{x}}\}$. The terms quadratic in the velocity in the action (19) induce the metric on mini-superspace by

$$(\text{proper distance}) = \int d\lambda \left(\frac{1}{2} M \underset{\sim}{\mathbf{U}} \cdot \underset{\sim}{\mathbf{U}} + \frac{1}{2} \mu \gamma(r) \underset{\sim}{\mathbf{u}} \cdot \underset{\sim}{\mathbf{u}} \right)^{1/2} . \tag{28}$$

In this expression, λ is a parameter along the path in mini-superspace, and $\underset{\sim}{\mathbf{U}} = d\underset{\sim}{\mathbf{X}}/d\lambda$ and $\underset{\sim}{\mathbf{u}} = d\underset{\sim}{\mathbf{x}}/d\lambda$ are velocities with respect to this parameter. The metric in the center of mass sector is flat. The metric in the relative coordinates can be read directly from (28). We note that the geodesics are confined to a plane, so, choosing $\theta = \pi/2$ as before, the metric reduces to a metric on a two dimensional surface:

$$d\sigma^2 = \gamma(r)(dr^2 + r^2 d\phi^2). \tag{29}$$

This surface can be embedded in three dimensions. A plot of the surface is shown in Figure 3. This surface is asymptotically flat as $r \to \infty$ (at the top of the Figure). As $r \to 0$ the geometry becomes conical, with a deficit angle of π. Geodesics on this surface correspond to orbits of the two black holes. Scattering orbits correspond to geodesics which start at infinity and return to infinity. Coalescence orbits are geodesics which start at infinity and go to $r = 0$, down on the cone. It is also clear from the figure that there is only one circular orbit, at the throat, since any other circle of constant separation is not a geodesic.

4.4 Radiation

We now address the question of the validity of our approximation. From the form of the potential U_{eff} we see that the slow motion approximation

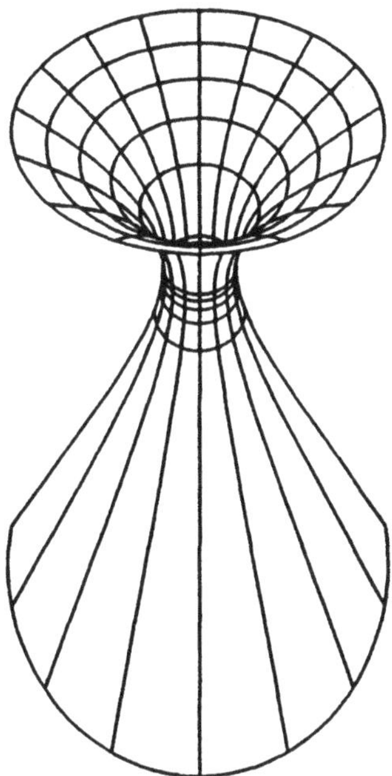

Figure 3. The mini-superspace geometry for two equal mass black holes ($m_1 = m_2$). A geodesic on this surface represents a trajectory for the two black hole system. Circles represent lines of constant separation of the black holes, $r = |\underset{\sim}{x}_1 - \underset{\sim}{x}_2|$. The throat is at $r = M\left(\left((b_{crit}/M)^2/3\right)^{\frac{1}{2}} - 1\right) \simeq 0.366M$. Circles above the throat are separated by $\Delta r = 1.46M$, below the throat by $\Delta r = 0.0731M$. The point $r = 0$ lies infinitely far down the cone.

is self-consistent, in the sense that velocities which are small at large separation remain small throughout the interaction. This is an indication that the forces really are small, as we had anticipated. However, it is the square of the spatial part of the 4-velocity which must be small compared to unity, and this is approximately $\psi^4 v^2$, not v^2. Of course, ψ is infinite at each black hole, so we cannot really discuss the 4-velocity of a black hole. This is why, in our derivation of the effective Lagrangian, we had to smear out the black holes into a smooth distribution of dust and consider the 4-velocity of each dust particle. If we think for a moment about the motion of two small, compact clumps of dust rather than two black holes we can understand when small v is not enough to satisfy the slow motion requirement. If the two clumps are well separated, then the field felt by a dust particle in clump 1 will have contributions from the other dust particles in clump 1, and also from the dust particles in clump 2. The contribution from other particles in clump 1 can be kept as small as desired by smearing out clump 1. In addition, if the two clumps are sufficiently far apart, the field due to clump 2 will not be too large. However, if the two clumps are too close together, say the distance between their "centers of masses" is r, then the field due to clump 2 will be large. We cannot control this by regularization. Thus, the square of the spatial part of the 4-velocity of a particle in clump 1 will be like $(1+M/r)^4v^2 \sim (M/r)^4v^2$ where M is some characteristic mass. The slow motion approximation is that this is small, $(M/r)^4v^2 \ll 1$. Since $v^2 = v_\infty^2/\gamma$ we expect breakdown of the approximation, therefore, when $r \sim v_\infty^2 M$.

For scattering orbits, the separation is large throughout the whole range of motion, since $r > r_{circ}$, so small $\psi^2 v$ amounts to small v. For scattering orbits, therefore, we speculate that there is little energy radiated in either electromagnetic waves (EMW) or in gravitational waves (GW). We expect that $\Delta E_{EMW} \sim \Delta E_{GW} \sim v_\infty^5 M$.

In contrast, orbits which result in coalescence necessarily have $r \to 0$, and we expect that the final state is a nearly extremal Kerr-Newman type rotating black hole. A Kerr-Newman black hole must satisfy the inequality

$$(\text{Total mass})^2 \geq (\text{Total charge})^2 + (\text{Total angular momentum})^2/(\text{Total mass})^2,$$

otherwise there is no event horizon. The total mass of the final black hole will contain contributions from the masses of the two black holes, and also a small contribution form the kinetic energy, $\frac{1}{2}\mu v_\infty^2$. There will also be some angular momentum if $b \neq 0$. Orbits which are characterized by $b \leq b_{max} \equiv M(M/\mu)^{1/2}$ satisfy the above inequality. We speculate that for these orbits there is little radiation released, $\Delta E_{EMW} \sim \Delta E_{GW} \sim v_\infty^5 M$. On the other hand, orbits with $b_{crit} > b > b_{max}$ would form a naked singularity unless some of the angular momentum is removed. We speculate that orbits in this category radiate away enough angular momentum to avoid formation of a naked singularity. In this case, we expect that $\Delta E_{radiation} \sim v_\infty^2 M$ after the slow motion approximation breaks down at $r \sim v_\infty^2 M$. In Figure 4 we plot both b_{crit} and b_{max} as functions of the mass ratios. From this figure it is apparent that most orbits are "radiation quiet".

5. FUTURE WORK

For this work to be of direct astrophysical interest, we must extend it to black holes which are not maximally charged and which are rotating (Kerr-Newman black holes). It is not immediately clear how to do so, since the starting point of this work was a static solution, and there is no known static solution for a system of Kerr-Newman black holes. Nevertheless, the mini-superspace picture of black hole interactions is a powerful aid to the understanding of the dynamics of several Reissner-Nordstrøm black holes, and this picture may be similarly helpful for black hole interactions in more general circumstances.

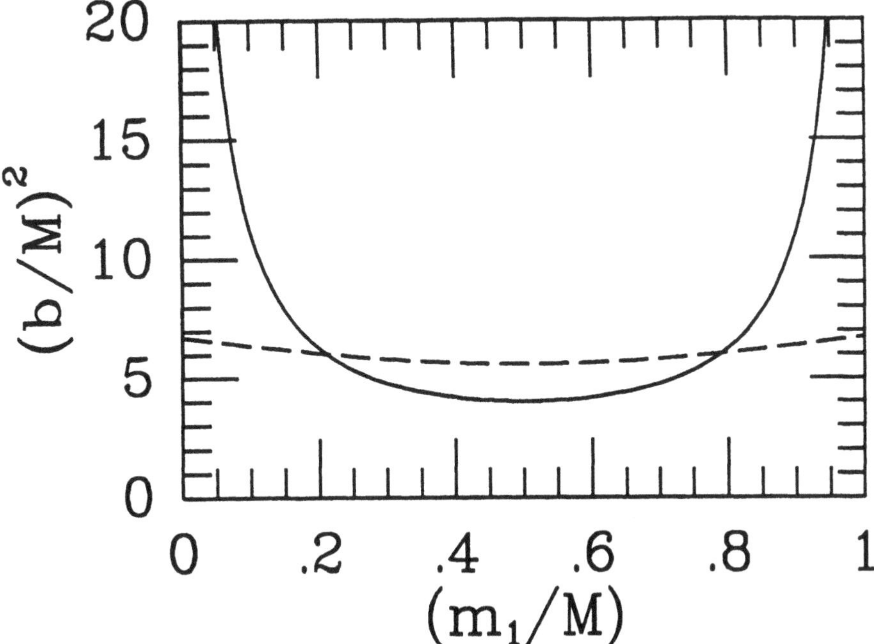

Figure 4. The dashed line shows $(b_{crit}/M)^2$ of (26) as a function of m_1/M. Only values of b^2/M^2 below the line result in black hole coalescence. The solid line is a plot of $(b_{max}/M)^2 = (M/\mu)$. As discussed in the text, only values of b below this line can result in a Kerr-Newman black hole. The parameter values above b_{max} and below b_{crit} yield "radiation-loud" coalescences.

Part of this work was described in Ferrell and Eardley (1987), and a longer paper is in preparation. This research was supported in part by the National Science Foundation under Grants No. PHY85-06686 and PHY82-17853, supplemented by funds from the National Aeronautics and Space Administration, at the University of California at Santa Barbara.

REFERENCES

Arnowitt, R., Deser, S., & Misner, C. W. (1960). *Physical Review* **120**, 313.

Arnowitt, R., Deser, S., & Misner, C. W. (1962). The Dynamics of General Relativity. *In: Gravitation, an Introduction to Current Research*, ed. L. Witten. New York: Wiley.

Damour, T. (1983). Gravitational Radiation and the Motion of Compact Bodies. *In: Rayonnement Gravitationnel (Gravitational Radiation), NATO Advanced Study Institute, Les Houches 1982*, ed. N. Deruelle & T. Piran, pp. 59–144. Amsterdam: North Holland Publishing Co.

DeWitt, B. S. (1967). *Physical Review* **160**, 1113.

Eardley, D. M. (1986). Approximate Methods for Black Hole Collisions. *In: Dynamical Spacetimes and Numerical Relativity*, ed.J. Centrella, pp. 347–364. Cambridge: Cambridge University Press.

Ferrell, R. C. & Eardley, D. M. (1987). *Physical Review Letters* **57**, 1617.

Gibbons G. W. & Ruback P. J. (1986). *Physical Review Letters* **57**, 1492.

Landau, L. D. & Lifshitz, E. M. (1975). *The Classical Theory of Fields, pp. 165–169.* New York: Pergamon Press.

Majumdar, S. D. (1947). *Physical Review* **72**, 39.

Manton, N. S. (1982). *Physics Letters B* **110**, 54.

Misner, C. W., Thorne, K. S. & Wheeler, J. A. (1972). *Gravitation.* San Francisco: W.H. Freeman.

Papapetrou, A. (1947). *Proc. Irish Acad. Sci., Sec. A* **51**, 191.

KEPLER'S THIRD LAW IN GENERAL RELATIVITY

Steven Detweiler
University of Florida, Gainesville Florida 32611

Abstract. A new approach to the study of the two black hole problem is outlined. Standing wave boundary conditions in the wave zone and near the throats are used to keep a system of two black holes in steady circular orbits, with no loss of energy or angular momentum. The resulting geometry is both time dependent and non-axisymmetric; however, there is a Killing vector field of the form $\partial/\partial t+\Omega\partial/\partial\phi$. A variational principle is found for the total mass of the binary system and is used to derive Kepler's third law in the Newtonian limit. This approach requires only modest computational resources and yet can be used to study problems such as the nature of the gravitational radiation and the stability of orbits of binary black holes in a fully nonlinear regime.

1 *INTRODUCTION*

A binary system of two black holes emits gravitational radiation and loses energy and angular momentum. As the system evolves it is more and more tightly bound with increasing angular velocity, amplitude of gravitational radiation and rate of evolution. Eventually, the distance scale of the system is comparable to its total mass, M, the time scale of evolution is comparable to the orbital period and, presumably, the system emits some significant burst of gravitational radiation in a very nonlinear fashion. If the total angular momentum, J, is less than M^2, then the system can settle down into a single, axisymmetric Kerr black hole. The details of this evolutionary process present an important unsolved problem in classical general relativity.

There have been a number of different approximations and approaches to the binary black hole problem. The post-Newtonian approximation suffices when the holes are far apart and slowly moving. Damour (1987) reviewed a variety of approaches which focus upon the emission and effects of gravitational radiation. Perturbation approximations of

Davis et al. (1971), Detweiler (1978) and Oohara & Nakamura (1983) examined the resulting gravitational radiation when one black hole is much less massive than the other. Misner (1960, 1963) and Gibbons (1972) considered the initial value problem for two holes in a moment of time symmetry. And the initial value problem for moving black holes was analyzed by Bowen & York (1980). The pioneering fully numerical attack on the two black hole problem was Smarr's (1978) analysis of the head on, equal mass collision; this clearly demonstrated the feasibility of the numerical approach.

Orbiting black holes may be contrasted with an orbiting electron-proton pair in classical electrodynamics. Electrostatics is a linear theory; the coulomb field of two charged particles is just the algebraic sum of the individual fields. Even the emission of radiation is essentially linear. Nonlinearities enter vacuum electrodynamics only in the effects of radiation reaction, but radiation reaction is often small and can be treated perturbatively. In the end, physicists feel comfortable with electrodyamics despite the effects of the radiative nonlinearities. The electron-proton pair presents a tractable problem. Schild (1963) presents an interesting approach to this electrodynamic two body problem. For a pair of orbiting charged particles he finds the "time symmetric" field consisting of half advanced and half retarded fields. This effectively imposes standing wave boundary conditions for the electromagnetic field at infinity and prevents the system from radiating a net amount of energy or angular momentum. Hence, the system does not evolve. The standing wave boundary condition is, of course, non physical. But its presence allows for an analytic solution to the equations of electrodynamics. As long as the amplitude of the standing wave at infinity is not large, this solution should accurately model many aspects of the interesting physical problem with only outgoing waves.

The major difficulty in solving the binary black hole problem is not nonlinearities but, rather, the lack of symmetry. The system is neither axisymmeteric nor time independent. And both the orbital parameters and the black holes evolve while gravitational radiation carries energy and angular momentum out at infinity and down the holes. But, a symmetry may be imposed upon the system by the choice of standing wave boundary conditions at infinity and at each throat in the geometry. This choice allows for orbital motion but stops the evolution of the orbital parameters. In intertial coordinates, there would be a Killing vector field of the form $\partial/\partial t+\Omega\partial/\partial\phi$. As long as the secular effects of the gravitational radiation are small, the standing wave system ought to be a good approximation to the physical system, with outgoing waves, for at least a few orbital periods.

Standing wave boundary conditions cause some difficulties which arise from the nonlinearity of Einstein's equations. At large distance, the contribution to the total energy of the system from a thin shell of standing wave approaches a constant, so the total contribution of the wave grows linearly with distance. Hence, the boundary conditions cannot really be imposed in the limit of $r\rightarrow\infty$. This difficulty may be circumvented with two different interpretations. First, in the wave zone the radiation has amplitude Ar^{-1} and the energy density is proportional to A^2r^{-2}. The total energy in the wave zone is proportional to A^2r. If A is small enough, then a value of r, r_∞, can be found such that $A^2r_\infty \ll M \ll r_\infty$. The place to impose the boundary conditions is at r_∞. Second, in this paper we consider solutions of the Einstein equations over some finite region of a space-time manifold. In fact, with the $\partial/\partial t+\Omega\partial/\partial\phi$ Killing vector field, we really only look at initial data which satisfies the initial value equations, and the restrictions imposed by the existance of the Killing vector, over a partial Cauchy surface. This data will evolve inside the future Cauchy development in a manner consistent with the Killing vector. Outside of the development, of course, the geometry is not determined by the initial data. Thus, the initial data we consider here has a meaningful, although not so transparent, interpretation even when A is large.

A similar problem exists near the event horizon. Radiation going down the hole from the distant past piles up near the apparent horizon of the black hole in the same way that outgoing radiation from the distant past piles up at spatial infinity. If the inner boundary of the partial Cauchy surface gets too close to the apparent horizon then the energy content of the radiation will dominate the geometry. The resolution of this problem is the same as that at infinity; the boundary conditions are imposed at a throat located some distance outside the apparent horizon.

In section 2, we briefly review the initial value formulation of general relativity. In section 3 this is specialized to the case where there exists one Killing vector field which can be identified with the evolutionary time vector t^a. The resulting equations we call the steady state equations. In section 4, we present a variational principle for the steady state equations and discuss the consequent boundary conditions far from the system and near the throats. In section 5, the variational principle is applied to a system of two black holes, far apart, and Kepler's third law is found as a consequence.

We use units with $\kappa=8\pi G$ and $c=1$.

2 THE INITIAL VALUE FORMALISM AND DYNAMICS OF GENERAL RELATIVITY

York (1978) presents a fine pedagogical treatment of the initial value formalism and the dynamics of general relativity. Here we summarize some of his results and notation.

A four dimensional spacetime with a metric, g_{ab}, may be foliated into constant t, spacelike hypersurfaces, with a unit normal vector, n^a, and a metric γ_{ab},

$$\begin{aligned} ds^2 &= g_{ab}dx^a dx^b \\ &= -\alpha^2 dt^2 + \gamma_{ab}(dx^a+\beta^a dt)(dx^b+\beta^b dt). \end{aligned} \tag{1}$$

The quantity α is the lapse function, and β^a is the shift vector. The indices $a, b, \ldots$ run from 0 to 3; however, in the spirit of the initial value formalism nearly all tensors except n^a are orthogonal to n^a. The three dimensional metric, γ_{ab}, has a derivative operator, D_a, and Ricci tensor R_{ab}. The extrinsic curvature, K_{ab}, measures the rate of change of γ_{ab} from hypersurface to hypersurface and is defined in equation (7) below.

In this paper we are primarily interested in the vacuum equations; but the stress energy tensor is included in many equations as an aid in understanding from where these equations come. The four dimensional stress energy tensor is decomposed into parts parallel and perpendicular to n^a by

$$T_{ab}=S_{ab}+2j_{(a}n_{b)}+\rho n_a n_b. \tag{2}$$

and we define

$$S=\gamma^{ab}S_{ab}. \tag{3}$$

Now the constraint equations are restrictions only on the data γ_{ab} and K_{ab} on a given hypersurface. With

$$K=\gamma^{ab}K_{ab}, \tag{4}$$

these are the Hamiltonian constraint,

$$2G_{nn}=R+K^2-K_{ab}K^{ab}=2\kappa\rho . \tag{5}$$

and the momentum constraint,

$$-\perp G_{an}=D_b(K^b{}_a-\gamma^b{}_aK)=\kappa j_a; \tag{6}$$

the symbol G_{ab} is the four dimensional Einstein tensor and the index n denotes contraction of a spacetime index with the vector n^a.

The contravariant vector $t^a\partial/\partial x^a=\partial/\partial t=(\alpha n^a+\beta^a)\partial/\partial x^a$ points in the direction of increasing t with all spatial coordinates held fixed. So the Lie derivative with respect to t^a, $\mathcal{L}_t$, of different geometric quantities describes their evolution in time. The evolution of the metric is essentially just the definition of K_{ab},

$$\mathcal{L}_t\gamma_{ab}=-2\alpha K_{ab}+\mathcal{L}_\beta\gamma_{ab}. \tag{7}$$

The dynamical part of the Einstein equations gives the evolution of K_{ab},

$$\mathcal{L}_tK_{ab}=-D_aD_b\alpha$$
$$+\alpha[R_{ab}-2K_{ac}K^c{}_b+KK_{ab}-\kappa(S_{ab}-\tfrac{1}{2}\gamma_{ab}S)-\tfrac{1}{2}\kappa\rho\gamma_{ab}]+\mathcal{L}_\beta K_{ab}. \tag{8}$$

Equations (5) - (8) are the four conerstones of the initial value formalism of general relativity.

When the spacetime is asymptotically flat, the mass and angular momentum of the geometry may be found from certain surface integrals over a large sphere. York (1980) gives the mass integral in terms of $h_{ab}=g_{ab}-f_{ab}$, where f_{ab} is the flat background metric, in any coordinate system, which g_{ab} asymptotically approaches, and ∇_a, its flat derivative operator. The total mass as measured at spatial infinity is

$$M=\frac{1}{16\pi}\oint_\infty \nabla^a(h^b{}_a-\delta^b{}_ah^c{}_c)d\Sigma_b, \tag{9}$$

The total angular momentum is a little more difficult to describe. However, if the geometry and coordinate system go flat fast enough at infinity then the total angular momentum is

$$J=\frac{1}{8\pi}\oint_\infty \xi^a(K^b{}_a-\delta^b{}_aK)d\Sigma_b, \tag{10}$$

where ξ^a is the axial Killing vector at infinity.

3 *THE STEADY STATE EQUATIONS*

In this paper we are concerned with the Einstein equations restricted to the vacuum and with t^a a Killing vector field. The right hand sides of the cornerstone equations all vanish, as well as the terms with ρ, j^a, and S_{ab}. We call the resulting equations the steady state equations:

$$R+K^2-K_{ab}K^{ab}=0, \tag{11}$$

$$D_b(K^b{}_a-\gamma^b{}_aK)=0, \tag{12}$$

$$\alpha K_{ab}-D_{(a}\beta_{b)}=0, \tag{13}$$

and

$$D_aD_b\alpha-\alpha(R_{ab}-2K_{ac}K^c{}_b+KK_{ab})-\beta^cD_cK_{ab}-2K_{c(a}D_{b)}\beta^c=0. \tag{14}$$

Geometry at infinity

For the rotating system which we envision, in inertial coordinates at large distances the Killing vector is $\partial/\partial t+\Omega\partial/\partial\phi$, with Ω the rotation speed. Also β^a goes to zero, and γ_{ab} approaches a flat metric.

But a coordinate transformation, $\phi_{old}=\phi_{new}+\Omega t$, to a system which rotates with the black holes is often convenient. Then the Killing vector is just $t^a\partial/\partial x^a=(\alpha n^a+\beta^a)\partial/\partial x^a=\partial/\partial t$. And $\beta^a\partial/\partial x^a=\Omega\partial/\partial\phi$ at infinity. The quantities α, K_{ab} and γ_{ab} do not change under this coordinate transformation. We use this rotating coordinate system henceforth.

Geometry at the throat

The formal description of the geometry at the black hole is difficult. A black hole has an event horizon, a null surface which is the boundary of the past of future null infinity. The generators of this surface are null geodesics. But the event horizon is a global concept and cannot be found without a global description of the geometry. A related concept is that of the apparent horizon on a given spacelike hypersurface. This is the marginally outer trapped surface, *i.e.* that two dimensional surface whose outward pointing null geodesics have zero convergence. The apparent horizon is much easier to find because its existence depends only upon the behavior of nearby null geodesics, a local concept.

The space-time manifold we consider has a spacelike hypersurface, $\mathcal{S}$, with two asymptotically flat regions connected by a bridge for each black hole, much like the Einstein-Rosen bridge of the Schwarzschild geometry. Each black hole will have an apparent horizon in $\mathcal{S}$ on the bridge. At or very near this surface the boundary conditions for the steady state equations are specified. The two dimensional surface on which we specify the boundary conditions we will call the throat, $\mathcal{T}_i$ for the ith hole, and either it is the apparent horizon in $\mathcal{S}$, or it is just outside of the apparent horizon. This distinction between the throat and the apparent horizon in $\mathcal{S}$ is important. Standing waves from the distant past pile up on the apparent horizon. To keep divergences from occuring it is important to cut them off at a boundary, the throat, some distance away from the apparent horizon in $\mathcal{S}$.

The apparent horizon of a Schwarzschild black hole is a dynamic object even with a $\partial/\partial t$ Killing vector. To evolve the constant t surfaces into each other it is necessary to choose $\alpha=0$ there. Similarly, with the steady state equations, α must vanish on the apparent horizon in $\mathcal{S}$, and also β^a must either vanish or lie in the apparent horizon in $\mathcal{S}$ to guarantee that the boundary conditions for the steady state equations are Lie derived by the Killing field.

If β^a does not vanish at the apparent horizon then it must have no component parallel to $D_a\alpha$. Equation (13) then implies that β^a is a Killing vector of the two dimensional manifold which is the apparent horizon in $\mathcal{S}$. Presumably, the apparent horizon is axisymmetric and rigidly rotating with $\beta^a\partial/\partial x^a=(\Omega-\Omega_i)\partial/\partial\tilde{\phi}$, where Ω_i is the angular speed of the hole as measured by an inertial observer at infinity and $\tilde{\phi}$ is a local axial coordinate which runs from 0 to 2π around the hole. In this case the angular momentum of the ith throat, J_i, comes from

$$8\pi(\Omega-\Omega_i)J_i=\oint_{\mathcal{T}_i} \beta_a(K^{ab}-\gamma^{ab}K)d\Sigma_b, \tag{15}$$

in the limit that the throat approaches the apparent horizon. Note that equation(15) gives the standard result for angular momentum for any axisymmetric object when the entire geometry is axisymmetric (with β^a proportional to the axial Killing vector), and it plays the role of angular momentum in the current context, as seen below.

If β^a vanishes on the apparent horizon in $\mathcal{S}$, then the hole always presents the same face to its companion. This is an interesting restriction as it implies that there is no radiation at the throat. This situation is just like sending radiation towards a rotating black hole at just the critical frequency for superradiance; the radiation is perfectly reflected. But having β^a vanish also allows the apparent horizon to be nonaxisymmetric. In general if the holes start out very far apart then the time scale for the evolution of the orbital parameters is much less than the time needed for the tidal distortion to effect the spin of a black hole (Hartle, 1974). Hence, we would not expect to find tidally locked black holes in the real world.

A consequence of the full set of equations is

$$D_a(D^a\alpha-\beta^bK^a{}_b)=\tfrac{1}{2}\alpha\kappa(\rho+S)+\kappa\beta^bj_b+\mathcal{L}_tK+\tfrac{1}{2}K^{ab}\mathcal{L}_t\gamma_{ab}. \tag{16}$$

In the vacuum, steady state case this gives a divergence free vector field. The integral of this divergence yields

$$\oint_\infty(D^a\alpha-\beta^bK^a{}_b)d\Sigma_a=\sum_i\oint_{\mathcal{T}_i}(D^a\alpha-\beta^bK^a{}_b)d\Sigma_a. \tag{17}$$

In the gauge described in Chapter 19 of Misner et al. (1973) the flux integral at infinity is $4\pi M-8\pi\Omega J$, and it must be gauge invariant because it always equals the right hand side. Equations (15) and (17) give the mass formula for two black holes orbiting in the steady state,

$$4\pi M=8\pi\Omega J-\sum_i\left[8\pi(\Omega-\Omega_i)J_i-\oint_{\mathcal{T}_i}D^a\alpha d\Sigma_a\right]. \tag{18}$$

For a single rotating black hole this equation reduces to Smarr's (1973) mass formula. In that case, the surface integral is κA, where κ is the surface gravity which is constant over the surface of the hole. Such a simple interpretation for the surface integral does not seem possible for the binary black hole problem.

4 *A VARIATIONAL PRINCIPLE FOR THE STEADY STATE EQUATIONS*

A consequence of equation (9) and the constraint equations gives the total mass as

$$4\pi M=\tfrac{1}{4}\oint_\infty\alpha\nabla_b(h^{ab}-f^{ab}h^c{}_c)d\Sigma_a$$

$$-\tfrac{1}{4}\int[\alpha(R+K^2-K_{ab}K^{ab})-2\beta_aD_b(K^{ab}-\gamma^{ab}K)]dV. \tag{19}$$

This provides a variational principle for the total mass of the system.

Consider a trial geometry (α, β^a, K^{ab}, γ_{ab}) and its infinitesimal variation ($\delta\alpha$, $\delta\beta^a$, δK^{ab}, $\delta\gamma_{ab}$). The quantity M defined in equation (19) changes by an amount δM which is found, after a lengthy calculation, to be given through first order in the variation by

$$4\pi\delta M=\int(\delta\alpha[\![11]\!]+\delta\beta^a[\![12]\!]_a+\delta K^{ab}[\![13]\!]_{ab}+\delta\gamma_{ab}[\![11,\ 12,\ 13,\ 14]\!]^{ab})dV$$

$$+4\pi\Omega\delta J-\sum_i 4\pi(\Omega-\Omega_i)\delta J_i$$

$$+\frac{1}{4}\oint_{\mathcal{T}}[-\alpha D^a\delta\gamma^b{}_b+\delta\gamma^b{}_b D^a\alpha+\alpha D_b\delta\gamma^{ab}-\delta\gamma^{ab}D_b\alpha]\,d\Sigma_a \tag{20}$$

The double square bracket denotes a linear combination of the equations numbered inside and vanishes if and only if these equations are satisfied for the trial geometry. And $\mathcal{T}$ is the union of the individual disjoint throats $\mathcal{T}_i$. The terms with the δJ come from surface terms. If the throat, $\mathcal{T}_i$, is chosen to be precisely where α vanishes then the integrand of the surface integral at $\mathcal{T}_i$ is proportional to

$$(\delta\gamma^{ab}-\gamma^{ab}\delta\gamma^c{}_c)s_a s_b=-\delta\gamma_{ab}(\gamma^{ab}-s^a s^b) \tag{21}$$

where s_a is the outward unit normal to a constant α surface. This integrand is proportional to the trace of the part of $\delta\gamma_{ab}$ which is in the constant α surface; and, therefore, vanishes precisely when the variation does not change the magnitude of an element of area of the throat. With a variation of γ_{ab} the shape of an area element may change, but again, its magnitude may not. If the throat is chosen to be just outside the $\alpha=0$ surface, then the additional terms in the integrand are just what is needed to extrapolate the integrand back to the $\alpha=0$ surface.

We can now state the variational principle for M. Consider the class of arbitrary infinitesimal variations in the geometry which hold constant both the total angular momentum, defined in equation (10), and the angular momentum of each individual throat, defined in equation (15), and which also do not change the magnitude of an infinitesimal area element of the throat. The total mass M of the system, as found from equation (19), is an extremum ($\delta M=0$) under this class of variations if and only if the trial geometry satifies the steady state equations exactly. This is a powerful result which in a short sentence reads: *For fixed areas of the holes and angular momenta, the mass of the system is an extremum at solutions of the steady state equations.*

For two nearby solutions of the steady state equations, equation (20) gives the difference in their masses in terms of the differences in angular momenta and area integrals,

$$4\pi\delta M = 4\pi\Omega\delta J - \sum_i 4\pi(\Omega-\Omega_i)\delta J_i$$

$$+\frac{1}{4}\oint_{\mathcal{T}}[-\alpha D^a\delta\gamma^b{}_b + \delta\gamma^b{}_b D^a\alpha + \alpha D_b\delta\gamma^{ab} - \delta\gamma^{ab}D_b\alpha]\, d\Sigma_a \qquad (22)$$

There is a considerable amount of dancing around the question of holding the areas of the apparent horizon constant during the variation. This would be alleviated if the surface gravity, essentially the magnitude of $D_a\alpha$, were a constant over the surface, which is true for single stationary rotating black holes. I have not been able to prove this for the steady state black holes, although it seems likely, to me, to be true.

Another interesting version of the variational principle follows from equation (19). Out of all *solutions to the constraint equations* with fixed area and angular momenta, the total mass, as determined in equation (9) as a flux integral at infinity, is an extremum if and only if all of the steady state equations are satisfied. This is handy because no volume integrals need to be performed to utilize the variational principle, given solutions to the contraint equations.

Hawking (1973) formulated a variational principle for single, rotating, axisymmetric black holes. His is similar to but differs from this principle in fundamental ways even in the single hole limit.

5 *THE NEWTONIAN LIMIT*

To test the variational principle in the Newtonian limit, we need a trial geometry which looks like two black holes far apart, moving in opposite directions. For one of the black holes located, at the moment, at $x=x_1$, $y=0$, and $z=0$ and moving in the positive y direction with a speed u, and similarly for the second black hole moving with a speed v, a useful metric is

$$ds^2 = -(\alpha_1{}^2-\psi_1{}^4)(dt-u\,dy)^2-(\alpha_2{}^2-\psi_2{}^4)(dt-v\,dy)^2$$
$$+\psi^4(d\vec{x}\cdot d\vec{x}-dt^2) \qquad (23)$$

where

$$\psi = 1+\frac{m_1}{2\rho_1}+\frac{m_2}{2\rho_2}, \qquad (24)$$

$$\psi_1 = 1+\frac{m_1}{2\rho_1}+\frac{m_2}{2\rho_2(1)}, \qquad (25)$$

$$\alpha_1 = \frac{1-\frac{m_1}{2\rho_1}}{1+\frac{m_1}{2\rho_1}}, \tag{26}$$

and

$$\rho_1{}^2 = (x-x_1)^2 + z^2 + y^2/(1-u^2), \tag{27}$$

with $\rho_2(1)$ meaning evaluate ρ_2 at the position of the center of hole 1, and with similar definitions for the second hole.

This geometry is not an exact solution to the Einstein equations. However, it is an interesting geometry. A constant t hypersurface has an asympotically flat region connected by bridges to two other distinct asympotically flat regions, so it looks like two black holes. For simplicity, assume that the masses are equal and the velocities are equal in magnitude but opposite in direction. Then, in the limit that the black holes are far apart, $R = x_1 \gg m_1$, this geometry is an exact solution to all of the Einstein equations, for any speed u. Each part of the geometry looks like a boosted black hole moving with constant speed in isotropic coordinates. In particular, with this approximation the two constraint equations are satisfied up to terms of order m_1/R. A different interesting limit is when $u^2 \ll 1$. In this slow motion limit the initial value equations are satisfied for any value of m_1/R up to terms of order u^2. All together then, it is clear that this trial geometry satisfies the constraint equations through terms of order u^2 and m_1/R and up to terms of order $m_1 u^2/R$.

To use this geometry in the Newtonian limit, when all calculations are carried out through first order in either u^2 or m_1/R but not both, we need the mass, angular momentum and areas of the holes. The total mass of this geometry from equation (9) is

$$M = m_1(1+u^2/2) + m_2(1+v^2/2). \tag{28}$$

The total angular momentum from equation (10), with K_{ab} found from equation (7) with the velocities assumed constant, is

$$J = m_1 u x_1 + m_2 v x_2. \tag{29}$$

The area of a throat in this geometry is not easy to find, but it is

$$A_1 = 16\pi m_1{}^2(1 - m_2/4R)^2 . \qquad (30)$$

Note that this makes the irreducible mass of the first hole

$$m_{1\ irr} = m_1(1 - m_2/4R) . \qquad (31)$$

Now, it is straightforward to apply the variational principle by finding the extreme value for M when J, A_1 and A_2 are held constant. In the equal mass, equal speed case this gives $v^2 = m_1/2R$, which finally gives the angular velocity of the system,

$$\Omega^2 = m_1/2R^3 . \qquad (32)$$

This last result is Kepler's third law. This is not the briefest derivation. But it demonstrates the ability of the variational principle to produce meaningful results. The application to strong field geometries is in progress.

6 *CONCLUSIONS*

Consider the sequence of steady state geometries which have fixed irreducible masses and spin angular momenta for the black holes and is parametrized by the separation. When the holes are far apart, the amplitude of the radiation is small and the steady state approximation should be a good one. The geometry can be used to estimate the amplitude of gravitational radiation which will then determine the rate of evolution along the sequence. With the entire sequence of solutions, the frequency and amplitude of radiation can be determined as a function of time as may be observed some day. In fact, the steady state approximation will work up until that time when the amplitude of the radiation is large. The nature of the orbit when this happens is perhaps the most observationally important piece of information about the black hole binary system. It will determine the likelihood of one day observing such an event with gravitational wave detectors.

Another interesting feature to look for in the sequence is the minimum of the total energy. This will occur at the marginally stable orbit. After that point the evolution ought to occur on a dynamical timescale. And this may even occur before the amplitude of the radiation gets too big for the approximation.

All of this information is accesible without a complete integration of the dynamical equations. In fact much of it needs only application of the variational principle.

Many of the talks at this conference made reference to the binary black hole problem; most implied that the time for its solution has not yet arrived. Now, more than a decade after Smarr's ground breaking work on the head on collision, it appears that the direct numerical integration of the field equations for sprialing infall remains beyond the reach of available computational power. When will direct numerical methods be successful with this problem? I suspect about the time that gravitational waves are detected—in five to ten years.

I have benefitted greatly by conversations with Doug Eardley and Jim York at the conference, and with Jim Ipser, David Garfinkle and Kent Blackburn at the University of Florida.

REFERENCES

Arnowitt, R., Deser, S. & Misner, C.W. (1962). The dynamics of general relativity. In Gravitation, ed. L. Witten, pp. 227-265. Wiley, New York.

Bowen, J.M. & York, J.W. (1980). Time-asymmetric initial data for black holes and black-hole collisions. Phys. Rev. D21, no. 6, 2047-56.

Damour, T. (1987). An introduction to the theory of gravitational radiation. In Gravitation in Astrophysics: Cargèse 1986, ed. B. Carter & J.B. Hartle. Plenum, New York.

Davis, M., Ruffini, R., Press, W.H. & Price, R.H. (1971). Gravitational radiation from a particle falling radially into a schwarzschild black hole. Phys. Rev. Lett., 27, 1466-1469.

Detweiler, S. (1978). Black holes and gravitational waves: perturbation analysis. In Sources of gravitational radiation, ed. L. Smarr, pp. 211-230. Cambridge: Cambridge University Press.

Gibbons, G.W. (1972). The time symmetric initial value problem for black holes. Commun. Math. Phys., 27, 87-102.

Hartle, J.B. (1974). Tidal shapes and shifts on rotating black holes. Phys. Rev., 9, 2749-59.

Hawking, S.W. (1973). A variational principle for black holes. Commun. Math. Phys., 33, 323-34.

Misner, C.W. (1960). Worm hole initial conditions. Phys. Rev., 18, 1110-1.

Misner, C.W. (1963). The method of images in geometrostatics. Ann. Phys., 24, 102-17.

Misner, C.W., Thorne, K.S., & Wheeler, J.A. (1973). Gravitation. Freeman, San Fransico.

Misner, C.W. & Wheeler, J.A. (1957). Classical physics as geometry: Gravitation, electromagnetism, unquantized

charge and mass as properties of curved empty space. Ann. Phys., 2, 525-603.
Oohara, K. & Nakamura, T. (1983). Energy, Momentum and angular momentum of gravitational waves induced by a particle plunging into a Schwarzschild black hole. Prog. Th. Phys., 70, 757-771.
Schild A. (1963). Electromagnetic two-body problem. Phys. Rev., 111, 2762-66.
Smarr, L. (1973). Mass formula for Kerr black holes. Phys. Rev. Lett., 30, 71-73.
Smarr, L. (1978). Gauge conditions, radiation formulae and the two black hole collision. In Sources of gravitational radiation, ed. L. Smarr, pp. 245-274. Cambridge: Cambridge University Press.
York, J.W. (1978). Kinematics and dynamics of general relativity. In Sources of gravitational radiation, ed. L. Smarr, pp. 83-126. Cambridge: Cambridge University Press.
York, J.W. (1980). Energy and momentum of the gravitational field. In Essays in general relativity: a festscrift for Abraham Taub, ed. F.J. Tipler, pp. 39-58. Academic, New York.

BLACK HOLE SPACETIMES: TESTING NUMERICAL RELATIVITY

David H. Bernstein, David W. Hobill, and Larry L. Smarr
National Center for Supercomputing Applications
605, East Springfield Avenue
Champaign, IL 61820

Abstract. Spherically symmetric black hole spacetimes are studied numerically using the standard 3+1 decomposition of the Einstein equations. Different finite difference methods are employed and the results are compared with the known analytic behavior of the Schwarzschild solution. A discussion of coordinate and gauge conditions is included and it is demonstrated that, given the appropriate combinations of finite difference methods and gauge choices, free evolution can accurately propagate the constraint equations.

I. INTRODUCTION

The numerical construction of black hole spacetimes provides an opportunity to study the purely geometrodynamical effects of the gravitational field. Along with vacuum cosmologies and pure gravitational radiation, these spacetimes yield important information concerning the nonlinear behavior of gravitational fields that have been freed from the influences of matter or other fields. One of the advantages of studying black hole spacetimes is that analytic initial data sets exist describing various configurations for N black holes (see the article by York in this volume). The solutions are obtained by modelling the black holes with Einstein-Rosen bridges connecting two isometric sheets. The boundary separating the two sheets (known as the throat) provides unambiguous inner boundary conditions for the construction of numerical solutions. Further simplifications arise if the initial data is assumed to be time symmetric.

Since one of the major goals of this project is to re-calculate the original Smarr-Eppley (1979) axisymmetric two black hole collision using modern numerical methods and the power of a supercomputer, we have begun with the careful construction of a code that is able to reproduce a number of well known analytic results concerning the behavior of black hole spacetimes. The most obvious verification of any black hole code would be that it reproduce the Schwarzschild spacetime. In the two black hole collision, for example, this spherically symmetric spacetime can be thought of as the solution at late times when all of the radiation has propagated away, or the solution in the limit that the mass of one of the black hole vanishes. Therefore, reproducing the Schwarzschild solution using the methods that will be employed for the two-black hole collision is of fundamental significance and will act as an important test-bed calculation.

In the next section different choices of gauges and coordinates useful for numerically constructing spherically symmetric spacetimes will be presented. In Section III we will discuss the different finite difference techniques that are employed, and the fourth section presents a brief review of the analytic behavior associated with the Schwarzschild solution and its relevance to the numerical results. Finally we discuss the propagation of the constraints by the finite differenced version of the Einstein equations and include a discussion of the next phase of this project and the tests that will have to be met before continuing on to the two black hole collision.

II. COORDINATE AND GAUGE CHOICES FOR BLACK HOLE SPACETIMES

In constructing analytic solutions of the Einstein equations whether they be in exact, closed form or the result of a perturbative scheme, one frequently chooses coordinates and gauges that simplify certain aspects of the field equations. The same point holds true for the construction of numerical solutions of the Einstein equations. Different gauge choices can be made either to simplify the equations or relate them more intimately to the geometry of the spacetime. We begin with the standard 3+1 (or 1+1 for spherical symmetry) decomposition of the Einstein equations.

The most general line element for spherical symmetry is

$$ds^2 = (-\alpha^2 + A\beta^2)\, dt^2 + 2A\beta\, dt\, dr + A\, dr^2 + Br^2\, d\theta^2 + Br^2 \sin^2\theta\, d\phi^2 \qquad (2.1)$$

where where α is the lapse function, $\beta = \beta^r$ is the shift vector (one component with spherical symmetry), and $\gamma_{ij} = \mathrm{diag}(A, Br^2, Br^2\sin^2\theta)$ is the 3-metric on a t=const. hypersurface. All the metric coefficients are functions of t and r. For the construction of numerical solutions of the Einstein equations, many authors have followed the suggestion of Wilson (1979) that a simplification of the evolution equations can be brought about through *algebraic* coordinate conditions on the 3-metric. Historically two choices have dominated research in both spherical and axial symmetry. Radial gauge used by Estabrook et. al.(1973), Bardeen & Piran (1983) and Choptuik (1983) sets B=1. The well known Schwarzschild coordinates are a prime example of this gauge choice where:

$$\alpha = \left(1 - \frac{2m}{r}\right)^{1/2} \qquad A = \left(1 - \frac{2m}{r}\right)^{-1} \qquad B=1 \qquad (2.2)$$

The other coordinate system, the quasi-isotropic or isothermal gauge, has been used by Evans (1986), Shapiro & Teukolsky (1986) and Choptuik (1983) and is distinguished by setting A=B. The Schwarzschild solution is also well known in these coordinates:

$$\alpha = \frac{\left(1 - \frac{m}{2r'}\right)}{\left(1 + \frac{m}{2r'}\right)} \qquad A = B = \left(1 + \frac{m}{2r'}\right)^4 \qquad (2.3)$$

where $r = (1+m/2r')^2 r'$ relates the Schwarzschild and isotropic coordinates. In eqs. (2.2) and (2.3) the solution is explicitly time independent and the shift vanishes.

In the 3+1 decomposition of a spacetime, the extrinsic curvature of the spatial hypersurfaces

$$K_{ij} = (D_i\beta_j + D_j\beta_i - \partial_t\gamma_{ij})/2\alpha \qquad (2.4)$$

expresses how these 3-surfaces are embedded in the four dimensional spacetime. Here D_i is the covariant derivative on the 3-surface. Substituting the conditions placed upon the 3-metric by the radial and isothermal gauge choices into eq. (2.4), one obtains the following conditions on the shift: in radial coordinates $\beta = \alpha K^\theta{}_\theta / r$ while in isothermal coordinates $(\beta/r)_{,r} = \alpha(K^r{}_r - K^\theta{}_\theta)/r$.

In this work, we shall not impose any conditions on the 3-metric, but rather maintain it in its most general form, thereby leaving the choice of the lapse function and shift vector free to explore gauge conditions that are more geometric in nature. This point of view has also been taken by Thornburg (see his contribution to this volume) and is likely to become more popular as the power supercomputers increase. We are now in the position to write out the vacuum Einstein equations in the standard 3+1 form: the constraints may be written as

$$R + (\mathrm{Tr}\,K)^2 + K_{ij}K^{ij} = 0$$
$$D_i(\mathrm{Tr}\,K) - D_k K_i{}^k = 0$$

and the evolution equations are:

$$\partial_t \gamma_{ij} = -2\alpha\, K_{ij} + D_i\beta_j + D_j\beta_i$$
$$\partial_t K_{ij} = -D_i D_j \alpha + \alpha\, [R_{ij} + (\mathrm{Tr}\,K)K_{ij} - 2K_{ik}K_j{}^k] + \beta^k D_k K_{ij} + K_{kj} D_i\beta^k + K_{ik} D_j\beta^k \tag{2.5}$$

where R_{ij} and $R = \gamma_{ij}R^{ij}$ are, respectively, the Ricci tensor and the scalar curvature intrinsic to the the 3-space.

Choosing the coordinate system to be in keeping with those used by Cadez (1975) for the two black hole collision problem, we introduce η, a radial coordinate that at t=0 is related to the isotropic radial coordinate by $\eta = \ln(\frac{2}{m}r')$, and $\xi = \theta$. Following the York (1979) prescription for obtaining solutions to the initial vale problem, a conformal factor ψ' (where $A = \psi'^4 a$ and $B = \psi'^4 b$) is factored out of the 3-metric. We then write the line element for the 3-space as $d\sigma^2 = \psi^4 (a\, d\eta^2 + b\, d\Omega^2)$ where $d\Omega^2 = d\xi^2 + \sin^2\xi\, d\phi^2$ and $\psi = \psi' e^{\eta/2}$. The 3+1 form of the field equations in these coordinates along with the expressions for the intrinsic (Ricci) curvature of the spatial hypersurfaces are given in the Appendix. The variables H_a and H_b that appear in the evolution are obtained from the extrinsic curvature written in the conformal form: $K_{ij} = \psi^4 \mathrm{diag}(H_a, H_b, H_b \sin^2\xi)$.

The use of auxiliary variables should be mentioned at this point. The Ricci tensor of the 3-space contains second derivatives of γ_{ij} with respect to η. The field equations as written in the Appendix are in a form that is similar to writing the 1-dimensional wave equation; $\partial_t{}^2 f - \partial_x{}^2 f = 0$ as the system: $\partial_t f = h$, $\partial_t h = \partial_x{}^2 f$ where h is a function that plays the role of the extrinsic curvature. Alternatively the wave equation may also be written in the advective form: $\partial_t g = \partial_x f$, $\partial_t f = \partial_x g$. In this case the temporal and spatial derivatives are both of first order and are treated in the same manner.

For the Einstein equations, the role of the auxiliary variable g could be taken by either the Christoffel symbols [Hahn & Lindquist (1963)] or the first derivative of γ_{ij} with respect to η [Smarr (1979)]. In both of these cases the 3-Ricci tensor can be written in terms of first spatial derivatives and products of the extra variables. As a test of the concept of introducing auxiliary variables, we added the evolution equations for the metric derivatives and the Christoffel symbols to the system (2.5). Unfortunately we found that it was difficult to maintain the stability of the solution beyond 4000 time steps with the Christoffel symbols and 18000 time steps with the metric derivatives (our minimum requirement was 20000 time steps). This result should not be taken as a condemnation of the use of auxiliary variables since only one test using simple leapfrog finite differencing was made. It is likely that more sophisticated methods will have to be employed in order to make the use of auxiliary variables successful.

Finally, there is the freedom to choose the lapse and the shift. We also choose to provide inner boundary conditions at the throat and this requires that the evolution cover part of the interior of the black hole. This rules out some choices of time slices, e.g. polar slicing which cannot cover the spacetime inside the horizon. We also want to avoid the central singularity by an appropriate gauge choice. The best known of the 'crushing' singularity avoiding time slices is maximal slicing ($\mathrm{Tr}\, K= 0$) and its properties have been elucidated by Eardley & Smarr (1979). Taking the trace of the extrinsic curvature evolution equation yields:

$$\partial_t(\mathrm{Tr}\, K)= \alpha[(\mathrm{Tr}\, K)^2+R] - D^iD_i\alpha. \tag{2.6}$$

In the case $\mathrm{Tr}\, K=0$, eq. (2.6) reduces to an elliptic equation for the lapse that gives the maximal spacelike slices. The problem with a singularity avoiding lapse is that the time slices wrap around the singularity causing large amounts of shear in the coordinate grid. This results in a radial metric coefficient that increases without bound.

The shift vector choices we make explore the differences between Eulerian (normal or $\beta=0$) gauge and shifts that minimize the deformations of the geometry from one slice to the other. Smarr and York (1978) describe two such conditions, 1) 'minimal strain', which leads to a linear elliptic equation for β^i:

$$D^i(D_i\beta_k + D_k\beta_i) = D^i(2\alpha\, K_{ik}) \tag{2.7}$$

and 2) 'minimal distortion' where the distortion tensor is related to the shear of the unit normals to the 3-surfaces.

$$D^iD_i\beta_k + \frac{1}{3} D_k(D_i\beta^i) + R_{ik}\beta^i = D^i[2\alpha\, (K_{ik} - \frac{1}{3}\gamma_{ik}\mathrm{Tr}\, K)]. \tag{2.8}$$

III. FINITE DIFFERENCING METHODS

This section is concerned primarily with finite differencing methods for the hyperbolic evolution equations. The one elliptic equation that appears in this section comes from eq. (2.6). In the spherical coordinates (η,ξ), the lapse equation yields a two-point boundary value problem for a single second-order ordinary differential equation for α:

$$\partial_{\eta\eta}\alpha + \partial_\eta\alpha \left[\frac{2\partial_\eta\psi}{\psi} + \frac{\partial_\eta b}{b} - \frac{\partial_\eta a}{2a}\right] = \alpha\, (R_{\eta\eta} + \frac{2a}{b} R_{\xi\xi}) \tag{3.1}$$

where $\partial_\eta\alpha = 0$ at the throat and $\alpha = \alpha_{\,\mathrm{Schwarzschild}}$ at the outer boundary.
The method of choice for solving such an equation was shooting based upon a fourth order Runge-Kutta method and it since has become the test-bed calculation against which other methods (to be used for the full two-dimensional calculation) are compared.

If one finite differences the lapse equation with a standard second order differencing for the first and second derivatives with respect to η, the result is a linear system of algebraic equations which can be written in matrix notation as $\mathbf{M}\alpha=\mathbf{b}$. Here the square matrix $\mathbf{M}$ is a tridiagonal matrix that contains the coefficients resulting from the finite differencing of the lapse equation, α is a vector whose components are the values of the lapse at the grid points, and $\mathbf{b}$ is also a vector of the same length that contains the values of the lapse at the boundaries. Methods used to solve the matrix equation and tested against the shooting method have been 1.) simultaneous over relaxation (SOR), 2.) direct Gaussian elimination with pivoting and interchange of rows 3.) an iterative conjugate gradient method with different preconditionings. All of the matrix methods except SOR (which eventually took so long to converge it could not make the required cut off of 20000 time steps) gave the same results as the shooting method (to within one tenth of one percent or less) on radial grids with 1000 and 2000 nodes.

The differencing of the hyperbolic equations produced significantly different results depending upon the schemes employed. Rather than write out the full finite differenced versions of the Einstein equations appearing in the Appendix, we shall write the evolution equations in the schematic form: $\partial_t g = -2\alpha K$ and $\partial_t K = \alpha(K^2+R) + \Delta\alpha$. N.B. the symbols K and R are *not* the contracted extrinsic and intrinsic curvatures, $\Delta\alpha$ will represent the covariant Hessian of the lapse and for the sake of simplicity, we have eliminated the terms involving the shift since they may be treated in the same manner as the lapse. The grid is unstaggered in space but depending upon the finite difference scheme, may be staggered in time.

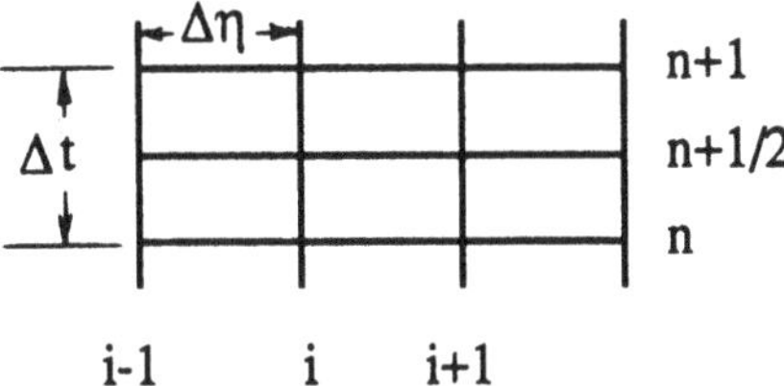

Figure 1. The numerical grid used for the finite differencing methods discussed in the text.

In all, nine different methods for finite differencing the evolution equations were utilized. The first order schemes used were a standard forward-time, centered-space method (FTCS) and the Lax method. The FTCS scheme is

$$g_i^{n+1} = g_i^n - 2\alpha_i^n K_i^n \Delta t \quad \text{and} \quad K_i^{n+1} = K_i^n + \{\alpha_i^n [(K_i^n)^2 + R_i^n] - (\Delta\alpha)_i^n\}\Delta t$$

and the Lax method, by averaging the variables on the nth time step, introduces an implicit artificial viscosity:

$$g_i^{n+1} = \tfrac{1}{2}(g_{i-1}^n + g_{i+1}^n) - 2\alpha_i^n K_i^n \Delta t\,; \quad K_i^{n+1} = \tfrac{1}{2}(K_{i-1}^n + K_{i+1}^n) + \{\alpha_i^n[(K_i^n)^2 + R_i^n] - (\Delta\alpha)_i^n\}\Delta t$$

In both of these cases, the first and second spatial derivatives of the metric that appear in the Ricci tensor and the Hessian of the lapse, $\Delta\alpha$ are all central differenced about the point, i, on the time step, n; i.e. $\partial_\eta g = (g_{i+1}^n - g_{i-1}^n)/(2\Delta\eta)$ and $\partial_{\eta\eta} g = (g_{i+1}^n - 2g_i^n + g_{i-1}^n)/(\Delta\eta)^2$.

Variations on the leapfrog scheme were implemented as well. The metric evolution was straight forward, but different methods were used to treat the nonlinear term in K for the extrinsic curvature evolution equation. The first method simply used the value of K on the previous time step and this added first order errors to the method:

$$g_i^{n+1/2} = g_i^{n-1/2} - 2\alpha_i^n K_i^n \Delta t \quad \text{and} \quad K_i^{n+1} = K_i^n + \{\alpha_i^n [(K_i^n)^2 + R_i^{n+1/2}] - (\Delta\alpha)_i^n\}\Delta t$$

In order to improve upon the method and make it completely second order in both temporal and spatial derivatives, the terms in the K evolution equation that are determined on the nth time step by the evolution equations need to be evaluated on the time step, n+1/2. This change makes the scheme a fully staggered leapfrog without introducing first order errors. To accomplish this one replaces the evolution equation for the extrinsic curvature with:

$$K_i^{n+1} = K_i^n + \{ {}^*\alpha_i^n[({}^*K_i^{n+1/2})^2 + R_i^{n+1/2}] - ({}^*\Delta\alpha)_i^{n+1/2}\}\Delta t$$

where the starred quantities were calculated from a $\Delta t/2$ step extrapolation defined by ${}^*u_i^{n+1/2} = \frac{3}{2}u_i^n - \frac{1}{2}u_i^{n-1}$.

Three two-step methods were used and these eventually led to true predictor-corrector schemes. While these schemes have the disadvantage of requiring more storage, they are provide better stability and can be run longer. Perhaps the best known of the two-step methods is the Lax-Wendroff scheme which consists of a Lax method for a one-half time step and then a leapfrog for a full time step:

(1) $g_i^{n+1/2} = \frac{1}{2}(g_{i-1}^n + g_{i+1}^n) - 2\alpha_i^n K_i^n (\Delta t/2);$

$K_i^{n+1/2} = \frac{1}{2}(K_{i-1}^n + K_{i+1}^n) + \{\alpha_i^n[(K_i^n)^2 + R_i^n] - (\Delta\alpha)_i^n\} (\Delta t/2)$

(2) $g_i^{n+1} = g_i^n - 2{}^*\alpha_i^{n+1/2} K_i^{n+1/2}\Delta t$

$K_i^{n+1} = K_i^n + \{{}^*\alpha_i^{n+1/2} [(K_i^{n+1/2})^2 + R_i^{n+1/2}] - ({}^*\Delta\alpha)_i^{n+1/2}\}\Delta t$

In this scheme the extrapolation of the values for K is unnecessary since they are determined by the Lax method. The value of the lapse on the (n+1/2)th time step is still determined by half-step extrapolation. Two variations on this scheme are 1.) replace the Lax method with a FTCS scheme and 2.) implement a full time step "double" leapfrog which can be considered as a simple predictor-corrector scheme. This latter two step method can be written in the form:

(1) $g_i^{n+1/2} = g_i^{n-1/2} - 2\alpha_i^n K_i^n \Delta t$

$K_i^{n+1/2} = K_i^{n-1/2} + \{\alpha_i^n[(K_i^n)^2 + R_i^n] - (\Delta\alpha)_i^n\} \Delta t$

(2) $g_i^{n+1} = g_i^n - 2\alpha_i^{n+1/2} K_i^{n+1/2}\Delta t$

$K_i^{n+1} = K_i^n + \{\alpha_i^{n+1/2}[(K_i^{n+1/2})^2 + R_i^{n+1/2}] - (\Delta\alpha)_i^{n+1/2}\}\Delta t$

After performing the first leapfrog calculation, the maximal slicing equation is solved on the n+1/2 time step and the result is used in the next leapfrog calculation to obtain the value of R for a subsequent calculation of the lapse. This method has the disadvantage of having to solve the lapse equation twice at each time level.

Finally two fully-second order predictor-corrector methods were employed. These two methods are now commonly employed in hydrodynamics codes, but as far as we know they have not been used extensively in numerical relativity, if at all. The better known of the two is the method of MacCormack:

(1) $\tilde{g}_i^{n+1} = g_i^n - 2\alpha_i^n K_i^n \Delta t$ predictor

$\tilde{K}_i^{n+1} = K_i^n + \{\alpha_i^n[(K_i^n)^2 + R_i^n] - (\Delta\alpha)_i^n\} \Delta t$

(2) $g_i^{n+1} = \frac{1}{2}[\tilde{g}_i^{n+1} + g_i^n - 2\alpha_i^n \tilde{K}_i^{n+1} \Delta t]$ corrector

$K_i^{n+1} = \frac{1}{2}[\tilde{K}_i^{n+1} + K_i^n + \{\alpha_i^n[(\tilde{K}_i^{n+1})^2 + \tilde{R}_i^{n+1}] - (\tilde{\Delta}\alpha)_i^{n+1}\}\Delta t]$

The MacCormack method differs from all other schemes that have been described up to now. It is distinguished by the fact that in the predictor equations, the first spatial derivatives of the metric appearing in the Ricci tensor and Hessian of the lapse are determined by forward differencing: $\partial_\eta g = (g_{i+1}^n - g_i^n)/\Delta\eta$. In the corrector equations these derivatives are calculated using backwards differencing: $\partial_\eta g = (g_i^n - g_{i-1}^n)/\Delta\eta$. An alternative method is to backward difference at the predictor level and forward difference the corrector equation. Second derivatives are always centered differenced.

On the other hand, the method of Brailovskaya, uses a centered differencing scheme for spatial derivatives in both the predictor and corrector equations.

$$(1)\quad \tilde{g}_i^{n+1} = g_i^n - 2\alpha_i^n K_i^n \Delta t \qquad \text{predictor}$$

$$\tilde{K}_i^{n+1} = K_i^n + \{\alpha_i^n[(K_i^n)^2 + R_i^n] - (\Delta\alpha)_i^n\}\Delta t$$

$$(2)\quad g_i^{n+1} = g_i^n - 2\alpha_i^n \tilde{K}_i^{n+1} \Delta t \qquad \text{corrector}$$

$$K_i^{n+1} = K_i^n + \{\alpha_i^n[(\tilde{K}_i^{n+1})^2 + \tilde{R}_i^{n+1}] - (\tilde{\Delta}\alpha)_i^{n+1}\}\Delta t$$

and is somewhat easier to code.

IV. ANALYTIC BEHAVIOR OF THE SCHWARZSCHILD SOLUTION APPROPRIATE TO NUMERICAL RELATIVITY

As every student of general relativity knows, the Schwarzschild solution is static. However if one slices the 4-dimensional spherical black hole space time by time slices that do not preserve the isometry of the 3-spatial geometry from one time slice to the next, the spacetime will appear to be time dependent since the spatial geometry changes from slice to slice. This is precisely what can happen in numerical relativity. A good illustration of how the spacelike surfaces depend upon the choice of constant time slices is found in Chapter 31 (p 839) of Misner, Thorne and Wheeler (1973), where different slices across the Kruskal diagram for the Einstein-Rosen bridge produce strikingly different results.

The initial data is for a single black hole is simple and its extension to multiple black holes is discussed by York in this volume. Using the York prescription for solving the initial value problem, imposing time-symmetry (K_{ij}=β=0) and conformal flatness (a=b=1 which implies the vanishing of the scalar curvature), one need only solve the flat space Laplacian for ψ' to obtain $\psi' = (1+\frac{m}{2r'})$. This solution describes the static Einstein-Rosen bridge. While this can be represented in a number of different coordinate systems (Kruskal-Szekeres, Novikov, and Eddington-Finkelstein to name but a few), we shall use the isotropic coordinates given in eq. (2.3). At the throat r'=m/2 (r=2m), one can identify the $m/2>r'>0$ with $m/2<r^*<\infty$ with the relationship $r^* = m^2/(4r')$ which gives the static two sheeted version of the Schwarzschild solution where the interior is mapped onto an asymptotically flat spacetime.

The slices that represent the static time translations should be reproduced easily since there is no evolution. In terms of the numerical metric the isotropic gauge can be written in the following form:

$$\alpha = \tanh(\eta/2) \qquad \psi = 2\sqrt{m/2}\cosh(\eta/2) \qquad a=b=1 \qquad (4.1)$$

These results are used as an initial test (albeit a rudimentary one) of the code. As such it tests mainly for simple errors (typographical, missing terms, etc.) since the lapse in eq. (4.1) must keep the metric and extrinsic curvature from evolving from one slice to the next. This will only result if the Ricci tensor (depending only upon derivatives of the conformal factor) exactly cancels the Hessian of the lapse.

Another well-known behavior associated with the Schwarzschild solution is that one can analytically calculate the radial geodesics and the motion of a test particle falling along them. If the test particle is at a position $R=r_0/2m$ where r_0 is the usual Schwarzschild coordinate, at time t=0 (proper time $\tau=0$) then the resulting position in Schwarzschild coordinates (r,t) after proper time,τ, has elapsed is given in the parametric form (see e.g. MTW):

$$\tau = mR^{3/2}(\lambda + \sin\lambda), \qquad r = mR(1 + \cos\lambda)$$

$$t/2m = \ln\left(\frac{\sqrt{R-1}+\tan(\lambda/2)}{\sqrt{R-1}-\tan(\lambda/2)}\right) + \sqrt{R-1}\left[\lambda + \tfrac{1}{2}R(\lambda+\sin\lambda)\right] \qquad (4.2)$$

Clearly the after an elapsed proper time $\tau=\pi mR^{3/2}$, the test particle has fallen into the central singularity ($\lambda=\pi$). This result can prove to be an important check on the numerical results of a code. If one sets the gauge to geodesic normal slicing ($\alpha=1$, $\beta=0$), the grid points free fall radially into the black hole. The initial data surface at the moment of time symmetry is described by the normal Einstein-Rosen bridge configuration where the throat is located at the Schwarzschild radius, r=2m. According to eqs. (4.2) this grid point will fall into the central singularity after an elapsed time of πm, at which time the code crashes. Making a crash test of the code during the initial stages of development, like the crash testing of automobiles, yields important information before a prototype is turned into a finished product. Oddly enough some codes failed even at this early a stage in development. The first order methods and the leapfrog without time step extrapolation did not crash until a few time steps after $\tau=\pi m$.

Finally, the most important analytic results concern the behavior of the maximally sliced (TrK=0) Schwarzschild solution. The static isotropic coordinates in eq. (2.3) are an example of a maximally sliced Schwarzschild solution. The lapse function is a solution of eq. (2.6) (K_{ij} vanishes due to the time symmetry and R_{ij} is determined solely from the conformal factor) given the Dirichlet boundary conditions $\alpha=0$ at the throat. If on the other hand the boundary conditions are Neumann conditions, with $\partial_\eta\alpha=0$ at the throat, the lapse is time dependent and must collapse to zero in order to prevent the grid from falling into the Schwarzschild singularity. The time dependent behavior of the lapse function at the throat was calculated by Smarr and York (1978) using a simplified model and was found to be $\alpha_{min}=\exp(-\tau/\tau_e)$ where the e-folding time $\tau_e \equiv 1.82$m. It has been argued by some that since the lapse collapses rapidly in the interior of the black hole, the physics is occurring at such a slow rate, that one may be able to simply replace the values of the lapse at the throat and its neighboring grid points with zero after some predetermined time in the evolution. However, had we performed such a procedure right from the start we would not have been able to determine the appropriate differencing schemes and minimum step size needed to reproduce the analytic behavior.

Another important result was reached in the work of Estabrook, et. al. (1974), who proved that for the dynamic maximal slicing of the Schwarzschild solution, the grid points asymptotically fall to the surface defined by the Schwarzschild radius, $r=\frac{3}{2}m$ which is itself maximal. This behavior can be understood qualitatively as follows: The initial hypersurface has the metric of a 3-parabaloid. (If it were a 2-parabaloid it would have

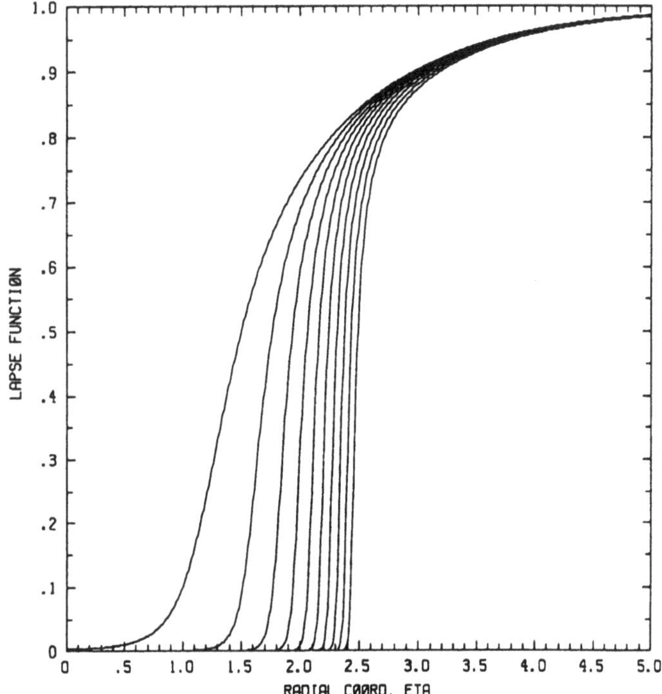

Figure 2. The lapse function for a dynamical maximally sliced Schwarzschild spacetime at time intervals of 10m from t=0 to t=100m. Most of the collapse of the lapse occurs within the first 10m of the evolution. Problems with maximal slicing occur when the spatial derivatives of the lapse grow at late times.

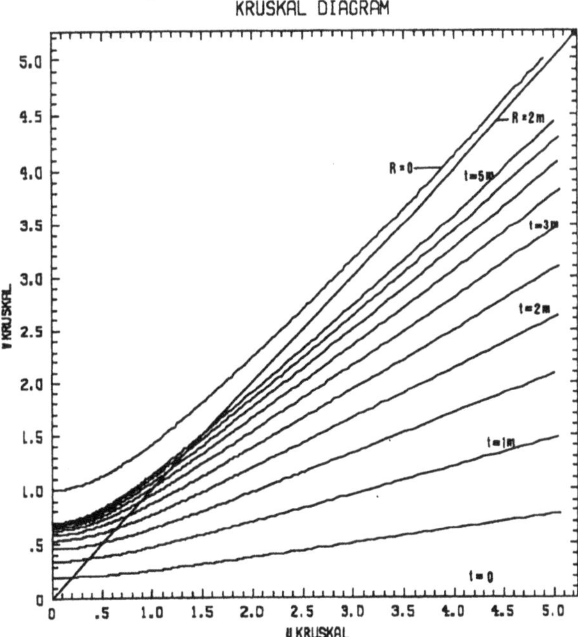

Figure 3. A Kruskal diagram showing the t=const maximal slices for the Schwarzschild spacetime. The slices at late times pile up at r=3m/2 in the inner region but still have to reach outside of the horizon at r=2m. This causes the grid stretching problems associated with maximal slicing. The slices are shown at time intervals of .5m from t=0 to t=5m. After 5m the inner region does not evolve significantly.

negative scalar curvature). As soon as the hypersurface begins to evolve, the bridge begins to stretch and and its central portion assumes the shape of a 3-cylinder (which has positive scalar curvature: essentially that of its generating 2-spheres). Since R is now positive, the lapse becomes concave upwards with increasing η in the region near the throat (see eq. (2.6)). Therefore it starts to collapse to zero, which eventually stabilizes the bridge in the throat region. The 3-scalar curvature of a cylindrical surface is $R=2/r^2$. Therefore at $r=3m/2$ one has $R=\frac{8}{9m^2}$. One may also use the results of eqs. (3.2) for the behavior of the freely falling test particles to calculate the total proper time for the throat to collapse from $r=2m$ to the $r=3m/2$ ($\lambda=\pi/3$) and obtain $\tau_{max}=\int_0^\infty \alpha_{min}\, d\tau = (\frac{\pi}{3}+\frac{\sqrt{3}}{2})m \cong 1.9132m$. It must be remembered that those points which are initially outside of the black hole event horizon have time-like trajectories that are affected by the effective acceleration produced by the lapse and therefore will not obey eqs. (3.2).

Figures 2 and 3 typify the behavior of the maximally sliced spherically symmetric solution. The collapse of the lapse for each 10m in time from t=10m to t=100m are shown in Figure 2. The outer boundary was placed at $\eta=10$ (or a Schwarzschild radius of about 11000m) and the step sizes were Δt =.005m and $\Delta\eta=0.01$. In order to avoid the singularity, the value of the lapse at the throat collapses to 10^{-2} by 5m in coordinate time. Initially the lapse is set to unity for all values of η. The behavior of the spatial slices can be represented by a Kruskal diagram (Figure 3) where the maximal t=const slices clearly halt at r=3m/2 and wrap themselves about the central r=0 singularity. The time slices in Figure 3 are shown at intervals of .5m from t=0 to t=5m.

The analytic results of Smarr and York (1978) for the 'collapse of the lapse' can be checked by plotting the logarithm of the lapse function at the throat versus time and determining the e-folding time. This measurement is also dependent upon the size of the time step as well. Figures 4a and 4b show the results for some of the finite difference methods described in the previous section. Calculations were performed that used time steps of .01m and .005m. Clearly the numerical dispersion that results from larger step sizes will ruin the accuracy of the computation at late times. The effect is particularly apparent for the Lax and Lax-Wendroff methods which introduce an implicit artificial viscosity into the code. The second order methods themselves are subject to dispersion and after 40000 time steps (t=200m for Δt =.005m) the strictly exponential behavior begins to die off with $\alpha \approx 10^{-94}$. Finally the e-folding time calculated for the second-order methods from Figure 4b yields $\tau_e=1.821$ compared to the Smarr-York value of 1.82.

Checking the results concerning the singularity avoidance of maximal slicing we calculate the total proper time elapsed at the throat during the evolution from t=0 to t=100m (with a time step of .005m) using a Simpson's rule integration procedure. We also determined the value of the Schwarzschild radius at the throat (for t=100m) and compare that to the analytic value of 1.5m. Since the limiting hypersurface at late times is a 3-cylinder we must also verify that the points at $\eta>0$ lie close to that cylinder with radius 3m/2. The numerical evolution proceeds only to a finite value of time. Therefore one would expect that the 3-surfaces have not yet reached the asymptotic cylindrical topology.

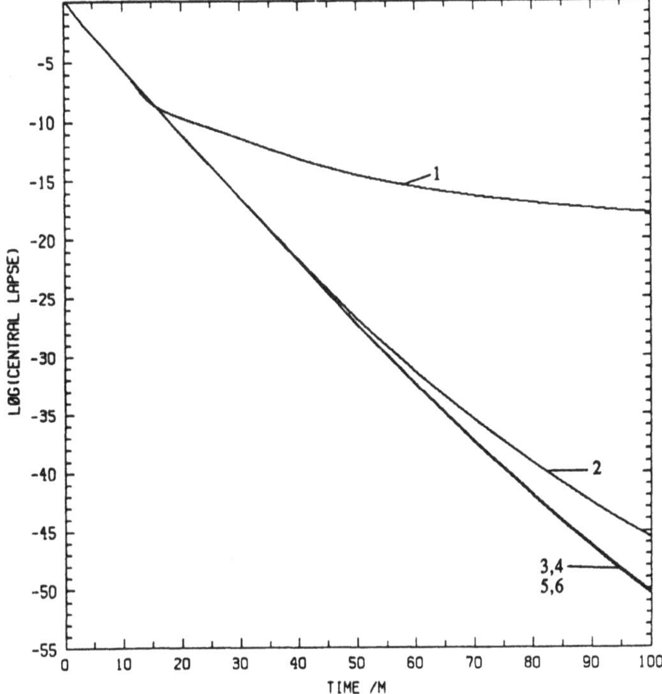

Figure 4a. The time dependent behavior of the lapse function at the throat calculated with different finite difference methods with Δt=.01m. The methods used are 1. Lax; 2. Lax-Wendroff; 3. Leapfrog with 1/2 time step extrapolation; 4. two step FTCS-Leapfrog; 5. MacCormack; 6. Brailovskaya.

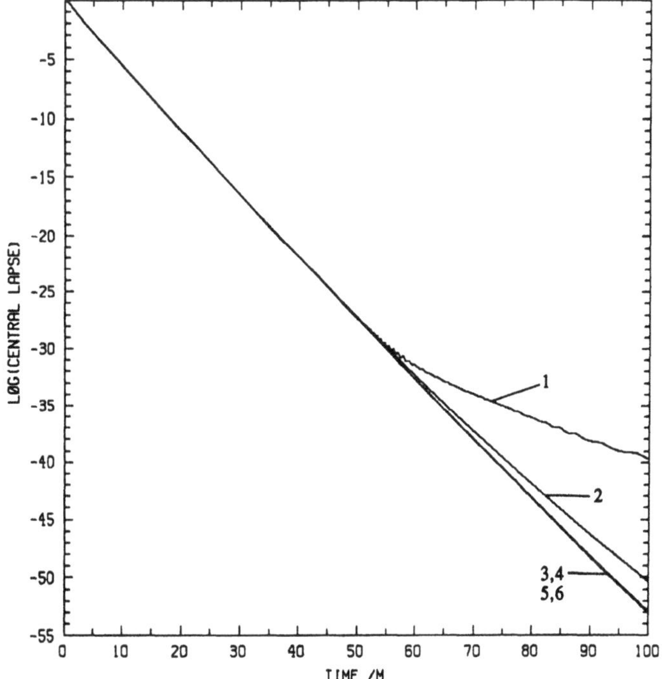

Figure 4b. The same graph as in Figure 4a but with Δt=.005m. Better agreement with the analytic results are obtained with the finer resolution. The e-folding time is determined to be τ_e=1.821m.

Those grid points that at t=0 were originally outside of the event horizon ($\eta>0$) should at late (but finite) times be located at Schwarzschild radii somewhat greater than 3m/2. If there is no crossing of the grid points then the value of the Schwarzschild radius will monotonically increase with the coordinate η. For many of the finite differencing schemes this did not occur. Some points located at $\eta>0$ were found to have a of the Schwarzschild radius less than the value of the Schwarzschild radius at the throat ($\eta=0$) for t > 50m in the evolution. The results of the test of proper time and monotonicity of the Schwarzschild radius for different finite differencing schemes are presented in Table 1.

finite differencing scheme	total proper time elapsed @ throat	R Schwarz. $\eta=0$ (t=100m)	R Schwarz. monotone @ $\eta>0$
Analytic	**1.9132m**	**1.50000m**	**yes**
Lax	1.9261m	1.48992m	no
FTCS	1.9223m	1.49548m	no
Leapfrog (1st ord. error)	1.9204m	1.49835m	no
Leapfrog (1/2 step extrap.)	1.9132m	1.49999m	no
Two time step Leapfrog	1.9132m	1.49998m	no
FTCS-Leapfrog	1.9174m	1.49968m	no
Lax-Wendroff	1.9212m	1.49727m	yes
McCormack	1.9132m	1.49999m	yes
Brailovskaya	1.9132m	1.49999m	yes

Table1. The results of the collapse of throat and points just outside of the horizon during the evolution to t=100m for the maximally sliced Schwarzschild solution. The step sizes were chosen as Δt=.005m and $\Delta\eta$=.01. The outer boundary was placed at η=10.

The results of Table 1 and Figure 4b can be summarized as follows. For those differencing methods that are more dispersive, the lapse does not collapse as rapidly as it should thereby allowing the evolution to proceed longer than expected in the region of the throat. This results in both a longer elapsed proper time and a limiting Schwarzschild radius of less than 3m/2. While four of the second order methods agree quite closely with the analytic results, it is only the two predictor-corrector methods that passed all tests. It should be noted that at t=100m, the limiting value of the Schwarzschild radius is less than 3m/2 for all methods. Part of the reason for this is that no method is completely free of dispersive effects and even at t=100m there is some deviation from exponential drop off of the lapse at the throat.

Our present philosophy is that if a numerical relativity code is expected to accurately reproduce known analytic results, then one result that the code should also be expected to reproduce is the propagation of the constraints by the evolution equations. If free evolution is properly carried out then the constraints should act as a check on how accurately one is solving the entire Einstein system. Analytically we know that the Einstein equations form a self consistent system and that self consistency should be maintained in the finite differenced version as well. For the believers in constrained evolution, where the constraint equations are used to determine some of the dependent variables, the self consistency check involves a verification that the extra evolution equations are satisfied.

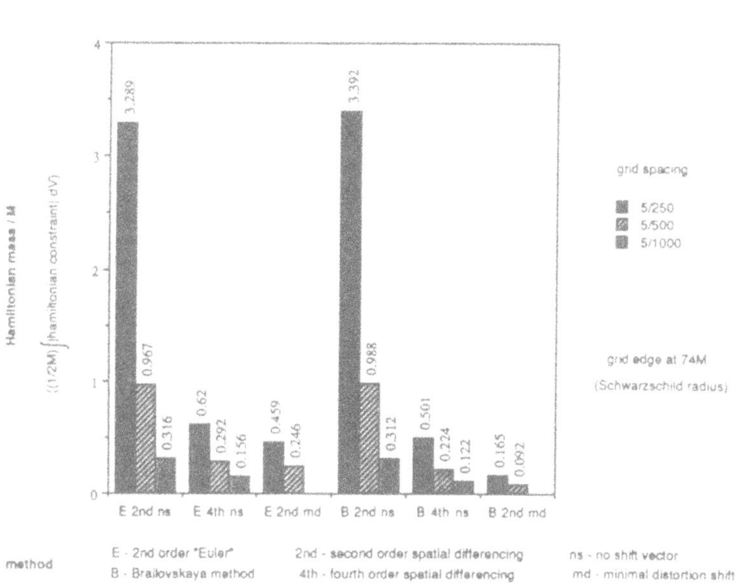

Figure 5. The Hamiltonian constraint integrated over the entire computational spacelike hypersurface. This provides a measure of how large the errors are in propagating the constraints with a free evolution. The minimal distortion shift vector combined with a higher order predictor-corrector method is necessary to maintain the constraints.

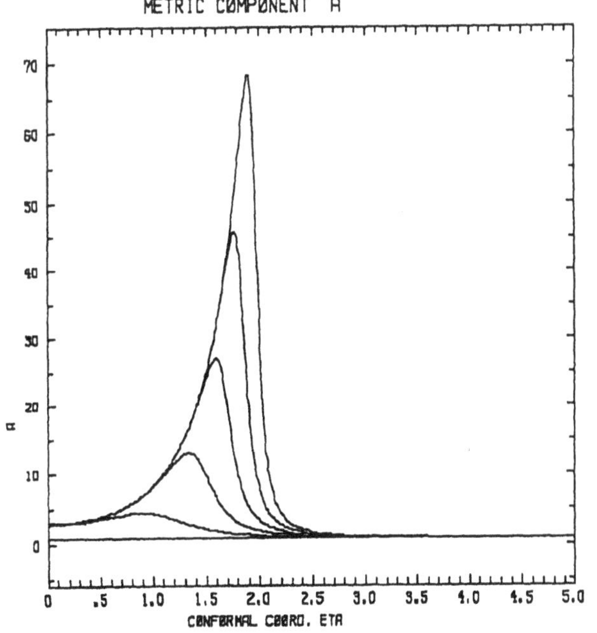

Figure 6. The radial metric coefficient from t=0 to t=40m with maximal slicing and vanishing shift vector. The coordinate shear at late times causes the metric to grow without bound in the vicinity of the event horizon. This eventually causes that code to halt at times greater than t=200m.

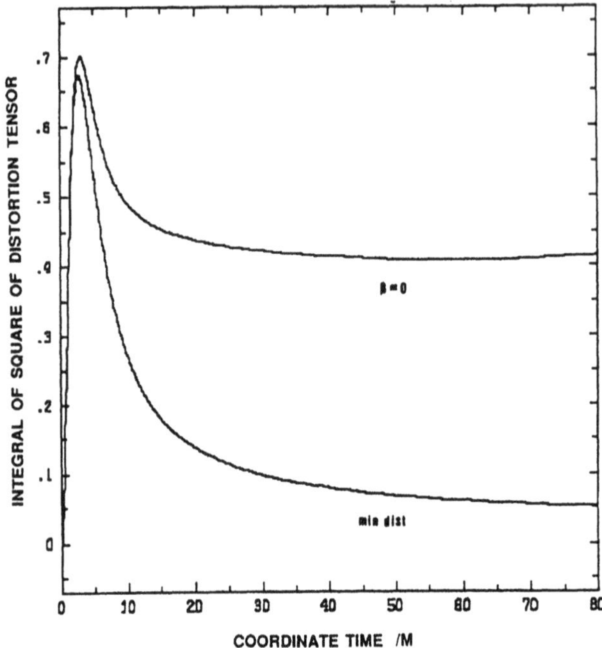

Figure 7. The integrated value of the square of the distortion tensor as a function of coordinate time. This demonstrates that the minimal distortion shift vector does reduce much of the coordinate shear.

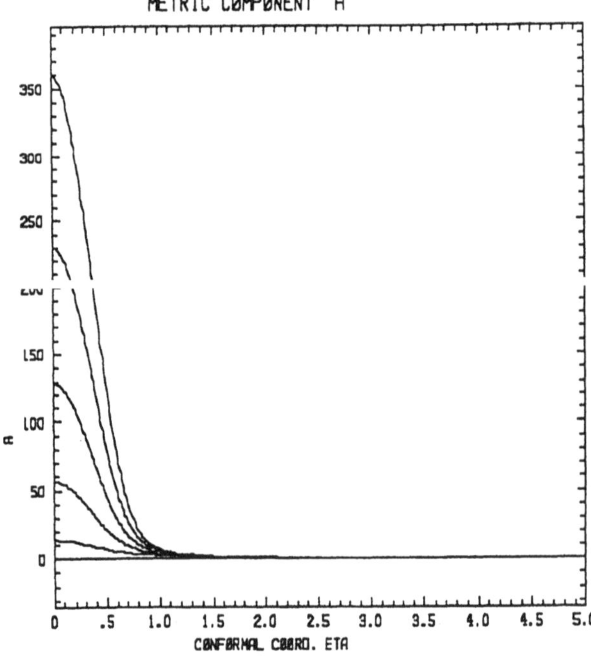

Figure 8. The radial metric coefficient from t=0 to t=40m with maximal slicing and minimal distortion shift vector. The coordinate shear at late times is shifted to the throat and the growth of the radial metric at this location occurs so rapidly that the evolution is halted soon after t=40m.

Since our intended purpose is to develop better methods of handling the hyperbolic equations, we have chosen to use the constraints as a check on free evolution. Figure 5 shows the results of a selected number of tests where the integrated value of the hamiltonian constraint is compared to the total mass of the system. While the absolute values of the constraints may actually be small, the integrated value over the hypersurface will give a better measure of how much 'energy' is associated with inaccuracies in numerical methods.

The important results of Figure 5 are that fourth order spatial differences are better at propagating the constraints than the simple second order difference methods. The reason why the spatial differencing is so important is that as the lapse collapses during the late time evolution, its spatial derivatives along with the spatial derivatives of the 3-metric and extrinsic curvature need to be calculated with ever increasing accuracy. The gradients of all the metric and extrinsic curvature components become very large in the vicinity of the event horizon if the shift vector vanishes (see Figure 2). Unless one has accurate methods of taking derivatives in the vicinity of these steep gradients, the code will quickly fail. The improved accuracy of the predictor-corrector methods clearly win out in this case.

While the leapfrog method has become the workhorse for finite differencing schemes in numerical relativity, its disadvantages outweigh its advantages when compared to the Brailovskaya or MacCormack predictor-corrector methods. The extra amount of effort that is needed to encode these latter two-step methods is negligible compared to the accuracy and stability that they provide.

At very late times ($t > 200m$) even the most accurate maximal slicing methods failed. As can be seen in Figure 6, the radial metric component develops a sharp peak in the vicinity of the event horizon and the finite difference approximation of the steep gradients eventually breaks down, forcing the code to halt. Smarr and York (1978) proposed the minimal distortion shift condition in order to overcome these problems. While the minimal distortion shift vector does reduce a measure of the distortion the (see Figure 7 which plots the square of the distortion tensor integrated over the 3-surface as a function of time) and removes the sharp gradients of the radial metric component in the region of the event horizon, the coordinate shear is transferred to the throat (where the volume elements are small) and produces even larger gradients. This behavior is shown in Figure 8 and is to be compared to figure 6. As a result, the minimal distortion shift condition caused the code to terminate at about 80m.

It is clear that numerically simulating black hole spacetimes with dynamic, singularity avoiding spacelike slices is not a trivial problem and more work remains to be done if longer and more accurate time evolutions are to be performed. Part of the problem may be blamed upon the use of a logarithmic radial variable. While it conveniently covers the region far outside the black hole it inadequately covers the region near the horizon. A partial cure may be found in the use of adaptive grids that allow the horizon (or any other region where large gradients form) to be covered with a greater degree of accuracy.

Our efforts have come closer to obtaining the Holy Grail of numerical relativity (Shapiro and Teukolsky (1986)). By definition the black hole codes we have written avoid singularities and treat black holes. With some combination of minimal principles, adaptive gridding, and perhaps higher order predictor-corrector methods, higher accuracy and longer term stability may finally be achieved.

V. FUTURE WORK

A two dimensional code that will be used to calculate the head-on collision of two black holes has been written using the methods described in this article. Deriving and coding a two-dimensional code based upon the most general three-metric possible is a project that can certainly introduce many errors. However using Macsyma we were able to derive the equations and translate many of the expressions into optimized Cray Fortran code. This meant altering the source code for the Macsyma tensor packages somewhat but since they are written in the Macsyma language, this proved to be a simpler job than had originally been expected.

The first test of the two-dimensional code, outside of reproducing all of the results presented here, will be to reproduce the normal mode analysis of an oscillating black hole. Analytic initial data exists for a single black hole superimposed on any number of gravitational multipole moments. While some of the contributions to the initial distortion of the gravitational field can be considered to result from incoming gravitational radiation, it is likely that most of this will leave the system quickly [see Schutz (1980) and Anderson and Hobill (1988) for arguments] and leave behind an oscillating black hole that loses its distortion through quasi-normal mode 'ringing'. It is hoped that the accuracy and stability of the code will be such that the numerical solution will be able to pick out the gravitational wave tails that exist long after the first burst of radiation is emitted.

If the amplitudes of the initial multipole moments are large enough, the pertubative results should break down. One goal is to determine when this occurs. A single oscillating black hole is expected form after two colliding black holes have merged. Therefore the next stage on the way to the development of a two black hole collision code will also provide an important test-bed calculation to guide the development of future codes designed to study black hole spacetimes.

REFERENCES

Anderson, J.L., and Hobill, D. W., (1988), *J. Comp. Phys.*, **75**, 283.

Bardeen, J., and Piran, T., (1983), *Phys. Reports.*, **96**, 205.

Cadez, A., (1971) Ph. D. Thesis, University of North Carolina

Choptuik, M., (1986), Ph. D. Thesis, University of British Columbia.

Eardley D., and Smarr, L., (1979) *Phys. Rev. D.*, **19**, 127.

Estabrook, F., Wahlquist, H., Christensen, S., DeWitt, B., Smarr, L., and Tsiang, E., (1973), *Phys. Rev. D.*, **7**, 2814.

Evans, C., (1986), In *Dynamical Spacetimes and Numerical Relativity*, edited by Centrella, J., Cambridge University Press, Cambridge.

Hahn, S. G. and Lindquist, R. W., (1964), *Ann. Phys.*, **29**, 304.

Misner, C.W., Thorne, K.S., and Wheeler, J. A., (1973), *Gravitation*, W. H. Freeman, San Francisco.

Shapiro, S. and Teukolsky, S., (1986), In *Dynamical Spacetimes and Numerical Relativity*, edited by Centrella, J., Cambridge University Press, Cambridge.

Schutz, B., (1980), *Phys. Rev. D.*, **22**, 249.

Smarr, L., (1979), In *Sources of Gravitational Radiation*, edited by Smarr, L., Cambridge University Press, Cambridge.

Smarr, L., and York, J., (1978), *Phys. Rev. D.*, **17**, 2529.

Wilson. J., (1979), In *Sources of Gravitational Radiation*, edited by Smarr, L., Cambridge University Press, Cambridge.

APPENDIX: VACUUM EINSTEIN EQUATIONS FOR SPHERICAL SYMMETRY

$$R_{\eta\eta}=-\frac{4\frac{\partial^2\psi}{\partial\eta^2}}{\psi}+\frac{4\left[\frac{\partial\psi}{\partial\eta}\right]^2}{\psi^2}-\frac{2\frac{\partial b}{\partial\eta}\frac{\partial\psi}{\partial\eta}}{b\psi}+\frac{2\frac{\partial a}{\partial\eta}\frac{\partial\psi}{\partial\eta}}{a\psi}-\frac{\frac{\partial^2 b}{\partial\eta^2}}{b}+\frac{\left[\frac{\partial b}{\partial\eta}\right]^2}{2b^2}+\frac{\frac{\partial a}{\partial\eta}\frac{\partial b}{\partial\eta}}{2ab}$$

$$R_{\xi\xi}=-\frac{2b\frac{\partial^2\psi}{\partial\eta^2}}{a\psi}-\frac{2b\left[\frac{\partial\psi}{\partial\eta}\right]^2}{a\psi^2}-\frac{3\frac{\partial b}{\partial\eta}\frac{\partial\psi}{\partial\eta}}{a\psi}+\frac{\frac{\partial a}{\partial\eta}b\frac{\partial\psi}{\partial\eta}}{a^2\psi}-\frac{\frac{\partial^2 b}{\partial\eta^2}}{2a}+\frac{\frac{\partial a}{\partial\eta}\frac{\partial b}{\partial\eta}}{4a^2}+1$$

Hamiltonian constraint

$$0=\frac{2R_{\xi\xi}}{b\psi^4}+\frac{R_{\eta\eta}}{a\psi^4}+\frac{2H_b^2}{b^2}+\frac{4H_aH_b}{ab}$$

momentum constraint

$$0=\frac{4H_b\frac{\partial\psi}{\partial\eta}}{b\psi}-\frac{4H_a\frac{\partial\psi}{\partial\eta}}{a\psi}+\frac{2\frac{\partial H_b}{\partial\eta}}{b}-\frac{\frac{\partial b}{\partial\eta}H_b}{b^2}-\frac{\frac{\partial b}{\partial\eta}H_a}{ab}$$

evolution equations

$$\frac{\partial a}{\partial t}=-\frac{4\beta\frac{\partial\psi}{\partial\eta}}{\psi^5}+\frac{2\frac{\partial\beta}{\partial\eta}}{\psi^4}-\frac{\frac{\partial a}{\partial\eta}\beta}{a\psi^4}-2\alpha H_a$$

$$\frac{\partial b}{\partial t}=\frac{4b\beta\frac{\partial\psi}{\partial\eta}}{a\psi^5}+\frac{\frac{\partial b}{\partial\eta}\beta}{a\psi^4}-2\alpha H_b$$

$$\frac{\partial H_a}{\partial t}=\frac{\alpha R_{\eta\eta}}{\psi^4}-\frac{4\beta H_a\frac{\partial\psi}{\partial\eta}}{a\psi^5}+\frac{2\frac{\partial\alpha}{\partial\eta}\frac{\partial\psi}{\partial\eta}}{\psi^5}+\frac{\beta\frac{\partial H_a}{\partial\eta}}{a\psi^4}+\frac{2\frac{\partial\beta}{\partial\eta}H_a}{a\psi^4}-\frac{2\frac{\partial a}{\partial\eta}\beta H_a}{a^2\psi^4}-\frac{\frac{\partial^2\alpha}{\partial\eta^2}}{\psi^4}+\frac{\frac{\partial a}{\partial\eta}\frac{\partial\alpha}{\partial\eta}}{2a\psi^4}+\frac{2\alpha H_aH_b}{b}-\frac{\alpha H_a^2}{a}$$

$$\frac{\partial H_b}{\partial t}=\frac{\alpha R_{\xi\xi}}{\psi^4}+\frac{4\beta H_b\frac{\partial\psi}{\partial\eta}}{a\psi^5}-\frac{2\frac{\partial\alpha}{\partial\eta}b\frac{\partial\psi}{\partial\eta}}{a\psi^5}+\frac{\beta\frac{\partial H_b}{\partial\eta}}{a\psi^4}-\frac{\frac{\partial\alpha}{\partial\eta}\frac{\partial b}{\partial\eta}}{2a\psi^4}+\frac{\alpha H_aH_b}{a}$$

THREE DIMENSIONAL INITIAL DATA OF NUMERICAL RELATIVITY

Ken-ichi Oohara
National Laboratory for High Energy Physics
Oho, Tsukuba-shi, Ibaraki-ken, 305, Japan
Takashi Nakamura
Department of Physics, Kyoto University, Kyoto, 606, Japan

Abstract : We present a method of solving 3 dimensional constraint equation of Einstein equation in (3+1)-formalism. A comparison is made between SOR and ICCG methods for solving non-linear elliptic equations. The initial data for colliding neutron stars is solved adopting the Cartesian coordinate system. No symmetry is assumed and therefore arbitrary density and momentum distributions can be considered. As an example, initial data for two neutron stars just contacting and rigidly rotating around the center of mass with the Keplerian angular velocity is presented.

1 Introduction

Evolutions of axially symmetric systems such as a head-on collision of black holes (Smarr 1977) and a gravitational collapse of a rotating star (Nakamura 1981, Piran & Stark 1986) were studied in the last decade in the field of numerical relativity. Gravitational waves to be emitted were successfully calculated there and it was shown that the resulting waves are very similar to the waves obtained by the perturbation calculation (Detweiler & Szedenits 1979, Nakamura, Oohara & Kojima 1987). On the other hand, the perturbation calculation revealed that three-dimensional (3-D) processes will be a more effective source than 2-D processes. Of 3-D processes, a non-head-on collision of compact objects such as black holes or neutron stars must be the most promising source of a strong gravitational radiation. One scene of such a collision will be found at the final stage in the evolution of a close binary system of black holes or neutron stars. It will collapse after the loss of the angular momentum due to a continuous emission of gravitational waves and then become a burst source of gravitational waves (Schutz 1986). The other may be found in the way of 3-D collapse of a rotating star. The multiple neutron stars will be formed there, if the angular momentum of the system is large enough, in accordance with a "collapse, pursuit and plunge scenario" (Misner, Thorne & Wheeler 1973). Therefore we will focus on colliding neutron stars.

As a first step of 3-D simulations, we constructed a code for solving the constraints of Einstein equations in (3+1)-formalism. This code can be used, in general, for arbitrary density and momentum distributions. (No symmetry is necessary.) As for the coordinate system, we adopt the Cartesian one. Since no symmetry is assumed, it is of little advantage to adopt the polar coordinate system. Indeed, larger memory might be needed in the numerical calculation if one uses the Cartesian coordinate system, but one does not have to suffer troubles with coordinate singularities. Moreover the expression of the equations becomes simpler and therefore it is likely to be solved faster numerically.

The constraint equations are the system of non-linear elliptic partial differential equations (PDE). Recently progress has been made in the technique of solving the elliptic PDE numerically by supercomputers. First we review shortly one of them, the ICCG method, and compare it with the classical SOR method in Section 2. In Section 3 we consider the constraint equations and the boundary conditions of the quantities to be solved. As a code testing, the comparison between the exact solution and the the numerically generated one will be presented in Section 4. In Section 5 we show the results for colliding neutrons stars just contacted and rotating around the center of the mass.

2 Methods of Solving Elliptic Equations

As shown in next section, the constraint equations are the system of non-linear elliptic equations. Now we consider 3-D non-linear or quasi-linear elliptic PDE;

$$\Delta\psi + F(\psi, \frac{\partial\psi}{\partial x}, \frac{\partial\psi}{\partial y}, \frac{\partial\psi}{\partial z}, x, y, z) = 0. \tag{1}$$

Here F is an arbitrary function of ψ, its first derivatives and x, y, z. By means of the finite difference method, Eq.(1) becomes the system of non-linear equations

$$A\boldsymbol{x} = \boldsymbol{b}(\boldsymbol{x}), \tag{2}$$

where $\boldsymbol{x}$ is a vector each of whose element is given from the value of ψ at each grid point. Usually matrix A is a symmetric, banded and, moreover, positive definite matrix, *i.e.*,

$$A = A^T \quad \text{and} \quad (\boldsymbol{x}, A\boldsymbol{x}) \geq 0 \quad \text{for all } \boldsymbol{x}. \tag{3}$$

As for the inversion of a symmetric, positive-definite and banded matrix, Preconditioned Conjugate Gradient (PCG) method (Meijerink & van der Vorst 1977) has become popular in these days. The conjugate gradient method is based on the theorem:

If a matrix A is symmetric and positive definite, and $\boldsymbol{b}$ is a constant vector, then the following statements are equivalent.

"The vector $\boldsymbol{x}_s$ is the solution of the equation $A\boldsymbol{x} = \boldsymbol{b}$"

and

"$f(\boldsymbol{x})$ takes its minimum at $\boldsymbol{x} = \boldsymbol{x}_s$,"

where

$$\begin{aligned} f(x) &= (r, A^{-1}r) \\ &= (b - Ax, A^{-1}b - x) \end{aligned} \tag{4}$$

Therefore, the solution of the equation $Ax = b$ will be obtained via an iteration

$$x^{(i+1)} = x^{(i)} + \alpha p^{(i)} \tag{5}$$

$$p^{(i+1)} = r^{(i+1)} + \beta p^{(i)}, \tag{6}$$

where $r^{(i)} = b - Ax^{(i)}$ and $p^{(0)} = r^{(0)}$. The value of α is determined such that $f(x^{(i+1)})$ be as small as possible and β is such that

$$\left(p^{(i+1)}, Ap^{(k)}\right) = 0 \qquad \text{for} \quad k = 1, 2, \ldots, i. \tag{7}$$

That is,

$$\alpha = \left(r^{(i)}, r^{(i)}\right) / \left(p^{(i)}, Ap^{(i)}\right) \tag{8}$$

$$\beta = \left(r^{(i+1)}, r^{(i+1)}\right) / \left(r^{(i)}, r^{(i)}\right) \tag{9}$$

Since $p^{(i)}$'s become linearly independent of each other if they are determined by Eqs.(5) - (9), the conjugate gradient method gives the exact solution after N times iteration in an analytical sense, where N is the rank of the matrix A. For a large value of N, acceleration of the convergence is needed in numerical calculation. For example, A is decomposed as

$$A = LL^T + R, \tag{10}$$

say, incomplete Cholesky decomposition, then the equation $Ax = b$ can be rewritten as

$$\tilde{A}\tilde{x} = \tilde{b}, \tag{11}$$

where $\tilde{A} = L^{-T}AL^{-1}$, $\tilde{x} = Lx$, $\tilde{b} = L^{-T}b$ and $L^{-T} = (L^T)^{-1}$. If $\tilde{A}$ is close to the unit matrix, the convergence is expected to be fast in numerical calculation. Such methods are called PCG methods and some kinds of PCG methods are proposed depending on the choice of L. In these ones, Incomplete Cholesky decomposition and Conjugate Gradient (ICCG) method by Meijerink and van der Vorst (1977) is famous and Modified ICCG (MICCG) method, which was proposed by Gustafsson (1978) as an improved version of ICCG method, is known to converge very fast. It has been shown that ICCG or MICCG is usually more stable than classical SOR or SLOR, that is, the former will converge by a smaller number of iteration than the latter and moreover ICCG converges for some of problems for which SOR and SLOR dose not converge. Figures 1 and 2 are the results of the benchmark test for

$$\Delta\psi = -2e^{-r^2} \tag{12}$$

with the boundary condition

$$\psi \to \frac{c}{r} \qquad \text{for} \quad r \to \infty, \tag{13}$$

Figure 1: Number of iteration in which ICCG, SOR or SLOR converges as a function of required tolerance for a solution of $\Delta\psi = -2\exp(-r^2)$. The mesh size is $21 \times 21 \times 21$ (labeled xxx21) or $51 \times 51 \times 51$ (labeled xxx51).

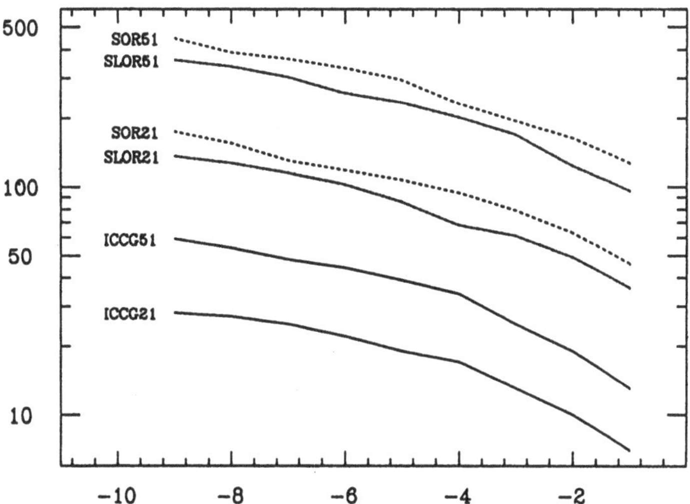

Figure 2: CPU time in which ICCG, SOR or SLOR converges as a function of required tolerance for the same problem of Fig.1. Labels xxxx-s and xxxx-v denote computation with scalar(M680H of HITAC) and vector(S810/10 of HITAC) processors, respectively. The results of mesh size $51 \times 51 \times 51$ only are shown.

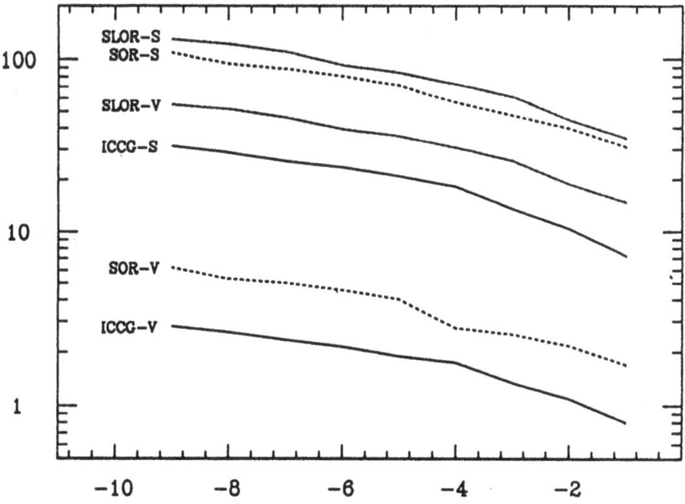

and $x = 0$ as an initial guess. Note that much smaller number of iteration is needed for ICCG than SOR or SLOR but the CPU-time for SOR is only about twice of that for ICCG.

For a quasi-linear elliptic PDE, in turn, we need a non-linear iteration to solve Eq.(2);

$$x^{(n+1)} = A^{-1} b(x^{(n)}) \tag{14}$$

The iteration is repeated until $\left| x^{(n+1)} - x^{(n)} \right|$ becomes sufficiently small. ICCG or SOR (or SLOR) method is applied to invert the matrix A. In the case of SOR, however, the way of iteration can be modified. That is, Eq.(2) will be solved by means of the iteration

$$\begin{aligned} \hat{x}_k^{(i+1)} &= \left\{ -\sum_{\ell=1}^{k-1} a_{k,\ell}\, x_\ell^{(i+1)} - \sum_{\ell=k+1}^{N} a_{k,\ell}\, x_\ell^{(i)} + b_k^{(i)} \right\} \Big/ a_{k,k}, \\ x_k^{(i+1)} &= x_k^{(i)} + \omega \left\{ \hat{x}_k^{(i+1)} - x_k^{(i)} \right\}, \end{aligned} \tag{15}$$

where ω is the over-relaxation factor, $a_{k,\ell}$ is the (k,ℓ) element of the matrix A and $x_k^{(i)}$ is the k-th element of the vector x at the i-th SOR iteration. The source term $b(x)$ is evaluated by $x_k^{(i)}$'s at each SOR iteration in this case, while in the case of Eq.(14) it is evaluated at each non-linear iteration and ICCG (or SOR) is applied regarding b as a constant vector. A modified SOR (MSOR) may not converge even if a ordinary SOR does, but MSOR usually converges faster than SOR if both of them converge. Figures 3 and 4 are the same figures as Figs. 1 and 2 for

$$\Delta\psi = -\frac{K(x)}{\psi^7}, \tag{16}$$

where $K(x)$ is a given function of x. In this case MSOR converges faster than ICCG! For actual problems, however, it is recommended to use ICCG rather than MSOR because the stability of MSOR is not clear and it is not easy to find the best value of the over-relaxation factor ω of SOR. It takes usually almost the same CPU time as solving the equation to find the best value of ω. Therefore we will use ICCG to solve the constraint equations.

3 Constraint Equations and Boundary Conditions

The constraint (or initial-value) equations in (3+1)-formalism are (York 1979)

$$\begin{aligned} {}^{(3)}R + K^2 - K_{ij}K^{ij} &= 16\pi\rho_{\rm H}, \qquad (17) \\ \nabla^j K_{ij} - \nabla_i K &= 8\pi J_i. \qquad (18) \end{aligned}$$

where $\rho_{\rm H}$ is the energy density of sources, J_i is the three-momentum of sources, ${}^{(3)}R$ is the scalar curvature of the spacelike three-metric γ_{ij}, K_{ij} is the extrinsic curvature, K is its trace $K = K^i_{\ i} = \gamma^{ij}K_{ij}$ and ∇_i is the covariant derivative with respect to γ_{ij}. Now we assume the Cauchy slice is conformally flat, $\gamma_{ij} = \psi^4\tilde{\gamma}_{ij}$, where $\tilde{\gamma}_{ij}$ is the three-metric

Figure 3: Same figure as Fig.1 for $\Delta\psi = -K(x)/\psi^7$.

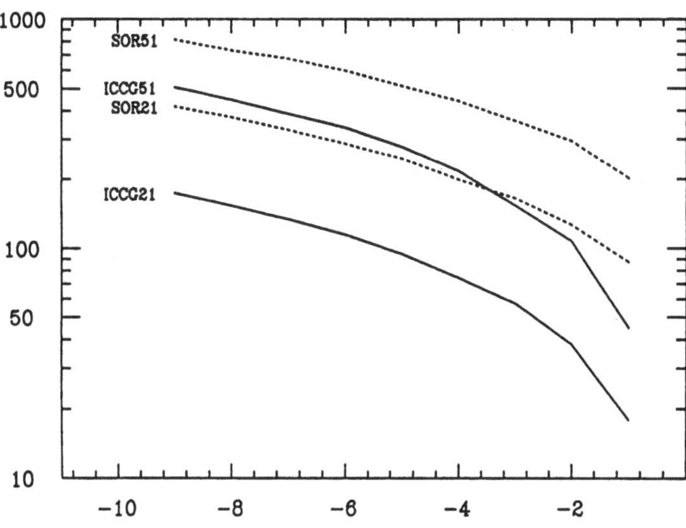

Figure 4: Same figure as Fig.2 for $\Delta\psi = -K(x)/\psi^7$.

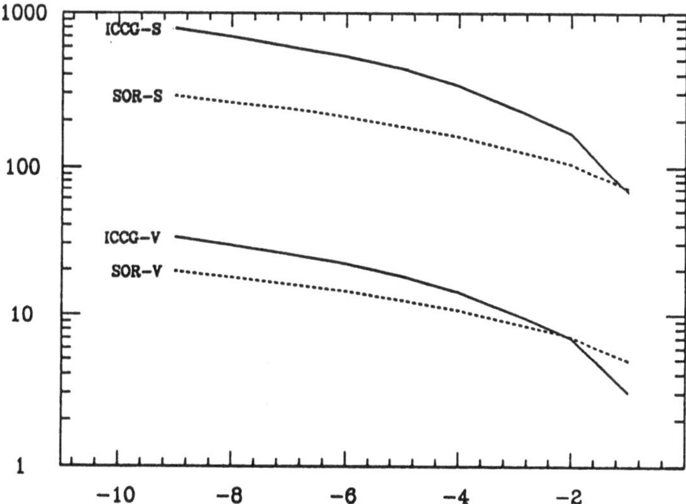

of the flat space, and the trace $K = 0$ in the Cauchy slice. Then Eqs.(17) and (18) can be expressed as (York 1979)

$$\Delta\psi = -\frac{1}{8}\tilde{K}_{ij}\tilde{K}^{ij}\psi^{-7} - 2\pi\rho_{\rm B}\psi^{-1} \tag{19}$$

$$\tilde{\nabla}^j\tilde{K}_{ij} = 8\pi\tilde{J}_i, \tag{20}$$

where $\rho_{\rm B} = \rho_{\rm H}\psi^6$, $\tilde{J}_i = J_i\psi^6$, $\tilde{K}_{ij} = K_{ij}\psi^2$, $\tilde{\nabla}_j$ is the covariant derivative with respect to $\tilde{\gamma}_{ij}$ and Δ is the flat-space Laplacian, $\nabla = \tilde{\nabla}^i\tilde{\nabla}_i$. The traceless extrinsic curvature can be decomposed with the transverse traceless part and the longitudinal traceless part (York 1973)

$$\tilde{K}_{ij} = \tilde{K}_{ij}^{\rm TT} + \left(\tilde{L}\tilde{W}\right)_{ij} \tag{21}$$

with

$$\left(\tilde{L}\tilde{W}\right)_{ij} = \tilde{\nabla}_i\tilde{W}_j + \tilde{\nabla}_j\tilde{W}_i - \frac{2}{3}\tilde{\psi}_{ij}\tilde{\nabla}^\ell\tilde{W}_\ell. \tag{22}$$

In addition, we assume $\tilde{K}_{ij}^{\rm TT} = 0$, then Eq.(20) becomes

$$\Delta\tilde{W}_i + \frac{1}{3}\tilde{\nabla}_i\tilde{\nabla}^j\tilde{W}_j = 8\pi\tilde{J}_i. \tag{23}$$

Now giving $\rho_{\rm B}$ and $\tilde{J}_i$, we will solve the system of linear elliptic equations for $\tilde{W}_i$ and then a non-linear elliptic equation for ψ by means of MICCG method with the boundary condition $W_i \to 0$ and $\psi \to 1$ for $r \to \infty$.

To solve Eqs.(19) and (23) numerically, we need to specify particular boundary conditions of W_i and ψ at the numerical boundaries. Clearly $\psi - 1$ should decrease as $M/2r$ for large r. To find the asymptotic form of W_i, in turn, we will consider them in the spherical coordinate system (r, θ, ϕ). Using tensor harmonics (Zerilli 1970), W_r, W_θ and W_ϕ are decomposed as

$$\begin{aligned} W_r &= \sum A_{\ell m}(r)Y_{\ell m}(\theta,\phi), \\ W_\theta &= \sum\left\{B_{\ell m}(r)\frac{\partial Y_{\ell m}}{\partial\theta} + \frac{C_{\ell m}(r)}{\sin\theta}\frac{\partial Y_{\ell m}}{\partial\phi}\right\}, \\ W_\phi &= \sum\left\{B_{\ell m}(r)\frac{\partial Y_{\ell m}}{\partial\phi} - C_{\ell m}(r)\sin\theta\frac{\partial Y_{\ell m}}{\partial\theta}\right\}. \end{aligned} \tag{24}$$

Setting $\tilde{J}_i = 0$ in the asymptotically flat region and substituting Eqs.(24) into Eq.(20), we obtain for each (ℓ, m)

$$\begin{aligned} 4r^3\frac{d^2A_{\ell m}}{dr^2} + 8r^2\frac{dA_{\ell m}}{dr} - \{3\ell(\ell+1)+8\}\,rA_{\ell m} & \\ -\ell(\ell+1)\left(r\frac{dB_{\ell m}}{dr} - 8B_{\ell m}\right) &= 0, \end{aligned} \tag{25}$$

$$r^2\frac{dA_{\ell m}}{dr} + 8rA_{\ell m} + 3r^2\frac{d^2B_{\ell m}}{dr^2} - 4\ell(\ell+1)B_{\ell m} = 0 \tag{26}$$

and

$$r^2\frac{d^2C_{\ell m}}{dr^2}-\ell(\ell+1)C_{\ell m}=0. \tag{27}$$

Assuming $A_{\ell m}=a_{\ell m}r^p$, $B_{\ell m}=b_{\ell m}r^{p+1}$ and $C_{\ell m}=c_{\ell m}r^s$, we have as vanishing solutions for $r\to\infty$

$$p=-\ell \qquad \text{with} \quad b_{\ell m}=-\frac{\ell-8}{\ell(\ell+7)}a_{\ell m} \qquad \text{for} \quad \ell\geq 1 \tag{28}$$

or

$$p=-\ell-2 \qquad \text{with} \quad b_{\ell m}=-\frac{1}{\ell+1}a_{\ell m} \qquad \text{for} \quad \ell\geq 0 \tag{29}$$

and

$$s=-\ell. \qquad \text{for} \quad \ell\geq 1 \tag{30}$$

If terms of $\ell=1$ are set to be non-zero as dominant terms and the others zero, we have

$$\begin{aligned}\tilde{W}_x &= \sum_{m=-1}^{1}\frac{1}{r}\left[\alpha_{1m}\sin\theta\cos\phi\,Y_{1m}-\beta_{1m}\left(\cos\theta\cos\phi\frac{\partial Y_{1m}}{\partial\theta}-\frac{\sin\phi}{\sin\theta}\frac{\partial Y_{1m}}{\partial\phi}\right)\right]\\ &+ \sum_{m=-1}^{1}\frac{c_{1m}}{r^2}\left[\frac{\cos\theta\cos\phi}{\sin\theta}\frac{\partial Y_{1m}}{\partial\phi}+\sin\phi\frac{\partial Y_{1m}}{\partial\theta}\right],\end{aligned} \tag{31}$$

$$\begin{aligned}\tilde{W}_y &= \sum_{m=-1}^{1}\frac{1}{r}\left[\alpha_{1m}\sin\theta\sin\phi\,Y_{1m}-\beta_{1m}\left(\cos\theta\sin\phi\frac{\partial Y_{1m}}{\partial\theta}+\frac{\cos\phi}{\sin\theta}\frac{\partial Y_{1m}}{\partial\phi}\right)\right]\\ &+ \sum_{m=-1}^{1}\frac{c_{1m}}{r^2}\left[\frac{\cos\theta\sin\phi}{\sin\theta}\frac{\partial Y_{1m}}{\partial\phi}-\cos\phi\frac{\partial Y_{m}}{\partial\theta}\right],\end{aligned} \tag{32}$$

$$\begin{aligned}\tilde{W}_z &= \sum_{m=-1}^{1}\frac{1}{r}\left[\alpha_{1m}\cos\theta\,Y_{1m}+\beta_{1m}\sin\theta\frac{\partial Y_{1m}}{\partial\theta}\right]\\ &- \sum_{m=-1}^{1}\frac{c_{1m}}{r^2}\frac{\partial Y_{1m}}{\partial\phi},\end{aligned} \tag{33}$$

where

$$\alpha_{\ell m} = a^{(1)}_{\ell m}+\frac{a^{(2)}_{\ell m}}{r^2} \tag{34}$$

$$\beta_{\ell m} = \frac{(\ell-8)a^{(1)}_{\ell m}}{\ell(\ell+7)}+\frac{a^{(2)}_{\ell m}}{(\ell+1)r^2}. \tag{35}$$

Changing coefficients properly, it is shown that this solution coincide with the solution of Bowen and York (1982);

$$\begin{aligned}\tilde{K}_{ij} &= \frac{3}{2r^2}\left[P_in_j+P_jn_i-(\tilde{\gamma}_{ij}-n_in_j)P^kn_k\right]\\ &\mp \frac{3a^2}{2r^4}\left[P_in_j+P_jn_i+(\tilde{\gamma}_{ij}-5n_in_j)P^kn_k\right]\\ &+ \frac{3}{r^3}\left[\epsilon_{ki\ell}J^\ell n^k n_j+\epsilon_{kj\ell}J^\ell n^k n_i\right]\end{aligned} \tag{36}$$

using Cartesian coordinates, where $n^i = x^i/r$ and ϵ_{ijk} is the unit alternating tensor. Therefore $a_{\ell m}$ and $c_{\ell m}$ of $\ell = 1$ respectively represent the linear momentum and the angular momentum. In the result, the boundary condition for $r \to \infty$ is $W_i = a_i/r$ if the system has a linear momentum or $W_i = c_i/r^2$ if the system has only an angular momentum, where a_i and c_i are constants.

4 Code Testing

The exact solution of Eq.(23) has been found by Bowen(1979) for a general spherically symmetric linear momentum source. Giving the spherically symmetric source as

$$\tilde{J}_i(r) = P_i \rho(r); \qquad \rho(r) = 0 \quad \text{if} \quad r > r_0, \tag{37}$$

where P_i is a constant vector and the density $\rho(r)$ is an arbitrary function of r for $r \leq r_0$ and normalized so that

$$\int \rho(r) dV = 1. \tag{38}$$

Then the solution of Eq.(23) is given by

$$W_i = P_i \left[-2F(r) + \frac{G(r)}{2r} \right] + \frac{1}{2} n_i n_j P^j \left[\frac{\partial G(r)}{\partial r} - \frac{G(r)}{r} \right], \tag{39}$$

where

$$\begin{aligned} F(r) &= \frac{1}{r} Q(r) + S(r) \\ Q(r) &= \int_0^r 4\pi \rho(r') r'^2 dr' \\ S(r) &= \int_r^{r_0} 4\pi \rho(r') r' dr' \\ G(r) &= \frac{1}{r^2} \int_0^r r'^2 F(r') dr'. \end{aligned} \tag{40}$$

On the other hand, we found the exact solution for the spherically symmetric angular momentum source given by

$$\tilde{J}_i = \epsilon_{ik\ell} J^k x^\ell \rho(r); \qquad \rho(r) = 0 \quad \text{if} \quad r > r_0, \tag{41}$$

where J_k is a constant vector and $\rho(r)$ is normalized so that

$$\int \frac{2}{3} \rho(r) r^2 dV = 1. \tag{42}$$

Then we have

$$\tilde{W}_i = -\epsilon_{ik\ell} J^k x^\ell F(r), \tag{43}$$

where

$$\begin{aligned} F(r) &= \frac{1}{r^3} Q(r) + S(r) \\ Q(r) &= \int_0^r \frac{8\pi}{3} \rho(r') r'^4 dr' \\ S(r) &= \int_r^{r_0} \frac{8\pi}{3} \rho(r') r' dr' \end{aligned} \tag{44}$$

Figure 5: Accuracy of extrinsic curvature K_{xy} on the x-axis. The solid line is the exact solution $K_{xy}^{(\mathrm{ex})}$ and the dashed line is the relative error $|(K_{xy}^{(\mathrm{num})} - K_{xy}^{(\mathrm{ex})})/K_{xy}^{(\mathrm{ex})}|$.

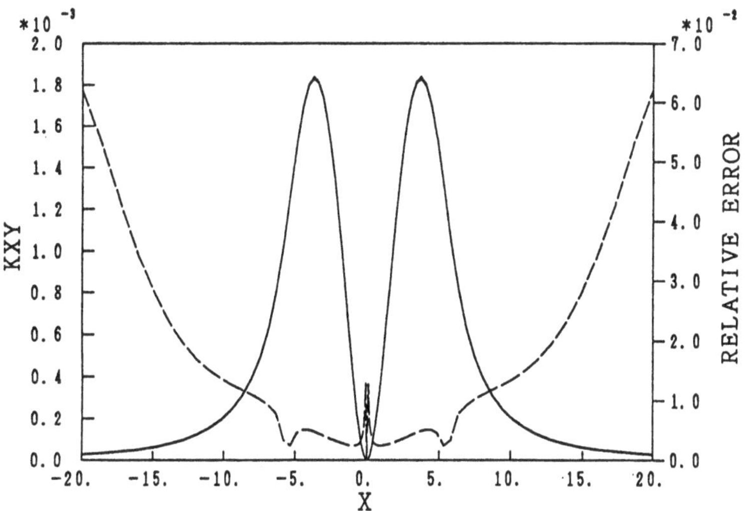

Figure 6: Accuracy of conformal factor ψ on the x-axis. The solid line is the exact solution $\psi^{(\mathrm{ex})} - 1$ and the dashed line is the relative error $|(\psi^{(\mathrm{num})} - \psi^{(\mathrm{ex})})/(\psi^{(\mathrm{ex})} - 1)|$.

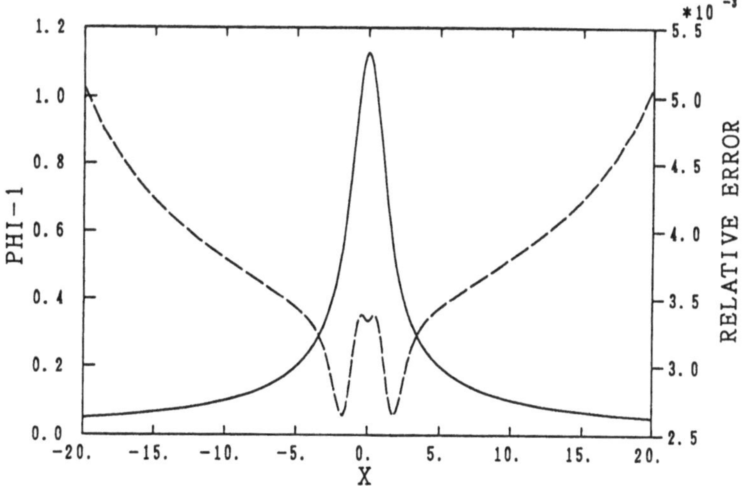

The extrinsic curvature $\tilde{K}_{ij}$ is given by

$$\tilde{K}_{ij}(r) = \frac{3}{r^3}\left[\epsilon_{ik\ell}J^k n^\ell n_j + \epsilon_{jk\ell}J^k n^\ell n_i\right] Q(r). \tag{45}$$

We have tested our code against Eq.(45) and found that K_{ij} can be reproduced in an accuracy of several per cents, or less than 1% in the central region, of relative error with 81 × 81 × 81 grid. Figure 5 shows the accuracy of K_{xy} on the x-axis for $\rho(r) = (315/8\pi r_0^5)(1-r/r_0)^2$ and $J_x = J_z \neq 0, J_y = 0$, with $r_0 = 3$. The solid line represents the exact solution given by Eq.(45) and the dashed line does the relative error $|\Delta\tilde{K}_{xy}/\tilde{K}_{xy}|$.

As for the conformal factor ψ, we compared the exact solution

$$\psi(r) = 1 + \frac{1}{r}\frac{2}{\sqrt{\pi}}\int_0^r e^{-r^2}dr \equiv 1 + \frac{1}{r}\mathrm{Erf}(r) \tag{46}$$

with the numerical result for the density

$$\rho_{\mathrm{B}} = -\frac{1}{2\pi}\left\{1+\frac{1}{r}\mathrm{Erf}(r)\right\}\left\{-\frac{4}{\sqrt{\pi}}e^{-r^2} + \frac{1}{8}\tilde{K}_{ij}\tilde{K}^{ij}\left(1+\frac{1}{r}\mathrm{Erf}(r)\right)^{-7}\right\}, \tag{47}$$

where $\tilde{K}_{ij}$ is the extrinsic curvature given by Eq.(45) and Erf(r) is the error function. As a result, we can obtain ψ in an accuracy less than 0.5%. The accuracy of ψ on the x-axis is shown in Fig.6.

5 Initial Data for Colliding Neutron Stars

Now we consider colliding neutron stars. Binary neutron stars get closer losing the orbital angular momentum due to a continuous emission of gravitational waves. Then they will collide with each other and coalesce into a single neutron star or black hole. The main part of gravitational waves will be emitted in this stage. Therefore, we consider neutron stars just contacting with each other (Fig.7). Here we assume two neutron stars are identical and the density ρ_{B} of each star is of a polytrope of $N = 3$ with mass M and radius $6M$;

$$\rho_{\mathrm{B}} = \rho_3(\boldsymbol{x} - \boldsymbol{x}_0) + \rho_3(\boldsymbol{x} + \boldsymbol{x}_0), \tag{48}$$

where $\boldsymbol{x}_0 = (0, 6M, 0)$. The history before this stage is ignored, that is, we assume that the stars have no linear momentum but they are rotating around z-axis with a Keplerian velocity at the center of the star;

$$\tilde{J}_x = -y\Omega\rho_{\mathrm{B}}, \quad , \tilde{J}_y = x\Omega\rho_{\mathrm{B}}, \text{ and } \tilde{J}_z = 0, \tag{49}$$

where

$$\Omega = \frac{1}{2}\sqrt{\frac{GM}{|\boldsymbol{x}_0|^3}}. \tag{50}$$

Now we will solve Eq.(23) and then Eq.(19) to obtain $\tilde{W}_i$ and ψ, respectively. Figures 8–9 show ρ, $\tilde{W}_x$ and $\tilde{W}_y$ on the equatorial plane. Here we have used 81 × 81 × 81 inhomogeneous grid. The computation space is from $-20M$ to $20M$ for each direction. Since Ω is

Figure 7: Colliding Neutron Stars

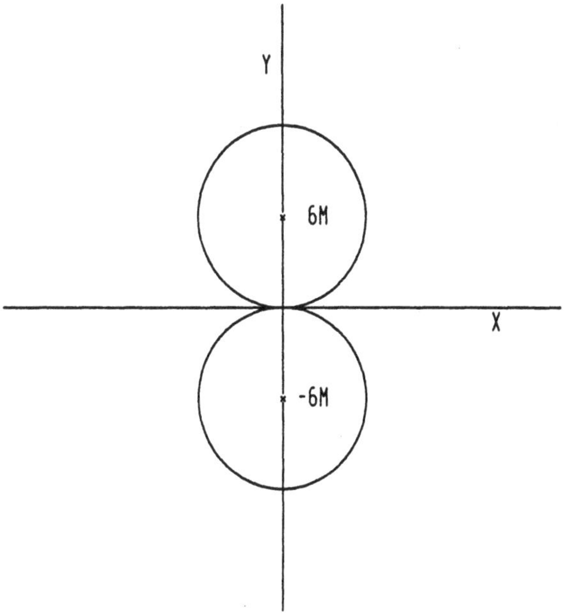

Figure 8: The conformal factor ψ on the equatorial plane. Grid used in numerical calculation is shown but some of grid lines at small x and small y are removed.

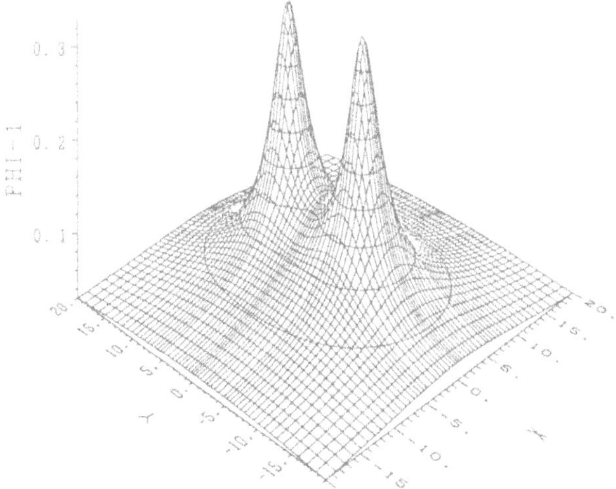

Figure 9: The vector potential $\tilde{W}_x$ on the equatorial plane.

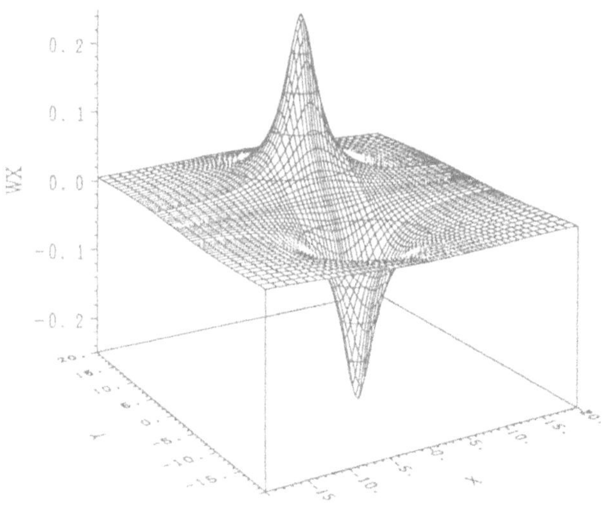

Figure 10: The vector potential $\tilde{W}_y$ on the equatorial plane.

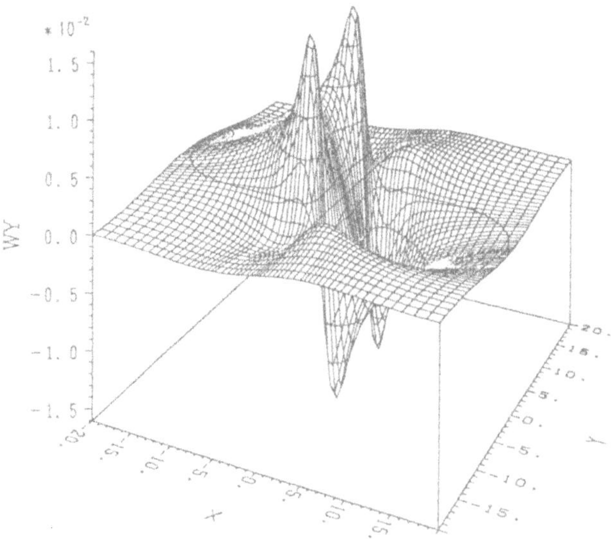

constant and

$$\rho(x,y,z) = \rho(-x,y,z) = \rho(x,-y,z) = \rho(x,y,-z), \tag{51}$$

it can be proven that

$$\begin{aligned}
\tilde{W}_x(x,y,z) &= \tilde{W}_x(-x,y,z) &= -\tilde{W}_x(x,-y,z) &= \tilde{W}_x(x,y,-z), \\
\tilde{W}_y(x,y,z) &= -\tilde{W}_y(-x,y,z) &= \tilde{W}_y(x,-y,z) &= \tilde{W}_y(x,y,-z), \\
\tilde{W}_z(x,y,z) &= -\tilde{W}_z(-x,y,z) &= -\tilde{W}_z(x,-y,z) &= -\tilde{W}_z(x,y,-z).
\end{aligned} \tag{52}$$

Furthermore $\tilde{W}_z$ is very small compared with $\tilde{W}_x$ and $\tilde{W}_y$, since $\tilde{\nabla}^j \tilde{W}_j \approx 0$ in this case.

6 Concluding Remarks

We have constructed the code for solving the constraint equations with no assumption of symmetry. Benchmark tests of the code shows the accuracy of the numerical solutions are very good. As we adopted the Cartesian coordinate system, there is no problem on coordinate singularities. In the Cartesian coordinate, however, the code is somewhat memory consuming. As we need a fine mesh in the central region, we have unnecessarily fine meshes at the outer region in the present code. In a future development of our code, a multi-grid method is indispensable to overcome the above problem. However, the maximal memory we can use at present(200MBytes) is large enough to evolve initial data with $81 \times 81 \times 81$ grid. We are now constructing a dynamical code to evolve initial data obtained here.

This work was partly supported by a Grant-in-Aid for Scientific Research of Ministry of Education, Science and Culture (62540188).

References

Bowen, J.M. (1979) Gen. Rel. Grav. **11**, 227.

Bowen, J.M. & York Jr., J.W. (1982) Phys. Rev. **D21**, 2047.

Detweiler, S.L. & Szedenits Jr., E. (1979) Astrophys. J. **231**, 211.

Gustafsson, I. (1978) BIT **18**, 142.

Meijerink, J.A. & van der Vorst, H.A. (1977) Math. Comp. **31**, 148.

Misner, C.W., Thorne, K.S. & Wheeler, J.A. (1973) *Gravitation* (W.H. Freeman, San Francisco).

Nakamura, T. (1981) Prog. Theor. Phys. **65**, 1876.

Nakamura, T., Oohara, K. & Kojima, Y. (1987) Prog. Theor. Phys. Suppl. No. 90

Piran, T. & Stark, R.F. (1986) in *Dynamical Spacetimes and Numerical Relativity*, ed. J. M. Centrella (Cambridge University Press) P.40.

Schutz, B.F. (1986) Nature **323**, 310.

Smarr, L. (1977), Ann. N.Y. Acad. Sci **301**, 569.

York Jr., J.W. (1973) J. Math. Phys. **14**, 456.

York Jr., J.W. (1979) in *Source of Gravitational Radiation*, ed. by L. Smarr (Cambridge University Press), P.83.

Zerilli, F.J. (1970) Phys. Rev. **D2**, 2141.

Initial Data for Collisions of Black Holes and Other Gravitational Miscellany

James W. York, Jr.
Institute of Field Physics
Department of Physics and Astronomy
University of North Carolina at Chapel Hill
Chapel Hill, NC 27599

Abstract. I review the method of constructing initial data for encounters among N black holes modeled by Einstein-Rosen bridges connecting two isometric sheets. A proof of uniqueness of the solution, depending explicitly on the inversion symmetry of the data, is given. Boundary conditions of the Robin type that can be employed to advantage at finite radii are given for the scalar and vector constraint equations and a proof of uniqueness is given in the latter case. The equation defining the apparent horizon of a black hole is derived in 3+1 form, including the minimal surface equation as a special case. Minkowski spacetime is described from the viewpoint of a curved spacelike hypersurface and a corresponding "test bed" numerical calculation is proposed. The fourteen algebraic invariants of the Riemann tensor are listed in a form suitable for calculations performed in the standard 3+1 format. These invariants are independent of the lapse and shift functions.

Introduction.

The problem of motion in pure general relativity is the problem of the encounters or collisions among N black holes. The corresponding spacetime can be free of matter, fields, and singularities if we model the holes by using topologically non-trivial initial data slices of certain types. One can obtain in this way four-geometries with direct relevance to astrophysics as well as being elegant and interesting in themselves. There should be no significant physically observable difference, classically, between the collision of black holes formed by the collapse of matter and those modeled by suitable topological structures such as Einstein-Rosen bridges. In particular, in either case, production of gravitational radiation will take place and coalescence of the holes can occur.

In this article, I shall review a method of constructing initial data for the collision of N black holes by using multiply-connected initial data surfaces. This method, which has been developed over the past five years or so, provides the necessary generalization of one proposed by Misner (1960, 1963) that produced the momentarily static data for the black-hole collision computed by Smarr and K.R. Eppley (Smarr 1979). I shall concentrate, for simplicity, on the cases N=1 and N=2 and shall give particulars of some points that have not appeared in the literature. Also, I shall treat several miscellaneous topics: (1) the algebraic invariants of the Riemann tensor, (2) Minkowski spacetime from the viewpoint of the numerical relativist, (3) derivation of the apparent horizon equation in 3+1 form, (4) writing mass multipole moments as surface integrals, and (5) some detail about vector Robin boundary conditions.

The background that I assume known is found in York (1979). See also York (1983). Remember what Samuel Johnson has told us, "Men more frequently require to be reminded than informed."

The Initial-Value Equations and the World's Simplest Way to Solve Them. The initial-value equations are

$$\nabla_j K^{ij}_{TF} - \tfrac{2}{3}\psi^6 \nabla^i K - 8\pi j^i = 0 \ , \tag{1}$$

$$8\Delta\psi - R\psi - \tfrac{2}{3}K^2\psi^5 + (K^{ij}_{TF})^2\psi^{-7} + 16\pi\rho\psi^{-3} = 0 \ , \tag{2}$$

where TF means "trace-free". Let us consider asymptotically flat (AF) spacetimes and choose K=0, *i.e.*, a maximal slice. (We know from Witt (1986) that not every AF spacetime admits a maximal slice because there can be topological obstructions. However, this seems to cause no problem in astrophysical applications, as opposed to quantum geometrodynamics, where interesting new phenomena could arise.) Take the metric to be flat (so the physical three-metric will be conformally flat: $g_{ij} = \psi^4 f_{ij}$) and take the transverse, trace-free part of K_{ij}, in the flat metric f_{ij}, to be zero, so that

$$K^{ij} = \nabla^i W^j + \nabla^j W^i - \tfrac{2}{3} f^{ij}\nabla_k W^k \ . \tag{3}$$

Then, the constraints become

$$\nabla^2 W^i + \tfrac{1}{3}\nabla^i(\nabla_j W^j) - 8\pi j^i = 0 \ , \tag{4}$$

$$\nabla^2\psi + \tfrac{1}{8}(K^{ij}_{TF})^2\psi^{-7} + 2\pi\rho\psi^{-3} = 0 \ . \tag{5}$$

Now choose ρ and j^i with compact support (to represent "stars") and $\rho \geq |j|^{1/2}$ (the stars aren't made of tachyons). Solve (4), then (5). In solving (4), note that Evans (1984) has constructed a suitable vector Green function based on (9) below. An alternative way to handle (4) is to set

$$W^i = V^i - \tfrac{1}{4}\nabla^i U \tag{6}$$

and (4) becomes

$$\nabla^2 V^i = 8\pi j^i \ , \tag{7}$$

$$\nabla^2 U = \nabla_i V^i \ . \tag{8}$$

This is especially convenient when Cartesian coordinates are being used. Bowen (1982) applied these equations systematically in constructing a form of the general solution of (4) on $\mathbf{R}^3$.

One Black Hole Made of Pure Geometry.

Set $\rho = j^i = 0$ in (4) and (5). Then, the basic monopole solution (Cartesian coordinates) for W^i is

$$W^i = -\frac{1}{4r}(7P^i + n^i n_j P^j), \tag{9}$$

where $n^i = x^i r^{-1}$ and the constants are chosen to make P^i the linear momentum in the Arnowitt-Deser-Misner (ADM) surface integral

$$P^i = \frac{1}{8\pi}\oint_\infty K^{ij}\, d^2S_j \ . \tag{10}$$

For K^{ij} we find from (3) and (9)

$$K^{ij} = \frac{3}{2r^2}\left[P^i n^j + P^j n^i - (f^{ij} - n^i n^j) n_k P^k\right] . \tag{11}$$

We now want a solution of (5), with $\rho = 0$, such that $\psi \to 1$ as $r \to \infty$. Then, we would have a moving (translating) black hole with linear momentum P^i. However, (11) is singular at the origin. This defect can be repaired by making an Einstein-Rosen bridge construction with a minimal surface at some $r = a > 0$. If, moreover, we can make the construction such that the inversion map $r \to \bar{r} = a^2 r^{-1}$ (generally, $x \to \bar{x} = J(x)$) is an isometry of the physical metric $\psi^4 f_{ij}$, then the sphere $r = a$, being a fixed point set with respect to the isometry, is a totally geodesic subspace of the three-surface (and of course then a minimal surface). The origin $r = 0$ becomes spatial infinity on another "sheet" of the universe. As explained in detail in Bowen and York (1980), these conditions are satisfied when

$$\psi(x) = \frac{a}{r}\psi(J(x)) \ , \tag{12}$$

$$K_{ij}(x) = -\left(\frac{a}{r}\right)^6 R_i^k R_j^l K_{kl}$$
$$\equiv (BK)_{ij}(x) \ , \tag{13}$$

where

$$R_i^k \equiv \delta_i^k - 2n^k n_i \tag{14}$$

and the Bowen (1979) operator B defined by (13) preserves the vanishing divergence and vanishing trace of any tensor with these properties that it acts upon. Note that BB = identity operator.

The tensor (11) does not satisfy (13), but applying the B operator to it shows that if it is augmented by the extra term

$$- \frac{3a^2}{2r^4}\left[P^i n^j + P^j n^i + (f^{ij} - 5n^i n^j) n_k P^k\right] , \tag{15}$$

then the sum satisfies (13). Let K_P^{ij} denote this sum. Note that (15) alone is also a solution of (1) with $K = j^i = 0$.

Another interesting tensor that satisfies (13) is the type that carries only angular momentum (spin) J_{ki}. This tensor needs augmentation by no further terms:

$$K_J^{ij} = \frac{3}{r^3}\left[n^i n^k J_k{}^j + n^j n^k J_k{}^i\right] . \tag{16}$$

Using either K_P^{ij} or K_J^{ij}, the scalar constraint is given by (5) with $\rho = 0$. Furthermore, we do not have to solve it on all of $\mathbf{R}^3 - \{0\}$. Instead we solve it on $\mathbf{R}^3$ with the sphere $r = a$ deleted. We invoke an inner boundary condition (minimal surface condition)

$$\frac{\partial \psi}{\partial r}(r{=}a) + \frac{1}{2a}\psi(r{=}a) = 0 , \tag{17}$$

and $\psi \to 1$ as $r \to \infty$. The physical data will now describe two isometric, asymptotically flat sheets joined by a generalized Einstein-Rosen bridge at the minimal surface $r = a$. Numerical studies of the above equation are found in York and Piran (1982), Choptuik (1982), and Bowen (1984). The isometry of two sheets guarantees that we know in advance *where* to put the inner boundary (at $r = a$) and *what* condition to impose there (minimal surface). We shall carry these features over to the N-body problem.

Uniqueness of Solution.

We observe that (17) looks superficially like a Robin boundary condition (see below), but it has the wrong relative sign between its two terms. Therefore, one cannot give a standard uniqueness argument for the problem we have posed, namely

$$\nabla^2 \psi = -\frac{1}{8}\,(K^{ij}_{J\,\mathrm{or}\,P})^2 \psi^{-7} , \tag{18}$$

$$\frac{\partial \psi}{\partial r} + \frac{1}{2a}\psi = 0 \quad \text{at } r = a , \tag{19}$$

$$\psi \to 1 \text{ as } r \to \infty . \tag{20}$$

(Also, M. Choptuik (1982) and others have noted that the "wrong sign" in (19) is awkward when used in conjunction with the SOR method of solving (19).)

Nevertheless, a "local" (in function space) uniqueness proof can be given. It relies explicitly on the use of inversion-symmetric K_{ij}'s (as described above) and that the minimal surface $r = a$ is the inversion sphere. (This approach can probably be extended to the N-body problem described below, while for the variants of our scheme suggested in Thornburg (1987), no uniqueness proof has been suggested.)

We work on $\mathbf{R}^3 - \{0\}$. The "exterior" problem for $0 < a \le r$ is given by (18), (19), and (20). The "interior" problem for $0 \le r \le a$, as shown in Bowen and York (1980), has a solution (by inversion properties)

$$\tilde{\psi} = \frac{a}{r}\psi(\frac{a^2}{r}) \tag{21}$$

of the equations

$$\nabla^2\tilde{\psi} = -\frac{1}{8}(K^{ij}_{J \text{ or } P})^2\tilde{\psi}^{-7} \ , \tag{22}$$

$$\frac{\partial\tilde{\psi}}{\partial r} + \frac{1}{2a}\tilde{\psi} = 0 \ \text{ at } r = a \ , \tag{23}$$

$$\tilde{\psi} \to \frac{C_1}{r} + C_2 + C_3 r + \ldots \ \text{ as } r \to 0 \ , \tag{24}$$

where the C's are finite constants. Note that (24) follows from the asymptotic behavior of ψ as $r \to \infty$ ($\psi \to 1 + \frac{E}{2r}$ + constant/r^2 + ...) and (21). Also note that (23) *is* a Robin condition for the interior problem.

Suppose there are two exterior solutions ψ_1 and ψ_2 of the inversion-symmetric problem. Then there are correspondingly two interior solutions $\tilde{\psi}_1$ and $\tilde{\psi}_2$, with $\tilde{\psi}_2 = \tilde{\psi}_1 + \tilde{u}$ and $\tilde{u} = (a/r)u(a^2/r)$. Let us suppose $\tilde{u}$ is small and keep its first powers only. Then $\tilde{u}$ solves

$$\nabla^2\tilde{u} = \frac{7}{8}(K^{ij}_{J \text{ or } P})^2\psi_1^{-8}\,\tilde{u} \equiv M^2\tilde{u} \ . \tag{25}$$

From what has been given, it follows that the "asymptotic" behavior of $\tilde{u}$ is given by

$$\lim_{r\to 0}\tilde{u} = d_1 \quad ; \quad \lim_{r\to 0}\frac{\partial\tilde{u}}{\partial r} = d_2 \ , \tag{26}$$

where the d's are finite constants. Now, take the region inside $r = a$ and surround $r = 0$ by a small ball of radius ε. Multiply (25) by $\tilde{u}$, integrate by parts, and evaluate the boundary terms to obtain

$$-\oint_{r=\varepsilon}\tilde{u}\frac{\partial\tilde{u}}{\partial r}\varepsilon^2 d^2\omega + \oint_{r=a}\tilde{u}\frac{\partial\tilde{u}}{\partial r}a^2 d^2\omega = \int[(\nabla\tilde{u})^2 + M^2\tilde{u}^2]\,dv \ . \tag{27}$$

Taking the limit of the first term as $\varepsilon \to 0$, we see that it vanishes. The second term on the left is negative, as follows from the Robin condition (23). Hence $\tilde{u} = 0$. Thus solutions $\tilde{\psi}$, and ψ, are (locally) unique.

N-Body Initial Data.

We shall take N=2 to illustrate the ideas. (See Bowen *et al.* (1984) for detailed calculations involving axisymmetric data for two spinning black holes.) To obtain inversion symmetry for *both* holes now requires an imaging method for the K_{ij}'s. (In practice, only a few image terms are needed to obtain excellent accuracy.) The payoff for using this method is that, by the symmetry we shall build in, we will know, without solving for ψ(!), *where* the inner boundaries are (at the non-overlapping inversion spheres with radii $r = a_\alpha$, $\alpha = 1,2$) and *what condition* applies there (minimal surface conditions like (19) for each of the spheres). It has been suggested by W. G. Unruh, as reported by Thornburg (1987), that one use the apparent horizon equation (discussed below) as the inner boundary condition. This is physically reasonable of course, but one does *not* know where the apparent horizon is until ψ is known. It seems unnatural to impose that an arbitrarily chosen two-surface be the apparent horizon. Furthermore, in the evolution the holes may merge. This means a further minimal surface - and apparent horizon - will suddenly appear, surrounding the two already present (which do not go away). Thornburg suggests searching for the new larger apparent horizon to use as an inner boundary in the continued evolution, an idea that obviously applies, *mutatis mutandis*, to minimal surfaces. However, such a surface will not be the level surface of any coordinate, while the original "inner" minimal surfaces are and remain so. Therefore, I do not think the idea worth the effort and that instructive physics and geometry might be lost. It seems best to keep the inner boundaries at the two initial minimal surfaces.

Before proceeding, let us make two observations. Firstly, Misner's (1963) method of images was applied to the linear equation for the scalar ψ that results when $K_{ij} = 0$ initially in the matter-free case. Our ψ equation is non-linear and no method of images is known. We apply an imaging method to the linear equation satisfied by the trace-free tensor K_{ij}. Secondly, the final imaged K_{ij}'s that we construct are not derivable entirely from vector potentials W^i, as shown by Rauber (1985).

The unphysical flat space in which we set the equations is shown in Figure 1.

Figure 1.

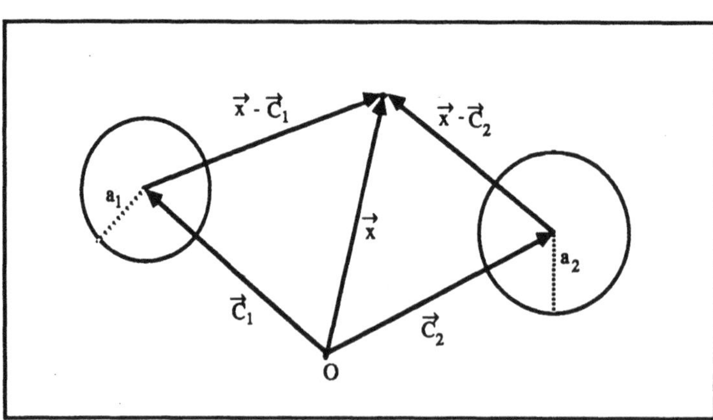

We desire the result

$$K^{ij}_{\text{physical}}(x) = -\left[\text{pull-back of } K^{ij}_{\text{physical}} \text{ at } J_\alpha(x)\right]^{ij} \tag{28}$$

for each α. To achieve this requires a multiple application of Bowen operators for each sphere (Kulkarni *et al.* 1983; Kulkarni 1984). We now have

$$x^i - C^i_\alpha \rightarrow [J_\alpha(x)]^i - C^i_\alpha = \frac{a^2_\alpha}{r_\alpha} n^i_\alpha \ , \tag{29}$$

$$n^i_\alpha = \frac{x^i - C^i_\alpha}{r_\alpha} \ ; \ r_\alpha = |\vec{x} - \vec{C}_\alpha| \ . \tag{30}$$

From (29) and (30) we construct operators B_α in an obvious way. The application of any such operator, or sequence of them, on a divergence-free, trace-free tensor preserves these properties. (For an excellent exposition of these techniques and their extension to the case of electrically charged black holes, see Bowen (1985).)

Starting with (say) a K^{ij}_J or a K^{ij} as in (11), or an appropriate sum of such, the complete sequence of images necessary to obtain the desired inversion symmetry is obtained by application of the operator

$$\boldsymbol{B} = \frac{1}{2}\left[I + \sum_{[\alpha_i]}^{(\text{all})} \prod_{i=1}^{m} B_{\alpha_i} \right] , \tag{31}$$

where the α_i's take values 1,2,...,N (N = the number of holes; here N=2) and the sum is over all sequences of α's of length m such that $\alpha_i \neq \alpha_{i+1}$. Recalling that $B_\alpha B_\alpha = I$, we have $B_\alpha \boldsymbol{B} = \boldsymbol{B}$ for any α. Our solution is denoted by

$$\mathcal{K}^{ij} = (\boldsymbol{B}K)^{ij} \ , \tag{32}$$

and we solve

$$\nabla^2\psi = -\frac{1}{8}(\mathcal{K}^{ij}\mathcal{K}_{ij})\psi^{-7} \ , \tag{33}$$

$$\frac{\partial\psi}{\partial r_\alpha} + \frac{1}{2a_\alpha}\psi = 0 \text{ at } r_\alpha = a_\alpha \ \forall\, \alpha \ . \tag{34}$$

At large r, we have $\psi \rightarrow 1$, or we use a Robin condition as discussed below. The geometry of the physical space is schematically indicated in Figure 2 for the case $a_1 = a_2$. Note that there are apparent horizons (dotted circles) in each "sheet" or "universe", located symmetrically with respect to the minimal surfaces (dark circles) (Bowen & York 1980). There can be, on each sheet, larger minimal surfaces, and apparent horizons, surrounding the two holes, but these are not shown. If these larger two-surfaces are present, the holes have merged. It is evident that only one sheet needs to be evolved.

Figure 2.

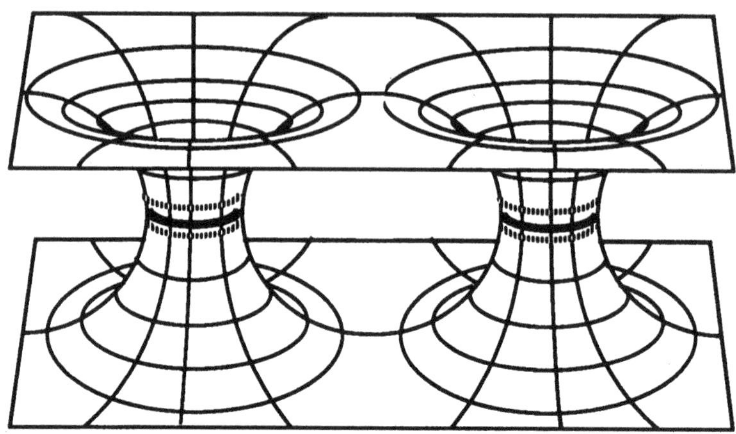

Two Identical Holes in an Initially Circular Orbit

The problem identified in the title of this section seems to be one of the most important of classical geometrodynamics. We consider the case when neither hole has intrinsic angular momentum initially. According to the method of estimation of Detweiler and Smarr (Detweiler 1979), a substantial conversion of initial energy into gravitational radiation might occur in this case. In this section, I only want to make a few elementary observations about the initial data.

In the unphysical flat space, let the two holes be equidistant from the origin. Let $\vec{d}$ be a vector from the origin to hole number one, whose linear momentum $\vec{P}$ is at right angles to $\vec{d}$. The linear momentum of hole number two is $-\vec{P}$. Let r and $n^i = x^i r^{-1}$ refer to the origin rather than the "center" of one or the other of the translating holes. Rewriting $K^{ij}(\vec{P}_1)$ and $K^{ij}(\vec{P}_2)$ with reference to the origin, starting with the form (11) for each, we see that each acquires new terms of $O(r^{-3})$. From (16) we see that the K^{ij}'s that carry angular momentum are also of $O(r^{-3})$. This suggested to Thornburg (1987) that the initial angular momentum therefore cannot, in general, be specified freely in this problem (that is, without first solving for ψ). However, this is not the case. When one adds the two K^{ij}'s (in a frame in which the total linear momentum is zero), there are two distinct types of terms of $O(r^{-3})$. One has the form (16) depending only on $\vec{J}$ (the three-dimensional dual of the antisymmetric three-tensor J_{ij}) and it incorporates the antisymmetric part of the tensor product (moment) of $\vec{x}$ and the effective current. The other $O(r^{-3})$ term carries no angular momentum and indeed depends on a symmetric three-tensor related to the "integrated" tensor product of $\vec{x}$ and the effective current generated by the moving Einstein-Rosen bridge. The two distinct $O(r^{-3})$ parts of K_{ij} were found by O'Murchadha and York (1976) and independently by Evans (1984).

Carrying out the indicated sum yields

$$K_{(1)}{}^{ij} + K_{(2)}{}^{ij} = K_J{}^{ij} + \frac{3}{r^3}[(\vec{d}\cdot\hat{n})(P^i n^j + P^j n^i) + (\vec{P}\cdot\hat{n})(d^i n^j + d^j n^i)$$

$$- (P^i d^j + P^j d^i) - (\vec{P}\cdot\hat{n})(\vec{d}\cdot\hat{n})(3f^{ij} - 5n^i n^j)] + O(r^{-4}), \qquad (35)$$

where the first term on the right hand side has the form (16) with $\vec{J} = 2\,\vec{d} \times \vec{P}$. One verifies that the angular momentum is found only in $K_J{}^{ij}$ by performing the surface integral (in Cartesian coordinates here)

$$J_i = \frac{1}{16\pi}\varepsilon_{ijk}\oint_\infty (x^j K^{lm} - x^l K^{jm}) d^2S_m. \qquad (36)$$

Because $\psi = 1 + O(r^{-1})$, (36) gives the physical or dressed value of $\vec{J}$.

Finally, it is important to note that the construction of the appropriate sequence of images to be added to (35), as described above, will not affect the value of $\vec{J}$ because they begin their contribution in $O(r^{-4})$, *i.e.*, they can only affect higher order moments of the effective current. However, this does show that topological models of black holes are indeed *physically* different depending on whether, for example, inversion symmetry is enforced (resulting in a two-sheeted manifold) or not (resulting in an (N+1)-sheeted manifold for N holes). The reason is that different topologies imply different multipole moments (of both momentum and mass) as determined on the top ("physical") sheet alone. (It is important to recall here that these data are not going to produce stationary or static spacetimes!) For example,this was demonstrated in the case of two black holes momentarily at rest, in the two-sheeted ("Misner-matched") *versus* three-sheeted ("Brill-matched") cases by Cantor and Kulkarni (1982). These authors normalized the two solutions by fixing the mass and the mass quadrupole moment of the two solutions. These quantities are clearly of interest in characterizing the initial data of the present problem, so I shall review them briefly.

It is well known that the ADM mass-energy integral can be written in terms of the conformal factor (Brill 1959; O'Murchadha & York 1974a):

$$E = -\frac{1}{2\pi}\oint_\infty \vec{\nabla}\psi\cdot d^2\vec{S}. \qquad (37)$$

The mass quadrupole moment in Newtonian gravity is defined by the volume integral

$$I^{ij}_{\text{Newtonian}} = \int \rho(x^i x^j - \tfrac{1}{3}r^2\delta^{ij})dV. \qquad (38)$$

Obviously we do not want a volume integral expression for the analog of (38) in our problem. However, we can use the correspondence between ψ and the Newtonian potential implied by (37), namely, $\Phi_N \to -2(\psi-1)$, and $\rho = (4\pi)^{-1}\nabla^2\Phi_N$, to write (38) as the surface integral (Cartesian coordinates)

$$I^{ij} = -\frac{1}{2\pi}\oint_{\infty}\Big[(x^i x^j - \tfrac{1}{3}\delta^{ij}r^2)\partial^k\psi - (x^i\delta^{ij} + x^j\delta^{ik} - \tfrac{2}{3}x^k\delta^{ij})(\psi-1)\Big]d^2S_k\,. \quad (39)$$

We cannot be sure such integrals exist in general relativity because, unlike in Newtonian theory, we cannot assume compact support for the effective energy density "ρ". However, the integral exists in the present problem because the "source" ~ $(K_{ij}K^{ij})\psi^{-7}$ is $O(r^{-6})$.

Robin Boundary Conditions

Robin boundary conditions are well known in the case of scalar elliptic equations (Duff 1959). If $\partial/\partial n$ denotes the derivative with respect to the outward-pointed unit normal of some region, then the Robin condition for a scalar ψ has the form

$$\frac{\partial\psi}{\partial n} + f\psi = g\ , \quad (40)$$

where f is a function non-negative on the boundary. The signs of the terms on the left-hand side of (40) are such as to make possible a uniqueness argument for an operator like the Laplacian employing integration by parts in the usual way.

The application of Robin conditions at a large but finite radius r_o to the scalar constraint is well known (York & Piran 1982). The principle behind the construction of such a condition is the same here as for the vector conditions below. One solves for the monopole in both the potential (here, $\psi = 1 + E/2r + \ldots$) and its normal derivative ($\partial\psi/\partial r = -E/2r^2 + \ldots$). Eliminating the monopole between the resulting expressions yields the result for large r_o:

$$\frac{\partial\psi}{\partial r}(r_o) + \frac{\psi(r_o) - 1}{r_o} = O(r_o^{-3}) \approx 0\,. \quad (41)$$

If we had set $\psi = 1$ at r_o, we would be making an error of $O(r_o^{-1})$. Here, by setting the right-hand side of (41) to zero, we are making an error of $O(r_o^{-3})$. We pick up two powers of r_o in accuracy.

The appropriate vector Robin condition for the operator

$$\nabla_j(\nabla^j W^i + \nabla^i W^j - \tfrac{2}{3}g^{ij}\nabla_k W^k) \equiv \nabla_j(LW)^{ij} \quad (42)$$

was also given in York & Piran (1982). Note that here we can consider the curved space form of the operator if the slice is asymptotically flat, for then the monopole part of the solution of the homogeneous equation looks just like (9). Solving (9) for the linear momentum (monopole) P^i gives

$$P^i = -\tfrac{4}{7}W^j(\delta^i_j - \tfrac{1}{8}n^i n_j)r\,. \quad (43)$$

Likewise, forming $(LW)^{ij}$ from (9) gives (11). Solving for P^i yields

$$P^i = \frac{2}{3} r^2 (LW)^{kj} n_j (\delta^i_k - \frac{1}{2} n^i n_k) . \tag{44}$$

Equating (43) and (44) yields the desired result

$$\left[(LW)^{kj} n_j (\delta^i_k - \frac{1}{2} n^i n_k) + \frac{6}{7r} W^k (\delta^i_k - \frac{1}{8} n^i n_k)\right]_{r=r_o} = O(r_o^{-3}) . \tag{45}$$

If the total linear momentum is not zero, the use of (45) with zero on the right-hand side gives an increase in accuracy of two powers of r_o compared to setting $W^i = 0$ at at r_o. However, it will often be the case that we set up the problem in such a way that the total linear momentum is in fact zero. In this case W^i will be $O(r_o^{-2})$. Nevertheless, using zero on the right-hand side of (45) is still an improvement of one power of r_o compared to setting $W^i = 0$ at at r_o.

A natural question to ask, when the total linear momentum is zero, is whether one can develop a Robin condition based on the elimination of the angular momentum J^i from W^i and $(LW)^{ij}$. It does not seem that this can be accomplished in general because there are (as mentioned earlier) nine harmonic W^i's of $O(r_o^{-2})$, of which three carry the angular momentum. The remaining six are related to the time rate-of-change of the mass-energy quadrupole moment tensor (O'Murchadha & York 1974b; Evans 1984). Evans (1984) has shown that when the latter are negligible, one can eliminate the angular momentum just as done above for the linear momentum, and one obtains

$$(LW)^{ij} n_j + \frac{3}{r_o} W^j (\delta^i_j + n^i n_j) \approx O(r_o^{-4}) . \tag{46}$$

The proof of uniqueness of a solution of

$$\nabla_j (LW)^{ij} = 8\pi j^i \tag{47}$$

with boundary condition (45) goes as follows. Suppose there are two solutions W^i_1 and W^i_2 for the given current j^i. The difference of these solutions is denoted V^i and it satisfies the homogeneous equation and the boundary condition (45). Then

$$0 = V_i \nabla_j (LV)^{ij} = \nabla_j [V_i (LV)^{ij}] - \frac{1}{2} (LV)_{ij} (LV)^{ij} . \tag{48}$$

Next we observe from (45), with zero on the right-hand side, that we can solve for

$$(LV)^{lj} n_j = -\frac{6}{7r_o} [V^l + \frac{3}{4} n^l (\vec{V} \cdot \hat{n})] \tag{49}$$

because the inverse of $(\delta^i_k - \frac{1}{2} n^i n_k)$ is $(\delta^l_i + n^l n_i)$. Integrating (48) and using (49) gives us

$$0 = -\frac{6}{7r_o}\oint_{r=r_o}[\vec{V}\cdot\vec{V} + \tfrac{3}{4}(\vec{V}\cdot\hat{n})^2]d^2S - \tfrac{1}{2}\int(LV)_{ij}(LV)^{ij}\, d^3V\,, \tag{50}$$

from which we infer $\vec{V} = 0$ and the solution is unique.

Evans (1984) has shown similarly that when the approximation he used in obtaining (46) is valid, a proof of uniqueness similar to that above can be given.

Apparent Horizon

Figure 3.

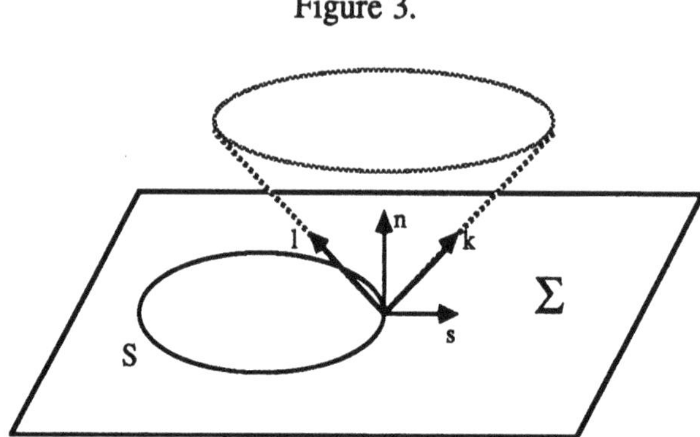

It is convenient in numerical relativity to have the equation satisfied by the apparent horizon in 3+1-dimensional form. Let S be the two-dimensional boundary of a region in a spacelike initial data surface Σ and consider the outgoing null geodesics orthogonal to S. Hypersurface Σ has unit normal n^μ ($n^\mu n_\mu = -1$) and s^μ is a spacelike unit vector orthogonal to S in Σ ($s^\mu s_\mu = 1$; $s^\mu n_\mu = 0$). Tangents to the outgoing null geodesics through S are denoted k^μ ($k^\mu k_\mu = 0$; $k^\mu s_\mu > 0$) and they are assumed to be affinely parameterized ($k^\mu{}_{;\nu} k^\nu = 0$). The corresponding ingoing null geodesics have tangents l^μ.

Clearly there is a null hypersurface η, the boundary of whose intersection with Σ is S and whose null generators are the k^μ. There is a second fundamental tensor (extrinsic curvature) associated with η. However, because η is null, to define its extrinsic curvature involves S in the construction as well. The outgoing null second fundamental tensor is then given, as usual, by a suitable projection of the covariant derivatives of its normal generators k^μ. However, in this case the projection operator P^α_μ is that associated with S. Writing

$$k^\mu = \frac{1}{\sqrt{2}}(n^\mu + s^\mu)\ ;\ \ l^\mu = \frac{1}{\sqrt{2}}(n^\mu - s^\mu)\,, \tag{51}$$

we have

$$P^{\alpha}_{\mu} = \delta^{\alpha}_{\mu} + k^{\alpha} l_{\mu} + l^{\alpha} k_{\mu}$$
$$= \delta^{\alpha}_{\mu} + n^{\alpha} n_{\mu} - s^{\alpha} s_{\mu} \ , \tag{52}$$

where

$$P^{\mu}_{\mu} = 2 \ ; \quad P^{\alpha}_{\mu} P^{\mu}_{\beta} = P^{\alpha}_{\beta} \ . \tag{53}$$

Now the extrinsic curvature in question is defined by

$$\kappa_{\mu\nu} = - P^{\alpha}_{\mu} P^{\beta}_{\nu} \, k_{\alpha;\beta} \ . \tag{54}$$

Recalling that

$$n_{\alpha;\beta} = - K_{\alpha\beta} - a_{\alpha} n_{\beta} \ , \tag{55}$$

where $K_{\alpha\beta}$ is the extrinsic curvature of Σ and $a_{\alpha} = n_{\alpha;\beta} n^{\beta}$, and using (51) in (54), we can obtain

$$\sqrt{2}\kappa_{\mu\nu} = X_{\mu\nu} + P^{\alpha}_{\mu} P^{\beta}_{\nu} K_{\alpha\beta} \tag{56}$$

where the first term on the right-hand side is the extrinsic curvature tensor of S defined by its embedding into the three-geometry of Σ:

$$X_{\mu\nu} = - P^{\alpha}_{\mu} P^{\beta}_{\nu} \nabla_{\alpha} s_{\beta} \ . \tag{57}$$

Here, ∇_{α} denotes (as throughout this article) the three-dimensional covariant derivative:

$$\nabla_{\alpha} s_{\beta} = \perp^{\mu}_{\alpha} \perp^{\nu}_{\beta} s_{\nu;\mu} \ , \tag{58}$$
$$\perp^{\mu}_{\alpha} = \delta^{\mu}_{\alpha} + n^{\mu} n_{\alpha} = P^{\mu}_{\alpha} + s^{\mu} s_{\alpha} \ . \tag{59}$$

The two-surface S is defined as the apparent horizon when the trace of $\kappa_{\mu\nu}$ vanishes:

$$-\sqrt{2}\kappa^{\mu}_{\mu} = k^{\mu}_{\ ;\mu} = \nabla_i s^i - K + K_{ij} s^i s^j = 0 \ . \tag{60}$$

Unlike the event horizon, the apparent horizon depends only on the Cauchy data defined on a spacelike slice. It does not depend on assumptions about the future evolution of the data, nor on the lapse and shift functions. (A more detailed description of the apparent horizon in the spacetime context is given by Wald (1984).) The vanishing of the quantity $\nabla_i s^i$ alone is the minimal (actually, "extremal") surface condition because

$$X^{\mu}_{\mu} = - \nabla_i s^i \ . \tag{61}$$

All the quantities in (60) refer, of course, to the physical initial data. Expressed in terms of the free data by using the conformal transformation rules for initial data, (60) has the form

$$\nabla_i s^i + 4 s^i \nabla_i \ln\psi + \psi^{-4} K_{ij} s^i s^j = 0 \ , \tag{62}$$

where $K = 0$ has been assumed.

Curvature Tensor.

How to write the spacetime Riemann tensor R_{abcd} in terms of the spatial Riemann tensor R_{abcd} and the extrinsic curvature tensor is well known. The resulting equations are the Gauss-Codazzi-Ricci equations. For the sake of convenience and future reference, I record them here together with a few additional quantities. The spacetime metric signature is (-+++) and $n^a n_a = -1$. All indices in this and the remaining section from the list a,b,...,h can be regarded as "abstract" in the well-understood sense. (See, for example, Wald (1984)).

The spatial metric and operator of projection onto the spacelike slice are

$$\gamma_{ab} = g_{ab} + n_a n_b \ ; \ \perp^a_b = \gamma^a_b = \delta^a_b + n^a n_b \ . \tag{63}$$

The symbol $\perp$ will be used to denote projection of every free index to its right using $\perp^a_b$. The Levi-Civita tensor ε_{abcd} (oriented by "0123 positive" and of weight zero) leads to the definition of the corresponding three-dimensional tensor

$$\varepsilon_{abc} = n^d \varepsilon_{dabc} \ , \tag{64}$$

and we have

$$\varepsilon_{abcd} = -(n_a \varepsilon_{bcd} + n_b \varepsilon_{adc} + n_c \varepsilon_{abd} + n_d \varepsilon_{acb}) \ . \tag{65}$$

The Einstein equation is

$$G_{ab} = \frac{1}{2} \kappa T_{ab} \ , \tag{66}$$

($\kappa = 16\pi G$, $c = 1$). We use the conventions of Misner *et al.* (1973) throughout. Note that κ in York (1979) is here (1/2)κ. The energy density, momentum density, and spatial stress tensor are, respectively

$$\rho = T_{ab} n^a n^b \ ; \ j^a = -\perp T^a_b n^b \ ; \ S_{ab} = \perp T_{ab} \ . \tag{67}$$

The Gauss equation is (with slanted *R*'s referring to the intrinsic three-geometry)

$$\perp R_{abcd} \equiv P_{abcd} = \mathit{R}_{abcd} + K_{ac} K_{bd} - K_{ad} K_{bc} \ , \tag{68}$$

with

$$P_{ab} \equiv P^{c}{}_{acb} = \mathcal{R}_{ab} + KK_{ab} - K_{ac}K^{c}_{b} , \tag{69}$$

$$P_{abcd} = - \varepsilon_{abe}\varepsilon_{cdf}(P^{ef} - \tfrac{1}{2}\gamma^{ef}P) , \tag{70}$$

$$P \equiv P^{a}{}_{ba}{}^{b} = \mathcal{R} + K^{2} - K_{ab}K^{ab} , \tag{71}$$

$$P_{abcd}P^{abcd} = 4P_{ab}P^{ab} - P^{2} . \tag{72}$$

The Codazzi equation is

$$\perp R_{abcd}n^{d} \equiv Q_{abc} = \nabla_{b}K_{ac} - \nabla_{a}K_{bc} , \tag{73}$$

with

$$Q_{a} \equiv Q_{ab}{}^{b} = \nabla_{b}(K^{b}_{a} - K\gamma^{b}_{a}) , \tag{74}$$

$$\begin{aligned} {}^{*}Q_{ab} &\equiv \tfrac{1}{2}\varepsilon_{a}{}^{cd}Q_{cdb} = {}^{*}Q_{(ab)} - \tfrac{1}{2}Q_{c}\varepsilon^{c}{}_{ab} \\ &\equiv B_{ab} - \tfrac{1}{2}Q_{c}\varepsilon^{c}{}_{ab} \quad (B^{a}_{a} = 0) , \end{aligned} \tag{75}$$

$$Q_{abc} = \varepsilon_{ab}{}^{d}(B_{cd} + \tfrac{1}{2}Q_{e}\varepsilon^{e}{}_{cd}) , \tag{76}$$

$$Q_{abc}Q^{abc} = 2B_{ab}B^{ab} + Q_{a}Q^{a} . \tag{77}$$

The Ricci equation is

$$R_{acbd}n^{c}n^{d} \equiv D_{ab} = \frac{1}{\alpha}[(\pounds_{t} - \pounds_{\beta})K_{ab} + \nabla_{a}\nabla_{b}\alpha] + K_{ac}K^{c}{}_{b} , \tag{78}$$

where α is the lapse function, β^{a} ($\beta^{a}n_{a} = 0$) is the shift vector, $t^{a} = \alpha n^{a} + \beta^{a}$, and £ denotes the Lie derivative operator. The Gauss-Codazzi-Ricci equations as given in (68), (73), and (78) are all geometric identities, as the field equation has not been used. Further useful relationships are

$$\begin{aligned} R_{abcd} = P_{abcd} &+ n_{a}Q_{cdb} - n_{b}Q_{cda} + n_{c}Q_{abd} - n_{d}Q_{abc} \\ &+ n_{a}n_{c}D_{bd} - n_{a}n_{d}D_{bc} + n_{b}n_{d}D_{ac} - n_{b}n_{c}D_{ad} , \end{aligned} \tag{79}$$

$$R_{ab} = P_{ab} - D_{ab} + n_{a}Q_{b} + n_{b}Q_{a} + n_{a}n_{b}D , \tag{80}$$

where $D = D^{a}{}_{a}$, and

$$R = P - 2D . \tag{81}$$

The field equations can be inserted into these relationships by using them in the form of constraints

$$P = \kappa\rho \ ; \ Q_a = \frac{1}{2}\kappa j_a \tag{82}$$

and equation of motion

$$D_{ab} = P_{ab} - \frac{1}{2}\kappa S_{ab} - \frac{1}{4}\kappa\gamma_{ab}(\rho - S) \ , \tag{83}$$

where $S = S^a{}_a$.

Flat Spacetime and Test-Bed Calculations

Consider vacuum AF spacetimes with AF Cauchy slices having the topology of $\mathbf{R}^3$. From the equations of the preceding section, it is easy to see that among the solutions of the constraints $P = 0$, $Q_a = 0$ there can be special ones that satisfy on some Cauchy slice Σ_o the stronger requirements

$$P_{ab} = R_{ab} + KK_{ab} - K_{ac}K^c_b = 0 \ , \tag{84}$$

$$Q_{abc} = \nabla_b K_{ac} - \nabla_a K_{bc} = 0 \ . \tag{85}$$

Equations (68), (70), (73), (79), and (80) then tell us that the Riemann tensor has the form

$$R_{abcd} = n_a n_d R_{bc} - n_a n_c R_{bd} + n_b n_c R_{ad} - n_b n_d R_{ac} \ . \tag{86}$$

But because the spacetime is "vacuum" (Ricci-flat) by hypothesis, we see from (86) that the Riemann tensor must be zero in the immediate vicinity of Σ_o. Suppose now that we evolve the data satisfying (84) and (85) using the vacuum equation of motion $P_{ab} = D_{ab}$. Will the full spacetime be flat? That is, will the spacetime Riemann tensor remain zero?

The question can be answered in the affirmative by appealing to the full Bianchi identity (York 1987), just as one can use the twice-contracted Bianchi identity $G^a_{b;a} = 0$ to demonstrate that the ordinary constraints $P = Q_a = 0$ are preserved in the evolution. Beginning with the Bianchi identity

$$R_{bcde;a} + R_{abde;c} + R_{cade;b} = 0 \ , \tag{87}$$

and exploiting the well known relations that hold between Lie derivatives and covariant derivatives, one can find after some calculation that

$$\begin{aligned} \pounds_t P_{abcd} = \alpha[&\nabla_b Q_{cda} - \nabla_a Q_{cdb} + 2\nabla_{a[c}K_{d]b} - 2\nabla_{b[c}K_{d]a} + 2P_{abe[c}K^e_{d]} \\ &+ 2Q_{ab[c}\nabla_{d]}\alpha + 2Q_{cd[a}\nabla_{b]}\alpha] + \pounds_\beta P_{abcd} \ , \end{aligned} \tag{88}$$

$$\pounds_t Q_{abc} = \alpha[2\nabla_{[b}D_{a]c} + 2D_{c[a}\nabla_{b]}\alpha + P_{abcd}\nabla^d\alpha - Q_{abd}K^d_c + 2Q_{dc[b}K^d_{a]}] + \pounds_\beta Q_{abc} \; . \quad (89)$$

The result follows by inspection of (88) and (89).

We conclude that (84) and (85) define the initial data for flat Minkowski spacetime with respect to evolution in the pure Einstein theory. Note that the data that satisfy (84) and (85) need not be the trivial data $\gamma_{ij} = f_{ij}$, $K_{ij} = 0$ of a standard spacelike hyperplane in Minkowski spacetime. The data could appear quite non-trivial and this fact forms the basis of what might define an interesting class of test-bed numerical evolutions. One can simply take Minkowski spacetime in its usual simple form and give it "by hand" a wild foliation by warped and lumpy spacelike slices of one's own choosing. One next evaluates the γ_{ij} and K_{ij} induced on one of these slices, the "initial" one. Next, crank up your favorite numerical evolution scheme and blast away. The test is to compute, after some "time" has elapsed, all the components of the Riemann tensor. They should all be zero.

Invariants of the Riemann Tensor

The scalars formed from the Riemann tensor are often said to be analogous to those formed from the Maxwell tensor, two familiar ones being

$$F_{ab}F^{ab} = 2(\vec{B}\cdot\vec{B} - \vec{E}\cdot\vec{E}) \; , \quad (90)$$

$$\tilde{F}_{ab}F^{ab} = 4\vec{B}\cdot\vec{E} \; . \quad (91)$$

But note two points. First, scalars formed from F_{ab} and its dual $\tilde{F}_{ab}$ involve only the initial data of the Maxwell field, that is

$$E^a = F^{ab}n_b \quad ; \quad B_a = \frac{1}{2}\varepsilon_{abc}F^{bc} \; . \quad (92)$$

Therefore, the scalars formed from the curvature tensor of the electromagnetic field are manifestly independent of the lapse and shift and depend only on the initial data without explicit use of the Maxwell equation of motion for $\pounds_t E^a$. The analogous statement in the case of gravity is different as we shall see below. Second, as we see from (90) and (91), the invariants can be zero (indeed, *all* the invariants) without the vanishing of the electromagnetic field. Thus, the invariants "filter out" plane waves. An analogous statement holds for the invariants of the Riemann tensor in the case of plane-fronted gravitational waves with parallel rays ("pp waves").

Petrov (1969) has listed the fourteen invariants of the Riemann tensor. We present them here in familiar notation. To convert them to 3+1 form can be done directly (a sample calculation of this type is presented below). However, the quantities are probably most easily computed by first reconstructing the projections P_{abcd}, Q_{abc}, and D_{ab} of the Riemann tensor, then forming the other curvature tensors, and then finally calculating the invariants (*i.e.*, numerically), rather than writing out all these invariants directly in terms of the Cauchy data as a first step. We stress that these quantities are independent of the lapse and shift, and can be expressed in terms of the Cauchy data of the gravitational and matter fields (including an equation of state for the latter), but that for these properties to be manifest, wherever the projection D_{ab} apears, it must be replaced using the equation of motion (essentially the equation for $\pounds_t K_{ab}$). One can write terms involving P and Q_a in terms of either gravity data or matter by using the constraints.

We list the invariants using Petrov's nomenclature, though not explicitly his method of breaking up the curvature tensor. We use the Riemann tensor, the Ricci tensor, the double-dual Riemann tensor ($\tilde{R}_{abcd}$), the left-dual Riemann tensor (${}^*R_{abcd}$), and the scalar curvature R.

$$B_1 = R \tag{93}$$

$$B_2 = \frac{1}{2}(R^{ab}{}_{cd} - \tilde{R}^{ab}{}_{cd})R^{cd}{}_{ab} \ , \tag{94}$$

$$B_2' = 2\,{}^*R^{ab}{}_{cd}R^{cd}{}_{ab} \ , \tag{95}$$

$$B_3 = (\frac{1}{4}R^{ab}{}_{cd} - \frac{3}{4}\tilde{R}^{ab}{}_{cd})R^{cd}{}_{ef}R^{ef}{}_{ab} \ , \tag{96}$$

$$B_3' = (\frac{3}{2}R^{ab}{}_{cd} - \frac{1}{2}\tilde{R}^{ab}{}_{cd})R^{cd}{}_{ef}\,{}^*R^{ef}{}_{ab} \ , \tag{97}$$

$$S_2 = R^a_b R^b_a - \frac{1}{4}R^2 \ , \tag{98}$$

$$S_3 = R^a_b R^b_c R^c_a - \frac{3}{4}RR^a_b R^b_a + \frac{1}{8}R^3 \ , \tag{99}$$

$$S_4 = R^a_b R^b_c R^c_d R^d_a - RR^a_b R^b_c R^c_a + \frac{3}{8}R^a_b R^b_a R^2 - \frac{3}{64}R^4 \ , \tag{100}$$

$$T_1 = \frac{1}{2}(R^{ab}{}_{cd} - \tilde{R}^{ab}{}_{cd})R^c_a R^d_b - \frac{1}{16}R^3 \ , \tag{101}$$

$$T_1' = 2\,{}^*R^{ab}{}_{cd}R^c_a R^d_b \ , \tag{102}$$

$$\begin{aligned} T_2 = {} & \frac{1}{8}(R^{ab}{}_{cd} - \tilde{R}^{ab}{}_{cd})(R^{cd}{}_{ef} - \tilde{R}^{cd}{}_{ef})R^e_a R^f_b - \frac{1}{4}\tilde{R}^{ea}{}_{cd}R^{cd}{}_{eb}(R^b_f R^f_a - RR^b_a) \\ & + \frac{1}{16}R_{abcd}R^{abcd}R^e_f R^f_e - \frac{1}{32}(R_{abcd} + \tilde{R}_{abcd})R^2 R^{abcd} \ , \end{aligned} \tag{103}$$

$$\begin{aligned} T_2' = {} & \frac{1}{8}\varepsilon^{ab}{}_{cd}(R^{cd}{}_{ef} - \tilde{R}^{cd}{}_{ef})(R^{ef}{}_{gh} - \tilde{R}^{ef}{}_{gh})R^g_a R^h_b \\ & + \frac{1}{2}\,{}^*R^{ab}{}_{ef}R^{ef}{}_{ah}(R^h_g R^g_b - RR^h_b) + \frac{1}{8}\,{}^*R^{ab}{}_{ef}R^{ef}{}_{ab}R^g_h R^h_g \ , \end{aligned} \tag{104}$$

$$\begin{aligned} T_3 = {} & \frac{1}{2}(R^{ab}{}_{cd} - \tilde{R}^{ab}{}_{cd})[4R^c_e R^e_f R^f_a R^d_b + 3R^c_e R^e_a R^d_f R^f_b - 6R^c_f R^f_a R^d_b R + \frac{3}{2}R^c_a R^d_b R^2] \\ & - \frac{1}{4}R^a_b R^b_c R^c_a R^2 + \frac{9}{32}R^a_b R^b_a R^3 - \frac{21}{256}R^5 \ , \end{aligned} \tag{105}$$

$$T'_3 = 2{}^*R^{ab}{}_{cd}[4R^c_eR^e_fR^f_aR^d_b + 3R^c_eR^e_aR^d_fR^f_b - 6R^c_fR^f_aR^d_bR + \tfrac{3}{2}R^c_aR^d_bR^2] \ . \quad (106)$$

To compare these with Petrov's results (after correction of a few misprints in Petrov (1969)), one can use the relations between Petrov's notations and ours. The "isotropic" part of the Riemann tensor is

$$\dot{R}^{ab}{}_{cd} = \frac{1}{2}(R^{ab}{}_{cd} - \tilde{R}^{ab}{}_{cd}) \ , \quad (107)$$

and the "simple" part is

$$\mathring{R}^{ab}{}_{cd} = \frac{1}{2}(R^{ab}{}_{cd} + \tilde{R}^{ab}{}_{cd}) \ . \quad (108)$$

Finally, we see that the four invariants that do not vanish identically when $R_{ab} = 0$ are B_2, B'_2, B_3, and B'_3.

Here I exhibit a characteristic sample of the reduction of a term, like some that appear above, to initial data plus matter variables:

$$\begin{aligned} R_{abcd}R^{abcd} = {} & 8\mathcal{R}_{ab}\mathcal{R}^{ab} - 16\mathcal{R}_{ab}K^a_cK^{bc} + 16K\mathcal{R}_{ab}K^{ab} + 8(K_{ab}K^{bc}K_{cd}K^{da}) \\ & - 16KK_{ab}K^a_cK^{bc} + 8K^2K_{ab}K^{ab} - 8\nabla_{[b}K_{a]c}\nabla^bK^{ac} \\ & + \kappa[4S_{ab}(K^a_cK^{cb} - KK^{ab}) - 4\mathcal{R}_{ab}S^{ab}] \\ & + \kappa^2[S_{ab}S^{ab} - \tfrac{1}{4}S^2 - \tfrac{9}{4}\rho^2 + \tfrac{3}{2}\rho S] \ . \end{aligned} \quad (109)$$

Both L. Smarr and I computed such formulas for invariants, in the cases where $R_{ab} = 0$, in the "early days" when numerical relativity (Smarr 1975) joined forces with the improved understanding of the initial-value problem (York 1972, 1973), but I think not much real use was made of them at that time. Since then, "Spacetime Engineering" has become a more nearly mature discipline and perhaps more serious tests of results using the full-blown curvature are in order.

Acknowledgments

I am very grateful to Gregory Cook and John Jaynes for their invaluable assistance in this work. I also thank Jeffrey Melmed, and especially Gregory Cook, for their help in preparing the manuscript. The research was supported by the National Science Foundation.

References

Bowen, J.M. (1979). Initial Value Problems on Non-Euclidean Topologies. University of North Carolina at Chapel Hill: Ph.D. dissertation.

Bowen, J.M. (1982). General solution for flat-space longitudinal momentum. Gen. Rel. and Grav., 14, 1183-1191.

Bowen, J.M. (1984). Graphic displays of gravitational initial data. J. Comp. Phys., 56, 42-50.

Bowen, J.M. (1985). Inversion symmetric initial data for N charged black holes. Ann. of Phys., 165, 17-37.

Bowen, J.M., Rauber, J.D. & York, J.W. (1984). Two black holes with axisymmetric parallel spins. Class. Quantum Grav., 1, 591-610.

Bowen, J.M. & York, J.W. (1980). Time-asymmetric initial data for black holes and black-hole collisions. Phys. Rev. D, 21, 2047-2056.

Brill, D.R. (1959). On the positive-definite mass of the Bondi-Weber-Wheeler time-symmetric gravitational waves. Ann. of Phys., 7, 466-483.

Cantor, M.R. & Kulkarni, A.D. (1982). Physical distinction between normalized solutions of the two-body problem of general relativity. Phys. Rev. D, 25, 2521-2526.

Choptuik, M. (1982). A Study of Numerical Techniques for the Initial Value Problem of General Relativity. University of British Columbia. Vancouver: M. Sc. thesis.

Detweiler, S.L. (1979). Black holes and gravitational waves: perturbation analysis. *In* Sources of Gravitational Radiation, ed. L.L. Smarr, pp. 211-230. Cambridge: Cambridge University Press.

Duff, G.F.D. (1959). Partial Differential Equations. Toronto: University of Toronto Press.

Evans, C.R. (1984). A Method for Numerical Relativity: Simulation of Axisymmetric Gravitational collapse and Gravitational Radiation Generation. The University of Texas at Austin: Ph.D. dissertation.

Kulkarni, A.D. (1984). Time-asymmetric initial data for the N black hole problem in general relativity. J. Math. Phys., 25, 1028-1034.

Kulkarni, A.D., Shepley, L.C., & York, J.W. (1983). Initial data for N black holes. Phys. Letters, 96A, 228-230.

Misner, C.W. (1960). Wormhole initial conditions. Phys. Rev., 118, 1110-1112.

Misner, C.W. (1963). The method of images in geometrostatics. Ann. of Phys., 24, 102-110.

Misner, C.W., Thorne, K.S. & Wheeler, J.A. (1973). Gravitation. San Francisco: Freeman.

O'Murchadha, N.S. & York, J.W. (1974a). Gravitational energy. Phys. Rev. D., 10, 2345-2357.

O'Murchadha, N.S. & York, J.W. (1974b). Mass, momentum, and multipoles. Unpublished notes.

O'Murchadha, N.S. & York, J.W. (1976). Gravitational potentials: a constructive approach to general relativity. Gen. Rel. and Grav., 7, 257-261.

Petrov, A.Z. (1969). Einstein Spaces. London: Pergamon.

Rauber, J.D. (1985). Initial Data for Black Hole Collisions. University of North Carolina at Chapel Hill: Ph.D. dissertation.

Smarr, L.L. (1975). The Structure of General Relativity with a Numerical Illustration: The Collision of Two Black Holes. The University of Texas at Austin: Ph.D. dissertation.

Smarr, L.L. (1979). Gauge conditions, radiation formulae and the two black hole collision. *In* Sources of Gravitational Radiation, ed. L.L. Smarr, pp. 231-244. Cambridge: Cambridge University Press.

Thornburg, J. (1987). Coordinates and boundary conditions for the general relativistic initial data problem. Class. Quantum Grav., 4, 1119-1131.

Wald, R.M. (1984). General Relativity. Chicago: The University of Chicago Press.

Witt, D.M. (1986). Vacuum space-times that admit no maximal slice. Phys. Rev. Letters, 57, 1386-1389.

York, J.W. (1972). Role of conformal three-geometry in the dynamics of gravitation. Phys. Rev. Letters, 28, 1082-1085.

York, J.W. (1973). Conformally invariant orthogonal decomposition of symmetric tensors on Riemannian manifolds and the initial-value problem of general relativity. J. Math. Phys., 14, 456-464.

York, J.W. (1979). Kinematics and dynamics of general relativity. *In* Sources of Gravitational Radiation, ed. L.L. Smarr, pp. 83-126. Cambridge: Cambridge University Press.

York, J.W. (1983). The initial value problem and dynamics. *In* Gravitational Radiation, eds. N. Deruelle & T. Piran, pp. 175-202. Amsterdam: North-Holland.

York, J.W. (1987). Bel-Robinson gravitational superenergy and flatness. *In* Gravitation and Geometry, eds. W. Rindler and A. Trautman, pp. 497-505. Naples: Bibliopolis.

York, J.W. & Piran, T. (1982). The initial value problem and beyond. *In* Spacetime and Geometry, eds. R.A. Matzner and L.C. Shepley, pp. 145-176. Austin: The University of Texas Press.

Analytic-Numerical Matching for Gravitational Waveform Extraction

Andrew M. Abrahams

Department of Astronomy, U. of Illinois and National Center for Supercomputing Applications

Abstract. We present a formalism designed to provide a consistent means for extracting gravitational radiation waveforms at finite radii during numerical relativity simulations of isolated sources. First we motivate and give complete solutions to the linearized vacuum Einstein equations in two gauges. We then show three procedures for matching these exterior solutions, good in the local wave-zone and the linear near-zone, onto a numerical solution for the source and strong-field region. The matching eliminates pure gauge terms, and an ODE integration over a time-like cylinder is employed to separate off the wave's near-zone field. The procedures are demonstrated with simulations of oscillating neutron star and Brill wave spacetimes.

INTRODUCTION

Ever since Smarr's (1975, 1977) pioneering study of black hole collisions, the question of how to identify the radiative component of a numerically generated spacetime has been of practical interest. As simulations on spacelike slices appear to be the most powerful for realistic astrophysical calculations, there is increasing effort aimed at developing techniques for treating gravitational radiation from compact sources on grids of finite spatial extent. The major objectives are 1) to have a rigorous way of extracting waveforms and energies and 2) to use matching information to improve the code's outer boundary conditions.

Smarr and York (1978a, b) suggest the minimal distortion gauge modelled on Coulomb gauge in electromagnetism. This gauge choice attempts to force as much of the non-physical coordinate wave as is possible into the shift vector components by minimizing the global rate of change of the conformal three-geometry. Smarr (1979) discusses three curvature-based and three connection-based quantities as candidates for a general relativistic Poynting vector. Bardeen (1983) performs weak-field analysis of several gauges and gives an energy flux expression for radial gauge good through order r^{-2}. Schutz (1986) employs a near-zone field expansion to derive a formula for extracting the quadrupole moment using surface integrals over projected Riemann tensor components. Anderson and Hobill (1986) describe a possible technique for matching an analytic solution on null cones to an interior numerical solution on spacelike slices. Damour (1986) gives a formula which may be useful for correcting extracted waveforms for backscattering off curvature, assuming an idealized asymptotically flat spacetime,

Perhaps the most important practical advance in the last few years was the identification by several authors (Dykema 1980, Bardeen and Piran 1983, Evans 1984) of special "radiative" variables. These variables, constructed from metric components of quasi-isotropic and radial gauges, have moments of the radiation field rather than the static mass-monopole as the leading term in their weak-field expansion. Thus, at large enough radii these could be used to obtain asymptotic waveforms. As we will see, however, unless the extraction two-surface lies well out in the local wave-zone ($r \gtrsim 4\bar{\lambda} \equiv 4\lambda/2\pi$), waveforms will be significantly contaminated by linear gauge effects and the wave's non-radiative near-zone field. The work described here represents the logical completion of the program of linear analysis of radiative spacetimes which produced the radiative variables. First we will motivate and present solutions to the linearized gravity equations expressed as multipole expansions. Then we will describe three methods for matching these exterior analytic solutions onto interior numerical solutions and show how to extract radiation information. Finally we will demonstrate using simulations of oscillating neutron star and Brill wave spacetimes how the gauge and near-zone field terms are separated off to yield asymptotic waveforms.

LINEAR ANALYSIS

Thorne (1980) discusses splitting the calculation of gravitational radiation from isolated sources into two parts: 1) generation and propagation of the waves through the near zone and into the local wave-zone and 2) propagation of the waves from the local wave-zone through the distant wave-zone to the observer. The interior solution is matched in the local wave-zone to a linearized gravity solution which in turn can be corrected for effects of propagation through the real universe using techniques of geometric optics (Thorne 1983). It is unnecessary to make the unrealistic assumption that the universe is asymptotically flat - merely that the source is sufficiently isolated. For our work, the interior solution for the source, near zone, and local wave-zone is calculated numerically and matched at some finite radius onto a linearized solution. This matching allows the physical wave to be separated from any (linear) pure-gauge effects inherent in the interior solution. As will be seen, our procedure actually provides for this matching to be accomplished in the weak-field near zone as long as the tails are unimportant. This can be guaranteed by putting the matching two surface at a large enough value of r/M [if $(r^3/M)^{1/2} \lesssim \lambda\!\!\!^{-}$ then tails *cannot* be neglected (Thorne 1980)]. The wave's (linear) non-radiative near-zone field can then be separated off by performing certain ordinary differential equation (ODE) integrations along a time-like cylinder. Performing the extraction at two radii allows for a consistency check (that is, one can verify that our approximations are good in a given radiation extraction region). These linear effects: the gauge-shearing of extraction two-surfaces and the non-radiative near-zone field terms in the metric are the most important hindrances to determination of the asymptotic waveforms from most sources of astrophysical interest, e.g., collapsing stellar cores, stellar collision, black hole formation and collapse of relativistic star clusters. It is likely that all such sources radiate less than a few percent of their rest energy in a typical wave-period. This weak-radiation condition should insure that the wave's non-linear self interaction (an effect not accounted for in our first-order perturbation theory approach) is negligible. In addition, the form of our matching solutions requires the source to have approximately stationary mass and angular-momentum moments; this is also guaranteed by the weak-radiation condition. It is important to mention that in many situations the errors in the waveform due to neglected (non-linear) terms in our approach will be of the same order as finite difference truncation errors in the calculation.

To implement this picture as a working scheme first we require complete solutions to the linearized Einstein equations in vacuum. It turns out that for our purposes it was useful to have solutions in two different gauges, each a subgauge of Lorentz gauge (unlike the gauges used for numerical purposes up to this point). The first, in transverse-traceless gauge, is merely a generalization of that given explicitly for quadrupole waves by Teukolsky (1982). For $l \geq 2$ the metric perturbation in a spherical orthonormal basis is

$$(h_{ij}^{TT})_{lm} = \left[\frac{(l+1)(l+2)(2l+3)}{(l-1)l(2l-1)}\right]^{1/2} \{I_{lm}^{(l)}\}_{l-2} T_{ij}^{2l-2,lm} + \left[\frac{6(l+2)(2l+1)}{l(2l-1)}\right]^{1/2} \{I_{lm}^{(l)}\}_{l} T_{ij}^{2l,lm} + \{I_{lm}^{(l)}\}_{l+2} T_{ij}^{2l+2,lm}$$
$$+ i\left[\frac{l+2}{2l+1}\right]^{1/2} \{S_{lm}^{(l+1)}\}_{l-1} T_{ij}^{2l-1,lm} + i\left[\frac{l-1}{2l+1}\right]^{1/2} \{S_{lm}^{(l+1)}\}_{l+1} T_{ij}^{2l+1,lm}. \qquad (1)$$

Here, I_{lm} and S_{lm} represent respectively even-parity mass and odd-parity current moments with total angular momentum quantum number l and azimuthal quantum number m. The brackets around the moments indicate that a functional of that moment (an arbitrary function of retarded or advanced time) is being taken. The subscript outside the bracket means that the functional satisfies a radial wave equation with that orbital angular momentum. The superscripted parentheses denote the highest derivative of the function that appears in the functional. These functionals can be operated on by raising and lowering operators to change their orbital angular momentum (see Burke 1971). The $T_{ij}^{\lambda l',lm}$ are pure-orbit tensor spherical harmonics where λ is the order of the representation group and l' is the orbital angular momentum. From (1) it follows that the two polarization states of gravitational waves have the asymptotic form $(h_{+})_{lm} \Longrightarrow (I_{lm}^{(l)}/2r)\, T_{\theta\theta-\phi\phi}^{E2,lm}$ and $(h_{\times})_{lm} \Longrightarrow (S_{lm}^{(l)}/r)\, T_{\theta\phi}^{B2,lm}$ in terms of pure-spin harmonics. Thus, the desired radiative part of the TT metric goes as the lth derivative of the l-pole moment over r. The other solution we use is due to Thorne (1980); he derived it in terms of symmetric-tracefree harmonics. It is written in a mass-centered coordinate system which permits the appearance of the static l=0 mass monopole and the stationary l=1 current dipole moments in the solution. We will show elsewhere that this mass monopole can be extracted and used as a quasi-local mass indicator accurate to about one part in 5000. The spatial part of the metric perturbation in the Lorentz-Thorne gauge for $l \geq 2$ is

$$(h_{ij}^{LT})_{lm} = -\left[\frac{3l(l-1)}{2(l+1)(l+2)}\right]^{1/2} \{I_{lm}^{(l)}\}_{l} T_{ij}^{0l,lm}$$
$$+ \left[\frac{2(2l-1)(2l+1)}{(l+1)(l+2)}\right]^{1/2} \{I_{lm}^{(l)}\}_{l-2} T_{ij}^{2l-2,lm} + i\left[\frac{2l+1}{l+2}\right]^{1/2} \{S_{lm}^{(l)}\}_{l-1} T_{ij}^{2l-1,lm}. \qquad (2)$$

We will now describe the three matching procedures and give the extraction equations for the mass quadrupole moment arising from each. We limit ourselves to consideration of this moment as it is the one for which the matching is tested numerically in the next section. Each method however generalizes easily to extract even and odd parity moments with any l and m. As these techniques will be of the most practical necessity for three-dimensional codes, this generality is important.

In the first, "template", matching procedure the analytic solution is transformed directly into the numerically useful gauge using

$$h'_{ij} = h_{ij} + 2\nabla_{(i}\zeta_{j)}, \qquad (3)$$

where ∇_i is a flat-space covariant derivative. The generator ζ is decomposed in terms of pure-orbit vector harmonics and the gauge conditions (for the numerical gauge) are solved for functions of time and radial coordinate. To complete the infinitesimal gauge transformation the generator is plugged into (3). We have performed this transformation for a solution consisting of l=0 and 2 mass moments and

l=1 and 3 current moments to both quasi-isotropic and radial gauges. Once the exterior solution in the desired gauge is obtained one could just calculate the code variables from this solution to accomplish the analytic-numerical matching. We have found it more fruitful, however, to decompose the metric perturbation in terms of multipole amplitudes (functions of t and r) and tensor spherical harmonics. As the multipole amplitudes are known analytically it is possible to form a linear combination of these amplitudes which excludes pure-gauge terms (we know from the gauge-invariant analysis discussed below that this can always be accomplished). The result of this procedure for quasi-isotropic gauge at some extraction radius r_0 is

$$r_0\{I^{(2)}(t-\varepsilon r_0)\}_2 \equiv I^{(2)}(t-\varepsilon r_0)+\frac{3}{r_0}I^{(1)}(t-\varepsilon r_0)+\frac{3}{r_0^2}I(t-\varepsilon r_0)=3r_0(A_{20,2}+\frac{1}{2}A_{02,2}), \tag{4}$$

where $\varepsilon=+1$ (outgoing waves) or $\varepsilon=-1$ (incoming waves). Here $I^{(2)}(t-\varepsilon r)$ is the second time derivative of the quadrupole moment $I=I_{20}(15/64\pi)^{1/2}$ - the desired asymptotic radiative part of the metric. The terms with lower derivatives of I are due to the wave's nearzone field. The multipole amplitude $A_{\lambda l',l}$ is the projection of the spatial metric onto the tensor harmonic $T_{ij}^{\lambda l',l0}$. Equation (4) can be used as an ODE for separating the radiative moment from the near-zone field. The right side source terms can be calculated as a numerical time series during the simulation. The ODE is integrated over a timelike cylinder yielding the asymptotic wave signal. It is interesting to note that the left side of (4) is precisely a solution to the radial wave equation with l'=2. This appears to be a generic property of successful matching equations.

Our second, "gauge-invariant", procedure is based on a technique first used by Moncrief (1974) for studying perturbations of the Schwarzschild geometry (we apply it to perturbations off flat space). One can calculate the action of Killing operator [see (3)] on a general generating function. Using this it is possible to construct combinations of multipole amplitudes which are invariant under transformations (3). There is one of these gauge-invariant variables for odd-parity perturbations and two for even-parity. Utilizing the Hamiltonian constraint results in one even-parity unconstrained gauge-invariant variable. Calculating this variable using either of the analytic solutions (1) or (2) yields the following extraction equation for the quadrupole mass moment

$$r_0\{I^{(2)}(t-\varepsilon r_0)\}_2 = r_0 Q_{e(l=2)} = \frac{r_0}{8}\left[6A_{02,2}-9A_{22,2}+5A_{24,2}-2r\partial_r A_{02,2}+r\partial_r A_{20,2}+r\partial_r A_{22,2}\right]_{r=r_0}. \tag{5}$$

The combination of amplitudes on the right side should be free of pure gauge terms in any gauge. The gauge-invariant matching equations for other l and m modes are basically as simple. For even-parity at most four multipole amplitudes and their radial derivatives are required. For odd-parity only two amplitudes and their radial derivatives are necessary. The left side is always automatically a solution to the radial wave equation so integration of an lth order ODE is required to extract an lm-pole moment.

Our third method for connecting analytic and numerical solutions is based on the Riemann tensor. Exploiting the well known relation between the second time derivative of the metric perturbation in TT gauge and the time-space-time-space components of the Riemann tensor and using the mass quadrupole part of the general solution (1) some manipulation yields

$$\{I^{(2)}(t-r_0)\}_2 = -\frac{5}{4}r_0^2\int_0^{\pi/2} d\theta \sin\theta P_2(\cos\theta) R_{trtr}. \tag{6}$$

As the driving term for the ODE is an integral over a two-sphere of a Riemann tensor component, this method is also gauge-invariant to linear order. The same result can be obtained from the Weyl tensor coefficient Ψ_2 and the linearized gravity solution for this quantity from Janis and Newman (1965). The real part of Ψ_2 gives the even-parity matching equation and the complex part gives an odd-parity matching equation. The odd-parity matching equation obtained in this fashion is not as suitable as that from the TT gauge solution, however as it uses the $tr\theta\phi$ rather than the $trt\phi$ component of the Riemann tensor the calculation of which requires taking derivatives between leaves of the foliation.

Finally we discuss the determination of initial values for the extraction ODEs. One can show with a Laplace transform that the effects on the solution of the initial values I_0 and $\dot{I}_0$ die off exponentially with an e-folding time approximately equal to the wave crossing time of the extraction radius. It is possible however to determine these values rigorously from standard initial data (γ_{ij}, K_{ij}) on a timeslice. This is done using the generalization of (5) and its first time derivative which is calculated employing the evolution equation for the three-metric (in terms of K_{ij} and $\nabla_i\beta_j$) to obtain time derivatives of the multipole amplitudes. To get the initial values for an order l extraction equation one applies l+1 angular momentum lowerering operators to the extraction equation and l lowering operators to its time time derivative yielding $2l$+1 equations which can be solved for the $2l$+1 unknown initial values (only l–1 of these are actually used). One nice thing is that this procedure can be performed at any time during the numerical simulation so with simple physical insight about the nature of the problem being studied, the accuracy of the radiation extraction can be increased.

NUMERICAL TESTS

We will now demonstrate the procedures of the last section on two test problems which represent in many ways the opposite extremes of radiative spacetimes. For our initial evaluations neutron star pulsations are highly useful. The neutron star serves both as a source of gravitational radiation and as a source for a strong field static background upon which the waves propagate as a perturbation. For this physical situation the near zone and the local and distant wave-zones are easily delineated. In addition, there has been extensive perturbation theory study of these oscillations providing an independent check of the accuracy of the simulations. Since one is numerically evolving a perturbed matter configuration there is no problem with incoming radiation from past asymptotic null-infinity. The outgoing gravitational wavetrain begins with the onset of oscillation. It should be emphasized that the purpose of these simulations is merely to provide a radiative spacetime for testing purposes, not to investigate the physics of neutron stars. We utilize tabulated, zero-temperature, nuclear equations of state which produce relativistic stellar models with masses and radii indicative of neutron stars. Only adiabatic perturbations are considered.

The simulations of the neutron star and the surrounding space were carried out with a 2-D (axisymmetric) code which solves the general relativity equations coupled with hydrodynamics. This code has been extensively described elsewhere (Evans 1984, 1986) but we will review here some of its essential features. The code uses spherical polar coordinates with the isotropic radial coordinate. The quasi-isotropic gauge is employed leading to a three-metric of the form

$$dl^2 = \phi^4 \, [e^{(2\eta/3)}(dr^2+r^2d\theta^2)+e^{(-4\eta/3)}r^2 \, (\sin\theta d\phi+\xi d\theta)^2]. \tag{7}$$

Here η is the even-parity and ξ is the odd-parity radiative variable. Asymptotically, $r\eta \Longrightarrow h_+$ and $r\xi \Longrightarrow h_\times$. Note that this code is restricted to axisymmetry, equatorial plane symmetry, and (in the version used for this work) no rotation [ξ=0]. In this setting we will be testing the extraction techniques only for even-parity quadrupole waveforms. Maximal slicing $K^i_i = 0$ provides an elliptic equation for the lapse on each timeslice. The gauge conditions lead to an elliptic system for the components of the shift vector β^r and β^θ. The conformal factor is determined by solution of the Hamiltonian constraint equation. The radiative variable η is evolved as are the matter variables. The neutron-star pulsation runs shown here were performed with the fully constrained version of the code described by Evans (this volume). Since the boundary conditions in the new version are still under development, some small amount of wave reflection off the outer boundary is picked up by the radiation extraction procedures. Our experience has been that the extraction techniques can play an important role in code diagnostics.

The initial data for the oscillating neutron star evolution are established as follows. First an equilibrium stellar model with the desired central energy density w_c is set up by solving the Tolman-Oppenheimer-Volkoff equation. The matter in the star is perturbed with a low amplitude quadrupole displacement. Initial data for the gravitational field includes the assumption K_{ij}=0 (time symmetry) with a conformally-flat three-metric. For the preliminary test runs a sinusoidal waveform was desired so that the effects of the wave's near zone field could be compared easily with predictions of the linear analysis. For sinusoidal oscillations we wish to excite only the f-mode, so an attempt is made to the shape of the f-mode eigenfunction using radial profiles of Detweiler and Ipser (1973). To test the matching procedure in a more realistic (and interesting) situation we also desired non-sinusoidal waveforms. For these oscillations we make a different guess of radial profile which we found strongly excited the p_1 mode as well as the fundamental mode. The equation of state used for both runs is that of Harrison, Wheeler, and Wakano (see Hartle and Thorne 1968) with a central mass-energy density of w_c=1.0×10^{15} $g cm^{-3}$. Our numerical grid had 180 radial zones and 18 angular zones (per π/2). The simulations were run for 650M - enough time for several wavelengths to cross the outermost extraction radius (150M).

Employing a method described by Evans (1986) we checked the accuracy of our simulations against perturbation theory calculations of Lindblom and Detweiler (1983). For f-mode evolution the frequencies and damping times agree with perturbation theory to about 2%.

Numerical implementation of the techniques is reasonably straightforward. The multipole amplitudes are calculated numerically as surface integrals of metric components multiplied by angular functions. In quasi-isotropic gauge the amplitude $A_{24,2}$ is identically zero; this was maintained numerically to a level of about 10^{-8}. In weak-field ${}^{(4)}R_{\hat{t}i\hat{t}j}={}^{(3)}R_{ij}$ so the necessary components for the Riemann method are easily calculated although they require two spatial derivatives of metric components in contrast to one derivative for the gauge-invariant method and none for the template method (mass quadrupole extraction in quasi-isotropic or radial gauges). Note that for testing the gauge-invariant and Riemann methods we assume nothing about the quasi-isotropic gauge other than the form of the three-metric, i.e., it is not necessary to do any detailed analysis to implement these two methods in a code. The second-order ODE's arising from the matching procedures are split into first order form and integrated with the code timestep using a second-order accurate scheme (Anderson and Hobill 1986). For the neutron-star pulsation evolutions we place numerical detectors at five radii: $\lambda\!\!\!^{-}/2$, $\lambda\!\!\!^{-}$, $2\lambda\!\!\!^{-}$, $4\lambda\!\!\!^{-}$, and $6\lambda\!\!\!^{-}$ where $\lambda\!\!\!^{-}$ is the reduced wavelength of the fundamental-mode of oscillation ($\lambda\!\!\!^{-}\sim 23M$ in code units $1M=1M_\odot$).

To illustrate the necessity of our radiation extraction techniques we will first show waveforms obtained directly from the radiative variable η. Figure 1a shows an equatorial slice of $r\eta$ at the five different extraction radii as a function of retarded time $t-r_0$ for the sinusoidal oscillations. Figure 1b is the same plot for the non-sinusoidal oscillations. Clearly there is radially dependent contamination of the waveforms both in amplitude and phase. It is difficult to see *any* convergence towards an asymptotic waveform even when looking at the time series from $r_0=4\lambda\!\!\!^{-}$ and $r_0=6\lambda\!\!\!^{-}$. If anything, there is an apparent divergence as one looks further out into the wave zone (at least in the range $\lambda\!\!\!^{-}/2\leq r_0\leq 4\lambda\!\!\!^{-}$)!

With the template method we can directly show how pure-gauge terms contaminate the numerical method. Quasi-isotropic gauge has such a term which arises as a homogeneous solution to the gauge transformation equation for $l=2$. This gauge term b_2 is independent of radius in the weak-field region around the source. We can calculate this term using the $A_{20,2}$ multipole amplitude and the first and second time derivatives of the extracted quadrupole moment. Figure 2 shows b_2 at the inner four extraction radii as a function of time. Obviously this part of the metric perturbation is purely a function of coordinate time and is independent of radius and therefore unphysical. The $b_2(t)$ gauge term is important because it appears in the radiative variable η in both quasi-isotropic and radial gauges (in radial gauge it has dependence on r as well). Using the template procedure we find that in quasi-isotropic gauge the variable η has the weak-field expansion (including only $l=0$–2)

$$\eta=\sin^2\theta\left[\frac{I^{(2)}}{r}+2\varepsilon\frac{I^{(1)}}{r^2}+\varepsilon\frac{b_2}{r^2}\right]. \tag{8}$$

Clearly the mass-monopole is absent but there is a near-zone field term and a gauge term. As we have a time series for $b_2(t)$ we can subtract off this term to leave only radiation and near-zone field terms. In Figure 3 we plot $r\eta/\sin^2\theta-b_2(t)/r$ as a function of retarded time for the five extraction radii. Now it is obvious that the most significant contamination in both phase and amplitude is at $r_0=\lambda\!\!\!^{-}/2$ and the waveforms converge as one moves out of the near zone and into the local wave-zone. It is also clear that, although the $r\eta$ waveforms for $r_0=\lambda\!\!\!^{-}/2$ in Figures 1a and 1b had fairly correct amplitudes (within 20% of the asymptotic value), that "accuracy" is coincidental: due to the coordinate wave not the physical one.

Now we will demonstrate how the extraction procedure eliminates the effect of the wave's near-zone field. Let us consider the gauge-independent amplitude $Q_{e(l=2)}(t;r)$. As can be seen from Eq. (5), it should tend asymptotically to the purely radiative part $I^{(2)}(t-r)/r$. For a sinusoidal waveform it is possible to analytically calculate the effect of the near-zone field terms on $Q_{e(l=2)}(t;r)$. If the asymptotic waveform is

Figure 1a. The mass quadrupole waveform from the sinusoidal neutron-star oscillation run is shown as a function of retarded time (in M) as obtained at five radii from the radiative variable η.

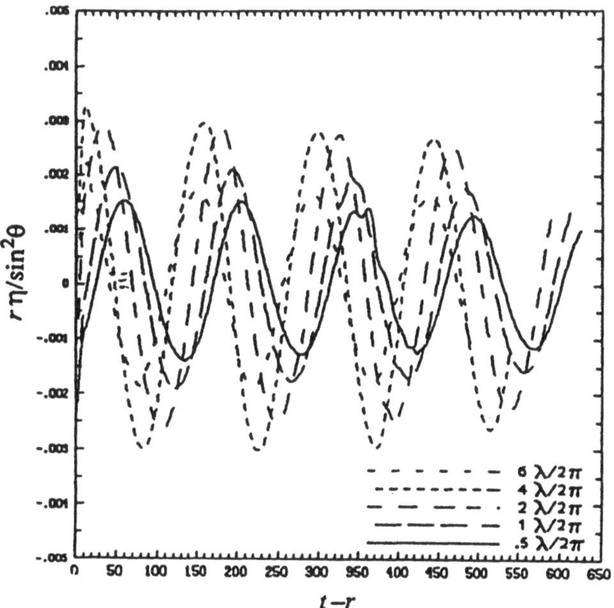

Figure 1b. Waveforms from the non-sinusoidal neutron star oscillations are shown as in 1a.

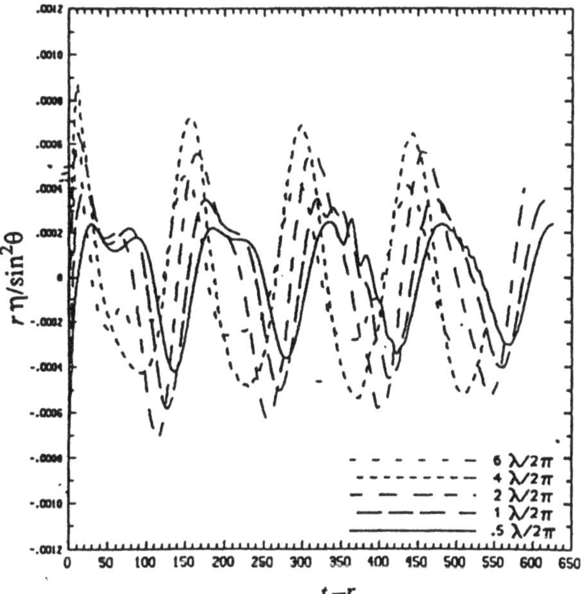

Figure 2. The mass quadrupole gauge term $b_2(t)$ is shown as a function of time as extracted with the template method.

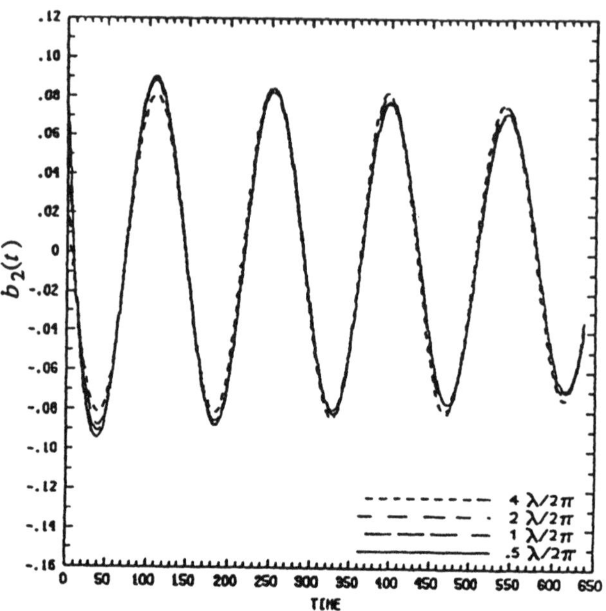

Figure 3. Here waveforms from the sinusoidal oscillation run corrected for the pure-gauge term are plotted as a function of retarded time.

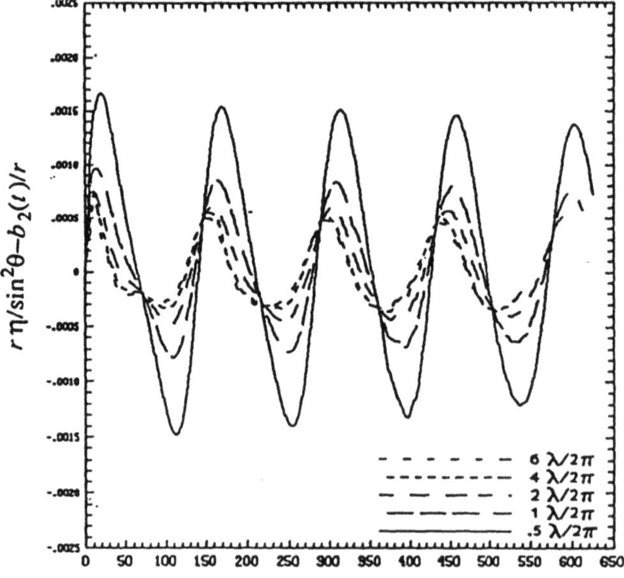

assumed to have the form $I^{(2)}/r = I_\infty \sin(k(t-r))/r$, and we plug this expression into (5) then $Q_{e(l=2)}(t;r)$ is expected to have the dependence

$$rQ_{e(l=2)}(t;r) = I_\infty \left[1+3(\frac{\lambdabar}{r})^2+9(\frac{\lambdabar}{r})^4\right]^{1/2} \sin(\frac{(t-r)}{\lambdabar}+\psi) \tag{9}$$

where the phase shift is given by $\psi = \tan^{-1}(\,3r\lambdabar[3\lambdabar^2-r^2]^{-1}\,)$. Figure 4 compares the analytic prediction for $r\,Q_{e(l=2)}(t;r)/I^{(2)}(t-r)$ as a function of r for three values of retarded time $(t-r)/\lambdabar$ against the same ratio calculated during the simulation at the five extraction radii. For the numerical comparison $Q_{e(l=2)}(t;r_0)$ is calculated directly from multipole amplitudes and $I^{(2)}(t-r_0)$ is the extracted value from solution of the ODE. Clearly, the agreement is excellent even at $r_0 = \lambdabar/2$ - deep in the near zone. This indicates that our exterior solution is a good approximation in the region $\lambdabar/2 \le r_0 \le 6\lambdabar$ and that our ODE integration routines are working correctly.

To complete our evaluation we now consider the ability of the template matching procedure to read off $I^{(2)}(t-r_0)\,/\,r_0$ consistently at all the extraction radii. The extracted waveforms should all agree in both amplitude and phase (once correction for retarded time is made). In Figure 5a we show the quadrupole waveform at four extraction radii extracted with the template method for the f-mode evolution. It is seen that there are only very small differences (< 5%) between the waveform extracted at $r_0=\lambdabar/2$ (where the near-zone field contribution to the gauge-independent amplitude is 93%) and that extracted well out in the wave zone at $r_0=6\lambdabar$. In Figure 5b we plot $I^{(2)}(t-r)$ for the nonsinusoidal oscillation run extracted with the gauge-invariant method. The agreement between different curves is comparable to that in 5a.

We turn briefly to an application to pure-radiative spacetimes. An interesting laboratory for studying non-linearities in general relativity is the implosion, self interaction and explosion of toroidal "Brill" waves (Brill 1959, Wheeler 1962). A incoming, pure l=2, weak Brill wave (solution from Teukolsky 1982 transformed into quasi-isotropic gauge by Evans 1986) is set up near the outer boundary and evolved using a vacuum version of Evans' code. For this problem the only modification of the extraction procedure is that those routines are not "turned on" until the wave has hit the origin at which point initial values are determined and integration commences. Figures 6a and 6b show a comparison of waveforms obtained with $r\eta/\sin^2\theta$ and with the Riemann procedure at five radii: λ, 2λ, 3λ, 4λ and 6λ, where $\lambda=35M$ is the characteristic wavelength of the sandwich wavepacket. This run is for a very non-linear Brill wave with an amplitude about 0.7 of the black hole forming amplitude. The central lapse goes down to ~.5 when the wave peak reaches the origin. Because of the spatial extent of the interaction region and the short wavelength of the waves, the near zone is very small so most of the wave contamination is due to pure-gauge terms. Both effects appear to have been eliminated correctly in 6b. The main error in the extracted waveforms is a phase discrepancy due to the time changing background from the waves. For the neutron-star oscillation problem we can remove this phase error (a consequence of the logarithmic divergence of our flat-space null cones from the true Schwarzschild ones) but that is much more difficult to do for the Brill wave case.

Figure 4. The accuracy of the analytic-numerical matching is demonstrated in this plot. The ratio $rQ_{e\,(l=2)}/I^{(2)}$ is plotted as a function of radial coordinate (in M) for three values of $(t-r)/\lambdabar$. The solid lines are analytic predictions and the points are numerical values obtained with the gauge-invariant method at the five extraction radii. At $r=\lambdabar/2$ and $t-r=\pi\lambdabar/4$ the near zone contribution to the waveform is about 14 times that of the asymptotic wave itself.

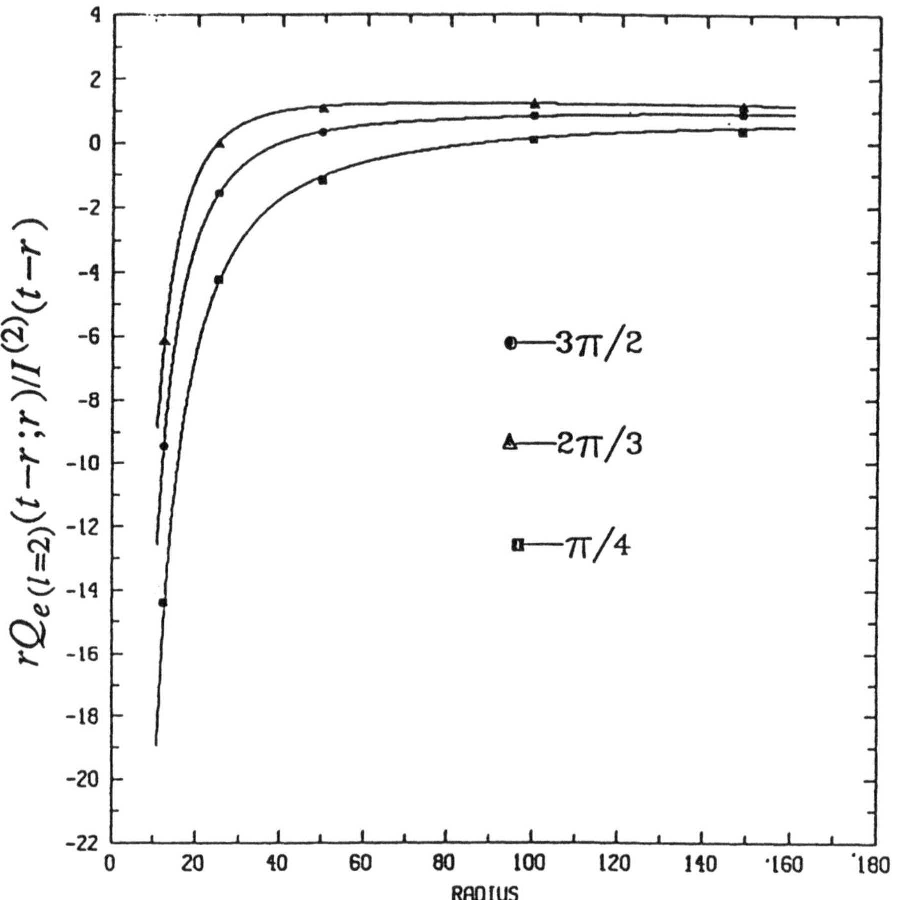

Figure 5a. The asymptotic mass quadrupole waveforms from the sinusoidal oscillation extracted with the template method at four radii are plotted. These should be compared with the waveforms in Figure 1a.

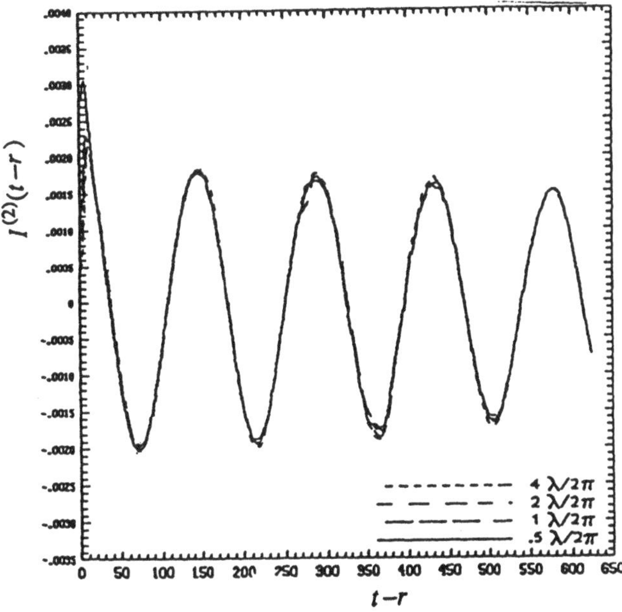

Figure 5b. The asymptotic mass quadrupole waveforms from the non-sinusoidal oscillation extracted with the gauge-invariant method are shown as a function of retarded time. These waveforms should be compared with those in 1b.

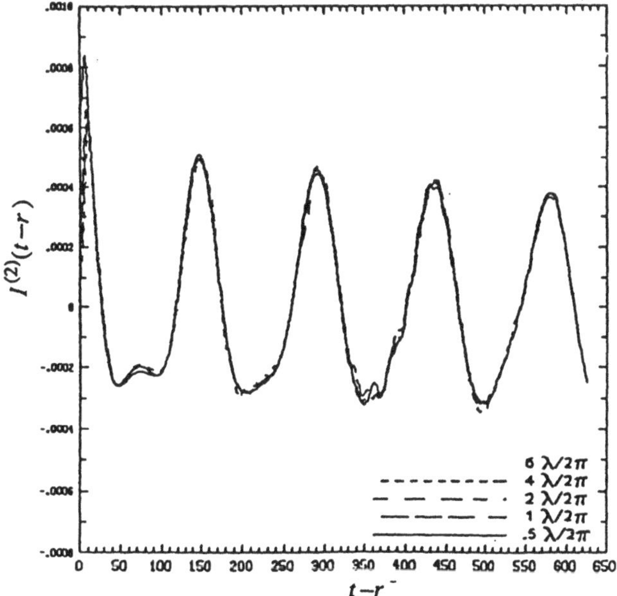

Figure 6a,b. Waveforms extracted with the radiative variable η (6a) and with the Riemann method (6b) for an imploding-exploding quadrupole Brill wave are compared.

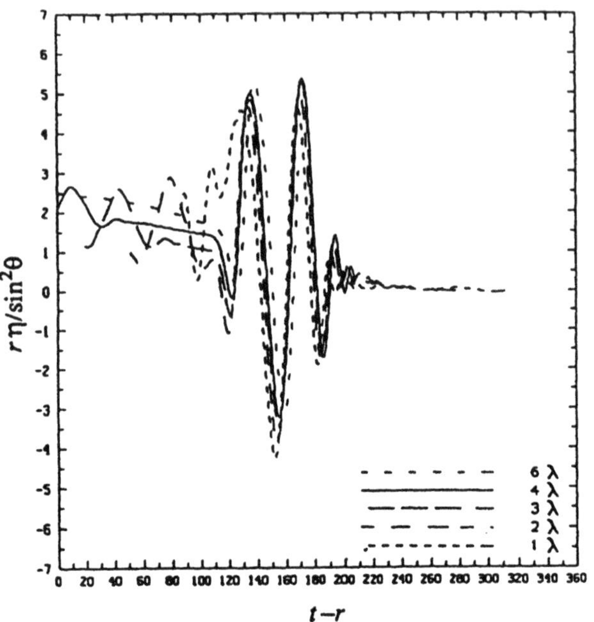

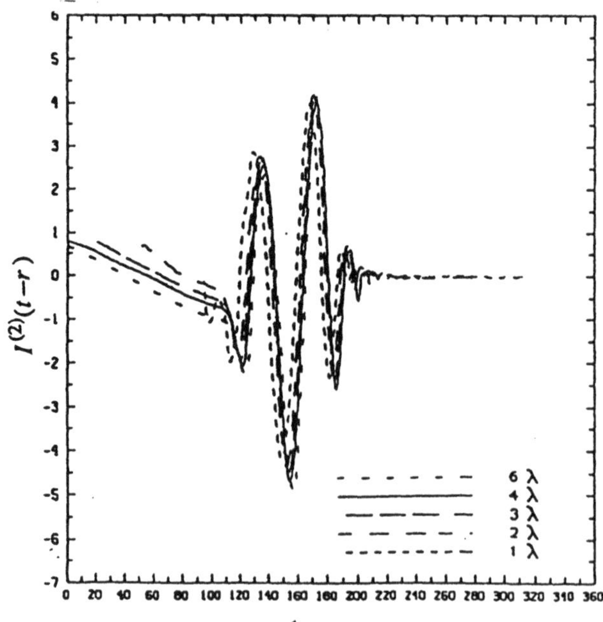

DISCUSSION

From a theoretical standpoint the logical direction to take now is to make the matching solutions more detailed. The framework described here has, we believe, closed the linear case for most practical purposes. To bring numerical detectors closer to the source - deeper into the near zone - will necessitate the inclusion of curved background and perhaps wave-wave interaction effects. Possible formalisms for accomplishing this include the Anderson-Hobill and Damour approaches discussed earlier. Moncrief's gauge invariant formalism provides a way to match an interior numerical solution onto solutions to the Zerilli or Regge-Wheeler equations at finite radii. The appropriate equation could be integrated numerically as was done by Cunningham, et al. (1978, 1979) and Seidel and Moore (1986) using the simulation data as an inner boundary condition.

The nice thing about the linearized gravity techniques, however, is their generality. No assumptions need be made about the detailed source structure. All that is required is that there be a weak field region around the source where space may be considered flat. It is not particularly clear that it is better to approximate the background of two coalescing black holes as Schwarzschild than it is to match on to flat space at a given radius. Additionally one must consider whether the beautiful results for ideal, asymptotically flat spacetimes are really relevant for the real universe. The practical approach seems to be to obtain a good approximate solution for your radiation field at a finite radius around a source and use the geometric optics approximation to propagate the waves to the earth.

Is it actually desirable to go any closer to the source? The purpose of these codes is to solve the full Einstein equations and we don't want to make any approximations that might lose or distort the physics in a critical region. The linear techniques are pretty safe because the extraction can be done in a consistent way to guarantee that the approximations are good in a given regime. The analysis is complete enough that there are several independent checks. Other nice features of the linear formalism may be lost by going to higher level techniques. Given initial data, all the information needed to extract radiation given be calculated on one time-slice from the spatial 3-metric. Integrals are performed over time-like cylinders but no derivatives across time slices are taken. We feel that this is aesthetically "right" for radiation extraction in the 3+1 picture. Another priority of this research is not only to be able to extract radiation with certainty but also to improve code boundary conditions to reflect information obtained through matching. It is not clear how this would be done with the null-cone matching techniques without performing full numerical, non-linear coordinate transformations between code and solution variables.

We look forward to the application of these techniques to some of the array of problems discussed in this meeting. Given the trend in numerical relativity towards precision and realism, we believe that analysis of this sort will be increasingly crucial to our understanding of geometrodynamical and astrophysical spacetimes. This research is described in more detail in Abrahams (1988) and Abrahams and Evans (1988a, b). The work was carried out in collaboration with C.R. Evans. It was completed as a doctoral dissertation at the University of Illinois under the advisorship of Larry Smarr. Support was provided by NSF grants PHY83-08826 and AST85-14911 and by the National Center for Supercomputing Applications.

Abrahams, A.M. (1988). Ph.D. Thesis, University of Illinois .

Abrahams, A.M. and Evans, C.R. (1988a). Phys. Rev. D, 37 ,318.

Abrahams, A.M. and Evans, C.R. (1988b). To be submitted to Phys. Rev. D. .

Anderson, J. and Hobill, D. (1986). in Dynamical Spacetimes and Numerical Relativity, ed. J. Centrella Cambridge: Cambridge University Press,

Bardeen, J.M. (1983). in Gravitational Radiation, eds. N. Dereulle and T. Piran, Amsterdam: North Holland.

Bardeen, J.M. and Piran, T. (1983). Physics Reports, 96 , 205.

Blanchet, L. and Damour, T. (1988). Phys. Rev. D., 37 , 1410.

Brill, D. (1959). Ann. Phys. 7 , 466.

Burke, W.L. (1971). J. Math. Phys., 12 , 401.

Chandrasekhar, S. (1983). The Mathematical Theory of Blackholes. New York: Oxford University Press.

Cunningham, C.T., Price, R.H., and Moncrief, V. (1978). Astrophys. J., 224 , 643.

Cunningham, C.T., Price, R.H., and Moncrief, V. (1979). Astrophys. J., 230 , 870.

Damour, T., in Proceedings of the Fourth Marcel Grossman Meeting on General Relativity, edited by R. Ruffini (in press).

Detweiler, S.L. and Ipser, J.R. (1973). Astrophys. J., 185 , 685.

Dykema, P.G. (1980). Ph.D. thesis, University of Texas.

Edmonds, A.R. (1974). Angular Momentum in Quantum Mechanics. Princeton: Princeton University Press.

Evans, C.R. (1984). Ph.D. thesis, University of Texas.

Evans, C.R. (1986). in Dynamical Spacetimes and Numerical Relativity, ed. J. Centrella, Cambridge: Cambridge University Press.

Hartle, J.B. and Thorne, K.S. (1968). Astrophys. J., 153 , 807.

Janis, A.I. and Newman, E.T. (1965). J. Math. Phys., 6 , 902.

Lindblom, L. and Detweiler, S.L. (1983). Astrophys. J. Supp., 53 , 73.

Mathews, J. (1981). Tensor Spherical Harmonics. Pasadena: Caltech Graphic Arts.

Moncrief, V. (1974). Ann. Phys., 88 , 323.

Schutz, B.F. (1986). in Dynamical Spacetimes and Numerical Relativity, ed. J. Centrella, Cambridge: Cambridge University Press.

Smarr, L. (1975). Ph.D. Thesis, University of Texas.

Smarr, L. (1977). Annals of the New York Academy of Sciences, 302 , 569.

Smarr, L. (1979). in Sources of Gravitational Radiation, ed. L. Smarr, Cambridge: Cambridge University Press.

Smarr, L. and York, J.W. (1978a). Phys. Rev. D., 17 , 1945.

Smarr, L. and York, J.W. (1978b). Phys. Rev. D., 17 , 2529.

Teukolsky, S.A. (1982). Phys. Rev. D., 26 , 745.

Thorne, K.S. (1980). Rev. Mod. Phys., 52 , 299.

Thorne, K.S. (1983). in Gravitational Radiation, ed. N. Dereulle and T. Piran, Amsterdam: North Holland.

Wheeler, J. (1983). in Relativity Groups and Topology, ed. B. De Witt and C. De Witt, New York: Gordon and Breach.

Zerilli, F.J. (1970). J. Math. Phys., 11 , 2203.

Supernovae, Gravitational Radiation, and the Quadrupole Formula

L. S. Finn
Newman Laboratory of Nuclear Science, Cornell University, Ithaca, New York 14851, and Institute for Theoretical Physics, University of California, Santa Barbara, California 93106

Abstract. Historically, supernovae have been considered among the most promising sources of detectable gravitational radiation; also historically, calculations of supernovae gravitational radiation efficiencies and waveforms have been subject to wide variation. Here we report on work in progress toward a better understanding of supernovae radiation efficiencies and waveforms. First, we review the supernova scenario, identifying those features of special importance to gravitational wave production. Next, we consider a Newtonian simulation of core collapse, and discuss three innovations that permit a more accurate determination of the quadrupole gravitational radiation than the standard quadrupole formula. Each of these variations is generally applicable to Newtonian sources of gravitational radiation. The standard quadrupole formula and each of the variations are tested and compared numerically on the problem of computing the gravitational radiation from rotating stellar core collapse. We find that for numerical finite difference calculations, the standard quadrupole formula is inferior to all three of the variations explored. In addition to their role in determining the gravitational radiation, the four waveform extraction formulae can be used to test the implementation and accuracy of the finite difference equations.

1. Introduction

1.1. Motivation

The next twenty years should see the birth of gravitational wave astronomy, and with it the potential for a revolution in the way we view the universe. Owing to its weak coupling to matter, a detectable burst of gravitational waves is produced only in violent events where high velocities and large matter densities predominate; and, once produced, these bursts propagate without dispersion, scattering, or attenuation. Consequently, the detection of gravitational waves will open an entirely new window onto the universe: one that complements existing windows in the information it will provide.

Type II supernovae provide an example of how gravitational wave observations can provide valuable and complementary information compared to observations made in other windows. Of the windows available to us, only neutrino radiation provides us with a way of exploring the interior of a collapsing stellar core; however, even neutrinos are scattered at the densities involved in core collapse. Consequently, we are really only able to see to the depth of the neutrino-sphere, and this leaves the innermost parts of the stellar core invisible to us. Furthermore, diffusion of neutrinos from the site of production in the inner core to the neutrino sphere takes place on a timescale longer than the collapse timescale, so what we observe in neutrinos does not allow us to resolve in time the actual events in the collapse or bounce of the core. Gravitational radiation, on the other hand, is generated by the coherent motion of the inner core and is quite sensitive to the maximum density achieved in the collapse and the timescale upon which the coherent motion of the inner core takes place. As it propagates to our detectors without further interaction, it provides a record undistorted in time of the bulk properties of the innermost regions of the stellar core.

In order to interpet the results of gravitational wave observations of stellar core collapse, we need accurate predictions of waveforms for various collapse models. Previous calculations of gravitational radiation efficiencies and waveforms have been subject to wide variation; in large part this variation is the result of refinements in the model of the collapsing star or the calculational techniques. Here we report on work in progress toward the generation of a waveform "catalog," to be based on the most accurate relevant model of stellar core collapse and state-of-the-art numerical simulation techniques.

The plan of this paper follows. The remainder of §1 describes the notations and conventions used throughout this paper. In §2, we briefly review the history of research on gravitational radiation from stellar core collapse. Section 3 describes the core collapse scenario — with particular emphasis on the features of the scenario important for the production of gravitational radiation — and summarizes the *physical* model upon which our calculations are built. In §4, we report on several new techniques for extracting quadrupole gravitational radiation from Newtonian or nearly-Newtonian sources. Conclusions are presented in §5.

1.2. Notations and Conventions

Generally we use the conventions of Misner, Thorne, & Wheeler (1973) when dealing with spacetime. In particular, the metric signature is $(-+++)$, greek indices run from 0–3, latin indices from 1–3, and we work in units where $G = c = 1$. For the most part, however, we work in flat space and in one of two orthonormal basis: either cartesian or spherical polar.

When we work in flat space and orthonormal bases, indices are raised and lowered with the Minkowskii metric, which takes the form $\eta = \mathrm{diag}(-1,+1,+1,+1)$. The Einstein summation convention is followed, with one modification: repeated latin indices, whether subscript or superscript are summed over as if the metric were present, so that

$$S_i T_i = S^i T^i = S_i T^i = S_i \delta^{ij} T_j \tag{1.1}$$

(as we are working in an orthonormal basis, the spatial part of the metric reduces to the Kronecker delta δ^{ij}).

In §4 we shall be considering multipolar expansions of various quantities. We shall express these expansions in terms of Symmetric and Trace Free (STF) tensors, using the formalism developed by Thorne (1979, and henceforth RMP). In that formalism, a STF tensor of order l is an order l spatial tensor that is symmetric under the interchange of, and vanishes when traced on, any pair of indices. We denote the operation of taking the symmetric and trace free part of any spatial tensor by a superscript STF, and the resulting STF tensor will be denoted by an uppercase script letter:

$$T^{STF}_{a_1 a_2 \cdots a_l} = \mathcal{T}_{a_1 a_2 \cdots a_l}. \tag{1.2}$$

For convenience, a sequence of (say) l indices on an arbitrary spatial tensor is abbreviated as

$$T_{a_1 \cdots a_l} = T_{A_l}. \tag{1.3}$$

In connection with our multipolar expansions, we will need to refer to the radial vector, denoted x_i, and the unit radial vector, denoted n_i. Tensor products of the radial vector and the unit radial vector are denoted

$$\begin{aligned} X_{A_l} &= x_{a_1} x_{a_2} \cdots x_{a_l}, \\ N_{A_l} &= n_{a_1} n_{a_2} \cdots n_{a_l}. \end{aligned} \tag{1.4}$$

2. Gravitational Radiation From Stellar Core Collapse

Ruffini & Wheeler (1971) provide one of the earliest qualitative investigations of stellar core collapse as a source of gravitational radiation. They identified several processes (non-symmetric collapse, fragmentation, collision and coalescence of fragments to form a neutron star) thought to take place in the collapse of a stellar core, and provided order of magnitude estimates of the gravitational radiation from each.

Thuan & Ostriker (1974) undertook the first quantitative calculation of the gravitational radiation from the axisymmetric collapse of a stellar core. They calculated the (quadrupole) radiation from the collapse of oblate dust spheroids in Newtonian gravity, both with and without rotation, and found a maximum radiation efficiency (ratio of gravitationally radiated energy to the rest mass energy of the star) of approximately 10^{-3} for a maximally rotating configuration. Epstein & Wagoner (1975) and Epstein (1976a,b) derived the post-Newtonian corrections to the dust collapse considered by Thuan & Ostriker (1974), and found that the effect of the post-Newtonian terms was to prolong the collapse, thereby *decreasing* the gravitational radiation luminosity (which is a strong function of the timescale of the collapse). The post-Newtonian effects become important, of course, only when the collapsing spheroid becomes quite compact.

Refer for a moment to figure 1, which represents a typical gravitational radiation waveform from a modern Newtonian stellar core collapse calculation. Three regions of the waveform are identified: the *infall*, *bounce*, and *ring-down* regions. The calculation of Thuan & Ostriker (1974) modeled core collapse as a free-fall of dust; in fact, during the early stages of collapse (the *infall* region in figure 1) the functional dependence of the waveform is quite well described by this radical approximation.

Novikov (1975) first considered the role of internal pressure in the collapse and the production of gravitational radiation, arguing that for rapidly rotating spheroids, the pressure-induced abrupt deceleration of a collapsing stellar core at nuclear densities would lead to an enhanced gravitational radiation luminosity. Shapiro (1977) demonstrated this point quantitatively, performing a series of calculations of the quadrupole gravitational radiation from the axisymmetric collapse of uniformly rotating, homogeneous, Newtonian spheroids with a polytropic internal pressure. An important result of these calculations was to establish the existence of a critical angular momentum J_C for each one-parameter sequence of spheroids (parameterized by angular momentum, with mass and equation of state held constant). Along any such sequence, the gravitational radiation luminosity is maximized for the model with the critical angular momentum, and falls to zero as the angular momentum moves away from the critical value in either direction. This result is understood as reflecting the role of both pressure and rotation in the collapse: for small angular momentum $J \ll J_C$ the pressure serves to symmetrize the spheroid, reducing the asymmetry responsible for the radiation, while for large angular momentum $J \gg J_C$, the centrifugal acceleration prevents the spheroid from undergoing any significant collapse, also reducing the radiation.

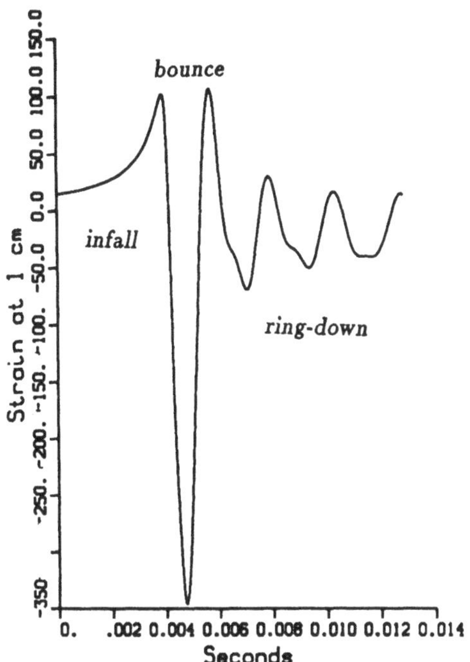

Figure 1. A typical example of a gravitational waveform that is expected to result from an axisymmetric type II supernova explosion. For the purposes of this discussion, the scale of the waveform is arbitrary.

Shapiro (1977) investigated the particular effects of two different equations of state then popular among supernova theorists. In the "cold" equation of state pressure support comes from degenerate electrons, and neutrinos from inverse beta decay during the collapse are assumed to freely stream out from the star. In the "hot" equation of state, the neutrinos are assumed to be trapped in the core during the collapse, leading to a significant thermal pressure component. For the cold collapse a maximum efficiency of approximately $2 \times 10^{-3}[M/(1.4\,M_\odot)]^{14/3}$ was found, while for hot collapse the maximum efficiency was approximately $4 \times 10^{-6}[M/(1.4\,M_\odot)]^{14/3}$. Saenz & Shapiro (1978) considered prolate configurations (representing the effects of a toroidal magnetic field) as well as oblate (owing to rotation and poloidal magnetic fields) configurations.

The initial investigations of Shapiro (1977) and Saenz & Shapiro (1978) described collapse and bounce in the absence of shocks and other dissipative mechanisms; consequently, the *ring-down* behavior of the gravitational waveform was not determined. Nevertheless, the inclusion of pressure is sufficient to provide an understanding of the second region of the gravitational waveform, the *bounce* region (*cf.* figure 1), where pressure forces and centrifugal accelerations become sufficient to slow and then halt the collapse.

Saenz & Shapiro (1979, 1981) continued to investigate the gravitational and neutrino radiation from collapsing spheroids with increasingly sophisticated equations of state and *ad hoc* prescriptions to model the effects of shocks and other dissapative mechanisms. Meanwhile, Turner & Wagoner (1979) pursued a different path: they developed a second order, time dependent perturbation theory to describe the evolution of a small rotational perturbation (and the resulting rotational deformation) on spherical core collapse. Applying this theory to the results of (then) state of the art numerical calculations of spherical core collapse (Wilson 1977, Van Riper 1978), they evaluated the gravitational radiation luminosity as a function of the small rotation. Since their results derived from a perturbation theory they were only accurate in the slow rotation limit, and could not, for example, determine

J_C and the corresponding maximum gravitational radiation luminosity. Nevertheless, their results (along with the later results of Saenz & Shapiro 1979 and 1981) began to present a picture of the *ring-down* waveform (*cf.* figure 1).

Müller (1982) evaluated the quadrupole gravitational radiaton from the Newtonian collapse models of Müller & Hillebrandt (1981). The Müller & Hillebrandt (1981) collapse calculations were fully two dimensional, included angular momentum, and had a realistic microphysical equation of state. Müller (1982) did not survey parameter space; however, his results indicated that differential rotation (as opposed to uniform rotation) could play an important role in gravitational collapse, enhancing by a factor of 10 or more the radiation efficiency of a collapsing star. Müller's (1982) calculated radiation efficiencies were also lower by a factor of 10 than had been previously calculated (Shapiro 1977, Saenz & Shapiro 1978).

All of the previous results have relied upon Newtonian (or post-Newtonian) calculations, with the gravitational radiation determined through the quadrupole formula. More recently, there have been several attempts at a general relativistic treatments of stellar core collapse. Stark & Piran (1986) investigated the collapse of pressure undercut rotating polytropes with a 2.5-D general relativistic code, focusing on the formation of a black hole and the resulting gravitational radiation. They found that, after the formation of the hole, the waveform could be very accurately described as a superposition of the first two black hole normal modes. For cases where a black hole did not form, the waveform was significantly more complicated and sensitive to their choice of initial model.

A different approach to a general relativistic treatment of core collapse and bounce has been taken by Seidel (1987) and Seidel & Moore (1988, this volume). They have developed a first order, general relativistic perturbation theory suitable for application to the spherically symmetric, dynamical general relativistic collapse models of May & White (1966, 1967). With this theory, they are able to evaluate the *current quadrupole* gravitational radiation for general relativistic collapse with slow differential rotation. Seidel & Moore are working toward extending their perturbation theory to encompass the more significant mass quadrupole radiation.

Axisymmetric collapse is a source of gravitational radiation; however, it is nowhere near as powerful a source as non-axisymmetric collapse is likely to be. Detweiler and Lindblom (1981) modeled the gravitational collapse of a stellar core as the evolution of a sequence of non-axisymmetric ellipsoids. By restricting themselves to ellipsoids whose angular momentum was about a principal axis of rotation, they lowered the total luminosity and bandwidth of the radiation; nonethelesss, they found that the energy radiated owing to non-axisymmetry was of the same order as that from axisymmetric collapse as described above. Since the radiation is confined to a very narrow bandwidth (and this would still be the case even if the rotation axis was not aligned with a principal body axis), the prospects for *detection* are considerably higher for this radiation, however (*cf.* Thorne 1987).

All other things being equal, the greater the radiation luminosity, the greater the detectability of the radiation. All other things are rarely, however, equal: as estimates of the radiation from supernovae are refined, not only does the total energy radiated change, but so does the frequency and bandwidth of the signal, and all of these factors affect the detection strategy and ultimate detectability of the radiation as measured by the signal-to-noise ratio in a real detector. Thus, even though the total luminosity expected from axisymmetric collapse has decreased as calculations and models have been refined, the detectability in laser interferometer gravitational wave detectors has remained roughly constant. For a further discussion of this point and other issues in the detection of gravitational radiation, as well as an excellent review of gravitational radiation, see Thorne (1987).

3. The Supernova Scenario

3.1. Introduction

Our eventual goal is an accurate determination of the gravitational radiation from supernovae explosions. In order to proceed toward that goal, we must consider the supernovae scenario, noting especially those aspects of the scenario that are relevant to the generation of gravitational radiation. An appropriate physical model for calculating the gravitational radiation from supernovae is one which includes all those details that are (either directly or indirectly) significant for the generation of the radiation, and excludes the rest. In this section, we examine the supernovae scenario, with particular attention paid to those details which are important for the generation of gravitational radiation. Most of the details of the supernova scenario are taken from the excellent review article by Woosley & Weaver (1986), to which the interested reader is directed for more detail.

Supernovae are observationally distinguished by their light curves into two classes. Generally, type I supernovae are believed to mark the thermonuclear explosion of white dwarf (Hoyle & Fowler 1960), and are distinguished by the absence of hydrogen spectral lines. Type II supernovae are believed to mark the death, by gravitational collapse, of more massive stars (Baade & Zwicky 1934), and are observationally distinguished by the presence of hydrogen spectral lines.

The exact nature of the thermonuclear burning that consumes the star in a type I supernova (detonation wave, deflagration wave, or some combination) is not yet determined (*cf.* Woosley & Weaver 1986 §3.1); however, the general energy budget of the combustion of the star is believed to be understood. The energy release in a type I supernova is thought to be about 10^{51} ergs (approximately 1MeV per nucleon). This energy release is very efficiently coupled to the fluid of the star, resulting in its disruption.

Type II supernova, on the other hand, are thought to result from the gravitational collapse of a massive stellar core. The collapse is either to a neutron star or a black hole, and results in the release of approximately 10^{53} ergs of binding energy. This energy is only weakly coupled to the fluid of the star: the bulk of the energy is released in neutrinos, with only a percent or less released in the optical display and the kinetic motion of the fluid.

The thermonuclear explosion that consumes the star in a type I supernova does not result in either the high densities or velocities that are typical of strong sources of gravitational waves. Additionally, the explosion is thought to proceed close to spherically symmetrically, which greatly lowers the production efficiency of gravitational radiation. In comparison, stellar core collapse does achieve high densities and velocities, and any initial magnetic fields or rotation in the collapsing stellar core will guarantee an aspherical collapse and so the production of gravitational radiation. Coupled with the greater energy available in core collapse as compared with thermonuclear combustion (10^{53} *vs.* 10^{51} ergs), it is apparent that type II supernovae are the more energetic sources of gravitational radiation. Since type II and type I supernovae have approximately the same event rate (Tammann 1982), we are much more likely to see a type II supernova with a gravitational wave detector. Henceforth, we will focus our attention on type II supernovae.

3.2. The Progenitor

A type II supernova begins its life as a massive star. The mass may range from approximately $8M_{\odot}$ to approximately $40M_{\odot}$ (Schild & Maeder 1983, Maeder 1984). Stars of mass less than $8M_{\odot}$ evolve either to form stable white dwarfs, or to a stage where carbon burns degenerately and explode in a type I supernova . Stars more massive than $40M_\odot$ may also undergo gravitational collapse; however, in the course of their evolution, these stars will have lost their hydrogen envelopes and so the resulting explosion will lack hydrogen spectral lines and will not be observationally identified as a type II supernovae.

It is useful to further resolve type II supernovae by mass. As the progenitor stars mass increases from 8 to $11M_\odot$, the evolving stellar core undergoes more and more complete nuclear burning: from carbon, through neon and oxygen, to silicon burning. The cores of stars in the mass range $11 \sim 20M_\odot$ undergo all the stages of nuclear burning and leave iron cores. Finally, progenitor stars with masses greater than $\sim 20M_\odot$ bypass carbon and neon burning, proceeding directly from helium to oxygen and silicon burning.

The degree to which nuclear burning proceeds in the evolving stellar core determines the cores entropy, and thus the mass of the collapsing core and the timescale for the collapse. Both of these are important in determining the gravitational radiation from stellar core collapse. Initially, a star has a uniform specific entropy of approximately 25 (in units of k_B). As the star evolves, first photons and then neutrinos carry entropy out from the core and into the envelope. By the time the star has evolved to the stage of collapse, the envelope entropy will be on order $40k_B$ while the core entropy will be on order $1k_B$. Consequently, the envelope (where the bulk of the mass of the star is) will be puffy and diffuse, while the core is quite condensed. For the purposes of studying the gravitational collapse of the star, and particularly the resulting gravitational radiation, we can ignore the envelope and focus attention on just the dense stellar core.

Since the later stages of nuclear burning take place on rapid timescales (of order days to hours for oxygen and silicon burning), the bulk of the entropy transfer takes places during the much longer early stages of nuclear burning (during the course of carbon and neon burning). Stars of greater than initially $20M_\odot$ bypass these stages of nuclear burning and have high entropy cores (specific entropy approximately 2.5–$3k_B$) compared to stars whose initial mass ranges from 8 to $20M_\odot$ (specific entropy approximately $1k_B$). The collapse scenario for these high entropy stellar cores is qualitatively different than for the low entropy stellar cores; henceforth we will focus attention on the collapse of low entropy stellar cores from progenitor stars in the mass range 8–$20M_\odot$.

3.3. The Pre-Collapse Stellar Core

A progenitor star whose mass ranges from 8–$20M_\odot$ will, at the end of its natural lifetime, have evolved a low entropy, degenerate stellar core. The core is supported primarily by degenerate electron pressure, with a small contribution from the finite entropy of the electrons. The specific entropy in the stellar core (on a per baryon basis) is approximately $1k_B$, and approximately half of that is in the degenerate electrons.

The maximum mass that the stellar core can achieve before collapsing is the Chandrasekhar mass. For a star supported by primarily by degenerate electrons and with a small electron entropy per baryon S_e and electron fraction Y_e, the Chandrasekhar mass is

$$M_{\rm Ch} = M^0_{\rm Ch}\left[1 + (S_e/\pi Y_e)^2\right], \tag{3.1}$$

where

$$M^0_{\rm Ch} = 1.255(Y_e/0.464)^2 M_\odot \tag{3.2}$$

is the Chandrasekhar mass for a cold white dwarf ($Y_e = 0.464$ for iron) (Bethe 1986). The corrections to the Chandrasekhar mass due to general relativity have been ignored here, since for iron white dwarfs other instabilities set in long before the general relativistic instability.

If the core were spherically symmetric, then the collapse would be spherically symmetric and there would be no gravitational radiation. Two mechanisms are primarily responsible for breaking spherical symmetry: rotation and magnetic fields. To date, neither rotation nor magnetic fields have been a part of any evolutionary calculations of stellar structure; consequently, it is not known what the magnetic dipole moment, the net angular momentum, or its distribution, is in the pre-collapse stellar core. Since the collapse of the stellar core occurs very rapidly, the angular momentum and the magnetic fields are conserved during the collapse; consequently, observations of young pulsars may provide us with some information on the pre-collapse rotation rates and magnetic fields. However, it is possible that some angular momentum is shed during the collapse and pulsar observations may only provide us with a lower bound. Regardless, there is still considerable controversy over the spin rates and magnetic fields characteristic of young pulsars to permit any kind of definitive answer to this question.

3.4. The Onset of Collapse

As nuclear burning proceeds, the ash of that combustion accumulates on the growing stellar core and the core density and temperature rises. Eventually, the temperature and density increase to the point that the core becomes marginally stable, and then unstable to collapse. One or both of two mechanisms are responsible for the instability to collapse: the first is inverse beta decay, and the second is nuclear photodissociation.

At a density of a few times $10^9\,\mathrm{g\,cm^{-3}}$, pressure support of the stellar core is provided almost entirely by the relativistic degenerate electron gas; consequently, the core is only marginally stable against collapse. With further compression, electron capture onto protons removes electrons and reduces the pressure support: thus, the core becomes unstable to gravitational collapse. This is the usual trigger mechanism for the collapse of stellar cores formed by progenitor stars in the mass range $8 \sim 15\mathrm{M}_\odot$, and shares the stage with nuclear photodissociation for stars in the mass range $15 \sim 20\mathrm{M}_\odot$. For stars more massive than $\sim 20\mathrm{M}_\odot$, nuclear photodissociation alone triggers the collapse.

The nuclear photodissociation instability sets in at a temperature of approximately $10^9\,{}^\circ\mathrm{K}$ (Bethe, Brown, Applegate, & Lattimer 1979). In the stellar core, photons are trapped and in thermal equilibrium with the electrons. As the temperature of the core increases, these trapped thermal photons dissociate an increasing number of nuclei into α's and neutrons, lowering the temperature of the photons, and thus in turn the electrons. This reduction in thermal pressure support lowers the Chandrasekhar mass, and when the Chandrasekhar mass falls below the core mass, the core becomes gravitationally unstable and collapses.

Though only one of these two mechanisms may be responsible for triggering the collapse, quickly after the collapse begins both inverse beta decay and nuclear photodissociation become important in continuing the collapse. At the time of collapse, the core mass for stars in the mass range $11 \sim 20\mathrm{M}_\odot$ is approximately the Chandrasekhar mass appropriate for an electron entropy of $\sim 1/2$ and an electron fraction appropriate to iron group nuclei (~ 0.42): $M_{\mathrm{Ch}} \simeq 1.2\mathrm{M}_\odot$. For less massive stars, the core mass is still the Chandrasekhar mass, but with a slightly larger core entropy and electron fraction, corresponding to the decreased time and incompleteness of the nuclear burning in less massive stars. For more massive stars, the core mass at the onset of the instability is higher (greater than or on order $2.2\mathrm{M}_\odot$, *cf.* Woosley & Weaver 1986) and is far from the low-entropy Chandrasekhar mass (the core entropy is much higher than in the lighter stars and thus so is M_{Ch}).

While the exact details of the initiation of the collapse are washed out during its course, it remains true that there is a qualitative difference between collapse initiation through the evolutionary development of an instability in the structure of the star and an *ad hoc* cut of the pressure support in a model polytrope. Goldreich & Weber (1980), in studying the homologous collapse of $\Gamma = 4/3$ polytropes, found that the mass of the homologous core was sensitive to the pressure deficit which initiated the collapse. Additionally, cutting the pressure of a rotating equilibrium configuration also excites oscillatory quadrupole and higher order pulsational modes, and in the ensuing collapse these modes can be adiabatically excited to significant amplitude, changing the course of the infall and the gravitational radiation therefrom.

3.5. *Collapse*

Once the core becomes gravitationally unstable, it collapses rapidly: on a timescale of 0.1 seconds. Regardless of whether nuclear photodissociation or inverse beta decay was responsible for triggering the collapse, very quickly both become important in determining the pressure support, and so the speed, of the collapsing core. The collapse occurs nearly adiabatically, with the specific entropy remaining nearly constant throughout. As the core contracts, it heats up; however, owing to the high heat capacity of excited nuclear states the temperature does not rise as rapidly as one would expect (Mazurek, Lattimer, & Brown 1979; Bethe, Brown, Applegate, & Lattimer 1979) and thermal pressure does not halt the collapse before nuclear densities are reached.

As the collapse proceeds, the presence of a finite sound speed and the development of an associated sonic point causes the core to segregate into an inner and an outer piece. The inner core collapses subsonically and homologously, with an inward radial velocity that is linear in r. The boundary of the inner core is the sonic point, where the inward radial velocity is equal to the sound speed of the fluid. Late in the collapse, the inner core encloses a mass of between 0.6 and $0.8 M_{\odot}$. In contrast, the outer core collapses supersonically and, because it is unable to maintain sonic communication with itself, does not collapse homologously. The radially inward fluid velocity of the outer core is proportional to $r^{-1/2}$, and the velocities are roughly half the free-fall velocities (the velocities in the outer core never reach free-fall because of the small but non-trivial remaining pressure support).

The gross details of the early stages of stellar core collapse are reasonably well approximated by the homologous collapse of a spherically symmetric $\Gamma = 4/3$ Newtonian polytrope (Goldreich & Weber 1980). Collapse in these polytropes is initiated by pressure reduction; corresponding to a pressure reduction typical of the instabilities that initiate collapse in real stellar cores (~26%, *cf.* Bethe, Brown, Applegate, & Lattimer 1979), Goldreich & Weber (1980) find the mass of the homologous core is approximately 70% of the initial mass of the star. This agrees quite well with the results of more detailed simulations. Accurate modeling of the collapse of the homologous core is quite important for determining the gravitational radiation: it is the homologous core that moves coherently throughout the collapse and bounce, and so is the main (gravitationally) radiating element in the supernova.

A second notable event in the core collapse is the trapping of inverse beta decay neutrinos. Initially, the neutrino products of inverse beta decay stream freely out of the core, decreasing slightly the core's specific entropy. At densities above approximately $10^{11}\,\mathrm{g\,cm^{-3}}$, however, the neutrino mean free path has decreased to the point that the neutrinos are effectively trapped in the core for the remainder of the collapse. One result of this neutrino trapping is that the collapse is now completely adiabatic: beta decay is now an equilibrium process, with the relative fractions of neutrinos, electrons, protons, neutrons, alphas, and nuclei determined by the Saha equation. Yet another result is that neutrinos become available as a source of pressure support; eventually, neutrinos provide approximately 15% of the pressure support in the collapsing core.

3.6. Bounce

The core continues to collapse, eventually reaching nuclear densities. At and above nuclear densities, the equation of state stiffens dramatically. Presuming the collapse does not proceed to the formation of a black hole, the inner core rapidly halts and, remaining in sonic communication throughout, coherently bounces, driving a shock wave out into the (still infalling) outer core. The shock begins as an accretion shock at the sonic point, and later begins to propagate outward through the outer core. Ultimately, the shock is believed responsible for the supernova optical display; however, initially the shock does not have sufficient energy to reverse the infall of the outer core. Several mechanisms are believed to play a role in re-energizing the shock (Colgate & White 1966; Wilson 1971; Wilson *et al.* 1975; Bethe & Wilson 1985; Baron, Cooperstein, & Kahana 1985; Wilson 1985), and our understanding of how the energy generated in the collapse is coupled to the shock is still incomplete. The passage of the shock through the outer outer core and the stellar envelope is unimportant in the production of gravitational radiation, however, and here we can ignore the controversy surrounding the shock. (If the shock stalls and is not revived, then subsequent accretion may cause the neutron star remanent to undergo further collapse to a black hole, with a corresponding burst of gravitational radiation. In this case, of course, the behavior of the shock is of critical importance; however, we shall not consider this possibility further here.)

Along with the details of the shape and matter distribution of the inner core, it is the depth of the bounce — the maximum central density reached during the collapse — and its timescale that are the important factors in determining the gravitational radiation from the supernova explosion. In turn, the maximum central density and timescale of the bounce depend sensitively on several factors: the speed of the collapsing core when nuclear density is reached, the nature of the nuclear equation of state, and the gravitational potential numbering prominently among them. Evans (1984) has shown that the difference between a Newtonian gravity calculation of spherical core collapse and a general relativistic calculation can be a factor of 3 or more in the maximum central density reached.

3.7. The Post-Bounce Stellar Core

After the collapse, the neutron star remanent will oscillate and eventually settle down to a stable configuration. This period of oscillation will have a characteristic gravitational radiation emission with several components. One component corresponds to the neutron star pulsational period and depends on the central density of the remanent core. A second component corresponds to any non-axisymmetric motion, whether it is due to a non-axisymmetric neutron star remanent, or a misalignment between the rotation axis and bodies symmetry axis. This second component has a frequency determined by the rotational frequency of the neutron star remanent.

3.8. Summary

To summarize, we conclude that type II supernovae, because they proceed through the mechanism of gravitational collapse, are more significant than type I supernovae as gravitational wave sources. The progenitors of type II supernovae are stars with mass in the range $8 \sim 40 M_\odot$. This mass range of progenitors can be further subdivided by the nature of the stellar core that evolves from the progenitor. Progenitors in the mass range $11 \sim 20 M_\odot$ evolve degenerate iron cores with low specific entropy; in progenitors of the mass range $8 \sim 11 M_\odot$ the combustion stops before iron is reached, while in progenitors of greater than $\sim 20M_\odot$ several stages of nuclear burning are bypassed and a high entropy iron core develops. The high entropy cores are significantly more massive, and collapse more slowly and by different mechanisms than the low entropy cores, and consequently are probably not as important sources of gravitational radiation as are the low entropy cores.

The low entropy cores are surrounded by high entropy, diffuse envelopes that, from the point of view of gravitational radiation, can be ignored in studying the gravitational collapse of the star. The core mass will grow until it approaches its Chandrasekhar mass (which depends on the core's composition through its electron fraction and specific entropy), whereupon the core becomes unstable to gravitational collapse. The gravitational radiation from the collapse is determined by the asphericity of the collapse, which is controlled by the core's angular momentum and magnetic fields; however, at present these are not known within wide limits.

Collapse proceeds rapidly and nearly adiabatically. The process of collapse causes the core to segregate into a coherently collapsing inner core and a supersonic outer core. Neutrino trapping during the collapse increases the pressure support significantly, but the core continues to collapse until nuclear densities are reached. The collapse is reversed by the sudden stiffening of the nuclear equation of state, and the inner core bounces coherently. It is the sudden deceleration and coherent motion of the inner core that is responsible for the bulk of the gravitational radiation. The details of the gravitational radiation are sensitive to the depth and timescale of the bounce, and so to the equation of state for densities above $\sim 10^9\,\mathrm{g\,cm^{-3}}$. Additionally, the calculation must ultimately be carried out using general relativistic gravity.

Finally, the post-collapse remanent will undergo oscillations which will be characteristic of the mass and angular velocity of the remanent.

4. Four Waveform Extraction Formulae

4.1. Introduction

The considerations of the previous section make it clear that any venture to determine accurate gravitational waveforms from theoretical models of supernovae is a complicated one. At minimum a two dimensional numerical simulation, capable of including at least one of angular momentum or magnetic fields, is indicated. The equation of state of the matter in the simulation must be sophisticated enough to model inverse beta decay and nuclear photodissociation, as well as the production of neutrinos. The low density equation of state must match smoothly onto a nuclear equation of state which is extendable up to super-nuclear densities. At least a primitive form of neutrino transport is required to properly model the effects of neutrino trapping. Finally, the simulation must be fully general relativistic.

Such a simulation is our ultimate goal; however, a project of this magnitude cannot be tackled all at once. To approach our goal, we have adopted a several step program. The first step in developing this simulation, now completed, is the construction of a two dimensional, axisymmetric, Newtonian gravity/special relativistic hydrodynamics simulation. Extensively tested and calibrated on problems with known, analytic solutions (for further details on the code, and an extensive list of code tests, see Finn & Evans 1988), this code serves several purposes: *i)* it provides a test bed for the development, testing, and calibration of the sophisticated equation of state required in the final simulation, *ii)* it provides a technological base upon which to build the more sophisticated general relativistic hydrodynamics and gravity code, and *iii)* it provides a series of test cases with which to calibrate the more sophisticated general relativistic code. Additionally, comparison of Newtonian and general relativistic collapse calculations may provide new insights into the nature of general relativistic gravity and gravitational wave production.

In order to use the Newtonian code in this final way, we must have one, and preferably several, reliable techniques for the extraction of gravitational radiation from Newtonian simulations. The usual method for determining quadrupole gravitational radiation waveforms in Newtonian simulations is the Standard Quadrupole Formula; however, as we shall see below, this is not the only technique, and it is certainly not the most reliable for determining the leading order quadrupole radiation from Newtonian sources. The remainder of this section describes several new techniques we have developed for determining the quadrupole gravitational radiation emitted from Newtonian sources. Each technique is based on a mathematical expression for the quadrupole radiation in the Newtonian limit, and is mathematically equivalent to the Standard Quadrupole Formula. Each technique is generally applicable, not just to axisymmetric sources or collapse problems, but to any Newtonian problem, with or without special symmetry, where the quadrupole gravitational radiation is desired.

4.2. The Standard Quadrupole Formula

We begin our discussion with a brief review of the use of the Standard Quadrupole Formula (henceforth, SQF) in finite difference simulations. The derivation and range of validity of the SQF has, in the past, been a matter of some controversy (*cf.* Damour 1987 for a recent review). That controversy has been settled, however, and a derivation of the SQF valid for sources with small but non-negligible internal gravity can be found in MTW §36.10. We quote here only the final result:

$$\begin{aligned} h_{ij}^{TT} &= \frac{2}{r}\frac{d^2}{dt^2}\left[\int d^3x\, T^{tt} r^2 \left(n_i n_j - \frac{1}{3}\eta_{ij}\right)\right]^{TT} \\ &= \frac{2}{r}\frac{d^2}{dt^2}\left[\int d^3x\, \rho r^2 \left(n_i n_j - \frac{1}{3}\eta_{ij}\right)\right]^{TT} \\ &\equiv \frac{2}{r}\frac{d^2}{dt^2}\, I\!\!\!-_{ij}^{TT}. \end{aligned} \tag{4.1}$$

In the above, h_{ij}^{TT} is the Transverse-Traceless projection of the gravitational radiation field in the asymptotically flat region far from the source, and retraded time is used throughout.

From a numerical standpoint, several features of the SQF deserve notice. The first is that, after evaluation of the trace-free quadrupole moment $I\!\!\!-_{ij}$, two numerical time derivatives are necessary to determine the gravitational radiation field; the second feature is the large moment arm (r^2) that weights the mass distribution of the source $dm = -\rho\, r^2 dr\, d\cos\theta\, d\phi$.

When viewed as a process in the frequency domain, it is clear that the two numerical time derivatives emphasize the high frequency variation of $I\!\!\!-_{ij}$. Owing to the discrete nature of the finite difference approximation, one expects that "numerical noise" will be present, peaking in strength at the average Nyquist frequency $< f_N > \sim \Delta t^{-1}$ corresponding to the (Courant limited) timestep Δt which is of order $\Delta r/\mathcal{V}$ (where Δr is the typical grid spacing, and $\mathcal{V}$ is the typical internal velocity in the source). This should be compared to the "numerical signal," which has a period of $\mathcal{L}/\mathcal{V}$ (where $\mathcal{L}$ is the typical extent of the source). The difficulty is that the two time derivatives amplify the noise by a factor of $(\mathcal{L}/\Delta r)^2$ relative to the signal.

The second feature of note in the SQF is the r^2 moment arm weighting the mass distribution dm. This moment arm gives large weight to the low density material far the the central regions of the source; so much so, in fact, that the low density material can provide the dominant contribution to the quadrupole moment. Nevertheless, it is the rapidly accelerating, high density regions of the source, *not* the low density, slowly moving material at large distances from the central regions, that is responsible for the gravitational radiation. The two time derivatives of the SQF (discussed in the previous paragraph) re-emphasize the high density regions of the source where the timescales are short. This involves a delicate cancellation of the large average value and rate of the quadrupole moment which, owing to the truncation error of the finite difference approximation, cannot be expected to be as reliable as a determination of the waveform from quantities that depend more directly on the high density, high velocity regions of the source.

4.3. The Momentum Divergence Formula

As our first variation on the SQF, we eliminate a single time derivative from equation (4.1). To do so, we make use of the continuity equation,

$$\frac{\partial \rho}{\partial t} = -\nabla \cdot (\rho \vec{v}) \,, \tag{4.2}$$

and reexpress the first time derivative of the trace-free quadrupole moment $\mathcal{I}_{ij}$ as

$$\begin{aligned} \frac{d\,\mathcal{I}_{ij}}{dt} &= \frac{d}{dt}\int d^3x\, r^2 \left(n_i n_j - \frac{1}{3}\eta_{ij}\right)\rho \\ &= \int d^3x\, r^2 \left(n_i n_j - \frac{1}{3}\eta_{ij}\right)\frac{\partial \rho}{\partial t} \\ &= -\int d^3x\, r^2 \left(n_i n_j - \frac{1}{3}\eta_{ij}\right)\nabla \cdot (\rho \vec{v}). \end{aligned} \tag{4.3}$$

The gravitational radiation field is thus given as a single time derivative of the divergence of the momentum density (henceforth QF2):

$$h_{ij}^{TT} = -\frac{2}{r}\frac{d}{dt}\left[\int d^3x\, r^2 \left(n_i n_j - \frac{1}{3}\eta_{ij}\right)\nabla \cdot (\rho \vec{v})\right]^{TT}. \tag{4.4}$$

Equation (4.4) for the radiation field has one fewer time derivative than equation (4.1), and this leads to an immediate advantage in determining the waveform and the power in gravitational radiation (*cf.* the discussion in the previous section), corresponding to an increase in the "signal-to-noise" ratio by the factor $\mathcal{L}/\Delta r$.

4.4. The First Moment of Momentum Density Formula

Our second alternative to the SQF reduces the weight of the moment arm in equation (4.3). Integrating equation (4.3) by parts and discarding the surface term, we obtain

$$\frac{d\,\mathcal{I}_{ij}}{dt} = 2\int d^3x\, \rho \left[v_{(i}x_{j)} - \frac{1}{3}\eta_{ij}\vec{v}\cdot\vec{x}\right] \tag{4.5}$$

(henceforth QF1). Here $v_{(i}x_{j)}$ represents the symmetric part of the tensor product $\vec{v} \otimes \vec{x}$: $v_{(i}x_{j)} = (v_i x_j + v_j x_i)/2$. By reducing the moment arm we decrease the emphasis on the momentum density at large radius, emphasizing more strongly the dense, high momentum fluid at small radius.

4.5. The Stress Formula

The two previous expressions for the gravitational radiation field have been based on manipulations of the SQF. The SQF is itself an approximation to a solution of the field equations, however. Let us return for a moment to a solution to the field equations. We assume our source of gravitational radiation is compact and at rest at the origin of an asymptotically flat spacetime; accordingly, we introduce an asymptotically Minkowskii coordinate system in which we may define the *gravitational field* $\bar{h}^{\alpha\beta}$:

$$\bar{h}^{\alpha\beta} \equiv -(-g)^{1/2} g^{\alpha\beta} + \eta^{\alpha\beta}. \tag{4.6}$$

If we impose the de Donder gauge condition,

$$\bar{h}^{\alpha\beta}{}_{,\beta} = 0, \tag{4.7}$$

then the (still exact) Einstein Field Equations (henceforth EFE) may be written in the form

$$\Box \bar{h}^{\alpha\beta} = -16\pi\tau^{\alpha\beta}. \tag{4.8}$$

In equation (4.8), $\Box$ is the flat space *scalar* wave operator and $\tau^{\alpha\beta}$ is the effective stress-energy tensor of matter and the gravitational field:

$$\tau^{\alpha\beta} \equiv (-g)(T^{\alpha\beta} + t^{\alpha\beta}_{\mathrm{LL}}) + (16\pi)^{-1}(\bar{h}^{\alpha\mu}{}_{,\nu}\bar{h}^{\beta\nu}{}_{,\mu} - \bar{h}^{\alpha\beta}{}_{,\mu\nu}\bar{h}^{\mu\nu}). \tag{4.9}$$

Here $t^{\alpha\beta}_{\mathrm{LL}}$ is the Landau-Lifshitz pseudo tensor (*cf.* MTW eq. 20.22) and $T^{\alpha\beta}$ is the stress-energy tensor of ordinary matter and fields (the source to the EFE when written in the conventional form $\mathbf{G} = 8\pi\mathbf{T}$). Far from the source (or in any region where $|\bar{h}^{\alpha\beta}| \ll 1$), the gravitational field $\bar{h}^{\alpha\beta}$ reduces to the trace-reversed metric perturbation:

$$\bar{h}^{\alpha\beta} = h^{\alpha\beta} - \frac{1}{2}\eta^{\alpha\beta} h^{\mu}{}_{\mu}, \tag{4.10}$$

where $h^{\alpha\beta} = -g^{\alpha\beta} + \eta^{\alpha\beta}$.

Formally, the solution to equation (4.8) may be written as

$$\bar{h}^{\alpha\beta} = 4\int d^3x' \frac{[\tau^{\alpha\beta}]_{\mathrm{ret}}}{|\mathbf{x} - \mathbf{x}'|}, \tag{4.11}$$

Equation (4.11) is a nonlinear integral equation, since the right hand side depends on $\bar{h}^{\alpha\beta}$ through equation (4.9). In the Newtonian limit, however, equation (4.11) provides an explicit solution for the quadrupole gravitational radiation field:

$$\begin{aligned} h^{TT}_{ij} &= \frac{4}{r}\left[\int d^3x\, T_{ij}\right]^{TT} \\ &= \frac{4}{r}\left[\int d^3x\, \rho(v_i v_j - \frac{1}{3}\eta_{ij}v^2) - (\sigma_{ij} - \frac{1}{3}\eta_{ij}\sigma^k_k) \right. \\ &\quad \left. + \frac{1}{4\pi}(\nabla_i\Phi\nabla_j\Phi - \frac{1}{3}\eta_{ij}\nabla\Phi\cdot\nabla\Phi)\right]^{TT} \end{aligned} \tag{4.12}$$

Note four important features of this expression for the radiation field: *i)* the integral expression (4.12) gives the gravitational radiation field directly — no time derivatives are required; *ii)* the integral for the radiation field does not involve any moment arms weighting the source: the contribution to the field comes from the regions of largest (aspherical) stress; *iii)* the integrand of equation (4.12) is not compact — the gravitational stress contribution extends out to infinity; and *iv)* the viscous stress contribution appears explicitly in the expression for the radiation field. These last two points require additional discussion.

Hydrodynamical shocks in an inviscid fluid are surfaces of discontinuity. Since a FD grid cannot resolve discontinuities, a FD hydrodynamical code must incorporate some mechanism to smooth shocks over several (usually two to three) zones. This mechanism is also responsible for seeing that the entropy is increased across the shock. In practice, the method used is to introduce an *artificial viscosity* into the equations of motion and energy conservation (*cf.* Smarr 1979; Wilson 1979; Hawley, Smarr, & Wilson 1984b; Richtmeyer & Morton 1967). The form of the viscosity is chosen so that it is dynamically negligible except at shocks, and so its effect at shocks is to smooth them just enough so that they can be resolved on the FD mesh. In the continuum limit, the artificial viscosity contribution to the equations of motion and energy conservation should vanish; however, they appear as accelerations in the FD Euler equations and as a source of in the FD internal energy equation.

The mathematical transition from the stress equation to the SQF is through the equations of motion and energy conservation; consequently, in order that a comparison can be made between the stress waveform and the SQF (or QF1 or QF2) there must be stress contributions corresponding to the artificial viscous forces and energy contributions. The standard formulations of artificial viscosity in two or more spatial dimensions (*cf.* Hawley, Smarr, & Wilson 1984b, Wilson 1979) are essentially *ad hoc* operator split generalizations of the 1D prescription of von Neumann & Richtmeyer (1950), which are not derivable from a viscous stress tensor. In order to resolve this problem, we have developed a formulation of artificial viscosity which derives from a viscous stress-energy tensor, thus permitting the use of the stress formula and its direct comparison with the SQF and its variations. For more details, see Finn & Evans 1988.

The remaining point that needs to be discussed is the non-compact integrand of the stress formula (eq. 4.12). In the integral expression for the SQF as well as the first two variations discussed above (QF1 and QF2) the integrand is compact; thus, as long as the FD mesh covers all the regions where there is matter the integral can be evaluated numerically and the formula applied directly. In the stress formula, however, the integrand includes gravitational field stress terms that *do not* vanish outside of the material body, but have a power law fall off. In order to apply the stress formula we must analytically evaluate the vacuum contribution to the gravitational radiation field from the regions beyond the extent of the FD mesh.

Assuming that the source is isolated and the region outside our simulation ($r > \mathcal{R}$) is vacuum, we can analytically evaluate the gravitational stress contribution to the stress formula from the region outside the FD grid. We find the contribution to the radiation field from the gravitational stress at radii larger than $\mathcal{L}$ is

$$\frac{1}{4\pi}\left[\int_{\mathcal{R}}^{\infty} d^3x\,\nabla_i\Phi\nabla_j\Phi\right]^{STF} = \sum_{l=0}^{\infty}\mathcal{R}^{-(2l+1)}\frac{l!l(2l-1)}{(2l+3)!!}\left[\mathcal{I}_{iA_{l-1}}\mathcal{I}_{jA_{l-1}}\right]^{STF}, \tag{4.13}$$

where

$$\mathcal{I}_{A_l} \equiv -\frac{(2l+1)!!}{l!(2l+1)}\left[\int d^3x\,\rho X_{A_l}\right]^{STF}. \tag{4.14}$$

Note that this expansion is quadratic in the moments of the gravitational field, and that the moments are not mixed. Equations (4.12) and (4.13) together can be used to evaluate the the quadrupole gravitational radiation field from a Newtonian source with arbitrary symmetry.

4.6. Discussion

Each of the four waveform extraction formulae described above has been specialized to axisymmetry and incorporated into our two dimensional Newtonian collapse code. Figure 2 shows, on the same chart, the waveform as measured by each of these diagnostics for a typical collapse calculation (mass: $1.4M_{\odot}$; angular momentum: 2.0×10^{49} erg s).

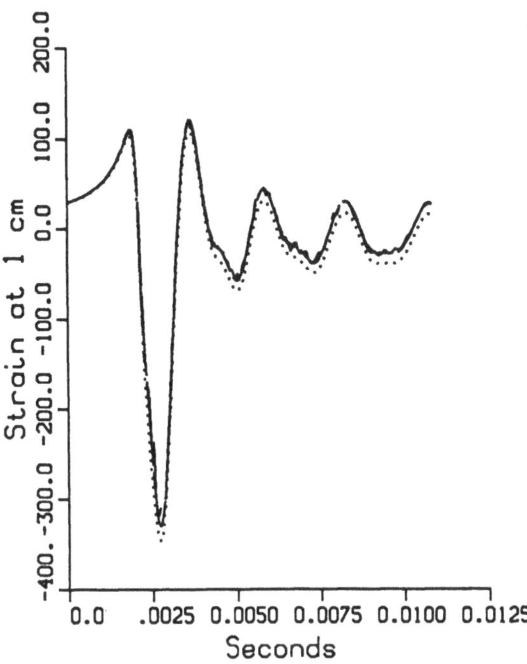

Figure 2. The four waveform diagnostics superimposed on the same chart. Three of these quadrupole waveform diagnostics (SQF, QF1, and QF2) are indistinguishable to the level of truncation error, while the fourth (the stress diagnostic, represented with a dotted line) shows a small DC offset. Time is given in seconds and the amplitude is given in strain normalized to a distance of 1 cm; however, for the purposes of this discussion, the scale of the waveform is arbitrary.

The first thing to notice is the remarkable agreement among the four diagnostics: though each is ultimately measuring the same quantity, the intermediate quantities used are all very different. Aside from high frequency noise, the SQF, QF1, and QF2 extraction formulae yield virtually the same results, while the stress formula yields a result of the same functional form but with a slight dc offset. This dc offset is directably attributable to finite difference truncation error: it is the result of a small difference between two (analytically) very nearly equal quantities that are determined numerically only to finite difference truncation accuracy. Numerical experiments have verified this understanding. A further discussion of this difference can be found in Finn & Evans (1988).

Consider now the high frequency component of the waveforms. Figure 3 shows the power spectral densities of the four waveforms of figure 2. Here the difference between the SQF and the other diagnostics is seen to be dramatic. The total integrated radiated energy as a function of time as determined from QF1 and from the stress formula do not differ significantly. Together, these differ from the radiated energy as determined from QF2 at just under the 10% level, while the SQF determined radiated energy is larger than all these measures by a factor of 5 or more. The difference in the radiated energy measures is clearly a function of the increased high frequency noise, owing to the additional numerical time

derivative, in the SQF.

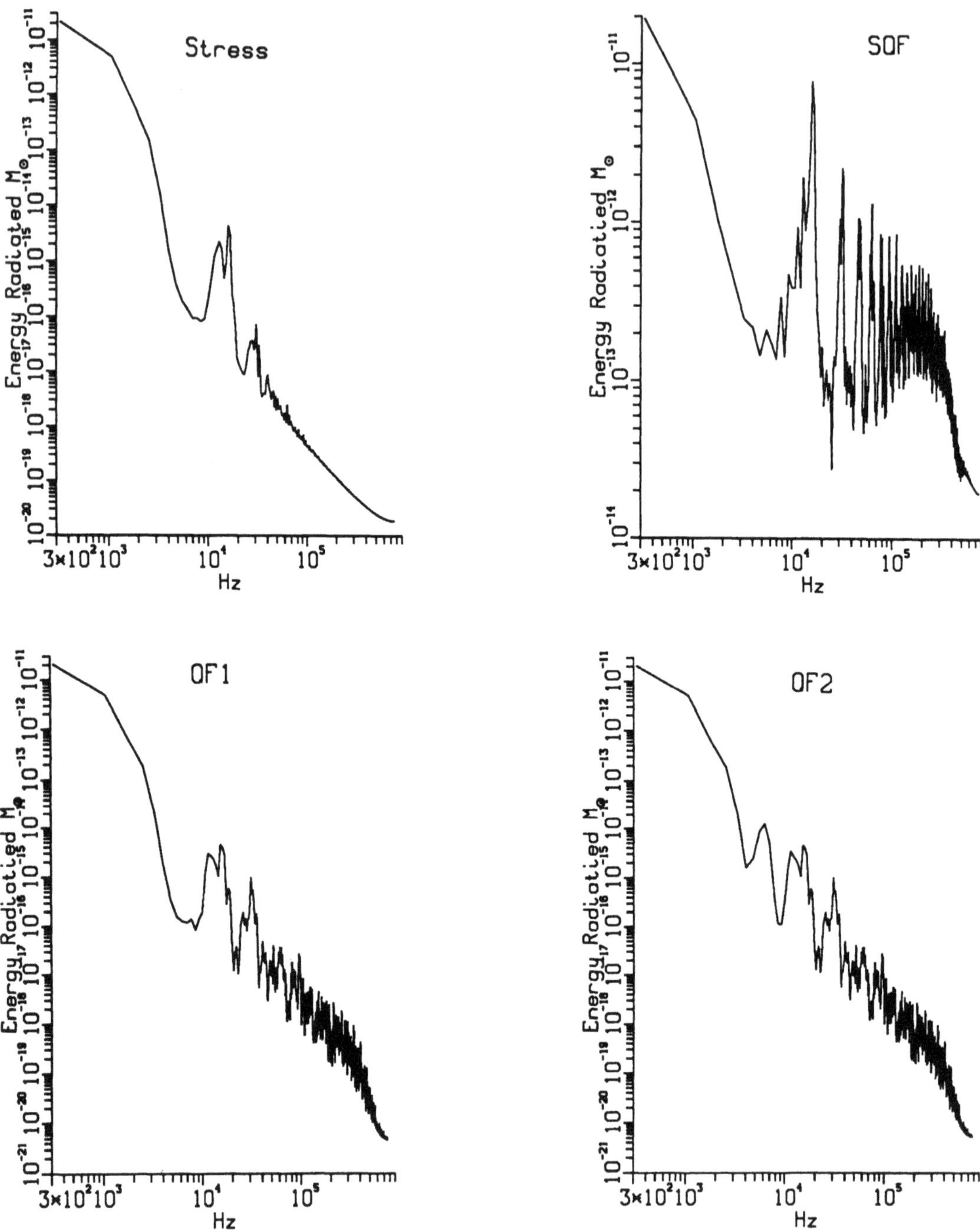

Figure 3. Power Spectral densities of the four waveform diagnostics.

One further point is worthy of note. The four diagnostics investigated here are, analytically, identical. The mathematical transformations that take one to another are all expressible in terms of applications of the continuity equation (which transforms between QF2 and the SQF), the Navier-Stokes equation and the internal energy equation (which transforms between the stress formula and the SQF), or an integration by parts (which transforms between QF2 and QF1). Consequently, any difference between these measure reflects directly on the accuracy and the implementation of the finite difference implementation of these equations. In this role, the four diagnostics provide an internal consistency check that the code is correctly solving the continuity, Navier-Stokes, and Euler equations; and, differences between pairs of diagnostics can be used to localize errors or inaccuracies to finite difference implementations of one or several of these equations.

5. Conclusions

We have examined the standard quadrupole formula (SQF) and three variations for the determination of the gravitational radiation waveform from Newtonian self-gravitating sources. In an application of the SQF, the quadrupole moment, which may be regarded as a functional of the density with a large moment arm, must be numerically differentiated twice with respect to time in order to reproduce the waveform. The three variations may be regarded as ways to reduce either the number of time derivatives or the weight of the moment arm required to recover the waveform.

To investigate the performance of the various waveform determination techniques, we examined numerically gravitational radiation from the collapse of rotating stellar core models. Our investigations show dramatic differences in the resolution of the waveform between the SQF and the three variations. Minimizing the number of numerical time derivatives that separate integrals over state variables from the gravitational waveform minimizes the numerical noise in the waveform. The "cleanest" determination of the gravitational radiation results from a direct calculation of the waveform as an integral over the stress tensor of the source; however, the waveform in this case emerges as the difference of two nearly equal quantities, and is more sensitive to the truncation error of the finite difference scheme than any of the other techniques.

The significance of these alternative waveform extraction formulae for code testing has been examined. The variations provide an opportunity to rigorously test the implementation of the continuity equation independently of the correctness of the implementation of the Euler or energy equations, and to test the implementation of the energy and Euler equations given the correctness of the implementation of the continuity equation.

6. Acknowledgements

It is a pleasure to acknowledge helpful conversations with David Chernoff, Charles Evans, John Hawley, and Bill Newman. Some of the work described in this paper has been carried out in collaboration with Charles Evans. I also wish to acknowledge the hospitality of the Institute for Theoretical Physics, where the manuscript for this paper was prepared. Computing in support of the research described here was carried out at the National Center for Supercomputing Applications and the Cornell National Supercomputing Facility. This research was supported by National Science Foundation Grants PHY 86-03284 and AST 87-14475.

References

Baade, W., & Zwicky, F. 1934. *Phys. Rev.*, **45**, 138.

Baron, E., Cooperstein, J., & Kahana, S. 1985. *Phys. Rev. Lett.*, **55**, 126.

Bethe, H. A., Brown, G. E., Applegate, J., & Lattimer, J. M. 1979. *Nucl. Phys.* , **A324**, 487.

Bethe, H. A., & Wilson, J. R. 1985. *Astrophys. J.,* **295**, 14.

Bethe, H. 1986, in *Highlights of Modern Astrophysics,* eds. S. Shapiro & S. Teukolsky (Wiley & Sons: New York).

Colgate, S. A., & White, R. H. 1966. *Astrophys. J.,* **143**, 626.

Centrella, J. M., Shapiro, S. L., Evans, C. R., Hawley, J. F., Teukolsky, S. A. 1986, in *Numerical Relativity and Dynamical Spacetimes,* ed. J. Centrella (Cambridge University Press: Cambridge).

Damour, T. 1986, in *Proceedings of the Fourth Marcel Grossmann Meeting on General Relativity,* ed. R. Ruffini (North Holland: Amsterdam).

Damour, T. 1987, in *Gravitation in Astrophysics,* eds. B. Carter & J. B. Hartle (Plenum: New York).

Evans, C. R. 1984. *A Method for Numerical Relativity: Simulation of Axisymmetric Gravitational Collapse and Gravitational Radiation Generation,* unpublished thesis, The University of Texas at Austin.

Evans, C. R. 1986, in *Numerical Relativity and Dynamical Spacetimes,* ed. J. M. Centrella (Cambridge University Press: Cambridge).

Finn, L. S. 1988, in *Proceedings of the NATO Advanced Research Workshop on Gravitational Radiation Data Analysis,* ed. B. Schutz, in press.

Finn, L. S., & Evans, C. R. 1988, in preparation.

Goldreich, P., & Weber, S. V. 1980. *Astrophys. J.,* **238**, 991.

Hawley, Smarr, & Wilson. 1984a. *Astrophys. J.,* **277**, 296.

Hawley, Smarr, & Wilson. 1984b. *Astrophys. J. Suppl.,* **55**, 211.

Hoyle, F., & Fowler, W. A. 1960. *Astrophys. J.,* **132**, 565.

Maeder, A. 1984, in *Observational Tests of Stellar Evolution Theory, IAU Symp. No. 105,* eds. A. Maeder, A.Renzini, (Reidel: Dordrecht), p. 299.

May, M. M., & White, R. H. 1966. *Phys. Rev.,* **141**, 1232.

May, M. M., & White, R. H. 1967. *Methods Comput. Phys.,* **7**, 219.

Mazurek, T. J., Lattimer, J. M., & Brown, G. E. 1979. *Astrophys. J.,* **229**, 713.

Misner, C. W., Thorne, K. S., & Wheeler, J. A. 1973, *Gravitation,* (San Francisco: Freeman).

Müller, E. 1982. *Astr. Astrophys.,* **114**, 53.

Norman, M. L., Wilson, J. R., & Barton, R. T. 1980. *Astrophys. J.,* **239**, 968.

Norman, M. L., & Winkler, K-H. A. 1986, in *Astrophysical Radiation Hydrodynamics,*, eds. K-H. Winkler & M. Norman, (Reidel: New York), p. 187.

Richtmyer, R. D. & Morton, K. W. 1967, *Difference Methods for Initial-Value Problems,* (Wiley: New York).

Ruffini, R., & Wheeler, J. A. 1971, in *Relativistic Cosmology and Space Platforms,* (ESRO: Paris), p. 140.

Saenz, R. A., & Shapiro, S. L. 1978. *Astrophys. J.,* **221**, 286.

Saenz, R. A., & Shapiro, S. L. 1979. *Astrophys. J.,* **229**, 1107.

Saenz, R. A., & Shapiro, S. L. 1981. *Astrophys. J.,* **244**, 1033.

Schild, R., & Maeder, A. 1983. *Astr. Astrophys.*, **130**, 237.

Seidel, E., & Moore, T. 1987. *Phys. Rev. D.*, **35**, 2287.

Seidel, E., & Moore, T. 1988, in *Frontiers in Numerical Relativity,* eds. C. R. Evans, L. S. Finn, & D. W. Hobill (Cambridge University Press: Cambridge).

Shapiro, S. L. 1977. *Astrophys. J.*, **214**, 566.

Stark, R. F., & Piran, T. 1986, in *Proceedings of the Fourth Marcel Grossmann Meeting on General Relativity,* ed. R. Ruffini (North Holland: Amsterdam).

Tammann, G. A. 1982, in *Supernovae: A Survey of Current Research,* eds. M. J. Rees & R. J. Stoneham (Reidel: Dordrecht).

Thorne, K. S. 1977, in *Topics in Theoretical and Experimental Gravitational Physics,* eds. V. D. Sabbata & J. Weber (Plenum: New York).

Thorne, K. S. 1980. *Rev. Mod. Phys.*, **52**, 299.

Thorne, K. S. 1987, in *300 Years of Gravitation,* eds. S. Hawking & W. Israel (Cambridge University Press: Cambridge).

Thuan, T. X., & Ostriker, J. P. 1974. *Astrophys. J. (Letters)* , **191**, L105.

Turner, M. S., & Wagoner, R. V. 1979, in *Sources of Gravitational Radiation,* ed. L. Smarr (Cambridge University Press: New York), p. 383.

von Neumann, J. & Richtmyer, R. D. 1950. *J. Appl. Phys.*, **21**, 232.

Wilson, J. R. 1971. *Astrophys. J.*, **163**, 209.

Wilson, J. R., Couch, R., Cochran, S., LeBlanc, J., & Barkat, Z. 1975. *Ann. NY Acad. Sci.*, **262**, 54.

Wilson, J. R. 1985, in *Numerical Astrophysics,* eds. J. Centrella, J. LeBlanc, & R. Bowers (Jones & Bartlett: Boston), p. 422.

Woosley, S. E., & Weaver, T. A. 1986. *Ann. Rev. Astr. Astrophys.*, **24**, 205.

Gravitational Radiation from Perturbations of Stellar Core Collapse Models

Edward Seidel
McDonnell Center for the Space Sciences, Physics Department, Washington University
St. Louis, MO 63130 USA

Thomas Moore
Department of Physics, Pomona College
Claremont, CA 91711 USA

Abstract Odd-parity gravitational perturbations of detailed spherically symmetric type II supernova models, which include realistic equations of state and neutrino transport, and even-parity perturbations of simplified core collapse models, are studied in the linearized Einstein theory. For the odd-parity case, we have performed simulations in which we vary the equation of state at supernuclear densities, the free proton mass fraction, and the precollapse iron core. These variations span much of the range of conditions currently thought to occur during stellar collapse and lead to the production of core-bounce shock waves that vary from the very strong to those that stall almost immediately. We find little difference in the gravitational radiation output among any of these models, though models that use stiffer equations of state or higher mass iron cores give somewhat more radiation output. For even-parity perturbations, we report on work in progress.

INTRODUCTION

The emission of gravitational radiation from stellar core collapse represents an interesting and increasingly important problem in General Relativity and astrophysics. The explosion of SN 1987A has given added excitement and importance to the study of supernova physics and has stimulated many groups to increase their efforts to develop satisfactory theoretical models of stellar core collapse and explosion, to extend the experimental search for gravitational radiation, and to develop better theoretical calculations of gravitational radiation from stellar collapse.

However, the complexity and non-linear nature of General Relativity make its application to realistic astrophysical systems very difficult, and much work remains to be done. In particular, detailed computer simulation is needed to take advantage of the predictive power of General Relativity in astrophysics. Full scale numerical codes to solve the non-linear Einstein's equations are still being developed after many years of work and only recently have attempts been made to tackle three dimensional problems. But, stars are essentially spherical objects, and one does not need to develop a three dimensional computer code to model a spherical object. In fact, most research to date on stellar collapse has been done under the assumption that the star is perfectly spherical throughout its evolution. Of course, if this were strictly true, there would be no need to calculate gravitational radiation emission due to stellar core collapse and supernovae (which is the goal of a great many of the participants of this conference), as a time dependent quadrupole (or higher) moment is required for the production of gravitational waves. But we know that stars rotate and have magnetic fields, both of which will serve to break the spherical symmetry. These effects are usually considered to be small, so that the detailed spherical models of stellar core collapse are thought to be very good approximations to real stellar core collapse events. Our approach to calculating gravitational radiation is to take advantage of the wealth of research done on spherical core collapse and incorporate it into a perturbational description of the star, with the perturbations representing the small deviations from spherical symmetry assumed to be present during the collapse.

In this alternative approach to non-linear codes we study the linearized Einstein's equations, taking advantage of a known (numerical) spherical solution as a background. The spherically symmetric system is evolved using a numerical scheme originally developed by May and White (1966, 1967) and can be used to model the collapse of a star with any equation of state. In addition, the code can include sophisticated microphysics, such as neutrino transport, which is known to be very important in determining whether or not the collapse will produce the optical display seen in the typical type II supernova. Perturbations of such a system are ideally suited to the study of gravitational wave emission from stellar collapse and supernovae. Relativistic perturbation theory of spherical systems has been developed extensively by many authors during the last 30 years, and we will bring these techniques to bear on the stellar collapse problem.

The current theories of the hydrodynamics of stellar collapse and type II supernovae have been described in numerous papers (see Woosley and Weaver 1986, and references therein). While our main emphasis in this paper is on the gravitational radiation emitted during a type II supernova, for completeness we provide a summary of the relevant supernova physics here. We are interested in the core collapse of stars whose total mass is in the range 10-25 $M_\odot$, the iron core masses for which may range from about 1.2-2.0 $M_\odot$. The cores of these stars are thought to destabilize and start collapsing as a result of either photodissociation of iron nuclei or by electron capture. It is the latter process that ultimately drives the collapse, since the removal of electrons deprives the core of its chief source of pressure. As the collapse proceeds, the core divides into two regions: an inner core in which the collapse is nearly homologous (velocity proportional to radius) and an outer mantle for which the collapse is supersonic and the stellar fluid is essentially in reduced free-fall. The size of the these regions is determined, for the most part, by the electron capture that takes place, with greater capture leading to a smaller homologous core. The neutrinos produced by the electron capture reactions initially stream from the core, but are eventually trapped and brought to equilibrium with the matter once the core reaches a density of greater than about 10^{12} g cm^{-3}.

As a result of nucleon degeneracy and the repulsive nuclear force, the collapse is halted once densities exceed that of nuclear saturation ($\cong 2.4 \times 10^{14}$ g cm^{-3}). The precise density at which this occurs is uncertain since the behavior of nuclear matter is not fully understood in the high-density regime. The signal announcing the core bounce travels as far as the sonic point, where the infall velocity equals the local sound speed (located about 0.6 to 0.8 $M_\odot$ from the center of the core and just beyond the limit of the infall homology). Near this point a pressure wave builds and quickly steepens to a shock wave that starts to propagate outward. It has been pointed out (Brown, et. al. 1982, Burrows and Lattimer 1985) that this shock wave efficiently dampens any subsequent oscillations of the core, thus reducing the gravitational radiation that would otherwise be emitted.

In conventional scenarios for type II supernovae, the explosion and resulting optical display are powered by the shock wave as it propagates out of the core and into the overlying envelope. The energy it delivers and the timescale over which this occurs, however, are in dispute since there are uncertainties in the initial shock strength, the magnitude of the losses it must sustain (through the dissociation of nuclei and neutrino losses), and possible sources of revitalizing energy (through neutrino energy deposition and resultant heating of material behind the shock). In the prompt scenario, the shock is initially energetic and suffers few subsequent losses so that it can propagate through the core in a few tens of milliseconds (Colgate and Johnson 1960, Bethe et. al. 1979, Baron et. al. 1985). In the delayed mechanism (Wilson 1985, Bethe and Wilson 1985), the prompt shock stalls, but is later revived by neutrino reheating and propagates off the core in a few hundreds of milliseconds. Fortunately, for the purposes of calculating gravitational radiation, both these timescales are longer than the radiation emission timescale, so the exact means of explosion is not directly related to thc results we present here. It turns out, however, that the nature of the equation of state at high-densities is important in determining the radiation emitted.

PERTURBATION TECHNIQUES

The perturbation approach for calculating gravitational wave emission from collapsing stars was pioneered by Cunningham, Price, and Moncrief (1978a, 1978b, 1980) (CPM I, II, and III) and will be followed in this work. Their idea was to treat the nonspherical aspects of core collapse as small perturbations of a spherically symmetric background, and expand the perturbations in terms of tensor spherical harmonics, following the early work of Regge and Wheeler on perturbations of the Schwarzschild spacetime. The angular dependence of the perturbations is expressed in terms of the tensor spherical harmonics, so the spatial part of the problem is reduced to one dimension, which makes the problem much more tractable. A particularly nice feature of their approach, which we will carry over into this work, is their use of gauge invariant quantities in the description of the perturbation fields. Physically observable quantities are gauge invariant, and a gauge invariant description of the problem will aid greatly in simplifying and clarifying the problem.

Their work, however, was limited to an Oppenheimer-Snyder description of the stellar interior. With this background solution, they were able to calculate gravitational wave emission from the formation of a perturbed black hole due to the collapse of a pressureless gas of dust. They found that one could expect up to 10^{-3} stellar masses to be radiated within the limitations of the perturbation approximation. The real advantage of such a background lies in its simplicity: collapse of a dust sphere can be calculated analytically. But real stars have pressure, which may enrich the variety of phenomena observed in the universe, but which very greatly complicates matters for a mathematical description of the problem.

The basic method we use in this project is like that of CPM: we perturb a spherically symmetric background system, whose evolution we know either through numerical or analytic methods, and then numerically evolve the equations describing the perturbations, which are derived using the Einstein equations, linearized in the perturbation quantities. We use a numerical computer code (an enhanced May-White code) to evolve the background spacetime inside the star, but when we calculate the gravitational wave spectrum and energy emitted by the perturbed, collapsing star, we must do this far from the source. Because the background spacetime inside the star is spherically symmetric, even though it is dynamic, the spacetime outside the star in the vacuum must be the (static) Schwarzschild geometry. So in the exterior of the star we know the background solution analytically at all times. In addition, perturbations of the Schwarzschild geometry have been studied for many years by Regge and Wheeler (1957), Zerilli (1970), and Chandrasekhar and Detweiler (1975), among others, and we can simplify our problem greatly by using the known perturbation equations of the Schwarzschild geometry. Regge and Wheeler discovered a wave equation describing odd-parity perturbations of Schwarzschild, Zerilli discovered a wave equation describing even-parity perturbations of Schwarzschild, and later Chandrasekhar (1975) was able to discover a transformation from the even-parity equation to the odd-parity equation, showing that for vacuum solutions, the two parity systems are equivalent. (Mathematically, odd-parity perturbations are those with parity $(-1)^{l+1}$ where l is the angular momentum quantum number of the spherical harmonic, whereas even-parity perturbations are those with parity $(-1)^{l}$.)

While in the stellar exterior the problem reduces to perturbations of vacuum spacetimes which have already been well studied, in the stellar interior the Einstein equations are coupled to the stellar matter, which we take to be a perfect fluid. In particular, gravitational perturbations of the metric tensor of either parity class couple to the corresponding perturbations of the stellar fluid. Physically, odd-parity perturbations of the stellar matter correspond to azimuthal fluid velocity perturbations, while even-parity perturbations correspond to convective fluid velocity perturbations, as well as to density and entropy perturbations. The fact that perturbations of scalar functions like density, pressure, etc., are even-parity can be seen from the fact that the scalar perturbation functions are decomposed in terms of scalar spherical harmonics, the usual Y_{lm}, which have parity $(-1)^l$. Because are able to consider such a large variety of different types of matter perturbations we expect to represent very well the physics of a slightly nonspherical collapsing star.

This separation of the problem into two parts, interior and exterior, is not without complications, however. We must derive equations which then match the two solutions at the stellar surface, which itself may be moving through spacetime as the star collapses or explodes. There is a well defined prescription for matching two regions of spacetime, namely that the induced metric (the first fundamental form) on the hypersurface joining the two regions must be continuous, and that the second fundamental form, which is related to the normal derivative of the metric across the surface, must be also be continuous. (See MTW section 21.13 for a discussion of this prescription.)

Finally, while one might choose a particular gauge to simplify the perturbation equations which result from the linearized Einstein equations, we have chosen to rewrite them entirely in terms of fully *gauge invariant* quantities. This considerable extra work makes the physics of the system more transparent in the end, since all physical quantities of interest are necessarily gauge invariant, and there is a decoupling of physical effects from pure gauge or coordinate effects. We are not interested in what the perturbation looks like in some particular gauge, but we are of course interested in certain gauge invariant quantities like the total energy radiated from the system. These gauge invariant functions in the stellar interior (there are six of them in the even-parity case and two in the odd-parity case) can be shown to match at the stellar surface to a single gauge invariant perturbation function in the stellar exterior, which is propagated by the Regge-Wheeler (odd-parity) or Zerilli (even-parity) equation on the fixed Schwarzschild background. Finally, the energy and spectrum of the radiation are computed across a surface of constant radius far from the source.

A particular computational advantage of our approach is that the perturbations are decomposed into tensor harmonics, so that the angular dependence of the perturbations is explicitly given by the angular momentum quantum numbers of the tensor harmonics, and the evolution equations for the perturbations are functions of radius and time only. Thus, we have the luxury to define a rather fine computer grid on which to solve the perturbation system. This means that some useful and accurate results can be obtained with a modest amount of computing power, although some supercomputer time is still desirable for a thorough investigation of the problem. We have run the code on various machines ranging from a VAX 750 to the Pittsburgh Cray X-MP.

The Regge-Wheeler Formalism

We write the total space-time metric of the system as a spherically symmetric background part, $\overset{o}{g}_{\alpha\beta}$, and add to it a small part which represents small deviations from spherical symmetry, $h_{\alpha\beta}$, so that the full metric becomes

$$g_{\alpha\beta} = \overset{o}{g}_{\alpha\beta} + \varepsilon h_{\alpha\beta}. \qquad 1$$

The perturbation metric $h_{\alpha\beta}$ could be any arbitrary tensor field, but following Regge and Wheeler, we will find it convenient to break it up into even- and odd-parity components. Of course any arbitrary perturbation would be a linear combination of even- and odd-parity tensor fields. The main reasons for studying the problem in this way are that we can take advantage of the mathematical properties of the even- and odd-parity tensor spherical harmonics developed by Regge and Wheeler for an expansion of perturbation fields on a spherical background, and in the exterior vacuum region the appropriate wave equation is already known (the Regge-Wheeler or Zerilli equation). We can further simplify the problem by restricting ourselves to axisymmetric perturbations, which leads to no loss of generality because the radial perturbation equations are independent of m, the z-component of the angular momentum quantum number l (Regge and Wheeler, 1957).

In the stellar interior we write the spherical background metric $\overset{o}{g}_{\mu\nu}$ in the form given by May and White with a background line element

$$d\overset{o}{s}^2 = -N^2\, d\tau^2 + A^2\, d\mu^2 + R^2\, (d\theta^2 + \sin^2\theta\, d\phi^2). \qquad 2$$

Here τ is a time coordinate, μ is a radial coordinate corresponding to proper rest mass within a sphere of radius μ, θ and ϕ are ordinary angular coordinates, and the background functions N, A and R are functions of μ and τ only, and are determined by the hydrodynamic portion of the computer code. For details, see Seidel and Moore (1987) (Paper I) or May and White. We also must specify the stress energy tensor $T_{\mu\nu}$. We will choose the form for a perfect fluid with an additional piece arising from the treatment of neutrinos in the star. To date the neutrino contribution has only been included in background used to drive the odd-parity perturbation equations. Then we can write the stress energy tensor as

$$T_{\alpha\beta} = (p + \eta)\, u_\alpha u_\beta + p\, g_{\alpha\beta} (+\, t_{\alpha\beta}) \qquad 3$$

where p is the pressure, η is the total energy density, u_α is the fluid four velocity and $t_{\alpha\beta}$ is a term describing the neutrinos.

For the odd-parity perturbations, the perturbation metric $h_{\mu\nu}$ is expanded in odd-parity, axisymmetric Regge-Wheeler harmonics, (see Paper I) so that the total line element takes the form:

$$ds^2 = -N^2\, d\tau^2 + A^2\, d\mu^2 + R^2\, (d\theta^2 + \sin^2\theta\, d\phi^2) + 2\varepsilon h_0 \sin\theta\, Y_{l0,\theta}\, d\tau d\phi \qquad 4$$

$$+\ 2\varepsilon h_1 \sin\theta\ Y_{l0,\theta}\, d\mu d\phi + 2\varepsilon\, [\tfrac{1}{2} h_2\, (\cos\ Y_{l0,\theta} - \sin\theta\, Y_{l0,\theta\theta})] d\theta d\phi,$$

where h_0, h_1, and h_2 are perturbation functions, dependent on μ and τ, and the Y_{l0} are the usual spherical harmonics with $m = 0$.

We must also consider perturbations of the stress energy tensor. There are off-diagonal terms $t_{\mu\tau}$ contributing to the stress-energy tensor that represent neutrino flux, and that should be perturbed in principle. However, because essentially all gravitational radiation is emitted at the time of the core bounce in the supernova, we need only consider the physics relevant to this time period in the evolution of the system. When the stellar fluid reaches densities of about 10^{13} gm cm^{-3}, the neutrinos have become strongly trapped and thermally equilibrated with the matter as a result of weak neutral currents between the neutrinos and the baryonic components of the matter. The core bounce itself occurs at densities well above nuclear saturation ($\rho_{sat} \cong 2.4 \times 10^{14}$ gm cm^{-3}). At such densities, the neutrino flux terms are negligible and the most important effect of the neutrinos is their contribution to the pressure and energy. Therefore, the effective stress-energy tensor which we perturb has the form for a perfect fluid:

$$T_{\alpha\beta} = (P + P_\nu + \eta + \eta_\nu) u_\alpha u_\beta + (P + P_\nu)\, g_{\alpha\beta}, \qquad 5$$

where η_ν and P_ν are the energy density and pressure from the neutrino component, and P and η are the pressure and total energy density of the rest of the fluid. Then the matter is perturbed by adding an odd-parity term to the fluid 4-velocity u_α having the following form:

$$(\delta u_\alpha) = (0, 0, 0, U(\tau,\mu) \sin\theta\, Y_{l0,\theta}). \qquad 6$$

Such a perturbation represents an axisymmetric fluid flow whose velocity lies entirely in the ϕ direction, and describes differential rotations of the stellar fluid about the z-axis. Note that bulk rotation about the z-axis cannot be studied using this method.

We can now expand the Einstein equations to first order in ε and δu_μ, but we also take advantage of the simplification which a gauge invariant description of the problem brings. If we define the quantity ψ as

$$\psi = l(l+1)\frac{4\pi\rho R^4}{N}\left(\frac{\partial}{\partial\tau}\left(\frac{h_1}{R^2}\right) - \frac{\partial}{\partial\mu}\left(\frac{h_0}{R^2}\right)\right) \qquad 7$$

then we find that it is *gauge invariant*, and the linearized Einstein equations show that it obeys the following wave equation in the interior of the star:

$$\frac{\partial}{\partial\tau}\left(\frac{A}{R^2N}\frac{\partial}{\partial\tau}(R^2\psi)\right) - \frac{\partial}{\partial\mu}\left(\frac{N}{AR^2}\frac{\partial}{\partial\mu}(R^2\psi)\right) + (l+2)(l-1)\,NA\frac{\psi}{R^2} \qquad 8$$
$$= 16\pi l(l+1)\frac{\partial}{\partial\mu}(NU(\tau,\mu)(P+P_v+\eta+\eta_v)).$$

By gauge invariant we mean that it is invariant under the gauge transformation generated by an arbitrary odd-parity vector field

$$(X_\mu) \equiv (X_\tau, X_\mu, X_\theta, X_\phi) = (0, 0, 0, C(\tau,\mu)\ \sin\theta\, Y_{l0,\theta}), \qquad 9$$

whereas the metric components themselves would transform as

$$\delta h_{\mu\nu} = X_{\mu;\nu} + X_{\nu;\mu}. \qquad 10$$

Because of the gauge invariant property of our function ψ (and $U(\tau,\mu)$), we can describe the interior perturbation with a simple wave equation whose meaning is clear: it describes a gravitational wave propagating at the speed of light inside the star. Without this description we would have a more complicated system of equations whose variables would not have a clear physical meaning.

The even-parity description is similar in principle, but substantially more complex in detail. The metric tensor and stress energy tensor are perturbed in an analogous way, with even-parity components. The perturbed metric tensor now contains seven independent functions instead of three, and the perturbed stress energy tensor now include density, pressure, and entropy perturbations, as well as a more complex even-parity form of the fluid four-velocity, which contains three functions instead of just one. This richer structure provides us with a larger set of gauge invariant functions to define. The perturbed metric tensor has the form:

$$(h_{\alpha\beta}) = \begin{pmatrix} N^2H_0Y_{l0} & H_1Y_{l0} & h_0\frac{\partial}{\partial\theta}Y_{l0} & 0 \\ H_1Y_{l0} & A^2H_2Y_{l0} & h_1\frac{\partial}{\partial\theta}Y_{l0} & 0 \\ h_0\frac{\partial}{\partial\theta}Y_{l0} & h_1\frac{\partial}{\partial\theta}Y_{l0} & R^2\left(K + G\frac{\partial^2}{\partial\theta^2}\right)Y_{l0} & 0 \\ 0 & 0 & 0 & R^2\left(K\sin^2\theta + G\sin\theta\cos\theta\frac{\partial}{\partial\theta}\right)Y_{l0} \end{pmatrix}. \qquad 11$$

We now follow the same procedure as above: namely, we plug the perturbed metric and stress energy tensors into the Einstein equations, define a set of gauge invariant quantities, and write the evolution equations in terms of them. We will spare the reader the gruesome details of the derivation and schematically display the resulting equations. One can find a set of six gauge invariant quantities which we call q, q_2, $\bar{\pi}$, π_2, π_4', and $\delta s'$, which satisfy the following (schematic) set of coupled PDE's:

Two equations form a system with the form:

$$\text{leading terms} = \begin{cases} q_{2,\tau} = \pi_2 + \ldots + \text{extra terms} \\ \pi_{2,\tau} = \left(\frac{\partial p}{\partial \eta}\right)_s \frac{q_{2,\mu\mu} N^2}{A^2} + \ldots + \text{many extra terms.} \end{cases} \qquad 12$$

These two equations clearly form a system which is a complicated wave equation for the variable q_2, which propagates at the speed of sound.

The sound wave quantity q_2 is very strongly coupled to the most important quantities of interest, q and $\bar{\pi}$, which form a system with the following structure:

$$\text{leading terms} = \begin{cases} q_{,\tau} = \bar{\pi} \\ \bar{\pi}_{,\tau} = \frac{q_{,\mu\mu} N^2}{A^2} + \ldots + \text{many extra terms .} \end{cases} \qquad 13$$

This system has the form of a wave equation for the variable q, which propagates at the speed of light. This function is clearly representing the propagation of a gravitational wave inside the star, and as such it carries information about the most important quantity of interest to us.

Finally, the function π_4' can be shown to describe the perturbed vorticity of the system, whose time derivative is equal to a function of q and q_2 and their gradients. An additional function $\delta s'$ is a gauge invariant measure of the perturbed entropy of the system, whose time derivative is equal to a function of all the other variables and some of their gradients.

For the sake of clarity (and to save space) we will not write out all of these equations, but as an example of the complexity of the system we provide one of the momemtumlike equations:

$$\begin{aligned} &\pi_{2,\tau} - \left(\frac{\partial p}{\partial \eta}\right)_s R_{,\tau} \rho^3 R^5 N \bar{\pi} - \frac{N^2}{A}\left(\frac{N_{,\mu}}{N} + \left(\frac{\partial p}{\partial \eta}\right)_s \left(4\frac{R_{,\mu}}{R} + \frac{\rho_{,\mu}}{\rho}\right)\right)\frac{q_{2,\mu}}{A} \qquad 14 \\ &- N^2\frac{R_{,\mu}}{RA}\left(1 + \left(\frac{\partial p}{\partial \eta}\right)_s\right)\frac{q_{,\mu}}{A} + \left(\frac{N_{,\tau}}{N} + 3\frac{R_{,\tau}}{R} - \left(\frac{\partial p}{\partial \eta}\right)_s \frac{\rho_{,\tau}}{\rho}\right)\pi_2 - 2\frac{R_{,\mu}}{A}\frac{(\rho R^3)_{,\tau}}{\rho R^3}\pi_4' \\ &+ \left(\left(\frac{\partial p}{\partial \eta}\right)_s \left(\frac{R_{,\tau}}{R}\frac{(\rho^3 R^6)_{,\tau}}{\rho^3 R^6} + 6\frac{N^2}{R^2} - 8\pi N^2 \eta\right) + 8\pi N^2 P + 3\frac{N u_{,\tau}}{R}\right) q_2 \\ &- \left(\left(\frac{\partial p}{\partial \eta}\right)_s \left(\frac{R_{,\tau}}{R}\frac{(\rho^3 R^6)_{,\tau}}{\rho^3 R^6} + 4\frac{N^2}{R^2} - 8\pi N^2 \eta\right) + 8\pi N^2 P + 3\frac{N u_{,\tau}}{R} - 2\frac{N^2}{R^2}\right) q \\ &+ 8\pi N^2\left(\frac{\partial p}{\partial s}\right)_\eta \delta s' - N^2\left(\frac{\partial p}{\partial \eta}\right)_s \frac{q_{2,\mu\mu}}{A^2} = 0, \end{aligned}$$

In the above equations, the quantities s, ρ, P, and η are the entropy, rest mass density, pressure, and total energy density, and along with the background metric functions A, R, and N, are supplied by the hydrodynamic portion of the code. All even-parity equations are specialized for $l = 2$.

To summarize, we expanded the Einstein equations to first order in the perturbation quantities, developed a complete set of gauge invariant variables which were then introduced into the partial differential equations resulting from the linearized expansion, and then we were able to interpret the physical significance of these variables simply from the structure of the equations which they satisfy. We are left with a rather complicated and strongly coupled system of partial differential equations which describes essentially four physical features of the perturbed system: 1) a quantity related to the evolution of the perturbed entropy of the system, 2) a quantity related to the evolution of the perturbed vorticity of the fluid perturbations, 3) a quantity describing density perturbations which propagate at the speed of sound throughout the star, which is very strongly coupled to 4) a quantity which represents the propagation of gravitational radiation through the star. The importance of q_2 (the soundlike quantity) in this scheme must not be overlooked: It is actually the quantity which is most strongly coupled to the exterior perturbation, which will be seen when we display the matching equations between the interior and exterior perturbations.

Perturbations of the Stellar Exterior

In the stellar exterior, the background geometry is given by the Schwarzschild solution to the Einstein equations. In this case the background metric $\tilde{g}_{\mu\nu}$ takes the standard Schwarzschild form. As discussed previously, the perturbations of this background solution have been investigated in great detail. The techniques for the perturbation expansion in the stellar exterior are the same as those used in the interior. The perturbed metric is expanded in the same manner as in the interior, the only difference being that the background coefficients need not be determined by a complex computer code; they are known exactly from the Schwarzschild metric above. We can take the perturbed metric and expand the Einstein equations to first order, define an analogous set of gauge invariant objects, and compute their equations of motion. This work has already been done by CPM I and II, and we quote their results here. A gauge invariant quantity $\tilde{\psi}$ can be defined in the exterior, which obeys the simple equation of motion given by

$$\frac{\partial^2\tilde{\psi}}{\partial t^2} - \frac{\partial^2\tilde{\psi}}{\partial r_*^2} + V(r)\,\tilde{\psi}=0 \qquad 15$$

$$\text{where } V(r)=\begin{cases}\left(1-\frac{2M}{r}\right)\left(\left(1+\frac{3M}{2r}\right)^{-2}\left(\frac{9M^3}{2r^5}-\frac{3M}{r^3}\left(1-\frac{3M}{r}\right)\right)+\frac{6}{r^2\left(1+\frac{3M}{2r}\right)}\right) & \text{(even)} \\ \left(1-\frac{2M}{r}\right)\left(\frac{l\,(l+1)}{r^2}-\frac{6M}{r^3}\right) & \text{(odd).}\end{cases} \qquad 16$$

where r_* is the "tortoise" radial coordinate defined by $r_* = r + 2M\ln(r/2M - 1)$. (We denote exterior quantities with a tilde.) Here the odd-parity potential was first discovered by Regge and Wheeler and the even-parity potential was first discovered by Zerilli.

Matching at Stellar Surface

The matching equations can be obtained by requiring that both the induced metric, or first fundamental form, on the hypersurface which bounds the interior and exterior spacetimes, and the second fundamental form, which is related to the covariant derivative of the normal vector to the surface, be continuous at the boundary. These conditions are similar in spirit to the analogous conditions on the electric field across a boundary surface in electromagnetism, and a discussion of these requirements may be found in MTW 21.13. We require that these conditions be satisfied for both the background and perturbed spacetimes. Once this is done, we find these matching equations between the interior and exterior perturbations for the odd-parity system,

$$R\,\psi = \bar{\psi} \quad \text{and} \quad n^{\mu}\frac{\partial \psi}{\partial x^{\mu}} - 16\pi\, l(l+1)\, \rho w U(\tau,\mu) = \bar{n}^{\mu}\frac{\partial(\bar{\psi}/r)}{\partial x^{\mu}}, \qquad 17$$

and the following matching equations for the even-parity system:

$$q_2 = \text{function of } (\bar{\psi}, \bar{\psi}_{,r}) \qquad 18$$

$$q_{2,\tau\tau} = \text{complicated function of } (q, q_{,\mu}, q_{,\tau}, \bar{\psi} \ldots) \qquad 19$$

$$\bar{\psi}_{,\tau\tau} = \text{complicated function of } (\bar{\psi}, q, q_2, \bar{\psi}_{,\tau}, \ldots) \qquad 20$$

A few remarks are in order about the matching equations. The even-parity system has six interior functions matching to one exterior function, while the odd-parity match equations are straightforward and intuitive, effectively matching the gravitational wave and its normal derivative at the surface. The even-parity system is also very odd in that the gravity wave quantity in exterior matches directly to a sound wave quantity in interior! This strange complication is required by the mathematics, but is a nonintuitive coupling of the gravitational wave perturbations from the interior to the exterior of the star.

NUMERICAL TECHNIQUES

The Stellar Interior

In the interior of the star, we have written the partial differential equations that describe the perturbations as finite difference equations, using the staggered leapfrog technique (Smarr, 1979). This technique is ideal for a system like ours where we have wave equations written as a system of two first-order-in-time PDE's. The "q" variables are placed on integer time slices, while the momentum variables "π" are placed on the staggered grid of half-integer time slices. For the odd-parity system, a naive difference scheme works well throughout the star, but in the even-parity case we have found that special care must be taken near the origin. As has been noted by many authors, a wave propagating toward the origin is likely to develop a numerical singularity as it reaches the inner grid point (Evans 1986). We have tried several techniques to deal with this problem, but have settled on the numerical regularity technique of Evans , which involves writing the difference operators for spatial derivatives in such a way that the finite difference scheme reproduces the exact analytic result for derivatives near the origin. This technique has worked extremely well, and we urge the interested reader to consult Evans' article cited above for details.

Another problem that is seen only in the even-parity system involves a numerical instability in the perturbation equations when strong shocks are present. In a typical core collapse, a shock wave develops that may be capable of blowing off the outer part of the star, resulting in a supernova explosion. If the shock is sufficiently strong, it will lead to very sharp gradients in the hydrodynamic quantities across the shock front. In addition, in the wake of the shock one finds a great deal of numerical noise superimposed on the physically reasonable profiles of the various hydrodynamic quantities such as density, pressure, fluid velocity, etc. The standard technique developed over the years for handling shocks involves using an artificial viscosity that one adds to the pressure. Its effect is to spread the shock discontinuity over several zones while maintaining the jump conditions across the shock. When properly chosen, it can minimize the noise level in the wake of the shock and smooth out sharp gradients across the shock, while still allowing the code to produce reasonable physical results. However, even when the artificial viscosity has done its job in keeping the background hydrodynamics under control, we find that sharp gradients and numerical noise are still strong enough to cause problems for perturbation equations. The primary source of the problem is that the even-parity perturbation equations contain terms with second derivatives of the background quantities, which can become numerical delta functions when the background contains numerical noise or very sharp gradients. These derivative terms then cause noise in the perturbation quantities that can dominate the numerical solution. Since we are interested in the radiation emitted during a supernova explosion which is caused by a strong core bounce, this can be a very serious problem.

The obvious solution might seem to be to artificially smooth the background, but we have found that this solution does not work well. One has to be extremely careful in smoothing the background solution, because this may lead to a background solution which does not satisfy the Einstein equations. Because the first order solution expects the background to be an exact solution to the unperturbed Einstein equations, an arbitrary smoothing procedure of the background can lead to trouble in the perturbation equations. This can be most easily seen if one appeals to the regularity of the perturbation equations near the origin. If one expands the equations out for small values the radial variable, using the known analytic behavior of the background and perturbation functions near the origin, one will find that, e.g., in equation (14) all terms in the equation are of order $\mu^{2/3}$ or higher, *except* the following terms:

$$\frac{6N^2}{R^2}\left(\frac{\partial p}{\partial \eta}\right)_s q_2 - N^2\left(\frac{\partial p}{\partial \eta}\right)_s 4\pi\rho R^2\,(4\pi\rho R^2 q_{2,\mu})_{,\mu} - 2\left(\frac{\partial p}{\partial \eta}\right)_s q_{2,\mu}(4\pi\rho R^2)^2\frac{R_{,\mu}}{R}, \qquad 21$$

which appear to be order 1! Since this equation must equal zero, there must exist a delicate cancellation among these terms. If one looks more closely, one will find that the regularity of the solution demands certain relationships among the various background coefficients of the perturbation quantity q_2, which will be destroyed if some arbitrary smoothing has been done on background quantities.

An alternative solution, which seems to work well when applied with care, is to add a small diffusion term to the wave equations, which has the effect of damping out short wavelength oscillations (on the order of the grid spacing) which we cannot model anyway, while leaving the longer wavelength part of the signal unaffected. The form of this term is the following:

$$q_{,\tau} = \overline{\pi} + \varepsilon\, q_{,\mu\mu} + \ldots \quad \text{where } \varepsilon \text{ is very small.} \qquad 22$$

This technique has been very effective in maintaining the stability of the perturbation equations, but of course one has to be very careful not to wash out the physically interesting part of the signal! In practice we have added such a term to both the gravitational and sound wave equations, and the value of the diffusion coefficient has been chosen by trial and error so that it has the desired effect.

The Stellar Exterior

In the exterior, we have found it convenient to use null coordinates u and v defined by the equations $u = t - r_*$ and $v = t + r_*$ instead of Schwarzschild coordinates. The wave equation for the exterior metric perturbation $\tilde{\psi}$ in these coordinates is given by:

$$4\frac{\partial^2 \tilde{\psi}}{\partial u \partial v} + V(r)\tilde{\psi} = 0. \qquad 23$$

In the numerical code, this exterior wave equation is solved on a grid in u and v.

The exterior grid is actually divided up into two regions, an "near exterior" and a "far exterior". The grid lines of the "near exterior" are chosen so that they intersect the irregularly spaced interior time slices at the stellar surface. This makes the task of numerically matching quantities across the stellar surface much less complicated. However, a regular grid (that is, a grid where $\Delta u = \Delta v =$ constant between grid lines) is much more efficient and accurate for computing the actual propagation of the gravitational waves and the energy flux. For this reason, we use a regular "far exterior" grid beyond the "near exterior" region to actually carry the wave well into the radiation zone before the energy flux is determined. The common boundary of these grid regions is the worldline of an ingoing light ray that intersects the stellar surface at the last interior time step. At this boundary, the metric perturbation function ψ is computed on the far exterior slices (lines of constant v) by simple interpolation using its values on the near exterior slices. For a schematic diagram of the interior and exterior grids, see Figure 1, Paper I.

INITIAL CONDITIONS

The hydrodynamic initial conditions for the odd-parity models were chosen using two contrasting pre-collapse models that represent what might be considered extreme possibilities for the initial conditions prevailing in the star at the time of collapse. The first of these possibilities is a low-mass, low-temperature core computed by Nomoto and Hashimoto (1987, 1988). Their model produces a 1.17 $M_\odot$ iron core from an original 3.3 $M_\odot$ helium core presumed to be embedded in a main-sequence star having a total mass of 13 $M_\odot$. The second possibility is the core of a 15 $M_\odot$ star modelled by Weaver, et. al. (1978). Though this model is an older calculation that no longer represents the state of the art in stellar evolution calculations, it was chosen because it contrasts sharply with the Nomoto-Hashimoto model, and it seems to be a good first approximation to the cores developed by more recent models for very massive stars The core in this model is more massive (1.6 $M_\odot$) and is hotter than the Nomoto-Hashimoto model. Because neutrino emission rates increase with temperature, this model suffers significant neutrino losses during the infall phase of the collapse and thus produces a weaker shock after the bounce. We find that the Nomoto-Hashimoto initial conditions, on the other hand, leads to the production of a robust shock after the bounce. Other parameters being equal, the strengths of the shocks produced by these contrasting initial conditions probably bracket the range of shock strengths to be considered physically reasonable.

The initial conditions for the velocity and metric perturbations are taken to be time-symmetric on the initial slice, as in our earlier work. In the exterior, this amounts to choosing a stationary solution to the wave equation having the asymptotic behavior

$$\psi \sim q_l \, (2M/r)^l \qquad 24$$

which is that of a multipole with moment q_l. All runs are standardized by setting the moment q_l to unity. Since the value of this perturbation cannot change at a given radius until a signal indicating that the star has begun to collapse reaches that radius, these values will remain constant in the near exterior from the initial slice of time-symmetry until the first outgoing light ray arrives at the radius in question. In our code, therefore, the initial values of the stationary solution are taken to as the initial conditions for the exterior perturbation along this first ray.

For the odd-parity case, the interior metric perturbation is chosen so as to have a vanishing time-derivative on the initial slice, to have the appropriate behavior at the origin, and to match the unit-moment perturbation in the exterior. The fluid perturbation that drives the metric perturbation also must have the correct behavior at the origin and the correct over-all magnitude to be consistent with the unit-moment exterior metric perturbation, but is otherwise unconstrained. In our models, we have used only fluid perturbation functions that vanish at the surface: this simplifies the matching of the metric perturbation across that surface. In the case of dust collapse, the gravitational radiation is independent of the details of the shape of this function. In earlier runs involving nonzero pressure, we have found that the gravitational radiation depends only very weakly on the shape of this function. The shape actually used for the present odd-parity runs has a gaussian-like peak well within the homologous part of the core and goes to zero at both the stellar center and the surface. For further details concerning the way that the odd-parity initial conditions were chosen and established, see Paper I.

The initial value problem for even-parity perturbations in the interior is, as one might expect, more complicated. A description of the details of how the initial conditions were chosen in the preliminary even-parity runs described here would make a long and tedious story, but basically these perturbations were taken to be as time-symmetric as possible on the initial slice. We found that self-consistent and nearly time-symmetric initial conditions can be determined if one takes all of the the perturbation momenta and *most* of the time-derivatives of perturbation functions to be zero on the initial slice. Details will be forthcoming in a later paper.

RESULTS

Both even and odd-parity results for the pressureless collapse ("dust collapse") problem have been presented in detail by CPM I and II. They found that the radiation from such a collapse is dominated by very distinctive quasi-normal ringing of the background spacetime, and that the radiation was completely independent of the details of the interior velocity perturbation. If we set the equation of state to be $P = 0$ during the entire collapse, our code should reproduce these results. This was used as a test of the current code.

In the odd-parity case, the agreement between the current code and CPM I is excellent. The gravitational waveforms generated and the results for the total energy radiated are nearly identical. The current even-parity model also reproduces quite well the both the waveforms and the results for radiated energy described in CPM II. See Figures 1a and 1b for a comparison of Cunningham, Price and Moncrief's results with those of the current even and odd-parity codes.

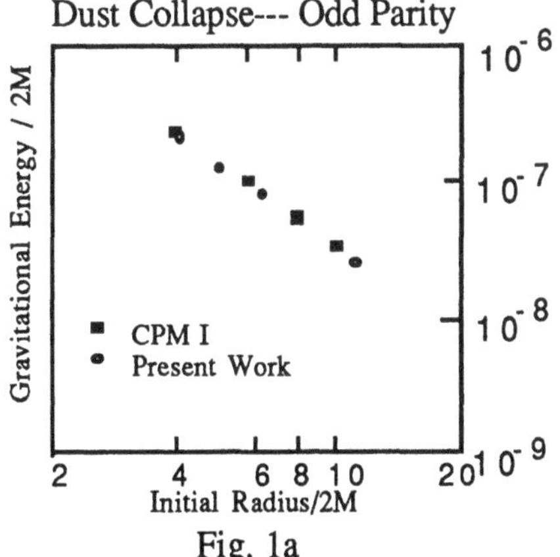

Fig. 1a

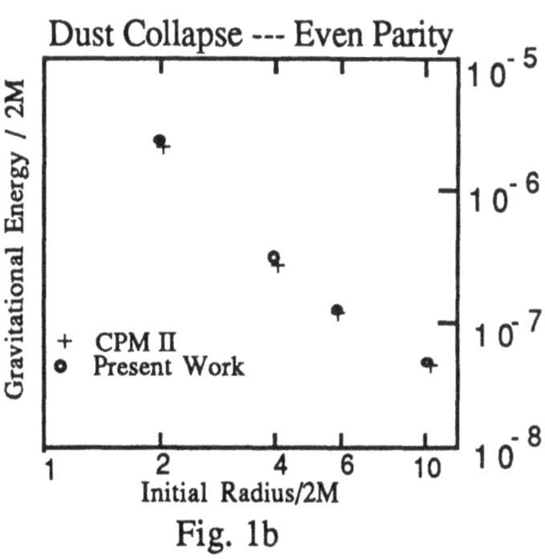

Fig. 1b

It is worth noting that gravitational radiation from collapse to a black hole, assuming that such a process reproduces most of the features seen in the dust collapse model, would be much easier to detect and identify than the gravitational radiation from a core bounce. For one thing, the magnitude of the energy radiated is large: the total energy radiated in the form of gravitational waves is about 10^{-3} M (where M is the mass of the black hole formed), which is at least three orders of magnitude more than that radiated by our current models that involve a core bounce. The quasi-normal ringing also makes for a very distinctive signature. If such processes turn out to be sufficiently common, they may represent excellent candidates for detection by the next generations of gravity-wave detectors.

Our earlier work on odd-parity radiation generated by core collapse, reported in Paper I, used very simple hydrodynamic model to provide the spherical background for the perturbation code. The model considered the collapse to be completely adiabatic, completely ignored neutrino transport effects, and employed a simplistic polytropic equation of state. The energy E radiated in the form of gravitational waves by these models lay within the range of 10^{-9} M < E < 10^{-5} M (where M is the mass of the collapsing core) with the most realistic model yielding $E \approx 10^{-7}$ M for a reasonable value of the perturbation strength.

In a set of recently completed odd-parity runs (Seidel et. al. 1988), we sought to bring the hydrodynamics model more in line with the current state of the art. A sophisticated neutrino transport model and a complex equation of state are currently considered to be essential components of a realistic Type II supernova model. But how do these features affect the gravitational radiation produced?

The hydrodynamics code used in our most recent runs is a fully "professional" model. It employs a multigroup flux limited neutrino transport scheme similar to the model described in detail in Myra et. al. (1987), but with the most important general relativistic correction added by consistently including a red-shift factor in all zones through which neutrinos are transported. The equation of state employed is the one described by Cooperstein (1985) with the supernuclear BCK parameterization described in Baron, et al. (1985a, b) This equation of state has three parameters that we have varied in this set of runs, one of which is important at sub-nuclear densities and the other two of which are important at trans-nuclear densities.

The first is the bulk symmetry coefficient W_s, which affects the ratio of free nucleons to nuclei as well as the ratio of free protons to neutrons. Its most important effect is its influence on the free proton mass fraction, with higher values of W_s corresponding to *fewer* free protons present. A smaller free proton fraction results in less electron capture, less energy radiated in the form of neutrinos, a larger homologous core at bounce, and ultimately a stronger shock wave after the bounce. The second and third parameters are the incompressibility K_0 of symmetric nuclear matter at saturation and the stiffness parameter γ (which amounts to an effective adiabatic index). Both of these parameters are primarily important at super-nuclear densities. Larger values of these parameters correspond to "stiffer" equations of state, which imply a smaller maximum density at the core bounce. The values for these parameters were varied to encompass the entire range of values currently considered to be likely, yielding equations of state that range from very soft to very stiff. It is likely that the correct values of these parameters lie somewhere in the middle of the ranges chosen. For more details concerning the hydrodynamics model for these runs, please refer to Paper II.

The hydrodynamics model only computes the evolution of the iron core of the star: because of the time scale of the collapse, the outer layers of the star do not contribute significantly to the details of the core collapse and therefore are ignored. Indeed, in the interests of simplicity, the core is considered by the model to be "bare", that is, surrounded by vacuum instead of the outer layers of the star. This assumption yields a tremendous simplification of the model code and is not likely to significantly affect the results. The model also chooses the pressure at the stellar surface so as to keep that surface at a fixed radius. This simplifies the matching equations and also ensures that the Schwarzschild mass of the star remains constant (if the surface is allowed to move with a non-zero pressure, work is done on the core that changes its effective mass, yielding an exterior background spacetime that is not strictly Schwarzschild). The outermost layers of the core do not really participate in the collapse, so this simplification also should not significantly affect the results. This boundary condition would lead to difficulties if the shock were to ever reach the surface, but since essentially all of the gravitational radiation goes through the surface long before the shock reaches it, we can stop the calculation before this becomes a problem.

For a given value of l, the amplitude of the gravitational waves produced by an odd-parity fluid perturbation depends only on a single parameter: the overall magnitude of that perturbation. It is difficult to choose an appropriate magnitude for this perturbation, as the details of the fluid motion in typical pre-supernova cores are not well known. We can put a physically reasonable upper limit on the magnitude of this fluid perturbation in the following manner. We follow the spatial maximum of the fluid perturbation throughout the collapse, find its minimum radius, and then compute the magnitude of the perturbation that would lead to rotational shedding of that layer (in the absence of external pressure). It is likely that well before this limit is actually reached, the perturbation can no longer be considered "small" and would significantly affect the hydrodynamic evolution of the star. The limit is therefore intended *only* to give a physical motivation for a choice of a perturbation strength. It does not meaningfully describe a physical upper limit on the energy that can be radiated (as highly aspherical cores may be able to radiate more), but probably does enclose the range of fluid perturbations that can be considered to be "small" in the sense of satisfying the constraints of this model.

As noted above, the initial shape of the fluid perturbation function does not appear to significantly affect the gravitational energy radiated. In our most recent runs, this shape was chosen to be a broad peak spanning most of the homologous core and falling to zero at the core surface. This shape was not varied from run to run. After choosing the magnitude and shape of the fluid perturbation as described above, we carried out a set of runs, varying both the parameters used in the equation of state and the model used to provide the initial hydrodynamic data. The results are summarized in Table 1.

Run	Unit Mom. ΔE	Max. ΔE	Prompt Expl.	γ	K_0^{sym}	W_s	Y_L	$\Delta v_{sh(cm/sec)}$
N09	1.4×10^{-14} M	3.5×10^{-8}M	yes	2.0	180	29.3	0.40	6.2×10^9
N07	2.5×10^{-14} M	6.8×10^{-8}M	yes	2.5	180	36.0	0.41	5.4×10^9
N10	3.9×10^{-14} M	1.1×10^{-7}M	no	4.5	220	29.3	0.40	3.9×10^9
W09	1.1×10^{-14} M	1.1×10^{-7}M	no	2.0	180	29.3	0.37	3.9×10^9
W11	1.8×10^{-14} M	1.7×10^{-7}M	no	2.5	180	29.3	0.37	3.4×10^9
W07	2.4×10^{-14} M	1.7×10^{-7}M	no	2.5	180	36.0	0.39	4.8×10^9
W10	3.6×10^{-14} M	3.5×10^{-7}M	no	4.5	220	29.3	0.37	2.4×10^9

Table 1. Y_L is the lepton to baryon number ratio averaged over the unshocked inner core. The smaller the value of Y_L, the larger is the amount of infall electron capture and neutrino loss for a given precollapse model. We give an indication of relative shock strengths, Δv_{sh}, by giving the change in velocity across the shock front once each shock wave has reached a radius of 150 km. This comparison, however, is only meaningful for calculations that use the same precollapse initial model. In fact, the weakest shock produced using the Nomoto and Hashimoto model (N10) is much stronger than the strongest shock produced using the Weaver, Zimmerman, and Woosley model (W07).

One can see that the energy radiated by all of these models at the perturbation magnitude chosen is on the order of magnitude of 10^{-7} M (where M is the core mass), with the Nomonto-Hashimoto hydrodynamic initial conditions generally yielding somewhat less than this and the Woosley initial models yielding somewhat more. A general trend that is present in runs involving both sets of initial conditions is that the models with stiffer equations of state are marginally more efficient in producing gravitational waves than models with "soft" equations of state. In collapses where the equation of state is soft, the self-similarity of the collapse is retained much longer into the collapse than in models where a stiff equation of state is used. Thus, a stiff equation results in a less orderly flow made evident by the greater generation of gravitational radiation. This result follows directly from the coupling between the matter and the gravitational perturbation found in the wave equation. If the matter terms are not smooth functions, or if they contain sharp gradients, they will drive strongly the gravitational perturbation function ψ. Furthermore, stiffer equations of state imply a shorter time scale for velocity and acceleration changes during the bounce, even though they ultimately produce a weaker shock. Since the energy radiated depends (very roughly) on the square of the third time-derivative of the mass quadrupole moment, the shorter time scale should lead to greater radiated energies.

Therefore, the kinds of soft equations of state that lead to prompt explosions may actually produce somewhat less gravitational radiation (other things being equal). If stiff equations of state do not lead to supernova explosions (This is a debate which we do not wish to enter!), then these results imply a very interesting conclusion: *optically silent core collapses will yield at least as much (and maybe somewhat more) gravitational radiation than Type II supernovae.* The stellar death rate in the Galaxy has been estimated to be ~ 0.1/year for stars with mass greater than 5 $M_{\odot}$, and perhaps half that rate for stars with mass greater than 20 $M_{\odot}$ (Bachall and Piran 1983). This rate is much larger than the observed rate. Presumably some fraction of these stellar deaths would lead to type II supernovae which would be obscured from view by dust in the Galaxy, while others may lead to core collapse events which will be optically silent. Our results indicate that to the gravitational wave astronomer, many of these events may be as luminous as typical type II supernovae.

Even-Parity Results and Work in Progress

The results described above for realistic core collapse models were for odd-parity perturbations. The even-parity code has been plagued with difficulties which were not encountered in the odd-parity system, but these problems have largely been solved and we expect to present results for this project soon. As shown in section above, the code reproduces the dust collapse calculations of CPM II as a test calculation. Runs that actually have been completed so far have used a hydrodynamic model like that of our earlier work on odd-parity perturbations (Paper I). This more basic model uses a simple polytropic equation of state and somewhat artificial initial conditions chosen to yield a reasonably realistic core bounce. A typical waveform for even-parity perturbations is shown in Figure 2, along with the corresponding collapse model.

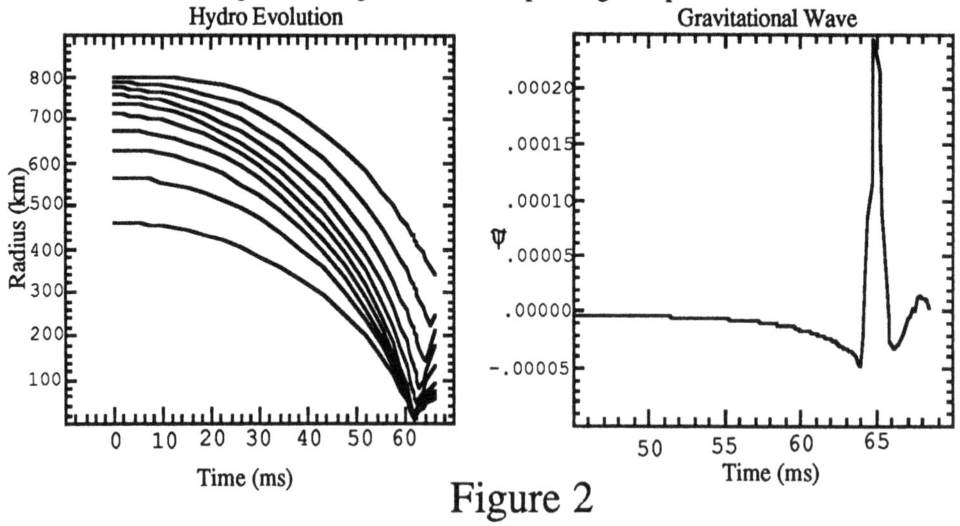

Figure 2

Both even and odd-parity perturbations can be calculated for a given hydro run so that they can be compared. Preliminary results indicate that even-parity perturbations radiate more energy than "comparable" odd-parity perturbations, but it is difficult to meaningfully compare the initial perturbations in the two cases because the nature of the driving perturbations involved are quite different. Furthermore, the initial choice of the perturbation fields is a much more complicated problem for the even-parity system, and it is possible that these differences in energy are due to differences in the initial conditions. We note that waveforms for the two cases are qualitatively similar. Work towards resolving this difficulty in the initial conditions is currently in progress.

A great deal of progress has been made on the even-parity perturbation problem, although some work remains to be done before results from realistic core collapse models can be obtained. We expect to continue the development of the even-parity perturbation code for several years as part of this research effort. This numerical code can be used as a continuing tool to test the effect of new ideas in supernova models on the production of gravitational waves. Any new equation of state for hot, dense matter can be easily incorporated in the code, and its impact on gravitational wave emission can then be evaluated. There is still great debate over the actual mechanism responsible for the supernova display, and as the various groups in the supernova model building industry converge on a common understanding of this mechanism, the perturbation code currently under development will be able to make quantitative predictions of the spectrum and amplitude of gravitational radiation produced in a supernova explosion.

A drawback of the odd-parity work already completed, and the even-parity code currently being developed, is that there is no particularly natural set of initial conditions for the perturbations. All models of supernovae are currently spherically symmetric, and it is difficult to estimate either the magnitude or even the nature of perturbations from spherical symmetry which might be found in an actual pre-supernova stellar configuration. For this reason, we have been limited so far to order of magnitude estimates based on physically reasonable restrictions on the strength of the fluid perturbation. This drawback can be removed in the even-parity problem, but one must continue to second order in the perturbation source. A first order, $l = 1$ odd-parity perturbation, which does not radiate, but which corresponds to a first order Kerr type perturbation on a non-rotating spherically symmetric background, can be used to drive an $l = 2$ even-parity perturbation mode. This slow rotation in first order provides natural initial conditions for the second order even-parity perturbation. Rotation is perhaps the most natural source of perturbations away from spherical symmetry in any stellar system, so this second order even-parity problem can be expected to give a reliable estimate of gravitational radiation from supernovae.

Another problem which we intend to tackle with this perturbation code is that of gravitational radiation from black hole collapse. Indications are that the quasi-normal ringing modes of a perturbed black hole may produce the strongest signal from stellar core collapse. However, the May-White metric coordinate conditions are not well suited to black hole collapse, so that the May-White code is more useful for core bounce models than for stellar collapse models which collapse to a black hole. Since the dust collapse results indicate that quasi-normal ringing radiation produced when a black hole is formed is several orders of magnitude stronger than radiation produced for realistic core bounce calculations, it is important to investigate black hole formation for more realistic equations of state. If the May-White code cannot be easily modified to follow black hole collapse in detail, then we may have to resort to a background code that uses a different choice of lapse and shift vector. This sort of background code would be ideal to follow black hole collapse, but it would require a major effort to rewrite the perturbation equations to accommodate a nonzero shift vector. Using a diagonal background metric, as in May-White, has the advantage of simplifying the perturbation equations, but a price is paid when trying to run the code near regions of singularity formation. Other choices of lapse and shift are much better suited for following the background evolution near singularities, though a considerable complication in the perturbation equations will result.

We would like to thank the conference organizers for an outstanding meeting. The calculations were performed at the Pittsburgh Supercomputer Center. Some of this work on odd-parity perturbations is the result of a collaboration of E. S., T. M., and Eric Myra at SUNY at Stony Brook, while the work on even-parity perturbations is the result of a collaboration of E. S. and Vincent Moncrief at Yale University. We would like to thank Vince Moncrief and Cliff Will for helpful discussions on a variety of issues related to this work. This research was supported by grants NSF-PHY 85-13953 and NASA NAGW-122 at Washington University, NSF-PHY85-03072 at Yale University.

References

Bachall, J. N., and Piran, T., *Astrophys. J.*, **267**, L77, (1983).
Baron, E. A., Cooperstein, J., and Kahana, S., *Nucl. Phys. A*, **440**, 744, (1985a).
Baron, E., Cooperstein, J., and Kahana, S., *Phys. Rev. Lett.*, **55**, 126, (1985b).
Bethe and Wilson *Ap.J.*, **295**, 14, (1985) .
Bethe, H.A., Brown, G.E., Applegate, J., and Lattimer, J.M., *Nuclear Physics A*, **324**, 487, (1979).
Brown, G.E., Bethe, H.A., and Baym, G., *Nuclear Physics A*, **375**, 481, (1982).
Burrows, A.S. and Lattimer, J.M., *Ap. J. Lett.*, **299**, L19, (1985).
Chandrasekhar S., and Detweiler, S., *Proc. R. Soc. Lond. A.*, **344**, 441, (1975).
Chandrasekhar S., *Proc. R. Soc. Lond. A.*, **343**, 289, (1975).
Colgate, S.A. and Johnson, H.J., *Phys. Rev. Let*, **5**, 235, (1960).
Cooperstein, J., *Nucl. Phys. A*, **429**, 527, (1985).
Cunningham, C.T., Price, R.H., and Moncrief, V. *Ap.J.* **224**, 643, (1978a).
Cunningham, C.T., Price, R.H., and Moncrief, V. *Ap.J.* **230**, 870, (1978b).
Cunningham, C.T., Price, R.H., and Moncrief, V. *Ap.J.* **236**, 674, (1980) .
Evans, C., "An Approach for Calculating Axisymmetric Gravitational Collapse", in Centrella, J., (ed), *Dynamical Spacetimes and Numerical Relativity.* (Cambridge University Press) (1986).
May, M.M. and White, R.H., *Methods Comput. Phys.* **7**, 219, (1967).
May, M.M. and White, R.H., *Phys. Rev.* **141**, 1232, (1966).
Myra, E. S., Bludman, S. A., Hoffman, Y., Lichtenstadt, I., Sack, N., and Van Riper, K. A., *Astrophys. J.*, **318**, 744, (1987).
Nomoto, K., and Hashimoto, M. 1988, *Physics Reports*, in press.
Nomoto, K., Shigeyama, T., and Hashimoto, M. 1987, in Proc. ESO Workshop on 'SN 1987A', ed. I. J. Danziger (ESO: Garching), p 325.
Regge, T., and Wheeler, J.A., *Phys. Rev.* **108**, 1063, (1957).
Seidel, E., and Moore, T., *Phys. Rev D.* **35**, 2287, (1987) (Paper I).
Seidel, E., Myra, E., and Moore, T., *Phys. Rev D.*, (1988), submitted (Paper II).
Smarr, L, "Basic Concepts in Differencing of Partial Differential Equations", in Smarr, L.L., (ed), *Sources of Gravitational Radiation.* (Cambridge University Press) (1979).
Weaver, T. A., Zimmerman, G. B., and Woosley, S. E., *Ap. J.*, **225**, 1021, (1978).
Wilson, J., *Numerical Astrophysics.* (ed. J. Centrella, J. LeBlanc, R. Bowers, Jones and Bartlett, Austin, p374 (1985)).
Woosley, S.E. and Weaver, T.A., *Ann.Rev. Astro. and Astrophys.*, **24**, 205 (1986).
Zerilli, F. J., *Phys. Rev. Lett.*, **24**, 737, (1970).

GENERAL RELATIVISTIC IMPLICIT RADIATION HYDRODYNAMICS IN POLAR SLICED SPACE-TIME

Paul J. Schinder
Physics Department, University of Pennsylvania, Philadelphia, PA 19104, USA

Abstract. The equations of general relativistic radiation hydrodynamics are presented under the condition of polar slicing of space-time. Adiabatic tests of a code developed using these equations are presented. This code will be used to compute the emitted spectra of neutrinos from various scenarios of stellar core collapse.

INTRODUCTION

The neutrino observations of Supernova 1987A (Hirata *et al.* 1987 ; Bionta *et al.* 1987) have dramatically confirmed the basic theory of Type II supernovae: a star of mass M > 8 $M_\odot$ undergoes collapse from an initially white dwarf-like state, stopping when the homologous (inner) core bounces at trans-nuclear densities. An outgoing shock is formed at the edge of the inner core, and propagates outwards. The shock may have sufficient energy to eject the outer layers of the star either directly, in about 100 ms (prompt mechanism) or after gaining additional energy from the neutrinos flowing through it (delayed mechanism) on a time scale of ~1 s. All during this dynamical phase, and for more than 20 s thereafter (Burrows & Lattimer 1986) the star radiates > 10^{53} ergs of neutrinos of all six types (e, μ, and τ neutrinos and antineutrinos), in roughly equal amounts. *Any* stellar core collapse, however, will be a copious producer of neutrinos, whether or not a supernova explosion occurs. There is no compelling reason to believe that every stellar core collapse results in the ejection of matter and a visible supernova. The occurrence or lack of mass ejection may well be sensitive to the detailed structure and composition of individual stars. All that can be said with confidence in general is that stellar collapse will result in either the formation of a remnant neutron star or a black hole and will radiate a large neutrino signal, with or without ejection of matter. Since stellar core collapse in our Galaxy is estimated to occur at a rate as high as 1/11 yr^{-1} (Bahcall & Piran 1983), and since neutrino astronomy is now well established, it is now desirable to construct codes which can predict the neutrino signal from various scenarios of stellar core collapse, both to neutron stars and to black holes. These codes will be useful in helping to interpret neutrino observations of Galactic core collapse. Neutrinos from Galactic core collapse may be able to tell us whether the prompt or delayed mechanism is correct. It is of course possible that both mechanisms play a role in nature, although it is becoming clear that it is difficult or impossible to get cores to explode by the prompt mechanism in numerical models with realistic equations of state and neutrino transport (Myra & Bludman 1988; Bruenn 1988). Galactic core collapse neutrinos may also provide a diagnostic for the equation of state of hot nuclear matter at trans-nuclear densities. The limited sample of 19 neutrinos from SN 1987A can be fit well by simple cooling models (Bludman & Schinder 1987), and gives no information on which supernova mechanism actually occurred or much information on the equation of state.

It is clear that both end points of any stellar collapse, either a neutron star or a black hole, require a fully general relativistic treatment of both gas dynamics and neutrino transport. Most modern supernova codes use fully relativistic, May and White (1966) type treatments of the gas dynamics, but range from fully relativistic to non-relativistic in their treatment of neutrino transport. While a non-relativistic or partially relativistic transport approximation may be adequate over time scales of ~100 ms, where neutrino transport affects the dynamics mostly by affecting the number of electrons per baryon Y_e (Myra and Bludman 1988, Bruenn 1988), it is unlikely to be adequate over the ~20 s during which the bulk of the neutrino signal from neutron star cooling occurs, or even over the ~1 s important for the delayed mechanism. May and White codes also suffer from the problem that once a singularity forms in the center, computation must stop, leaving a large amount of matter outside the developing horizon and still able to radiate neutrinos to infinity.

In this paper, I briefly describe our effort at constructing a fully general relativistic neutrino hydrodynamics code. A more detailed discussion may be found in Schinder *et al.* (1988) and Schinder (1988). We start with the Arnowitt *et al.* (1962; henceforth ADM) 3+1 formalism, and assume spherical symmetry throughout. We use Lagrangian (comoving) coordinates, and slice space time according to the polar slicing criterion of Bardeen & Piran (1983). Neutrino energies and directions are measured in comoving frames. The matter gas is assumed to be perfect. Neutrino transport will be described by moments of the fully general relativistic neutrino Boltzmann equation in polar slicing. One of the key features of this code is that the gas equations and neutrino transport equations are differenced *implicitly*, which enables us to escape the Courant restriction on the size of the timestep. The timestep is limited only by considerations of accuracy. With this time implicit code we have successfully integrated adiabatic stellar collapse calculations with simple equations of state to a time of 1000 s. We hope that the non-adiabatic code may reach times of 20-30 s with accurate neutrino transport of all six types of neutrinos in a reasonable amount of computer time. We are not at the moment striving for the accuracy necessary for a supernova code, which must accurately compute the 1% of the total liberated energy which powers the optical display and mass ejection, although we will in the future.

For a similar computational effort see Mezzacappa & Matzner (1988; this volume), who use maximal slicing, explicit hydrodynamics, a different coordinate system, and different techniques for solving the Boltzmann equation.

Gas Dynamics and Metric Equations

I start with the ADM metric

$$ds^2 = -(\alpha^2 - \beta_r\beta^r)dt^2 + 2\beta_r dr\, dt + (R'/\Gamma)^2\, dr^2 + R^2(d\theta^2 + \sin^2\theta\, d\phi^2) \tag{1}$$

where α is the lapse function, β^r is the shift vector, $\beta_r = g_{rr}\beta^r$, and R is the areal radius. The radial coordinate r can be any Lagrangian coordinate, such as the enclosed baryon number. I use geometrical units (G=c=1) throughout. Greek indices range over $\{t,r,\theta,\phi\}$. I use the notation $\dot{} \equiv \partial/\partial t$ and $' \equiv \partial/\partial r$. I require the gas four-velocity to be $u^\alpha = (u^t,0,0,0)$, so the coordinates move with the gas (Lagrangian coordinates). The condition $\mathbf{u}\bullet\mathbf{u} = -1$ requires $u^t = 1/\sqrt{\alpha^2 - \beta_r\beta^r}$. Note that u_α has non-zero components $u_t = -\sqrt{\alpha^2 - \beta_r\beta^r}$ and $u_r = \beta_r/\sqrt{\alpha^2 - \beta_r\beta^r}$.

Polar slicing requires $K^\theta_\theta = 0$, where K_{ij} is the extrinsic curvature tensor. This leads to a simple evolution equation for R

$$\dot{R} = \beta^r R' \tag{2}$$

I will define $U \equiv \dot{R}/\alpha$ and will use U to eliminate $\dot{R}$, β^r, and β_r.

The evolution equation for $K_{\theta\theta}$ becomes

$$\frac{\alpha'}{\alpha} = \frac{R'}{\Gamma^2}\left[\frac{M}{R^2} + 4\pi R\left(\frac{\Gamma^2(P+P_\nu) + 2U\Gamma F_\nu + U^2(\rho+E_\nu)}{\Omega^2}\right)\right] \tag{3}$$

Here $\Omega^2 \equiv \Gamma^2 - U^2$, $M \equiv \frac{1}{2}R(1-\Gamma^2)$, P is the gas pressure, and ρ the total energy density of the gas, related to the baryon rest mass density ρ_o by $\rho = \rho_o(1+e)$, where e is the internal energy per baryon mass. The gas is assumed to be a perfect fluid, so the gas stress energy tensor $T^{\alpha\beta} = (\rho+P)u^\alpha u^\beta + P\,g^{\alpha\beta}$. The quantities E_ν, F_ν, and P_ν are, respectively, the total neutrino energy density, radial flux, and radial pressure, and include all six neutrino types.

The Hamiltonian constraint equation is

$$M' = 4\pi R^2 R' \left(\frac{\Gamma^2(\rho+E_\nu) + 2U\Gamma F_\nu + U^2(P+P_\nu)}{\Omega^2}\right) \tag{4}$$

The energy equation is

$$\dot{e} + P\dot{v} = -\frac{2\pi}{h^3}\frac{\alpha\Omega v}{\Gamma}\int dE \int d\mu\, E^3 \sum_{\text{types}}(G_i - L_i f_i) \tag{5}$$

where $v \equiv 1/\rho_o$, E is the neutrino energy, μ the neutrino direction cosine from the radial, G the total source of neutrinos, L the total sink of neutrinos, and f the neutrino distribution function. The source (sink) includes any process which can move a neutrino into (away from) a given energy and direction.

The equation of baryon number conservation is

$$\frac{\partial\left(\frac{\rho_o R^2 R'}{\Omega}\right)}{\partial t} = 0 \tag{6}$$

The momentum equation is

$$\frac{P'}{R'} + \frac{\dot{P}U}{\alpha\Omega^2} + \frac{(\rho+P)\Gamma}{\alpha\Omega^4}(\Gamma\dot{U} - U\dot{\Gamma}) + \frac{(\rho+P)}{\Omega^2}\left(\frac{M + 4\pi R^3[P+P_\nu+(U/\Gamma)F_\nu]}{R^2}\right)$$
$$= -\frac{2\pi}{h^3\Omega}\int dE \int d\mu\, E^3\mu \sum_{\text{types}}(G_i - L_i f_i) \tag{7}$$

Conservation of lepton number (only electron lepton number need be explicitly conserved; μ and τ lepton numbers are conserved identically) may be written

$$n_B\frac{\partial Y_e}{\partial t} = -\frac{2\pi}{h^3}\frac{\alpha\Omega}{\Gamma}\int dE \int d\mu\, E^2(G_{\nu_e} - L_{\nu_e} f_{\nu_e} - \overline{G}_{\nu_e} - \overline{L}_{\nu_e}\overline{f}_{\nu_e}) \tag{8}$$

where n_B is the baryon number density. Here an overbar denotes antineutrino.

The boundary condition used below for the adiabatic tests may be found by matching our interior metric to a vacuum Schwarzschild metric at the surface of the star where $T^{\alpha\beta} = 0$. We are currently investigating whether matching to a Vaidya metric will provide an exact boundary condition for the case with neutrino radiation. The adiabatic boundary condition at the surface is $\alpha = \Gamma$.

Neutrino Boltzmann Equation

To derive the neutrino Boltzmann equation, I follow Lindquist (1966), who derived the Boltzmann equation using the coordinates and metric (equation 1 with $\beta^r = 0$) of May and White (1966). Choosing our comoving coordinates, comoving neutrino observers to measure neutrino energies and directions, and cranking through the formalism leads to the comoving neutrino Boltzmann equation

$$\Bigg\{\left[\frac{1}{\sqrt{\alpha^2-\beta_r\beta^r}}+\frac{\mu\beta_r\Gamma}{R'\alpha\sqrt{\alpha^2-\beta_r\beta^r}}\right]\frac{\partial}{\partial t}+\frac{\Gamma}{R'}\frac{\sqrt{\alpha^2-\beta_r\beta^r}}{\alpha}\frac{\partial}{\partial r}$$
$$-\frac{E}{\sqrt{\alpha^2-\beta_r\beta^r}}\left[\mu A+\mu^2 B+\frac{\dot{R}}{R}(1-\mu^2)\right]\frac{\partial}{\partial E} \tag{9}$$
$$-\frac{1-\mu^2}{\sqrt{\alpha^2-\beta_r\beta^r}}\left[A-C+\mu B-\mu\frac{\dot{R}}{R}\right]\frac{\partial}{\partial\mu}\Bigg\}f(E,\mu)$$
$$=G(E,\mu;f,\overline{f})-L(E,\mu;f,\overline{f})\,f(E,\mu)$$

where the notation $G(E,\mu;f,\overline{f})$ and $L(E,\mu;f,\overline{f})$ is used as a reminder that G and L are integral functions of both the distribution function of the neutrino and its corresponding antineutrino. The coefficients

$$A=\frac{\Gamma}{\alpha R'}\left(\alpha\alpha'-\frac{1}{2}(\beta_r\beta^r)'+\dot{\beta}_r-\beta_r\frac{\alpha\dot{\alpha}-\frac{1}{2}(\dot{\beta_r\beta^r})}{\alpha^2-\beta_r\beta^r}\right)$$
$$B=\frac{1}{\alpha^2-\beta_r\beta^r}\left(\beta_r\dot{\beta}^r-\beta_r\beta^r\frac{\dot{\alpha}}{\alpha}+\alpha^2\left[\frac{\dot{R}'}{R'}-\frac{\dot{\Gamma}}{\Gamma}\right]\right) \tag{10}$$
$$C=\frac{\Gamma}{\alpha R}\left(\beta_r\frac{\dot{R}}{R'}+\alpha^2-\beta_r\beta^r\right)$$

When $\beta^r = 0$, this agrees with the equation derived by Lindquist (1966).

Adiabatic Tests

We now set all neutrino distributions to zero, and consider two adiabatic tests which we made of our code (which we call GRIPOS). The first test is to start with an initially hydrostatic solution of the Tolman-Oppenheimer-Volkoff equation, and integrate it forward in time to make sure that the hydrostatic solution is maintained. Figure 1 shows what happens for the first 20 ms (the integration was run forward to 1000 s, but the rest is boring). Note that a slight amount of noise is introduced as the TOV solution was interpolated from the fine grid used by the TOV solver to the 100 equal baryon mass zones used here. However, the noise quickly disappears. Also note the small triangles on the time axis, showing the timesteps used. Near 20 ms the effects of the implicit differencing is noticeable, as the timestep increases since there is no dynamics occurring. At the end of this computation the timestep is much greater than the Courant time.

Figure 1. An initially hydrostatic star is integrated forward in time, testing whether or not the hydrostatic configuration is preserved. Only the first 20 ms of 1000 s is shown.

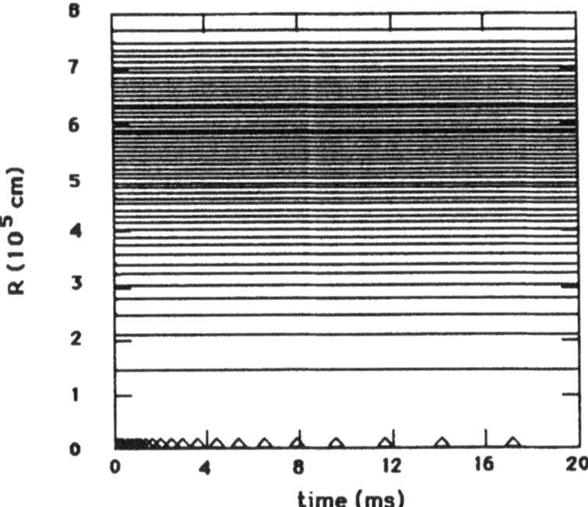

The second test which we perform is to compare our solution for the collapse of an initially uniform sphere of pressureless dust with the analytic solution of Petrich *et al.* (1986) for the same problem in polar slicing. I will compare here only two quantities, the lapse α and the baryon rest mass density ρ_0; other quantities agree equally well. Figure 2 shows a side by side comparison of the "collapse of the lapse" for the computed and analytic solution. Figure 3 shows a side by side comparison of ρ_0. As is evident, the computed solution is very close to the actual solution (the actual errors are on the order of 1%).

Figure 2. Side-by-side comparison of the lapse function α for the collapse of initially homogeneous dust computed by GRIPOS (left) and from the analytic solution (right). Shown is the value of α at 20 different times during the collapse.

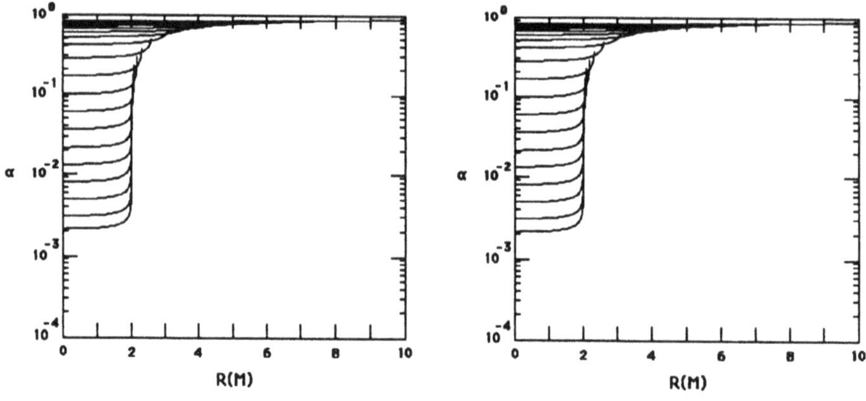

Figure 3. Side-by-side comparison of the baryon rest mass density ρ_0 for the collapse of initially homogeneous dust computed by GRIPOS (left) and from the analytic solution (right). Shown is the value of ρ_0 at 20 different times during the collapse.

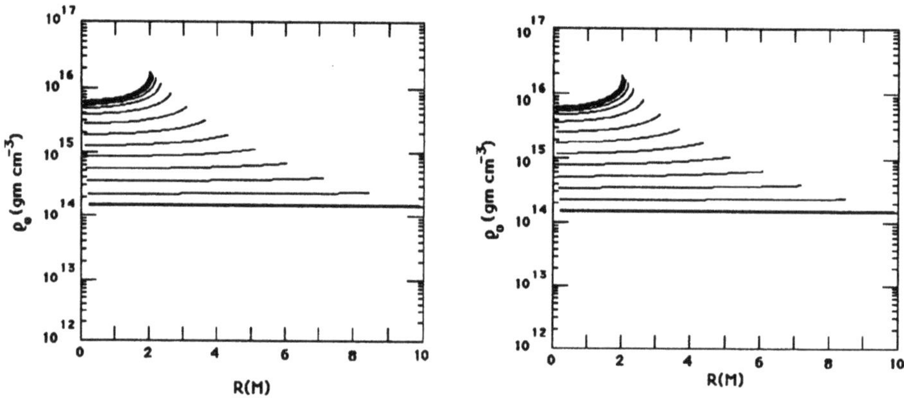

The results of these tests and others make us confident that the adiabatic version of our code is working correctly.

Future Plans and Conclusion

The construction of RGRIPOS, the neutrino transport version of GRIPOS, is in progress. For the details of how we will numerically handle neutrino transport, see Schinder (1988). Neutrino transport will be done fully implicitly, as the gas dynamics already is. With the completed code we expect to be able to compute to a time at least 20 s from the beginning of stellar collapse, and calculate accurate neutrino spectra for all six types of neutrinos. With these computed spectra, we will be able to predict what kinds of neutrino signals will be observed for various cases of stellar core collapse, with a wide range of input physics.

To conclude, I show the results of an adiabatic calculation of stellar core collapse, computed with a very simple equation of state designed to mimic the real equation of state but which will produce a prompt explosion. The initial condition is an n=3 relativistic polytrope with a central density of 4×10^9 gm cm^{-3}. The pressure is reduced by 1% throughout the star to start the collapse. This computation was run out to 1000 s, which took about 6000 timesteps.

Figure 4. A "mock supernova", a core collapse computed with a simple equation of state designed to give mass ejection. Shown are the areal radii of half of the 50 zones used in this computation over three different time intervals. Note that a hydrostatic core forms very quickly after bounce, while several zones are ejected from the star. The triangles on the time axis show the timesteps used.

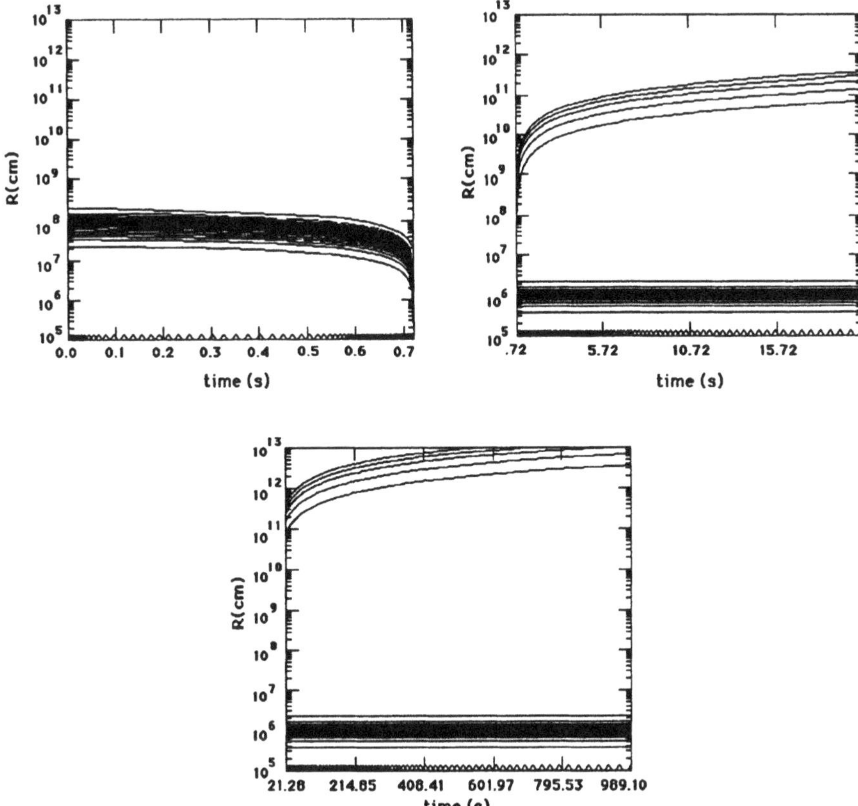

This is an example of the kind of calculation we hope to do in the future with fully relativistic neutrino transport and hydrodynamics.

This work was supported in part by DOE contract EY-76-C-02-3071 and in part by a grant of computer time at the John von Neumann National Supercomputer Center.

References

Arnowitt, R., Deser, S., & Misner, C. W. (1962), in *Gravitation: an Introduction to Current Research*, edited by L. Witten. Wiley.

Bahcall, J. N., & Piran, T. (1983), *Ap. J. (Letters)*, **267**, L77.

Bionta, R. M., *et al.* (1987), *Phys. Rev. Letters*, **58**, 1494.

Bludman, S. A., & Schinder, P. J. (1987), *Ap. J.*, **326**, 265.

Bruenn, S. (1988), submitted to *Ap. J.*

Burrows, A., & Lattimer, J. M. (1986), *Ap. J.*, **307**, 178.

Hirata, K., *et al.* (1987), *Phys. Rev. Letters*, **58**, 1490.

Lindquist, R. W. (1966), *Annals of Physics*, **37**, 487.

May, M., & White, R. H. (1966), *Phys. Rev.*, **141**, 1232.

Mezzacappa, A., & Matzner, R. A. (1988), this volume.

Myra, E. S., & Bludman, S. A. (1988), UPR-0352T, submitted to *Ap. J.*

Petrich, L. I., Shapiro, S. L., & Teukolsky, S. A. (1986), *Phys. Rev. D*, **33**, 2100.

Schinder, P. J., UPR-0351T, to appear in *Phys. Rev. D.*

Schinder, P. J., Bludman, S. A., & Piran, T. (1988), *Phys. Rev. D*, **37**, 2722.

GENERAL RELATIVISTIC RADIATION HYDRODYNAMICS IN SPHERICALLY SYMMETRIC SPACETIMES

A. Mezzacappa
Center for Relativity, University of Texas at Austin,
Austin, Texas 78712-1081

R. A. Matzner
Center for Relativity and Physics Department,
University of Texas at Austin, Austin, Texas 78712-1081

ABSTRACT. In this paper we present the basic equations on which our "3 + 1" general relativistic radiation hydrodynamics code for spherically symmetric spacetimes is based. We also present the results of a code test for the radiation transport and matter-radiation coupling.

During stellar-core collapse neutrinos are generated by electron capture onto nucleons and nuclei in the stellar core. Electron capture reduces the electron degeneracy pressure in the core, thereby destabilizing it to further gravitational collapse. Furthermore, electron capture in the core depends on the subsequent neutrino transport of the produced neutrinos. A careful treatment of the neutrino transport in the core is therefore essential to obtaining correct core dynamics. Most treatments up to now have used some approximation in handling the neutrino transport, for example, multigroup flux-limited diffusion (Bruenn 1985). In our code the neutrino transport is handled without approximation by using a general relativistic Boltzmann equation. A general relativistic Boltzmann equation was also used by Wilson (1971) in early studies of gravitational collapse. A careful treatment of the neutrino transport is also essential in computing accurate neutrino spectra from gravitational collapse. The neutrino spectrum provides a signature of the collapse, and when compared with an actual spectrum, a computed spectrum can be used as an important diagnostic tool for core collapse models and as a way to distinguish between the birth of a neutron star and the birth of a black hole. Mayle et al. (1987) have computed neutrino spectra from stellar collapse to stable neutron stars for different collapse models using multigroup flux-limited diffusion for the neutrino transport. We want to compute neutrino spectra from stellar collapse to neutron stars *and* to black holes, using Boltzmann transport for the neutrino transport in both cases. Because stellar-core collapse may result in the formation of a neutron star or in the formation of a black hole, it is essential to be able to compute at least up to the formation of an event horizon. Computations of spherically symmetric gravitational collapse without radiation that were taken beyond the formation of an event horizon have been carried out by a number of authors (Shapiro & Teukolsky

1980; Evans 1984; Schinder et al. 1988). Their computer codes were based on the "3 + 1" versions of the general relativistic hydrodynamics equations and Einstein equations. We have added general relativistic Boltzmann transport, the radiative contributions to the general relativistic hydrodynamics equations, and the radiative contributions to the Einstein equations to Evans' code for spherically symmetric general relativistic hydrodynamics. In our code the radiation phase-space distribution function is obtained by solving the "3 + 1" general relativistic Boltzmann equation. The spacetime radiation densities are obtained by calculating the various moments of the distribution function. The emergent radiation spectrum is also calculated from the distribution function. The Boltzmann equation is coupled to the hydrodynamics via the radiation 4-force density in the "3 + 1" general relativistic hydrodynamics equations. These equations are solved for the matter densities. Once the matter and radiation densities are known, they are used as sources in the "3 + 1" Einstein equations. The Einstein equations then determine the spacetime geometry.

In this paper we present the basic equations on which our radiation code is based. We also present the results of one code test we have successfully run. For more details regarding the theoretical foundation of our work and for details regarding the numerical techniques used in our code, we refer the reader to Mezzacappa & Matzner (1988). We also refer the reader to work by Schinder (1988 a,b). Schinder is developing a code based on the "3 + 1" general relativistic radiation hydrodynamics moment equations.

For spherically symmetric spacetimes the "3 + 1" metric in the isotropic gauge is

$$ds^2 = -(\alpha^2 - \beta^2 A^2)\,dt^2 + 2\beta A^2\,dr\,dt + A^2(\,dr^2 + r^2\,d\theta^2 + r^2 \sin^2\theta\,d\phi^2)\ . \quad (1)$$

The lapse function, α, is determined by the maximal slicing condition

$$\frac{1}{r^2}\partial_r\left[r^2\partial_r(\alpha A^{1/2})\right] = \alpha A^{1/2}\frac{1}{4}A^2\left[8\pi G(\rho_E + E_R) + 16\pi G(S + P_R) + \frac{21}{4}(K^r{}_r)^2\right] \quad (2)$$

where ρ_E, E_R, S, and P_R are the matter energy density, radiation energy density, trace of the matter stress tensor, and trace of the radiation stress tensor, respectively, measured by the Eulerian observers (those observers whose 4-velocities are the unit timelike normals to the spatial slices). G is the gravitational constant. The isotropic gauge condition gives

$$\beta = -\frac{3}{2}\,r\int_r^\infty \alpha K^r{}_r\,\frac{1}{r'}\,dr' \quad (3)$$

where ${K^r}_r$ is the radial component of the extrinsic curvature tensor. ${K^r}_r$ is determined by the momentum constraint

$$ {K^r}_r = 8\pi G A^{-3} r^{-3} \int_0^r A^3 r'^3 (S_{r'} + F_{r'})\, dr' \tag{4} $$

where S_r is the radial matter momentum density measured by the Eulerian observers and $F_r = A{F_R}^1$, where ${F_R}^1$ is the radial radiative flux measured by these observers. We then have ${K^\phi}_\phi = {K^\theta}_\theta = -\frac{1}{2}{K^r}_r$ for the other nonzero components of the extrinsic curvature tensor. The metric function, A, is determined by the Hamiltonian constraint

$$ \frac{1}{r^2}\,\partial_r\,(r^2 \partial_r A^{1/2}) = -\frac{1}{4}A^{5/2}\left[8\pi G(\rho_E + E_R) + \frac{3}{4}({K^r}_r)^2\right] . \tag{5} $$

We assume that the matter is a perfect fluid in its own rest frame so that the matter energy-momentum tensor may be written in terms of the matter 4-velocity, $U^a = (U^t, U^r, 0, 0)$, as $T^{ab} = \rho h U^a U^b + P g^{ab}$, where $h = 1 + \varepsilon + p/\rho$ is the specific relativistic enthalpy. ε is the specific internal energy density, ρ is the rest mass density, and P is the isotropic pressure. Define $U = \alpha U^t$, $V = U^r/U^t$, $D = \rho U$, and $E_M = \rho\varepsilon U$. In terms of these variables, a Lorentz transformation from the Eulerian frame to the fluid frame gives

$$ \rho_E = U(D + E_M) + (U^2 - 1)P \tag{6} $$

$$ S = U^{-1}(U^2 - 1)(D + E_M) + (U^2 + 2)P \ . \tag{7} $$

The hydrodynamics equations governing the evolution of D, E_M, and S_r are

$$ \begin{aligned} \partial_t D = &- D\partial_t\, \ell n\, A^3 \\ &- \frac{1}{r^2 A^3}\,\partial_r\,(r^2 A^3 D V) \end{aligned} \tag{8} $$

$$ \begin{aligned} \partial_t E_M = &- E_M \partial_t\, \ell n\, A^3 \\ &- \frac{1}{r^2 A^3}\,\partial_r (r^2 A^3 E_M V) \\ &- \alpha P \theta \\ &+ \alpha\gamma(G^0 - V^1 G^1) \end{aligned} \tag{9} $$

$$ \begin{aligned} \partial_t S_r = &- S_r \partial_t\, \ell n\, A^3 \\ &- \frac{1}{r^2 A^3}\,\partial_r\,(r^2 A^3 S_r V) \\ &- \alpha(D + E_M + PU)\left[U\partial_r\, \ell n\, \alpha + \left(\frac{1}{U} - U\right)\partial_r\, \ell n\, A\right] + S_r \partial_r \beta \\ &- \alpha\partial_r P \\ &+ \alpha A G^1 \end{aligned} \tag{10} $$

where $\theta = U^a{}_{;a}$ is the fluid expansion, $V^1 = AU(\beta + V)/\alpha$, and G^0 and G^1 are the rates of energy and momentum exchange per unit volume between the matter and the radiation, respectively, as measured by the Eulerian observers. These equations are supplemented by the following equations for U and V

$$U = \left[1 + \frac{S_r^2}{A^2(D + E + PU)^2}\right]^{1/2} \tag{11}$$

$$V = \frac{\alpha}{A} \frac{S_r}{\left[A^2(D + E + PU)^2 + S_r^2\right]^{1/2}} - \beta \tag{12}$$

and by an equation of state of the form $P = P(E_M)$.

We use spherical-momentum space coordinates so that the radiation 4-momenta measured by the Eulerian observers are given by $p^0 = E$, $p^1 = \mu E$, $p^2 = (1 - \mu^2)^{1/2} E \sin\phi_p$, and $p^3 = (1 - \mu^2)^{1/2} E \cos\phi_p$, where $\mu \equiv \cos\theta_p$. The general relativistic Boltzmann equation for the metric (1) and for maximal slicing is

$$\begin{aligned} D_r f &+ \frac{\mu}{(rA)^2} \frac{1}{A} \frac{\partial}{\partial r} \left[(rA)^2 f\right] \\ &+ \frac{1}{A}\left(\partial_r \ell n \frac{rA}{\alpha}\right) \frac{\partial}{\partial \mu} \left[(1 - \mu^2) f\right] + \frac{3}{2} K^r{}_r \frac{\partial}{\partial \mu} \left[\mu(1 - \mu^2) f\right] \\ &- \left[\frac{\mu}{A} \partial_r \ell n\, \alpha + \frac{3}{2}\left(\mu^2 - \frac{1}{3}\right) K^r{}_r\right] \frac{1}{E^3} \frac{\partial(E^4 f)}{\partial E} \\ &= \frac{1}{E}(e - of) - \left[\frac{2\mu}{A} \partial_r \ell n\, \alpha - \frac{3}{2}\left(\mu^2 - \frac{1}{3}\right) K^r{}_r\right] f \\ &\equiv \frac{1}{E}(e - o^* f) \, . \end{aligned} \tag{13}$$

e is the invariant emissivity given by the nonrelativistic emissivity in the Eulerian frame, η, as $h^2\eta/2E^2$, and o is the invariant opacity given by the nonrelativistic opacity in the Eulerian frame, χ, as $E\chi$. In terms of the invariant distribution function, f, the radiation moments measured by the Eulerian observers are

$$E_R = R^{00} = \frac{4\pi}{h^3} \int E^3 \, dE \, d\mu \, f \tag{14}$$

$$F_R{}^1 = R^{01} = \frac{4\pi}{h^3} \int E^3 \, dE \, d\mu \, \mu \, f \tag{15}$$

with $P_R = E_R$. We also have

$$G^0 = \frac{4\pi}{h^3} \int E^2 \, dE \, d\mu \, (of - e) \tag{16}$$

$$G^1 = \frac{4\pi}{h^3} \int E^2 \, dE \, d\mu \, \mu \, (of - e) \; . \tag{17}$$

We refer the reader to Mezzacappa & Matzner (1988) for details on coding the above equations. Here we present the results of a code test for the radiation transport and matter-radiation coupling. This test was first described by Lund (1985). It is based on physical conditions in a neutron star during the neutrino cooling phase. Consider a *static* star with density profile, $\rho = \rho \exp(-r/r_0)$, where $\rho_0 = 1 \times 10^{15}$ g cm^{-3} and $r_0 = 1 \times 10^6$ cm. The star is assumed to have a constant heat capacity, $C_V = 1 \times 10^{15}$ ergs g^{-1} MeV^{-1}. Hence, from $C_V = \rho^{-1}(\partial E_M/\partial T)_V$, we have

$$T = \rho^{-1} C_V^{-1} E_M \; . \tag{18}$$

The *initial* temperature profile is given by an adiabatic, polytropic law, $T = T_0(\rho/\rho_0)^{\gamma-1}$, with $\gamma = \frac{4}{3}$ and $T_0 = 24.4$ MeV. The radiation is initially isotropic and is given in terms of the local matter temperature by the Planck distribution, $[\exp(E/T) - 1]^{-1}$. The opacity, χ, is

$$\chi = \kappa\rho \tag{19}$$

with $\kappa = \kappa_0 E^2$ and $\kappa_0 = 1.24 \times 10^{-20}$ cm^2 g^{-1} MeV^{-2}. The emissivity, η, is

$$\eta = \chi B(T) \; , \tag{20}$$

where $B(T)$ is the Planck emission function, $B(T) = 2h^{-2}E^3[\exp(E/T) - 1]^{-1}$.

In the first figure we plot the momentum-space averaged mean free path. It can be seen that the radiation is very collisional near the center of the star where the average mean free path is much less than the scale height of the star (r_0) but essentially noncollisional in the outer parts of the star where the average mean free path is much greater than the scale height of the star.

In the second figure we plot the average value of μ at an elapsed time of 0.1 sec. The results indicate a smooth transition from the optically thick central region of the star where the radiation field is isotropic and the radiative transport is well described by diffusion to the optically thin outer region of the star where the radiation field is highly anisotropic and radiative transport is well described by free streaming. $\mu = -1$ corresponds to radial, inward propagation. $\mu = +1$ corresponds to radial, outward propagation.

In Fig. 3 we display the emitted spectrum at an elapsed time of 0.1 sec.

The results of the Lund test indicate that our code correctly computes the radiation transport in optically thick regions, in optically thin regions, and perhaps most importantly, in intermediate regions where diffusion and free streaming approximations break down. The results include a reliable spectrum of emitted radiation, computed by a correct description of the radiation transport.

The results of the Lund test give us confidence that our code will also yield accurate results when applied to gravitational collapse. Because of the singularity avoidance properties of maximal slicing (Shapiro & Teukolsky 1980; Evans 1984), the investigation of gravitational collapse to neutron stars *and* to black holes, including neutrinos and neutrino transport, will be possible and the determination of the resulting emitted neutrino spectra from these events will also be possible.

We want to thank Dr. Charles R. Evans for very extensive and useful discussions regarding the radiation code test presented in this paper and especially for providing a copy of his hydrodynamics collapse code for our use. We also want to thank Dr. Robert Harkness for useful discussions regarding code optimization. This work was supported by NSF grants PHY-8404931 and PHY-8506567.

Bruenn, S. W. (1985). Stellar core collapse: numerical model and infall epoch. Astrophysical J. Suppl., 58, 771–841.

Evans, C. R. (1984). A method for numerical relativity: simulation of axisymmetric gravitational collapse and gravitational radiation generation. Ph. D. dissertation, The University of Texas at Austin.

Mayle, R. et al. (1987). Neutrinos from gravitational collapse. Astrophysical J., 318, 288–306.

Mezzacappa, A. & Matzner, R. A. (1988). Computer simulation of time-dependent, spherically symmetric spacetimes containing radiating fluids: formalism and code tests. Astrophysical J., submitted.

Schinder, P. J. (1988a). General relativistic radiation hydrodynamics and neutrino transport in polar sliced space-time. University of Pennsylvania preprint UPR-0351T.

Schinder, P. J. (1988b). This volume.

Schinder, P. J. et al. (1988). General relativistic implicit hydrodynamics in polar sliced space-time. Phys. Rev. D37, 2722–2731.

Shapiro, S. L. & Teukolsky, S. A. (1980). Gravitational collapse to neutron stars and black holes: computer generation of spherical space-times. Astrophysical J., 235, 199–215.

Wilson, J. R. (1971). A numerical study of gravitational stellar collapse. Astrophysical J., 163, 209–219.

FIGURE 1

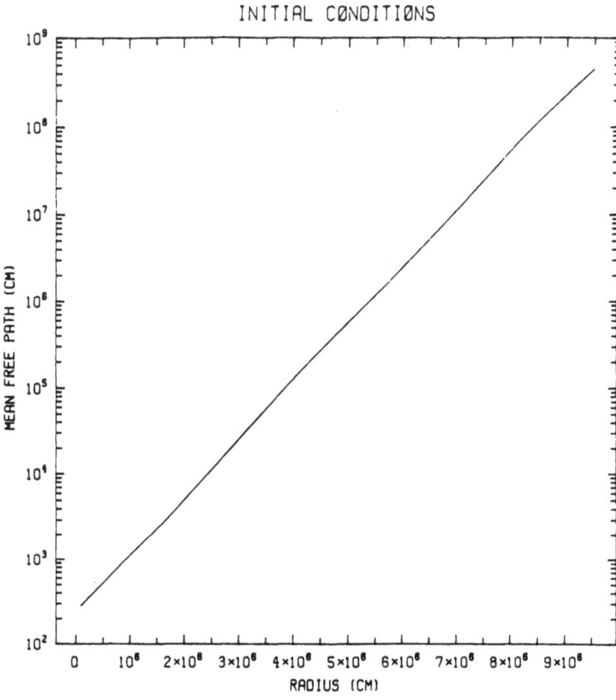

FIGURE 2

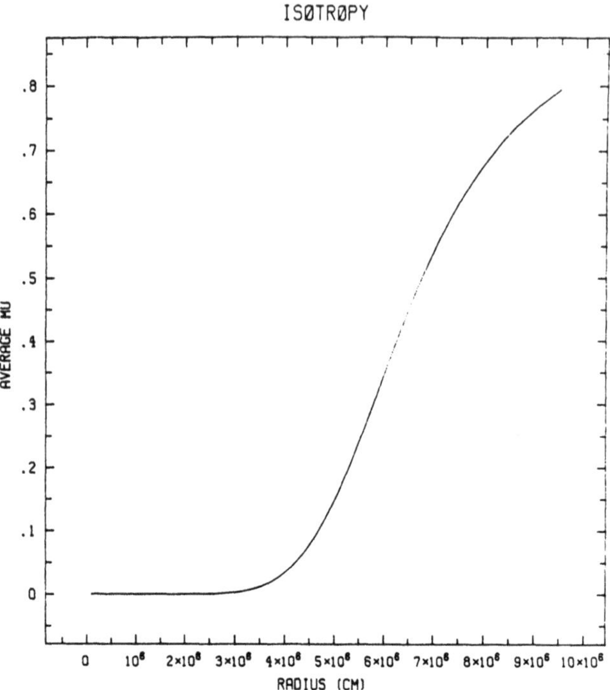

FIGURE 3

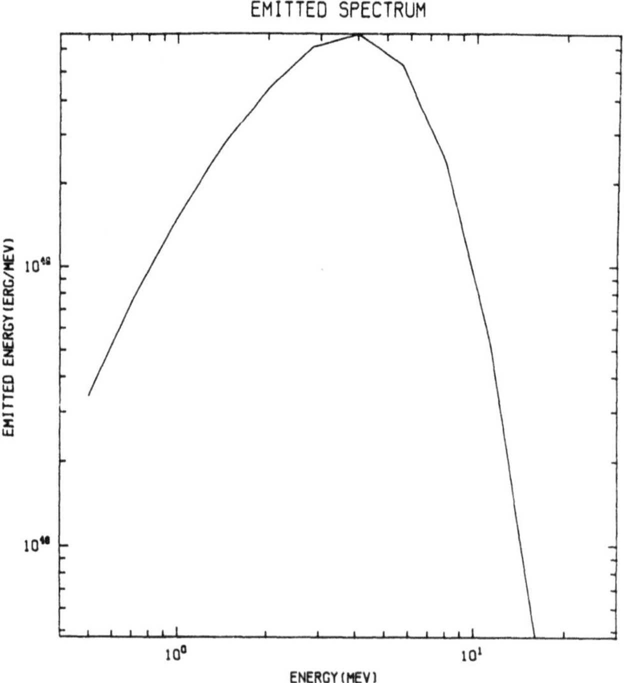

CONSTRAINT PRESERVING TRANSPORT FOR MAGNETOHYDRODYNAMICS

John F. Hawley
Dept. of Astronomy, University of Virginia, Charlottesville, VA 22903

Charles R. Evans
Theoretical Astrophysics 130-33, Caltech, Pasadena, CA 91125

ABSTRACT We present an optimal strategy for evolving the relativistic magnetohydrodynamic (MHD) field transport (induction) equation. The induction equation is shown to assume its simplest form when written in terms of the contravariant forms of velocity and magnetic vector density. The approach places the induction equation in integral form and uses the magnetic flux as a fundamental variable. We then describe a new numerical technique, called constrained transport (CT), for evolving the induction equation in a way that maintains vanishing divergence of the poloidal (constrained) field components to within machine round-off error. We present an example (nonrelativistic) MHD wind calculation that make use of constrained transport.

INTRODUCTION

If we were to attempt to identify an area of numerical astrophysics that is likely to see substantial development in the next few years, a strong case could be made for magnetohydrodynamics (MHD). Interest is rapidly increasing as researchers realize what magnetic fields can do (or perhaps what hydrodynamics alone *can't* do). Following the adage 'the strength of the magnetic field is proportional to our ignorance' we expect strong magnetic fields to be located at the black hole central engines of active galactic nuclei. Indeed, for some active galaxies magnetic fields may not just be important, but *fundamental.* The most promising process for powering radio galaxies and their jets is the magnetic extraction of energy from spinning black holes through the 'Blandford-Znajek mechanism' (Blandford and Znajek 1977; Rees, *et al.* 1982). Just as important may be the role of magnetic fields in directing outflowing winds into collimated jets, and influencing the evolution of accretion disks (e.g., Blandford and Payne 1982).

Although fully relativistic MHD simulations were pioneered by Jim Wilson as far back as 1975 (Wilson 1975) very little has been done in the interim. With the arrival of the national supercomputing centers, and the availability of machines with substantial speed and memory, large scale, high-resolution MHD simulations have become practical. To date, MHD simulations of active galaxy phenomena have largely been confined to questions of jet stability and propagation. Clarke, Norman and Burns (1986), and Lind *et al.* (1988) have evolved axisymmetric gas jets with both passive and active toroidal fields, and have since extended their work to include an active poloidal field (Norman, Clarke and Burns 1988). The complexity uncovered by even these idealized simulations forcefully demonstrates that while we expect MHD processes to be important, we do not fully appreciate either the solutions of the dynamic MHD equations or their consequences. This is especially true when one focuses not on the axisymmetric jet, but on the region near the black hole where, it is theorized, these jets are launched.

Although it is clear that there is an abundance of interesting and important MHD black hole processes to simulate, it is equally clear that the path to a fully relativistic MHD code will necessarily be a complicated one. The successful development of codes capable of investigating the complex physics of relativistic MHD will require meeting several formidable challenges. These include (1) development of accurate field advection schemes, (2) accurate and stable calculation of the Lorentz force including the relativistically important displacement current and space charge terms, and (3) development of techniques to handle both the plasma-dominated and force-free regions in the same model and with the same code.

In this paper we discuss the first of these challenges, the evolution of the magnetic field. In numerical MHD one seeks to obtain accurate fields from which to calculate the Lorentz force. Those fields should be divergenceless, i.e., $\nabla \cdot B = 0$. However this condition is, of course, an initial constraint not an evolution equation, and will in general not be maintained by the finite-difference MHD evolution equations. The constraint problem is a familiar one in numerical relativity. In past MHD simulations constrained evolutions have been done in special coordinate systems (e.g., Cartesian), or by the use of a vector potential in two dimensional simulations. In some cases researchers have settled for unconstrained field evolution. Below we show that by casting the relativistic magnetic induction equation in a covariant 3+1 form, it is possible to develop a coordinate-independent technique for evolving the magnetic field components while maintaining the field constraint to machine roundoff.

RELATIVISTIC MHD AND THE INDUCTION EQUATION

As is usually the case, what can be elegantly expressed in a four-dimensional covariant form must be recast into the so-called 3 + 1 form to make it suitable for a numerical simulation. The well-known advantages of the 3 + 1 formalism are that it (1) explicitly splits apart the time and space dependence, (2) clearly identifies the dynamical character of the equations and (3) expresses the general relativistic Maxwell's equations in terms that are completely analogous to the usual flat-space presentation. Several recent works on black hole electrodynamics (Thorne and MacDonald 1982; MacDonald and Thorne 1982) and relativistic MHD (Phinney 1983; Sloan and Smarr 1985) have discussed the 3 + 1 formalism. We adhere most closely to the work of Sloan and Smarr (1985), adopting 'numerical relativity' notation for the 3 + 1 decomposition of the gravitational field (York 1979; Evans 1984, 1986). Greek indices run from 0-3 and Latin indices from 1-3. Units are assumed such that $c = 1$, and the metric signature is $(-1, +1, +1, +1)$. We take $g_{\mu\nu}$ to be the spacetime metric. Spacetime is decomposed into time slices, each slice labeled by a coordinate time t. Locations on the slice are denoted by spatial coordinates x^i. The unit normal vector field to the slices is n^μ ($n^\mu n_\mu = -1$), and hence the three-metric of the slice is $\gamma_{\mu\nu} = g_{\mu\nu} + n_\mu n_\nu$. We denote the spatial covariant derivative by ∇_μ. Spatial coordinates x^i are propagated along $t^\mu = \alpha n^\mu + \beta^\mu$, where α and β^μ are the lapse function and shift vector. The extrinsic curvature is denoted by K_{ij}, and its trace by K.

Now consider the Faraday tensor $F^{\mu\nu}$ expressed in 3 + 1 form

$$F^{\mu\nu} = n^\mu E^\nu - n^\nu E^\mu + \epsilon^{\mu\nu\sigma} B_\sigma. \tag{1}$$

Here the electric and magnetic fields satisfy $E^\mu n_\mu = B^\mu n_\mu = 0$ and $\epsilon^{\mu\nu\sigma} = n_\kappa \eta^{\kappa\mu\nu\sigma}$. Similarly, the electromagnetic current four-vector $\mathcal{J}^\mu$ is rewritten as

$$\mathcal{J}^\mu = n^\mu \rho_e + J^\mu, \tag{2}$$

where ρ_e and J^μ are the charge density and (spatial) current density as seen in the frame at rest with respect to the time slice.

These can now be used to write the four-dimensional form of the (general relativistic) Maxwell's equations (see, e.g., Misner, Thorne and Wheeler 1973; hereafter MTW) in 3 + 1 form:

$$\nabla_i E^i = 4\pi \rho_e, \tag{3}$$

$$\partial_t E^i = \epsilon^{ijk} \nabla_j (\alpha B_k) - 4\pi\alpha J^i + \alpha K E^i + \pounds_\beta E^i, \tag{4}$$

$$\nabla_i B^i = 0, \tag{5}$$

$$\partial_t B^i = -\epsilon^{ijk} \nabla_j (\alpha E_k) + \alpha K B^i + \pounds_\beta B^i, \tag{6}$$

$$\partial_t \rho_e = -\nabla_i (\alpha J^i) + \alpha K \rho_e + \beta^i \nabla_i \rho_e, \tag{7}$$

where $\pounds_\beta$ is the Lie derivative along β^i. The trace, K, is related to changes in the three-determinant $\gamma \equiv \det(\gamma_{ij})$ by (Evans 1984)

$$\alpha K = -\partial_t \ln(\gamma^{1/2}) + \nabla_i \beta^i. \tag{8}$$

The Lie derivative terms $\pounds_\beta$ are the usual transport terms that account for spatial coordinate motion relative to the fiducial observers. For a Schwarzschild background we can take $\beta^i = 0$, but for the Kerr metric the nonvanishing gravitomagnetic potential, β^ϕ, plays an important role in the Blandford-Znajek (1977) process of extracting energy and angular momentum from a rotating magnetized black hole.

The expression for the Faraday tensor $F_{\mu\nu}$ in terms of the gradient of the four-vector potential $\mathcal{A}_\mu$ (see e.g., MTW) can be decomposed by using

$$\mathcal{A}_\mu = \phi\, n_\mu + A_\mu, \tag{9}$$

where A_μ is the (spatial) vector potential and ϕ is the electric potential. This then gives

$$\partial_t A_i = -\alpha E_i - \epsilon_{ijk}\beta^j B^k - \nabla_i(\alpha\phi - \beta^k A_k), \tag{10}$$

$$B^i = \epsilon^{ijk}\nabla_j A_k = \epsilon^{ijk}\partial_j A_k. \tag{11}$$

Note that Newtonian expressions can be recovered by setting $\alpha = 1$ and $\beta^i = 0$.

The MHD approximation results from adding the (resistive) condition

$$J_\mu - \tilde{\rho}_e U_\mu = \sigma F_{\mu\nu} U^\nu, \tag{12}$$

where σ is the conductivity, $\tilde{\rho}_e$ is the charge density in the *fluid rest frame*, and U^μ is the fluid four-velocity. This condition can be 3+1 decomposed using (2) to yield the two spatial equations

$$W\tilde{\rho}_e = \rho_e - \sigma U^\mu E_\mu, \tag{13a}$$

$$J_i - \tilde{\rho}_e U_i = \sigma(W E_i + \epsilon_{ijk} U^j B^k), \tag{13b}$$

where $W = -n_\mu U^\mu = \alpha U^t$ is the generalized Lorentz factor. Using the velocity definition $V^i \equiv U^i / U^t$ and the flux-freezing condition ($\sigma \to \infty$) gives

$$\alpha E_i = -\epsilon_{ijk} V^j B^k \tag{14}$$

which, except for the appearance of α, is quite familiar.

Now focus attention on Faraday's Law, equation (6). The Lie derivative term can be written

$$\pounds_\beta B^i = \beta^k \nabla_k B^i - B^k \nabla_k \beta^i. \tag{15}$$

We can rewrite this in the form of the curl of the cross product of β^i and B^k, which when combined with the K term in (6), using (8) and the vanishing of the divergence of B^i, gives

$$\alpha K B^i + \pounds_\beta B^i = \nabla_k(\beta^k B^i - \beta^i B^k) - B^i \partial_t \ln \gamma^{1/2}. \tag{16}$$

Using the ideal MHD relation (14), Faraday's Law becomes

$$\frac{1}{\gamma^{1/2}}\partial_t(\gamma^{1/2}B^i) = \nabla_j\left[(V^i-\beta^i)B^j-(V^j-\beta^j)B^i\right]. \tag{17}$$

The covariant divergence of an antisymmetric tensor has a simple reduction which brings the general relativistic magnetic induction equation (17) into the final form

$$\partial_t\mathcal{B}^i = \partial_j\left[(V^i-\beta^i)\mathcal{B}^j-(V^j-\beta^j)\mathcal{B}^i\right], \tag{18}$$

by using the *magnetic vector density*

$$\mathcal{B}^i \equiv \gamma^{1/2}B^i. \tag{19}$$

Note that the constraint on the magnetic field (5) is equally concise when written in terms of the magnetic vector density

$$\partial_i\mathcal{B}^i = 0. \tag{20}$$

Before we consider the significance of these results, we can look at the equivalent *Newtonian* magnetic induction equation

$$\partial_t\mathbf{B} = \boldsymbol{\nabla}\times(\mathbf{V}\times\mathbf{B}). \tag{21}$$

Written in component form this is

$$\partial_t B^i = \epsilon^{ijk}\nabla_j(\epsilon_{klm}V^l B^m), \tag{22}$$

and we can quickly show that it reduces to the simple expression

$$\partial_t\mathcal{B}^i = \partial_j(V^i\mathcal{B}^j - V^j\mathcal{B}^i) \tag{23}$$

using the same definition for the magnetic vector density. Thus the general relativistic and Newtonian forms of the induction equation are identical except for the addition of the shift vector. Further, we have the pleasing result that these equations have a Cartesian appearance yet they are valid for arbitrary choices of spacetime coordinates, and the relativistic equation is valid in a fully dynamic spacetime.

To obtain a finite-difference version of the induction equation we use a closely related integral formulation. An important byproduct is that it will then be obvious how to incorporate an adaptive mesh into the numerical scheme. The fundamental quantity about which to build a numerical scheme is the magnetic flux Φ_S piercing a two-surface S defined by

$$\Phi_S = \int_S B^i\, d\Sigma_i, \tag{24}$$

where $d\Sigma_i$ is the spatially covariant area element. Now suppose that the magnetic field is time dependent and that the finite area element is as well, i.e., $S = S(t)$. Let V_g^i be defined as the coordinate velocity describing the flow of points within $S(t)$ from time slice to time slice. The region $S(t)$ has a contour $\partial S(t)$ as its boundary. Just as in flat spacetime, the total rate of change of the flux piercing this moving surface, $d\Phi_{S(t)}/dt$, equals minus the total electromotive force (EMF), $\mathcal{E}$, around the contour $\partial S(t)$. The partial time derivative ∂_t gives the change in a quantity per unit coordinate time along fixed spatial coordinate values x^i. The convective (or adaptive mesh) time derivative δ_t is defined as giving the change per unit coordinate time along points that are moving with velocity V_g^i. We then have, for example, the rate of change of the magnetic field along points moving with $S(t)$,

$$\delta_t B^i = \partial_t B^i + V_g^k \partial_k B^i. \tag{25}$$

In particular the convective velocity V_g^i (a "grid" velocity for a finite-difference scheme) is itself defined by

$$\delta_t x^i \equiv V_g^i. \tag{26}$$

The total rate of change of flux through a moving surface is

$$\frac{d\Phi_{S(t)}}{dt} = \frac{d}{dt}\int_{S(t)} B^i \, d\Sigma_i = \int_{S(t)} (\delta_t B^i \, d\Sigma_i + B^i \, \delta_t d\Sigma_i). \tag{27}$$

We can express the surface area element $d\Sigma_i$ in terms of two coordinates ς and ξ that span the surface:

$$d\Sigma_i = \epsilon_{ijk} \frac{\partial x^j}{\partial \varsigma} \frac{\partial x^k}{\partial \xi} \, d\varsigma \, d\xi = \gamma^{1/2} [ijk] \frac{\partial x^j}{\partial \varsigma} \frac{\partial x^k}{\partial \xi} \, d\varsigma \, d\xi \equiv \gamma^{1/2} d\Sigma_i^{(c)}, \tag{28}$$

where $d\Sigma_i^{(c)}$ denotes the coordinate area element. To evaluate the convective derivative of the area element we assume the coordinates ς and ξ are fixed with respect to the motion V_g^i of the circuit element, i.e., $\delta_t \varsigma = \delta_t \xi = 0$. Then using the definition of the velocity of the moving circuit, we have

$$\delta_t \left(\frac{\partial x^j}{\partial \varsigma} \right) = \frac{\partial V_g{}^j}{\partial \varsigma} = \frac{\partial V_g{}^j}{\partial x^l} \frac{\partial x^l}{\partial \varsigma}. \tag{29}$$

We then find the change in the surface area element to be expressed by

$$\delta_t d\Sigma_i = \left(\partial_t \ln(\gamma^{1/2}) + \nabla_k V_g^k \right) d\Sigma_i - \frac{\partial V_g^k}{\partial x^i} d\Sigma_k, \tag{30}$$

which is the surface element (and relativistic) equivalent of the so-called "Euler expansion formula" for volume elements (Mihalas and Mihalas 1984, p. 58).

Combining this with the convective change in B^i we have

$$\delta_t B^i \, d\Sigma_i + B^i \, \delta_t d\Sigma_i = \frac{1}{\gamma^{1/2}} \partial_t \left(\gamma^{1/2} B^i \right) d\Sigma_i + \frac{1}{\gamma^{1/2}} \partial_k \left(\gamma^{1/2} V_g^k B^i - \gamma^{1/2} V_g^i B^k \right) d\Sigma_i, \tag{31}$$

using again the vanishing divergence of B^i. This yields the integral form of the field transport equation

$$\frac{d}{dt} \Phi_{S(t)} = \frac{d}{dt} \int_{S(t)} B^i \, d\Sigma_i = \frac{d}{dt} \int_{S(t)} B^i \, d\Sigma_i^{(c)} = -\mathcal{E} \tag{32}$$

$$= \oint_{\partial S(t)} (V^i - V_g^i - \beta^i) B^j \epsilon_{ijk} dx^k = \oint_{\partial S(t)} (V^i - V_g^i - \beta^i) B^j [ijk] dx^k.$$

In (32) $\mathcal{E}$ is the EMF about the contour $\partial S(t)$ due to the combined effects of fluid motion, gravitomagnetic potential and motion of the circuit. Equation (32) can also be derived from eq. (3.49) of Thorne *et al.* (1986) (see also Thorne and MacDonald 1982) using the ideal MHD condition. The usual nonrelativistic result arises by simply setting the shift vector β^i to vanish. This equation becomes the basis of our constraint-preserving adaptive mesh numerical transport algorithm.

CONSTRAINED TRANSPORT

We now consider the finite difference version of equation (32), beginning with a brief review of our notation. For simplicity we label coordinates by x, y, and z, but, it is most important to note, we do *not* mean to imply any restriction to Cartesian coordinates; equation (32) and its finite-difference analogue hold in arbitrary coordinate systems including those describing curved space. We use a staggered mesh that provides different centerings for various quantities, and label finite-difference cell corners with integer indices, i.e., (x_i, y_j, and z_k). The cell center is labeled by half-integer locations defined by

$$x_{i+1/2} = \frac{1}{2}(x_{i+1} + x_i), \quad y_{j+1/2} = \frac{1}{2}(y_{j+1} + y_j), \quad z_{k+1/2} = \frac{1}{2}(z_{k+1} + z_k).$$

The finite-difference cell increments are straightforwardly defined. Using the x direction as an example, we have

$$\Delta x_{i+1/2} = x_{i+1} - x_i,$$

and

$$\Delta x_i = x_{i+1/2} - x_{i-1/2}.$$

Discrete time levels are indicated by a superscript n. The time interval Δt is defined as $\Delta t = t^{n+1} - t^n$, and is typically a function of time as well.

The usual centering of hydrodynamic quantities places density ρ, energy density $\rho\epsilon$, pressure p and the generalized Lorentz factor W at zone centers and on the integer time levels n. The relativity quantities, the lapse function α and the spatial metric γ_{ij}, take the same centering. Velocities V^i, $V_g{}^i$, momenta S_i and the shift vector β^i are assigned face-centered locations and are considered located at half-integer time levels $n + 1/2$ (Hawley, Smarr, and Wilson 1984). Proper time- and space-centered finite-difference equations can be written by assigning the magnetic field the centerings

$$(\mathcal{B}^x)^n{}_{i+1/2,j,k}, \quad (\mathcal{B}^y)^n{}_{i,j+1/2,k}, \quad (\mathcal{B}^z)^n{}_{i,j,k+1/2}.$$

The magnetic field components are face centered, but we have chosen to center them in a different way from the velocity and momentum terms. In principle, other centerings are possible.

The magnetic field is really defined in terms of the flux Φ_S piercing a surface S. Using equation (24) we express the flux as an integral over the magnetic vector density $\mathcal{B}^i$ using the coordinate area element $d\Sigma_i{}^{(c)}$, a form that can be immediately finite-differenced. Take the surface S to be the (staggered) cell face located at the fixed value $x = x_{i+1/2}$ and spanned by zone increments Δy_j and Δz_k. The flux through this face is

$$\Phi_{S(i+1/2)} = \int_{\Delta y_j, \Delta z_k} \mathcal{B}^x \, dy \, dz, \tag{33}$$

with the orientation chosen so that Φ_S is positive for a positive-directed field component (right-handed rule). This defines the finite-difference value of the $\mathcal{B}^x$ component to be the face area averaged flux

$$(\mathcal{B}^x)_{i+1/2,j,k} = \frac{1}{\Delta y_j \Delta z_k} \Phi_{S(i+1/2)}. \tag{34}$$

Now integrate over the closed surface $S = \partial\mathcal{V}$ made up of the six faces bounding the cell volume $\mathcal{V}$ centered on (i, j, k) as indicated in Figure 1. We know analytically that the net flux through $S = \partial\mathcal{V}$ must vanish so a constraint on our numerical scheme is that

$$\Phi_{S(i+1/2)} - \Phi_{S(i-1/2)} + \Phi_{S(j+1/2)} - \Phi_{S(j-1/2)} + \Phi_{S(k+1/2)} - \Phi_{S(k-1/2)} = 0. \tag{35}$$

Since S is a closed surface, we can use Gauss' theorem (in a simply connected region) to write

$$\int_{S=\partial\mathcal{V}} \mathcal{B}^i \, d\Sigma_i^{(c)} = \int_{S=\partial\mathcal{V}} B^i \, d\Sigma_i = \int_{\mathcal{V}} \nabla_i B^i \, d\mathcal{V} = \int_{\mathcal{V}} \partial_i \mathcal{B}^i \, d^3x = 0. \tag{36}$$

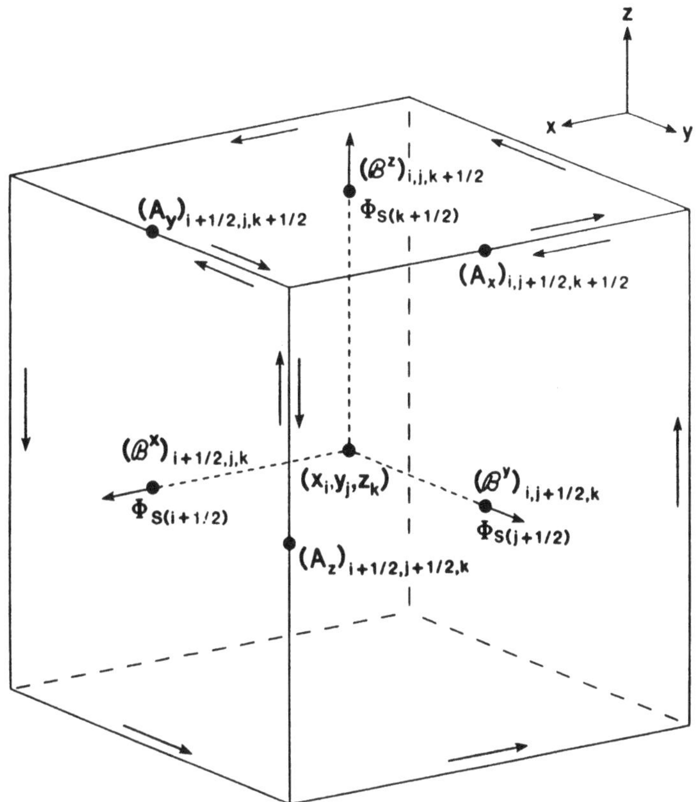

Figure 1: The three dimensional centering of electromagnetic quantities. In the constrained transport scheme the total flux piercing the surface of a finite-difference cell is always zero, achieved by balancing the EMF contributions taken around the cell edges. The arrows indicate the direction of the closed contour integral.

We see that the constraint that the total flux through S vanish is connected with a numerical requirement on the finite-difference form of the divergence operator. Substituting (33) into equation (35) and dividing by the cell coordinate volume $\Delta x_i \Delta y_j \Delta z_k$ we find

$$\frac{1}{\Delta x_i}\left((\mathcal{B}^x)_{i+1/2,j,k} - (\mathcal{B}^x)_{i-1/2,j,k}\right) + \frac{1}{\Delta y_j}\left((\mathcal{B}^y)_{i,j+1/2,k} - (\mathcal{B}^y)_{i,j-1/2,k}\right)$$

$$+\frac{1}{\Delta z_k}\left((\mathcal{B}^z)_{i,j,k+1/2} - (\mathcal{B}^z)_{i,j,k-1/2}\right) = 0, \tag{37}$$

as the required finite-difference analogue of the divergence constraint on the magnetic field.

Now consider the time evolution of the flux and hence the magnetic field. Each finite-difference cell face is bounded by a closed contour made up of four edges. The rate of change per unit coordinate time of the flux through a face is minus the total EMF around this contour; thus

$$\frac{d}{dt}\Phi_{face} = - \sum_{edges\ l=1}^{4} \mathcal{E}_l, \tag{38}$$

where $\mathcal{E}_l$ are the contributions to the total EMF *taken in the sense of a right-handed integration* about the bounding contour. There are twelve edges C to each finite-difference cell $V_{i,j,k}$. Edges are associated with constant values of two coordinates; for example the edge $C(i+1/2, j+1/2)$ is located at constant values $x = x_{i+1/2}$ and $y = y_{j+1/2}$ and has coordinate length Δz_k. We will define $\mathcal{F}_{(l)}$ to be the EMF contributions of the edges with the integrations *taken in the increasing coordinate direction.* For example the EMF contribution from the edge $C(i+1/2, j+1/2)$ is

$$\mathcal{F}_{i+1/2,j+1/2,k} \equiv - \int_{z_{k-1/2}}^{z_{k+1/2}} (u^x \mathcal{B}^y - u^y \mathcal{B}^x)\, dz, \tag{39}$$

where, for simplicity, we have written

$$u^i = V^i - V_g{}^i - \beta^i,$$

for the net velocity.

Assuming that the EMF contributions $\mathcal{F}_{(l)}$ can be derived by a suitable finite-difference approximation (to be discussed below), we can obtain finite-difference evolution equations for the fluxes. As an example, consider the evolution equation for the flux at the $i+1/2$ face

$$\begin{aligned}\frac{1}{\Delta t}(\Phi^{n+1}_{S(i+1/2)} - \Phi^{n}_{S(i+1/2)}) = &+ \mathcal{F}_{i+1/2,j,k+1/2} - \mathcal{F}_{i+1/2,j,k-1/2} \\ &- \mathcal{F}_{i+1/2,j+1/2,k} + \mathcal{F}_{i+1/2,j-1/2,k}.\end{aligned} \tag{40}$$

We see that by substituting in the magnetic field definition (34) we find the corresponding evolution equation for $\mathcal{B}^x$.

The new *total* flux through the closed surface of a mesh cell is found by adding together the individual updates obtained at the six faces of the cell. The total flux continues to vanish if it did originally, even though the cell position and shape may change. This numerical conservation law follows from the fact that each EMF edge contribution appears in the full sum exactly twice with opposing signs. This is illustrated in Figure 1 by arrows indicating the directions that the contour integrations follow in computing the EMF contributions in (40).

The fundamental point to recognize is the consistency required. The edge EMF contributions are calculated *once* even though each contribution is used in two separate evolution equations. This is analogous to the more familiar numerical implementation of the continuity equation, where mass conservation is insured by calculating the flux through each face and using it twice, once for mass leaving the volume and once when that mass enters the adjacent volume. Conservation schemes of this type for Cartesian meshes were originally described by Roberts and Weiss (1966).

In recognition of its constraint-preserving feature, we call this numerical approach *Constrained Transport* (CT). Most noteworthy is the scheme's generality. However, note that the CT scheme only guarantees that field evolution *changes* are divergence free. If the initial magnetic field fails to satisfy the constraint, it will continue to do so. A more serious issue arises with inflow and outflow boundary conditions. Here one must be careful to not under-specify or over-specify the field and flow at the boundary. Failure to prescribe free-flow boundary conditions properly can lead to subtle inflow of non-divergence-free magnetic field. Because boundary conditions are frequently changed, we recommend that a global measure of the constraint satisfaction be monitored as a diagnostic during simulations (Evans and Hawley 1988).

The CT approach to magnetic field evolution can be implemented with any of several choices of finite-difference schemes. In MHD an EMF arises from magnetic field advection, or transport, as the field is dragged along by the conducting fluid. This means that the scheme used to obtain the EMFs must be stable in the usual transport sense. Specifically, if we consider a two dimensional system a typical EMF contribution will be given by

$$\mathcal{F}_{i+1/2,j+1/2,k} = -(\overline{u}^x\overline{\mathcal{B}}^y - \overline{u}^y\overline{\mathcal{B}}^x)_{i+1/2,j+1/2}, \tag{41}$$

where bars indicate some appropriately centered average value. The evolution equation for a magnetic flux component looks like

$$\begin{aligned}(\mathcal{B}^x)^{n+1}_{i+1/2,j}\Delta y_j{}^{n+1} = (\mathcal{B}^x)^{n}_{i+1/2,j}\Delta y_j{}^{n} - \Delta t\Big((\overline{u}^x\overline{\mathcal{B}}^y - \overline{u}^y\overline{\mathcal{B}}^x)_{i+1/2,j+1/2} \\ - (\overline{u}^x\overline{\mathcal{B}}^y - \overline{u}^y\overline{\mathcal{B}}^x)_{i+1/2,j-1/2}\Big),\end{aligned} \tag{42}$$

The $\overline{\mathcal{B}}^x$ difference that appears on the right hand side of (42) forms a transport term, requiring upwind differencing for stability; the $\overline{\mathcal{B}}^y$ difference in the same equation is a shear term that requires no special treatment. However, the same two terms appear on the right hand side of the equation for $\mathcal{B}^y$ evolution with their roles reversed. What appears as a shear term in one equation is a transport term in the other. Constrained Transport requires that these terms must be differenced in a consistent manner. Hence upwind differencing is required for each such term. The bar over a $\mathcal{B}^i$ component then indicates upwind averaging in the direction of the velocity component with which it's paired. The averaging of velocity components can be done, as usual, by a direct equal-weight average. An explicit example of finite differencing these terms is given in Evans and Hawley (1988).

It can be seen that the CT scheme is not overly restrictive. Freedom still exists to modify the transport algorithm further, if desired. The CT scheme is also extendible to non-ideal MHD simulations. If the conductivity is finite, one should include the current density as part of the electric field, which is then used in the resistive induction equation that is the generalization of equations (18), and (30). Since Stoke's theorem still applies, properly constrained evolution equations can be written.

An elegant aspect of the CT scheme is that because the evolved field is divergence free, it is always possible to reconstruct the vector potential on each time slice. This can be done, up to an arbitrary constant, by integrating

$$dA_z = -\mathcal{B}^y dx + \mathcal{B}^x dy, \tag{43}$$

either by holding x constant and integrating $\mathcal{B}^x$ in the y-direction, or by holding y constant and integrating $\mathcal{B}^y$ in the x-direction. Since the magnetic vector density $\mathcal{B}^i$ is related directly to the vector potential, the CT scheme is related to a particular numerical approach for advecting the vector potential. In effect, we are simply obtaining a monotonic distribution of $\partial_i A_z$ rather than of A_z. In this way the first derivative is itself constrained to be monotonic, which means that the subsequent derivative used to calculate the Lorentz force will be better behaved near crucial sharp features in the flow.

AN EXAMPLE: MHD WIND

As always, the value of any proposed numerical scheme is revealed only through its actual use. A series of advection test problems (Evans and Hawley 1988) demonstrated some of the promising properties of CT, but to test the evolution of dynamically important fields we must include the Lorentz force terms. At present, the relativistic Lorentz force terms still pose formidable numerical difficulties of the type outlined by Sloan and Smarr (1985). The Newtonian Lorentz term is considerably simpler however, and we have begun using a Newtonian MHD code employing CT to examine the properties of MHD winds from rotating systems.

Interest in MHD winds has been stimulated by observations of the magnetized winds from stars, including our own sun, magnetized winds from pulsars, and the possibility of magnetic forces playing a role in the collimation and focusing of high energy jets from active galactic nuclei. To date, most studies of MHD winds and accretion flows have employed considerable simplification in order to obtain a solution. An example is the classic paper of Weber and Davis (1967) that solves a set of purely radial equations for a rotating magnetized solar wind. More recently Sakurai has relaxed the restriction of a purely radial solution to obtain steady-state, axisymmetric MHD winds from both a spherical (Sakurai 1985) and disk-like (Sakurai 1987) geometry. Phinney (1983) obtained general relativistic MHD wind solutions for axisymmetric flows along fixed field geometry; he lacked only a solution for force balance in the poloidal direction to arrive at a self-consistent time-independent solution. However, self-consistent solutions can be obtained by solving the full time-dependent MHD equations, and in fact this approach can reveal instabilities and nonstationarity that a time-independent analysis cannot. A study of MHD winds provides both the opportunity to validate the CT approach, and to investigate models of importance to astrophysics. For example, it would be worthwhile simply to generate a series of stationary outflows from rotating sources. Since they would be obtained from a fully dynamic MHD code the results would give self-consistent field configurations, information about the ability of fields to collimate a wind, mass outflow rates as a function of poloidal angle, and the location of critical surfaces.

As an example consider the wind solution pictured in Figure 2. This wind was generated from a split monopole radial field emerging from a uniformly rotating stellar surface. The boundary conditions for this model were set by calculating a particular Weber-Davis radial wind solution and using those values at the inlet. Pressure and density at the inlet are assumed to be constant. The illustrated calculation has evolved into a steady state wind. The split-monopole field geometry produces a current sheet on the equatorial plane. Angular rotation in the outplow shears the poloidal magnetic field component into a field that is dominated by its toroidal term. This produces an axis-directed tension that collimates the flow. This poleward bending of the field lines is evident in the figure. Also depicted are the slow and fast magnetosonic critical surfaces (dashed lines), and the Alfven surface (solid line).

Our investigation of MHD winds has only begun. However, based on the results of preliminary experiments such as that shown in Figure 2, we feel reasonably confident in stating that the CT scheme works well. In the near term we intend to continue this work by (1) improving the finite differencing of the Lorentz force term, (2) completing the calibration and testing of the code through detailed comparison with analytic critical point analysis, (3) developing more general boundary conditions (e.g., dipole fields), and (4) conducting a survey of rotating magnetized wind solutions.

CONCLUSION

Although there remains much to be done in the development of a fully relativistic MHD code, we believe that CT will prove to be the optimal approach to the magnetic field induction equation. Since this is a relativity conference it is appropriate to explicitly point out that the CT scheme was discovered through a relativistic covariant treatment of the equations. While conservative

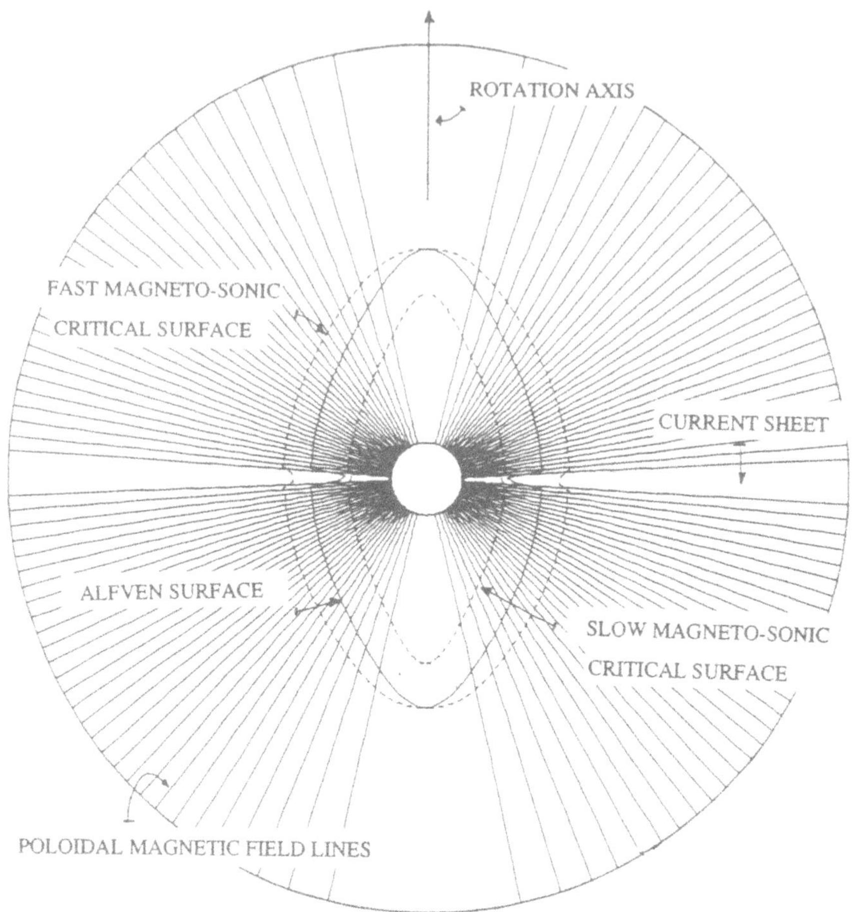

Figure 2. Contours of vector potential showing magnetic field coming off the surface of a uniformly rotating star. The field geometry is that of a split monopole. The locations of applicable critical surfaces are indicated. Note the slight bending of the field lines towards the axis.

numerical schemes have been developed in the past (e.g., for hydrodynamics), the ability to do this in MHD was overlooked. The single most important factor in developing CT was to search for the numerical conservation scheme while keeping the analysis as general as possible by not reducing to a coordinate-specific description early on.

The authors acknowledge the National Center for Supercomputing Applications for a portion of the computer resources used in this project. This work was supported in part by NSF Grants PHY88-02747 and AST85-14911.

REFERENCES

Blandford, R. D., and Payne, D. G. 1982, *M.N.R.A.S.*, **199**, 883.

Blandford, R. D., and Znajek, R. L. 1977, *M.N.R.A.S.*, **179**, 433.

Clarke, D. A., Norman, M. L., and Burns, J. O. 1986, *Ap. J. (Letters)*, **311**, L63.

Evans, C. R. 1984, Ph.D. Thesis, University of Texas at Austin.

Evans, C. R. 1986, in *Dynamical Spacetimes and Numerical Relativity*, ed. J. Centrella (New York: Cambridge University Press), pg. 3.

Evans, C. R., and Hawley, J. F., *Ap. J.*, **332**, in press.

Hawley, J. F., Smarr, L. L., and Wilson, J. R. 1984, *Ap. J. Supp.*, **55**, 211.

Lind, K., Payne, D. G., Meier, D. L., and Blandford, R. D. 1988, preprint.

MacDonald, D. A. and Thorne, K. S. 1982, *M.N.R.A.S.*, **211**, 345.

Mihalas, D. and Mihalas, B. W. 1984, *Foundations of Radiation Hydrodynamics* (New York: Oxford University Press), pg 58.

Misner, C., Thorne, K., and Wheeler, J. 1973, *Gravitation* (San Francisco: Freeman).

Norman, M. L., Clarke, D. A., and Burns, J. O. 1988, to appear in *Extragalactic Magnetic Fields.*

Phinney, E. S. 1983, Ph.D. Thesis, Cambridge University.

Rees, M. J., Begelman, M. C., Blandford, R. D., and Phinney, E. S. 1982, *Nature*, **295**, 17..

Roberts, K. V., and Weiss, N. O. 1966, *Math. Comp.*, **20**, 272.

Sakurai, T. 1985, *Astr. Ap.*, **152**, 121.

Sakurai, T. 1987, *Publ. Astron. Soc. Japan*, **39**, 821.

Sloan, J., and Smarr, L. L. 1985, in *Numerical Astrophysics*, ed. J. Centrella, J. LeBlanc, and R. Bowers (Boston: Jones and Bartlett), pg. 52.

Thorne, K. S., Price, R. H, and MacDonald, D. A. 1986, *Black Holes: The Membrane Paradigm* (New Haven: Yale).

Thorne, K. S., and MacDonald, D. A. 1982, *M.N.R.A.S.*, **198**, 339.

Weber, E. J., and Davis L. 1967, *Ap. J.*, **148**, 217.

Wilson, J. R. 1975, *Ann. N. Y. Acad. Sci.*, **262**, 123.

York, J. W. 1979, in *Sources of Gravitational Radiation* ed. L. Smarr (Cambridge: Cambridge University Press), p. 83.

ENFORCING THE MOMENTUM CONSTRAINTS DURING AXISYMMETRIC SPACELIKE SIMULATIONS

Charles R. Evans
Theoretical Astrophysics 130-33
Calfornia Institute of Technology
Pasadena, California 91125

ABSTRACT

The issue of enforcing the momentum constraints during a spacelike numerical relativity simulation is discussed. We show how the momentum constraints of general relativity can be solved for a subset of the components of the extrinsic curvature during a numerical simulation. We found little difference in the emitted waveforms that were calculated using (1) a code that freely evolves the extrinsic curvature and (2) a code based on employing the momentum constraints during the simulation. However, a slowly growing numerical instability was identified in long dynamical calculations using the free evolution code. This instability eventually caused the calculation to terminate. The new fully constrained code shows no evidence of such problems. The instability is due to a gradual drift in the evolved quantities away from satisfying regularity conditions near the origin of the coordinate system. By solving the momentum constraints on each time slice, these regularity conditions can be satisfied indefinitely to high accuracy. We also discuss how asymptotic, radiative boundary conditions are incorporated in the fully constrained code.

INTRODUCTION

This paper focuses on the issue of enforcing the momentum constraint equations of general relativity during a spacelike numerical simulation. We must at the outset clarify what we mean here by enforcing, or solving, the momentum constraints. First, we do *not* mean solving the initial value problem. In that problem, as York (*cf.* York 1979, Evans 1984, and York's article in this volume) has outlined, the aim is to determine a vector potential W^i for the longitudinal (constrained) part of the extrinsic curvature, K_{ij}. That is a separate, though important, question. Rather, we are concerned here with using the momentum constraints

$$D_j(K^{ij} - \gamma^{ij}K) = 8\pi S^i, \tag{1}$$

during the course of an evolution to reduce the number of components of K_{ij} that must be evolved to equal just the number of dynamical degrees of freedom and to fix the remaining components of K_{ij} through an elliptic or parabolic determination using equation (1) within a spacelike slice. Here γ_{ij} is the spatial metric, S_i is the momentum density, K is the trace of K_{ij}, D_j is the spatial covariant derivative and we use units such that $G = c = 1$.

It is clear that given three momentum constraint equations and six components of K_{ij} there is ambiguity in this process and with the choice of variables to constrain. We will see that the resulting mathematical problem is not always well posed. In any case, the process of using the general relativistic constraints during an evolution is not as elegant as, for example, the method of building in the constraint into the evolution equations as can be done in electromagnetic or MHD calculations (Evans & Hawley 1988, Hawley & Evans, this volume). There seems virtually no prospect for being able to do this in general relativity.

Numerous authors (*e.g.*, Smarr 1979, Wilson 1979, Dykema 1980, Bardeen & Piran 1983, Piran 1983, Stewart 1984, Evans 1984, 1986, Choptuik 1986, Shapiro & Teukolsky 1987 private communication) have faced the issue of either using the constraints during a simulation or performing a free evolution of the metric and extrinsic curvature fields while using the constraints merely as a consistency check on the numerical evolution. A third alternative is to make a partially constrained evolution that uses only some of the four constraint equations to replace evolution equations and the rest for consistency checks. In practice, all three approaches have been used, though heretofore the mathematical task of employing the momentum constraints in two dimensions has been on shaky ground.

In this paper we first show that the way thought most natural to use the constraint equations (1) during a simulation is mathematically ill posed. We then discuss an alternative method that is mathematically sound and easily solved numerically. Results from numerical simulations are shown that compare emitted gravitational waves calculated using both the new fully constrained approach and a previously used partially constrained approach (free evolution of K_{ij} but employment of the Hamiltonian constraint for the conformal factor). The waveforms are nearly identical during the period of time the simulations overlap, lending considerable weight to the results. Nonetheless, we show that the partially constrained approach suffers at late times from a slowly-growing numerical instability that ultimately destroys the simulation. No instabilities are evident in the fully constrained approach due to a careful treatment of the regularity conditions on the extrinsic curvature at the singular point ($r = 0$) in the coordinate system. We conclude that for many simulations there may be little advantage in one approach over the other, though the fully constrained evolution should help maintain stability in certain numerical implementations.

PRELIMINARIES

We use the quasi-isotropic (QI) spatial gauge (Smarr 1979, Wilson 1979, Dykema 1980, Bardeen & Piran 1983, Evans 1984, 1986, Abrahams & Evans 1988). In spherical-polar coordinates the line element is

$$ds^2 = A^2(dr^2 + r^2 d\theta^2) + B^2 r^2 (\sin\theta d\phi + \xi d\theta)^2, \tag{2a}$$

which stems from demanding that three gauge conditions on the metric

$$g_{r\phi} = 0, \tag{2b}$$

$$g_{r\theta} = 0, \tag{2c}$$

$$g_{\theta\theta}g_{\phi\phi} - (g_{\theta\phi})^2 = g_{rr}g_{\phi\phi}\, r^2, \tag{2d}$$

hold at all times. The metric variable ξ allows for the presence of rotation in the material source. For the questions we address in this paper, it is irrelevant whether the source has rotation or not, and so we will simplify and assume no rotation and that $\xi = 0$.

A more natural set of metric variables can be defined by

$$\Phi^6 \equiv A^2 B, \tag{3}$$

$$\eta \equiv \ln T \equiv \ln(A/B), \tag{4}$$

where Φ is the conformal factor and η is a measure of the anisotropy of the three-space. As discussed elsewhere (Evans 1984, 1986) Φ is determined during a simulation by solving the (elliptic) Hamiltonian constraint, making the method at least partially constrained, while η has useful properties as a dynamical variable.

Of the six components of the extrinsic curvature K_{ij}, two components, $K_{r\phi}$ and $K_{\theta\phi}$, vanish in the absence of rotation. One of the remaining four is usually used to fix the time slicing condition (setting $K = K^i{}_i = 0$ in the case of maximal time slicing and setting $K^\theta{}_\theta + K^\phi{}_\phi = 0$ for polar time slicing). The Einstein equations provide evolution equations for the remaining three components (we use the independent combinations $K^r{}_r$, $K^r{}_\theta/r$, and $\lambda \equiv K^r{}_r + 2K^\phi{}_\phi$, where λ, like η, has useful characteristics as a dynamical variable), yet the momentum constraints provide two instantaneous relations between these three quantities. Note that if rotation is included the remaining component of the momentum constraints gives a single equation relating the odd-parity components $K^r{}_\phi$ and $K^\theta{}_\phi$.

The essence of the problem considered here is to use the two momentum constraint equations during a simulation to determine two of the extrinsic curvature variables from a knowledge of the third. Thus, only one component need be time evolved and we are guaranteed that the evolved data will remain confined to the constraint hypersurface. As mentioned however, one problem with this notion is that not every choice for constrained variables leads to a well-posed set of equations.

AN APPROACH THAT FAILS

It is instructive to consider first a case that is *not* well posed. We can take (as is most natural) λ to be the time-evolved variable. The momentum constraints are then viewed as equations for $K^r{}_r$ and $K^r{}_\theta/r$ and in QI gauge and axisymmetry are found to be

$$\frac{T^{1/2}}{r^2}\partial_r\left[\frac{r^3}{T^{1/2}}\hat{K}^r{}_r\right]+\frac{1}{\sin\theta}\partial_\theta\left[\sin\theta\left(\frac{\hat{K}^r{}_\theta}{r}\right)\right]=8\pi r\hat{S}_r-\frac{1}{2}\hat{\lambda}r\partial_r\eta, \tag{5}$$

$$-\frac{1}{2T}\partial_\theta\left[T\hat{K}^r{}_r\right]+\frac{1}{r^2}\partial_r\left[r^3\left(\frac{\hat{K}^r{}_\theta}{r}\right)\right]=8\pi\hat{S}_\theta+\frac{T}{2\sin^2\theta}\partial_\theta\left[\frac{\sin^2\theta}{T}\hat{\lambda}\right], \tag{6}$$

where $\hat{K}^i{}_j=\Phi^6K^i{}_j$ refers to the conformally-related extrinsic curvature (York 1979, Evans 1984). Written in this way, the system is elliptic (Evans 1984). In some of what follows, we simplify the notation by setting $F\equiv\hat{K}^r{}_r$, $G\equiv\hat{K}^r{}_\theta/r$ and $H\equiv\hat{\lambda}$.

The mathematical problem with this approach can be demonstrated without loss of generality by considering the spatially flat limit ($\eta=0$ and $T=1$). Assuming this, the momentum constraints reduce to

$$\frac{1}{r^2}\partial_r\left[r^3F\right]+\frac{1}{\sin\theta}\partial_\theta\left[\sin\theta G\right]=8\pi r\hat{S}_r, \tag{7}$$

$$-\frac{1}{2}\partial_\theta F+\frac{1}{r^2}\partial_r\left[r^3G\right]=8\pi\hat{S}_\theta+\frac{1}{2\sin^2\theta}\partial_\theta\left[\sin^2\theta H\right]. \tag{8}$$

These equations can be decomposed into spherical harmonics by recognizing that the tensor components have dependencies $F(r,\theta)=f(r)P_l(\cos\theta)$ and $G(r,\theta)=g(r)\partial_\theta P_l(\cos\theta)$. The decomposition yields the following radial equations for the *homogeneous* part of equations (7) and (8):

$$\frac{1}{r^2}\partial_r\left[r^3f\right]-l(l+1)g=0, \tag{9a}$$

$$-\frac{1}{2}f+\frac{1}{r^2}\partial_r\left[r^3g\right]=0. \tag{9b}$$

The solutions of these equations are power laws, *e.g.*, $f(r) \propto g(r) \propto r^\alpha$, with the exponent given by

$$\alpha = -3 \pm \left[\frac{l(l+1)}{2}\right]^{1/2} . \tag{10}$$

It is interesting that the constraints when written in the form (5) and (6) give non-integer power homogeneous solutions, but the most striking feature is that the two solutions for $l = 2$ *both* decay in the outward radial direction. This fact was pointed out by Bardeen and Piran (1983) and emphasized by Shapiro and Teukolsky (private communication). The result is that if data is prescribed on an outer boundary of a spacelike slice, one requires an essentially perfect specification of the boundary data for equations (5) and (6) in order to obtain a solution regular at $r = 0$. Any errors in specifying the outer boundary data will lead to effects that *grow* as one moves inward, leading to an irregular behavior at $r = 0$. This is a classic example of an ill-posed elliptic system.

Figure 1 illustrates this effect by solving the radial differential system (9a,b) (including nonvanishing source) for $l = 2$. The dashed curve shows an assumed test solution, while the solid curve represents the numerical solution that was obtained after introducing a 2.5% perturbation in the *outer* boundary condition. A similar effect is seen near $r = 0$ when the two dimensional elliptic equations (5) and (6) are solved, though with the added complication of the angular dependence. There is essentially no recourse but to abandon this route for enforcing the momentum constraints during an evolution.

AN APPROACH THAT SUCCEEDS

Fortunately an alternate approach is possible. This involves using $\hat{K}^r{}_\theta/r$ (G) as the known, evolved source and using the constraints to instantaneously determine the variables $\hat{K}^r{}_r$ (F) and $\hat{\lambda}$ (H). Using F, G, and H once again to simplify the notation, the momentum constraints (5) and (6) can be rewritten as

$$\frac{T^{1/2}}{r^2}\partial_r\left[\frac{r^3}{T^{1/2}}F\right] + \frac{1}{2}Hr\partial_r\eta = 8\pi r\hat{S}_r - \frac{1}{\sin\theta}\partial_\theta\left[\sin\theta G\right], \tag{11}$$

$$-\frac{1}{2T}\partial_\theta\left[TF\right] - \frac{T}{2\sin^2\theta}\partial_\theta\left[\frac{\sin^2\theta}{T}H\right] = 8\pi\hat{S}_\theta - \frac{1}{r^2}\partial_r\left[r^3G\right]. \tag{12}$$

Let's consider the solution procedure if the spatial manifold is assumed to be flat ($\eta = 0$ and $T = 1$). The second term in equation (11) disappears, decoupling this equation from H. This equation can then be solved for F with a simple, outward radial integration. Once F is known, it can be substituted into equation (12), which then becomes a parabolic equation for H. H is determined by an integration in angle outward from the polar axis (the stable direction).

Figure 1. Plot of a solution to the radial equations (9a,b), modified to include a nonvanishing source, for $l = 2$. Dashed curve shows the expected solution. The solid curve is the result of solving (9a,b) with a 2.5% error in the outer boundary condition. Note the irregularity produced near the origin. A similar irregularity arises near $r = 0$ when the partial differential equations (5) and (6) are solved, though then there appear the additional complications of angular dependence.

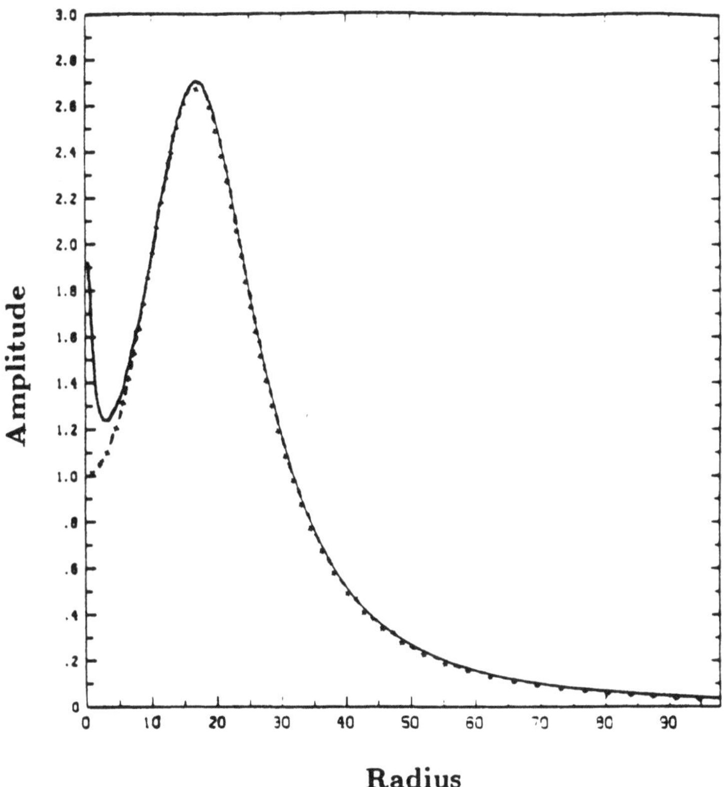

In general of course the three-space is not flat and so equations (11) and (12) are coupled. Nonetheless in this case we have found that the same integrations can be performed, with multiple iterations to obtain convergence to the full solution. The convergence was found to be linear and very rapid, with for example errors decreasing like

$$\epsilon^{n+1} \sim 10^{-4}\epsilon^{n}, \tag{13}$$

on a 60×18 mesh with $\max[\eta] \sim 0.3$. While I have no proof detailing a precise radius of convergence for this procedure, numerous test problems have been constructed with varying anisotropic amplitudes ($\eta \neq 0$) to test convergence. In all cases when $|\ \eta\ | \lesssim 1$ convergence was achieved. It has been observed that even in calculations involving copious emission of gravitational radiation, η rarely approaches unit amplitude.

REGULARITY

The clearest reason for using a fully constrained evolution, as we show below, turns out to be to maintain regularity of various tensor components at the singular point ($r = 0$) of the coordinate system. To understand this point we first need to explain what we mean by regularity. Because we use spherical-polar coordinates (and this would be true of other curvilinear coordinate systems), certain components of a tensor like K_{ij} have multivalued behavior at $r = 0$ (the singular points) and the differential operators that appear in the Einstein equations have singular behavior at $r = 0$ and along $\theta = 0$. Of course the physical fields themselves, in most situations, are in no way singular at these places. For this to be true the fields have certain local symmetries and in the differential equations cancellations must occur. The trick is to find ways of maintaining these crucial cancellations on a numerical level.

The most important example for the present discussion is the behavior of the extrinsic curvature tensor near $r = 0$. The local dependence of these components can be determined by assuming that the components of K_{ij} in a locally *Cartesian* coordinate system are each nonsingular and Taylor expandable. Numerous Taylor series terms are required to vanish to enforce axisymmetry (and, less crucially, vanishing rotation and equatorial plane symmetry if these are assumed). The tensor can then be transformed from Cartesian to spherical-polar (curvilinear) coordinates, uncovering the local dependence of the tensor's coordinate components at the singular points. Under the assumptions of axisymmetry, equatorial plane symmetry and vanishing rotation, the Cartesian tensor components of K_{ij} are expressible as (Evans 1984)

$$K_{xx} = \bar{k}_1(\rho^2, z^2) + \bar{k}_3(\rho^2, z^2)x^2, \tag{14a}$$

$$K_{yy} = \bar{k}_1(\rho^2, z^2) + \bar{k}_3(\rho^2, z^2)y^2, \tag{14b}$$

$$K_{zz} = \bar{k}_2(\rho^2, z^2), \tag{14c}$$

$$K_{xy} = \bar{k}_3(\rho^2, z^2)xy, \tag{14d}$$

$$K_{xz} = \bar{k}_4(\rho^2, z^2)xz, \tag{14e}$$

$$K_{yz} = \bar{k}_4(\rho^2, z^2)yz, \tag{14f}$$

where the functions $\bar{k}_i(\rho^2, z^2)$ are Taylor-expandable scalars reflecting the required symmetries. Note that vanishing rotation has reduced to four the number of independent functions determining K_{ij}. If maximal slicing is imposed, only three independent scalars will remain. Transforming to spherical-polar coordinates, raising one index and enforcing maximal slicing, we find

$$K^r{}_r = -k_1(\cos 2\theta + T^2\cos^2\theta) - k_3 r^2 \sin^2\theta\cos 2\theta + 2k_4 r^2\cos^2\theta\sin^2\theta, \quad (15a)$$

$$\frac{K^r{}_\theta}{r} = \left[k_1(T^2+2) + 2k_3 r^2\sin^2\theta + k_4 r^2\cos 2\theta\right]\cos\theta\sin\theta, \quad (15b)$$

$$\lambda = k_1\left[T^2(2-\cos^2\theta) - \cos 2\theta\right] - k_3 r^2\sin^2\theta\cos 2\theta + 2k_4 r^2\cos^2\theta\sin^2\theta, \quad (15c)$$

where $A^2 k_i = \bar{k}_i$ results from introducing the metric.

As the origin is approached ($r \to 0$, $T \to 1$), we find the following multivalued behavior of these components:

$$K^r{}_r \simeq -2P_2(\cos\theta)k_1{}^\circ, \quad (16a)$$

$$\frac{K^r{}_\theta}{r} \simeq -\partial_\theta P_2(\cos\theta)k_1{}^\circ, \quad (16b)$$

$$\lambda \simeq 3\sin^2\theta k_1{}^\circ, \quad (16c)$$

where $k_1{}^\circ = k_1(r=0)$ is a (limiting) constant value. More significantly, the three extrinsic curvature variables are no longer independent as the origin is approached. This is the crux of the problem with freely evolving all of the components of K_{ij}: even if initial data are prescribed such that (16a-c) are satisfied, independent values of $k_1{}^\circ$ can develop during the course of the evolution depending upon how the finite differences are constructed. Indeed, even with our best efforts to regulate the finite difference operators in the evolution equations for K_{ij}, we have found subtle, late-time growth of differences in the values of $k_1{}^\circ$. This lack of agreement slowly causes an irregularity to develop in the extrinsic curvature near $r = 0$, which ultimately causes the simulation to terminate. By being very careful with the finite differencing the onset of this irregularity has been pushed back. However, we eventually have required simulations to run for 2×10^4 to 4×10^4 time steps; too long to avoid trouble.

It was in hopes of getting around this limitation that a renewed examination was made of the issue of enforcing the momentum constraints. As we approach $r = 0$, we find that the momentum constraints (11) and (12) reduce to the conditions

$$3F = -\frac{1}{\sin\theta}\partial_\theta\left[\sin\theta G\right], \quad (17a)$$

$$-\frac{1}{2}\partial_\theta F - \frac{1}{2\sin^2\theta}\partial_\theta\left[\sin^2\theta H\right] = -3G. \quad (17b)$$

Figure 2. Gravitational radiation waveform emitted from a relativistic star that is undergoing a finite amplitude nonradial oscillation. We compare (a) the waveform obtained *- continued on next page*

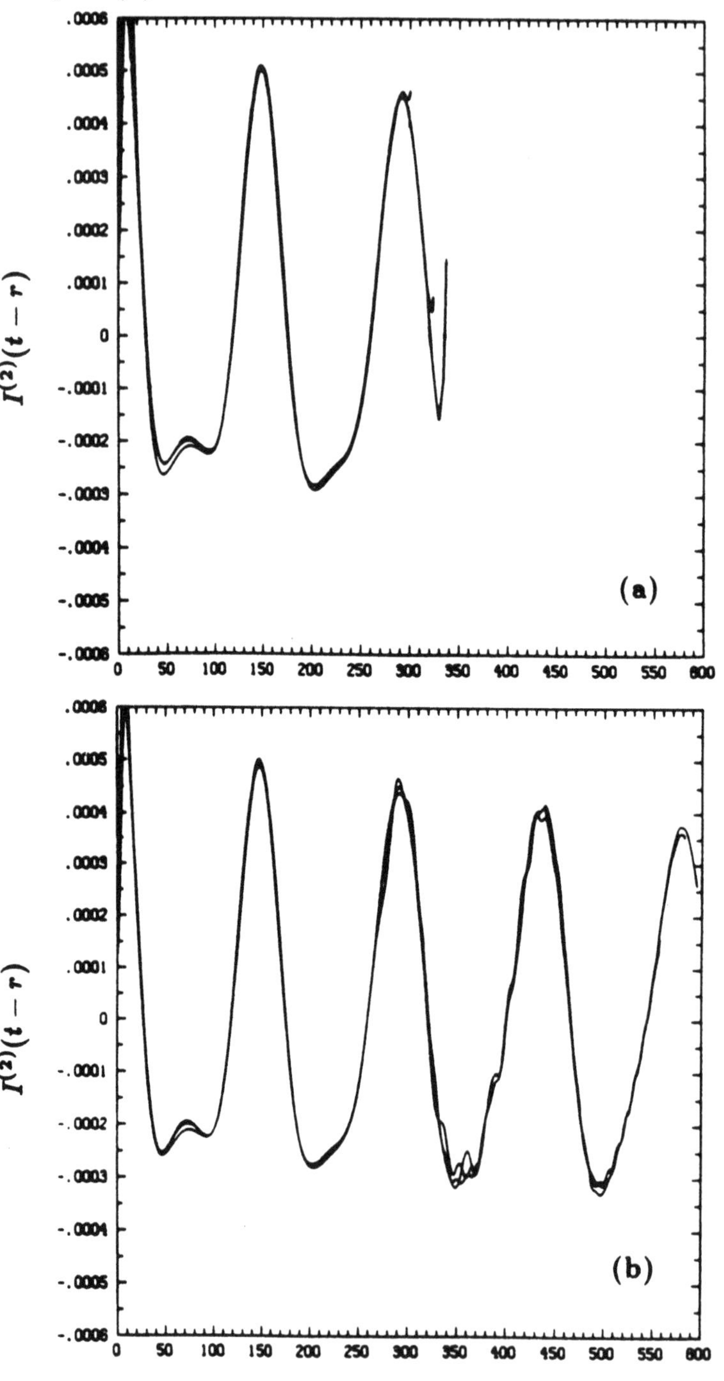

Figure 2. -cont'd by the unconstrained (K_{ij}) code to (b) the waveform obtained using the fully constrained version of the code. In each figure waveforms extracted from three different radii are plotted together, giving a sense of the accuracy and consistency of the simulation. There is essentially no significant difference between the waveforms calculated by the two versions of the code up until about $t \sim 350M$. At that point the unconstrained simulation terminated due to growth of a numerical irregularity at the origin, $r = 0$.

These conditions are trivially satisfied by (16a-c) and it is possible to achieve numerically the same satisfaction to high accuracy by using well regulated finite difference equations. In this way the near origin behavior of the variables $K^r{}_r$ and λ are determined on each time slice from the time evolved behavior of $K^r{}_\theta/r$ and irregularities have no chance of developing.

With both a constrained and unconstrained version of the axisymmetric code available, we were able to run side-by-side comparison calculations. The quantity of ultimate interest, the emitted waveform, was found to differ little in this comparison, even right up to the point the unconstrained simulation terminated. Figure 2 details this comparison.

ASYMPTOTIC BEHAVIOR

We must also be concerned with the asymptotic, large r properties of the solutions of the momentum constraints and requisite boundary conditions. We will assume in this discussion the boundary is outside the material source so that the matter energy density and stresses vanish as well as $S_i = 0$. In the local weak field (radii large enough that $| g_{\alpha\beta} - \eta_{\alpha\beta} | \ll 1$, but not so large a region as to allow significant cumulative nonlinearities) we can employ linearized gravity about Minkowskii space. Then the asymptotic gravitational field can be expressed in terms of static and dynamic (radiative) multipole moments (Thorne 1980, Abrahams & Evans 1988). The most important moments will be $l = 0$ and $l = 2$ and these manifest themselves in the extrinsic curvature in QI gauge as

$$K^r{}_r = -P_2(\cos\theta)\left[4\frac{I^{(1)}}{r^3} + 12\frac{I^{(0)}}{r^4} + 12\frac{I^{(-1)}}{r^5} + 12\frac{a_2}{r^5}\right] - 2\frac{a_0}{r^3}, \tag{18a}$$

$$\frac{K^r{}_\theta}{r} = -\sin\theta\cos\theta\left[2\frac{I^{(2)}}{r^2} + 6\frac{I^{(1)}}{r^3} + 12\frac{I^{(0)}}{r^4} + 12\frac{I^{(-1)}}{r^5} + 12\frac{a_2}{r^5}\right], \tag{18b}$$

$$\lambda = \sin^2\theta\left[\frac{I^{(3)}}{r} + 2\frac{I^{(2)}}{r^2} + 3\frac{I^{(1)}}{r^3} + 3\frac{I^{(0)}}{r^4} + 3\frac{I^{(-1)}}{r^5} + 3\frac{a_2}{r^5}\right], \tag{18c}$$

where $I = I(t-r)$ is the (radiative) quadrupole moment of the exterior gravitational field and $I^{(k)} \equiv d^k I(u)/du^k$, with $u = t-r$. Negative superscripts indicate integrals of $I(t-r)$ over past retarded time. These arise from the elliptic nature of QI gauge. Similarly, functions arise in this gauge, $a_0 = a_0(t)$ and $a_2 = a_2(t)$, that depend only on coordinate time t (not retarded time) and represent instantaneous gauge effects.

Note that the radiative part of the field shows up only in λ, and not in $K^r{}_r$ or $K^r{}_\theta$. The limiting appearance of the momentum constraints in the vacuum, weak field surrounding the source is given by equations (7) and (8) after setting $S_i = 0$. If we evolve only $K^r{}_\theta$, which does not contain a radiative component in its asymptotic structure, how does the radiative part we know must exist in the asymptotic behavior of λ come about? The answer is that even though a term like $I^{(3)}/r$ is not present in $K^r{}_\theta$, evanescent (nonradiative) terms that depend on the quadrupole moment are present and the radial differential operator in equation (8) has the effect of "digging" the radiative piece out of $K^r{}_\theta$. This can easily be verified by substituting the quadrupole wave solution (18a-c) into the source-free version of the constraints (7) and (8).

All of this is of more than just academic interest. When we finite difference the momentum constraints (11) and (12), we find that the last radial zone in which we carry a value for λ cannot be determined from the discrete constraint equations. This value must be supplied by imposing a boundary condition that expresses some condition on the gravitational field exterior to the mesh. What we want to assume is that there are only outgoing gravitational waves or static moments derived from the central source present on the outer boundary (*i.e.*, that there are no incoming waves and that the source is "isolated"). The interesting facet is that we achieve this by imposing a weak-field, outgoing-wave Sommerfeld condition on a quantity, λ, that is being derived from a constraint equation and not from an evolution equation. This condition is

$$\partial_t \lambda = -\frac{1}{r}\partial_r(r\lambda). \tag{19}$$

We find that for quadrupole waves this condition annihilates not only the leading outgoing radiative term in λ but also the next order nonradiative term, leaving a remainder in (19) of $\simeq -2I^{(2)}/r^3$.

CONCLUSIONS

The fact that nearly identical results are obtained for the emitted gravitational radiation whether we use a partially constrained evolution or a fully constrained evolution is powerful evidence that the long standing issue over employing the constraints is less critical than previously imagined. Nonetheless, we have pinpointed an example in which an unconstrained evolution goes awry, despite our best efforts, generating an irregularity that terminates the calculation. The constrained evolution shows no sign of such defects and, accordingly, a fully constrained evolution appears to be the method of choice.

ACKNOWLEDGEMENTS

I would like to acknowledge several conversations with Stu Shapiro and Saul Teukolsky that provided the assurance, from their own studies of the momentum constraints in two dimensions, that a solution algorithm could be found. It is my understanding that their method of employing the momentum constraints during simulations is similar, if not identical, to the one described here. I am also indebted to my colleague Andrew Abrahams whose thesis work lead to the apparent necessity to find a fully constrained approach and who provided several tests of the scheme. This work was supported by NSF grant AST85-14911. Calculations were performed on the Cray X-MP/48 at the National Center for Supercomputing Applications in Champaign-Urbana, Illinois.

REFERENCES

Abrahams, A. M. & Evans, C. R. 1988, *Phys. Rev.*, **D37**, 318.

Bardeen, J. M. & Piran, T. 1983, *Phys. Rep.*, **96**, 205.

Choptuik, M. W. 1986, Ph.D. Thesis, The University of British Columbia, unpublished.

Dykema, P. G. 1980, Ph.D. Thesis, The University of Texas at Austin, unpublished.

Evans, C. R. 1984, Ph.D. Thesis, The University of Texas at Austin, unpublished.

Evans, C. R. 1986, in *Dynamical Spacetimes and Numerical Relativity,* ed. J. Centrella (Cambridge University Press: Cambridge).

Evans, C. R. & Hawley, J. F. 1988, *Ap. J.*, **332**, 659.

Piran, T. 1983, in *Gravitational Radiation,* ed. N. Deruelle & T. Piran (North-Holland: Amsterdam).

Smarr, L. 1979, in *Sources of Gravitational Radiation,* ed. L. Smarr (Cambridge University Press: Cambridge).

Stewart, J. M. 1984, in *Classical General Relativity,* ed. W. B. Bonner, J. N. Isham & M. A. H. MacCallum (Cambridge University Press: Cambridge).

Thorne, K. S. 1980, *Rev. Mod. Phys.*, **52**, 299.

Wilson, J. R. 1979, in *Sources of Gravitational Radiation,* ed. L. Smarr (Cambridge University Press: Cambridge).

York, J. W. 1979, in *Sources of Gravitational Radiation,* ed. L. Smarr (Cambridge University Press: Cambridge).

EXPERIENCES WITH AN ADAPTIVE MESH REFINEMENT ALGORITHM IN NUMERICAL RELATIVITY

Matthew W. Choptuik
Laboratory of Nuclear Studies
Cornell University, Ithaca NY 14853–5001

Abstract. An implementation of the Berger/Oliger mesh refinement algorithm for a model problem in numerical relativity is described. The principles of operation of the method are reviewed and its use in conjunction with leap–frog schemes is considered. The performance of the algorithm is illustrated with results from a study of the Einstein/massless scalar field equations in spherical symmetry.

INTRODUCTION

A key issue in the numerical treatment of any set of partial differential equations (PDE's) is the choice of the discrete domain on which algebraic approximations of the PDE's are to be solved. When finite difference methods are used (Richtmeyer & Morton 1957; Kreiss & Oliger 1984), this is accomplished through the familiar technique of introducing a grid, or mesh, on the continuum domain of the problem. The discrete domain is then simply the set of all intersection points of the grid lines, approximations of the continuum unknowns are defined at these grid points, and algebraic equations for the unknowns are obtained from difference analogues of the various differential equations. Typically, the finite difference grid will be characterized by a *discretization scale*, h, such that distances between neighboring grid points are of order h. Computationally, of course, the choice of h has a direct bearing on both the cost of the calculation, and the accuracy of the results: for a d–dimensional problem (including time) on a domain with characteristic diameter L, the computation will involve at least $O((L/h)^d)$ operations and $O((L/h)^{d-1})$ storage, while the local error in the computed solution will, at best, be $O(h^p)$, if a difference equation with pth–order truncation error is used.

Historically, most finite difference calculations have been performed on *uniform* meshes—grids which have constant mesh spacings in each of the coordinate directions, or *graded* meshes, where the grid spacings are relatively slowly varying functions of position. For problems whose solutions have scales of variation which are roughly uniform over the solution domain, this approach is entirely adequate. However, the use of such meshes is more frequently motivated by considerations such as the ease with which finite difference equations can be formulated and solved on these structures. Many of the most interesting and challenging problems in computational physics in general, and numerical relativity in particular, involve solutions which are *significantly* harder to approximate in some regions than in others. For such systems, especially those involving two or three spatial dimensions as well as time, the use of a uniform grid which is sufficiently fine to adequately represent the smallest scale features of the solution will be extremely inefficient, if not computationally prohibitive. In addition, particularly for highly dynamic and/or non–linear problems, it will not always be possible to choose, a *priori*, a mesh spacing which will lead to a good solution and the computational physicist will often be lead to a trial–and–error process of experimenting with the grid until it is felt that the results are acceptable, or the calculation becomes too expensive.

Considerations such as these, as well as the success of classes of algorithms for other problems in numerical analysis (such as the solution of ordinary differential equations), have led to a considerable research effort directed towards the development of *adaptive* techniques for the finite difference treatment of PDE's. The basic idea behind these algorithms is to make the discretization scale a *local* quantity which will *automatically* vary from place to place in the solution domain in accordance with the demands that the solution places on the finite difference approximation. The purpose of this article is to report on some work I have done with an adaptive algorithm due to Berger & Oliger (1984, hereafter BO84), which is particularly suited for hyperbolic systems such as the sets of equations encountered in numerical relativity. The next section provides a brief overview of the operation of the algorithm; following this is a slightly more detailed discussion of the generation of error estimates and the use of leap–frog schemes. Finally some results are presented from an application of the algorithm to a model problem in numerical relativity. Although the discussion here is restricted to finite difference techniques, it should be stressed that adaptive methods are being studied with all of the principal techniques for solving time dependent PDE's—including finite–element methods (Baines & Waltham 1986; Löhner 1987), spectral methods (Kopriva 1986), and the method of lines (Babuska & Bieteman 1986). All of this research shares the common goal of producing the best possible solution for a given amount of computational resources, as has previous work in numerical relativity, where grid–adaptation techniques have been used to various degrees.

THE BERGER/OLIGER ALGORITHM

The Berger/Oliger algorithm is a general method for solving hyperbolic systems based on recursive (nested) local refinement of uniform meshes. Although the method is applicable to problems in any number of spatial dimensions, attention is restricted here to the case of *one* spatial coordinate—the interested reader is referred to BO84 for discussion of the non–trivial modifications necessary for the treatment of 2– and 3–d problems, as well as for algorithmic details which are omitted here. As a prototype system for illustrative purposes, I will adopt the simplest of all hyperbolic equations,

$$u_t = u_x, \tag{1}$$

for $u(x,t)$, on the half plane $t > 0$, which with the initial condition $u(x,0) = f(x)$, has the solution $u(x,t) = f(x+t)$, representing leftwards propagation of the initial disturbance, $f(x)$, with unit velocity. (The subscript notation for partial differentiation will be used throughout the following.)

The (finite) space–time domain on which the difference solution is to be constructed is covered by a *base grid*, G_0^0, which is uniform with mesh increments $\Delta x_0 \equiv h_0$ and $\Delta t_0 = \lambda h_0$, for constants h_0 and λ. (λ typically < 1.) The requisite notion of local mesh refinement may then be defined in brief by the recursive statement: any grid, G_i^l, may admit sub–grids (refinements), G_a^{l+1}, G_b^{l+1}, $\ldots$, such that: 1) each of the G_c^{l+1} is properly contained in G_i^l, and 2) $h_l = \rho_{l+1} h_{l+1}$ for some integer ρ_{l+1} Thus, the superscript on G labels the level of refinement, while the subscript indexes grids at the same level. Typically, the *refinement factors*, ρ_l, will be the same small integer (which will be a tuneable parameter of the algorithm), and refinement will be allowed up to some maximum level, l_m. Figure 1a depicts a simple example of a grid pattern constructed in this fashion with $l_m = 2$, $\rho = 2$. Note that the number of (spatial) grid points is, in general, a dynamic quantity, as is the finest level of refinement at a given space–time location. At any time, the spatial gridding pattern can be represented as a *tree*, with each "child" grid properly contained within its "parent". (Figure 1b). Any program which implements this algorithm must maintain a representation of this tree, and as described in BO84 and Berger (1986), use of appropriate data structures greatly facilitates this task. It is also important to note that the Berger/Oliger scheme treats the constituent grids as separate entities in the sense that difference equations are solved on each level and on each grid relatively independently of the equations on coarser or finer levels. Operationally, this produces a clean split between the problem–dependent portions of the algorithm (such as the specific scheme or equation solver which is being used) and the problem–independent parts (such as the inter–grid communications processes outlined below). Clearly, such a separation is essential for the method to be of a "general–purpose" nature.

The process described in BO84 for choosing a suitable local level of refinement is based on ideas dating back to the time of L.F. Richardson (1910) and his human computers, who performed some of the earliest finite difference computations. For concreteness, consider solving (1) using a Lax–Wendroff scheme on a uniform mesh with $\lambda \equiv \Delta t / \Delta x$. Using the standard notation $f_j^n \equiv f(j\,\Delta x, n\,\Delta t)$, the difference equation is

$$\hat{u}_j^{n+1} = \hat{u}_j^n + \frac{1}{2}\lambda\left(\hat{u}_{j+1}^n - \hat{u}_{j-1}^n\right) + \frac{1}{2}\lambda^2\left(\hat{u}_{j+1}^n - 2\hat{u}_j^n + \hat{u}_{j-1}^n\right), \qquad (2)$$

where the caret is used here and below to distinguish functions which satisfy difference equations from their continuum counterparts. Supressing the spatial grid indices, this can be written as

$$\hat{u}^{n+1} = \hat{Q}^{\Delta x}\,\hat{u}^n, \qquad (3)$$

Figure 1a. A simple example of the type of grid structure admitted by the Berger/Oliger algorithm. Here, there are three levels of grids, distinguished by the various line–types in the diagram. Note that refinement is in both space and time.

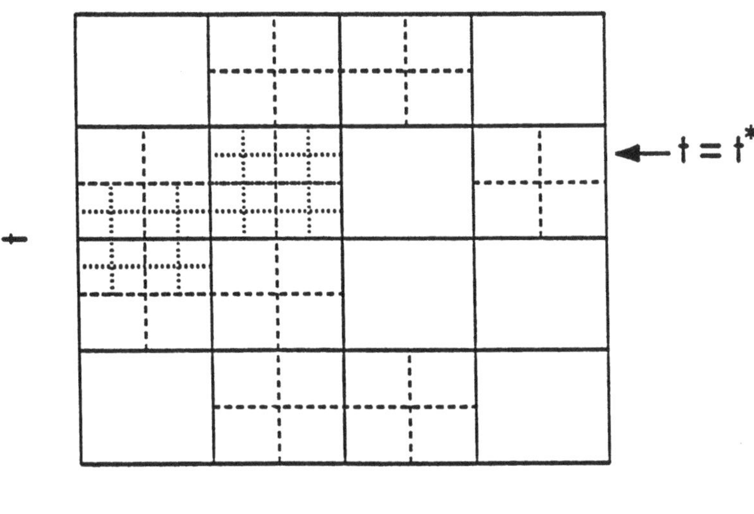

Figure 1b. The tree structure associated with the above gridding pattern at time t^*. Solid lines indicate parental–child links, while the dotted line shows a link between grids at the same level. Both types of links are maintained by the refinement algorithm for use in inter–grid communication.

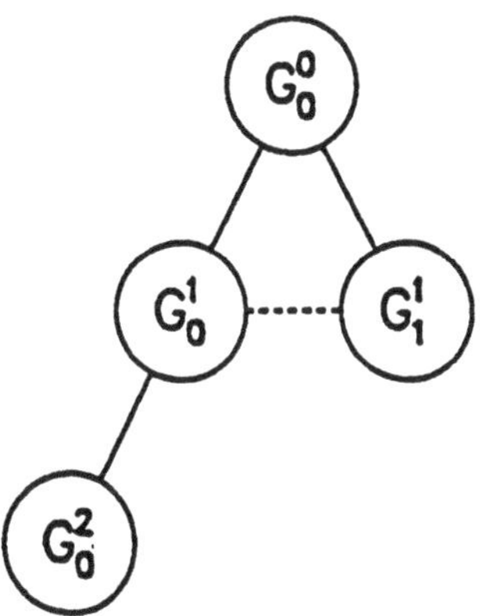

where the difference operator, $\hat{Q}^{\Delta x}$, can be shown to have the following expansion:

$$\hat{Q}^{\Delta x} = 1 + (\lambda \Delta x)\partial_x + \frac{1}{2}(\lambda \Delta x)^2 \partial_{xx} + \frac{1}{6}\lambda \Delta x^3 \partial_{xxx} + O(\Delta x^4) \tag{4}$$

The *local truncation error* (l.t.e.) of the difference equation is defined as follows:

$$\tau^n \equiv u^{n+1} - \hat{Q}^{\Delta x} u^n, \tag{5}$$

and is just the error that is incurred in propagating the *continuum* solution from $t = n\,\Delta t$ to $t = (n+1)\,\Delta t$ using the difference equation rather than the differential one. It is precisely this quantity which the Berger/Oliger algorithm attempts to monitor and control by introducing refinements in regions where τ is larger than some user–supplied criterion τ_m. Estimates of τ are generated using a procedure which, although outlined here for a specific equation and difference scheme, is quite generally applicable. Using Taylor series expansion, (1) and (4), we find

$$\tau_1 \equiv \tau^n = \frac{1}{6}\lambda\,(\lambda^2 - 1)\,u^n_{xxx}\,\Delta x^3 + O(\Delta x^4), \tag{6}$$

so the l.t.e. is third order in Δx, as is to be expected for one step of a second–order scheme. Similarly we can compute

$$\begin{aligned} \tau_2 \equiv u^{n+2} - \hat{Q}^{\Delta x}\left(\hat{Q}^{\Delta x} u^n\right) &= \frac{1}{3}\lambda\,(\lambda^2 - 1)\,u^n_{xxx}\,\Delta x^3 + O(\Delta x^4) \\ &= 2\tau_1 + O(\Delta x^4) \end{aligned} \tag{7}$$

Now consider the truncation error associated with a difference operator, $\hat{Q}^{2\Delta x}$ which would arise from Lax–Wendroff differencing on a mesh twice as coarse as the original:

$$\begin{aligned} \tilde{\tau}^n_1 \equiv \tilde{\tau}^n &= u^{n+2} - \hat{Q}^{2\Delta x} u^n \\ &= \frac{1}{6}\lambda\,(\lambda^2 - 1)\,(2\,\Delta x)^3\,u_{xxx} + O(\Delta x^4). \end{aligned} \tag{8}$$

Eliminating u^{n+2} from (7) and (8), we find

$$\tau^n = \frac{1}{6}\left(\hat{Q}^{\Delta x}\hat{Q}^{\Delta x} - \hat{Q}^{2\Delta x}\right)u^n + O(\Delta x^4). \tag{9}$$

If we assume that the difference solution admits a Richardson–expansion (Richardson 1910) of the form

$$u = \hat{u} + \Delta x^2 \hat{e}_2 + \ldots \tag{10}$$

where $\hat{e}_2(x,t)$ is a smooth function, it is straightforward to show that

$$\tau^n = \frac{1}{6}\left(\hat{Q}^{\Delta x}\hat{Q}^{\Delta x} - \hat{Q}^{2\Delta x}\right)\hat{u}^n + O(\Delta x^4), \tag{11}$$

so to leading order, provided $\hat{u}$ is smooth enough, we can estimate the l.t.e. by taking two steps on the original grid and combining the result with that obtained by taking one step on a coarse grid—but using the same difference equations and approximate solution. In the next section, I will examine this process in slightly more detail for the case of leap–frog differencing.

Truncation error estimation is only a component of the Berger/Oliger regridding algorithm and can be thought of as a representative black–box which, when given a grid and its mesh functions, returns a list of points which have been "flagged" as lying in regions requiring refinement. (This, in fact, is one of the most attractive features of the method from the point of view of experimenting with other regridding schemes.) At each existing level, $l < l_m$, the regridding process as a whole is performed periodically (typically every four or so time steps) through the clustering/buffering/merging process described in detail in BO84. Contiguous flagged points are formed into clusters which are then augmented at their extremities with buffer regions of some specified size. Finally, clusters which are within a given small distance of each other are merged, leaving a number of disjoint regions, each of which is to be covered with a grid at level $l+1$. The addition of buffer regions allows for a longer interval between regridding times at the expense of additional fine grid points. Note that the newly identified regions may intersect grids which exist at the fine level just prior to regridding. The initialization of the level $l+1$ mesh functions, then, will generally involve a combination of the interpolation of level l quantities, and the transferral of old level $l+1$ values. In addition, when regridding takes place at level l it is also performed *first* on all existing finer levels to ensure the proper nesting of refinements. Thus, regridding at level l essentially entails a bottom–up reconstruction of the entire tree structure at level $l+1$ and below. Nevertheless, the regridding overhead generally amounts to a small fraction of the total computational work expended. This is particularly true if there are a significant number of grid points at level l_m, since the regridding process is never initiated at this level, and the actual integration of the finest level difference equations is likely to dominate the computation.

Finally, there are two basic types of inter–grid/inter–level communcation which are performed by the mesh refinement algorithm. The first of these involves the use of level l quantities to provide boundary values for level $l+1$ grid functions. Extrapolation is avoided by advancing the solution in time on the coarse grid before taking the ρ_{l+1} steps on the fine grid which will bring the child grid into temporal alignment with its parent. Empirically, polynomial interpolation in the coarse mesh seems to work adequately, but it should be noted that this one area of the Berger/Oliger algorithm that needs additional research since many difficult questions connected with stability and accuracy arise (Berger 1985). The second type of grid–to–grid process involves the transferral of level $l+1$ values to corresponding parental quantities whenever the grids have been integrated to the same time. If this is *not* done, then coarse grid unknowns which lie within refined regions, and are inaccurate by assumption, could eventually lead to bad coarse grid values near the fine grid extremities. This in turn would result in poor boundary conditions for the fine grid functions and the entire solution could be corrupted. A straightforward assignment (injection) of level $l+1$ values to the level l quantity at the same space–time location is probably adequate for most problems, but is another area which needs further study.

LEAP–FROG SCHEMES AND L.T.E. ESTIMATES

One of the principal differences between the calculations described in the next section and other results obtained with the Berger/Oliger algorithm (BO84; Berger & Jameson 1985), is my use of a leap–frog method for integrating the scalar wave equation. Such a scheme is based on data at three time levels, whereas the previous work appears to have dealt exclusively with two–level methods. The choice of leap–frog is motivated by the desire to keep the differencing second order in space and time without unduly increasing the complexity of the mesh refinement algorithm. In particular, although two–level schemes based on "staggered" grid structures have proven very useful in many previous computations in numerical relativity, the incorporation of staggered meshes into the algorithm is not straightforward.

The leap–frog discretization of (1) is

$$\hat{u}_j^{n+1} = \hat{u}_j^{n-1} + \lambda\left(\hat{u}_{j+1}^{n} - \hat{u}_{j-1}^{n}\right). \tag{12}$$

while for a second order equation such as $\phi_{tt} = \phi_{xx}$, a leap–frog scheme can be constructed by introducing auxiliary variables, $\Phi \equiv \phi_x$ and $\Pi \equiv \phi_t$, rewriting the equation in first order form

$$\begin{aligned} \Phi_t &= \Pi_x \\ \Pi_t &= \Phi_x, \end{aligned} \tag{13}$$

and then differencing in the obvious fashion. Schemes such as (12) are subject to the usual stability criterion for explicit schemes, $\lambda \le 1$. In addition, these methods are non–dissipative: evolution with the difference equations preserves the amplitudes of (discrete) Fourier components, as does evolution with the continuous equations. The phases of the Fourier modes, however, do not do so well; equation (12) is dispersive, while (1) is not. If we look for normal modes of (12) of the form $\exp(i\omega(x + \mu t))$, we find

$$\mu = (\xi\lambda)^{-1}\tan^{-1}\left(\frac{\pm\lambda\sin\xi}{\sqrt{1-\lambda^2\sin^2\xi}}\right), \tag{14}$$

where $\xi \equiv \omega\,\Delta x$ assumes values in the range 0 to 2π. Of course, in the differential case, $\mu = 1$ for all ω. Considering the limit $\lambda \to 0$ of the above, which corresponds to treating time continuously, while retaining the spatial discretization of (12), we get

$$\mu = \pm\frac{\sin\xi}{\xi}. \tag{15}$$

Clearly the propagation of high frequency components (with respect to the mesh) of a grid function via (12) will be seriously in error—the shortest wavelength modes will not propagate at all! If the data is sufficiently smooth, this should not be too worrisome. It is, however, more significant when the l.t.e. estimating process described earlier is applied to the leap–frog scheme.

The local truncation error of (12) is

$$\tau = \frac{1}{3}\lambda(\lambda^2 - 1)u_{xxx}\,\Delta x^3 + O(\Delta x^5). \tag{16}$$

Is is easy to see that in Fourier space, a mode of frequency ω (of u or $\hat{u}$) will result in a mode of the same frequency in τ, but weighted by ω^3. Thus the l.t.e. estimate can easily become dominated by "noise" from short wavelength modes, particularly in the mesh refinement algorithm where the interpolation operations inherent to the regridding and the setting of boundary values tend to introduce high frequency components into the solution. In cases like this, it proves quite advantageous to add dissipative terms, which damp high frequency modes, to equations such as (12). As described in Kreiss & Oliger (1973, p. 35), if (12) is modified to

$$\begin{aligned}\hat{u}_j^{n+1} &= \hat{u}_j^{n-1} + \lambda\left(\hat{u}_{j+1}^{n} - \hat{u}_{j-1}^{n}\right)\\ &\quad - \frac{\epsilon}{16}\Delta x^4\left(\hat{u}_{j+2}^{n-1} - 4\hat{u}_{j+1}^{n-1} + 6\hat{u}_{j}^{n-1} - 4\hat{u}_{j-1}^{n-1} + \hat{u}_{j-2}^{n-1}\right),\end{aligned} \tag{17}$$

for $0 \le \epsilon \le 1$, then the per–step amplification factor, q_ω, of a mode with frequency ω is

$$q_\omega = 1 - \epsilon\sin^4\left(\frac{1}{2}\omega\,\Delta x\right), \tag{18}$$

so that, even for a fairly small value of ϵ, high frequency components can be effectively damped. At the same time, (17) is still a second order difference approximation of (1).

Another troublesome feature of the leap–frog scheme is the possible appearance of a stray family of modes, which propagate in the wrong direction as can be seen from the $\pm$ character of expressions (14) and (15). In this simple linear case, there will be no coupling between the left–moving modes and the spurious right–moving ones, so if the initial data is prepared carefully, this should not present a problem. Empirically, however, I have some evidence that such modes are introduced in the regridding part of the mesh refinement algorithm to the extent that they eventually dominate the l.t.e. estimates, but this needs further study.

At the conference, I expressed some concern about introducing refinements during an evolution at levels finer than those extant at the start of the calculation. Since then, I have studied the refinement process more carefully using the model system (1), as well as the spherically symmetric wave equation, and the indications are that, at least in principle, it *is* possible to go to arbitrarily fine grids without a serious breakdown of the error estimation process. Again, the key idea here is that a Richardson–expansion like (10) must hold, with $\hat{e}_2$ a *smooth* error function. When a new level is introduced, the grid functions must be interpolated from parental quantities, and this is likely to result in an error term which is not particularly smooth. However, as long as the evolution scheme is dissipative, after a few steps the error should, in general, be smoothed, and the expansion will be valid. Now, for a leap–frog scheme, two time levels of data must be generated when a new grid is introduced at level $l+1$. Suppose the regridding occurs at time t_l on a level l grid. Then the $t = t_l$ data on the new grid will always be obtained from interpolation of the $t = t_l$ quantities on parental grids. However, there are at least two possibilities for generating the needed $t = t_l - \Delta t_{l+1}$ values. The first of these is to interpolate in the coarse grid using $t = t_l,\ \ t = t_l - \Delta t_l,\ t = t_l - 2\,\Delta t_l \ldots$ data, which is how my current version of the refinement algorithm works. The second is to use the freshly interpolated $t = t_l$ data and differenced versions of the equations of motion to get the retarded quantities. For example,

$$\hat{u}_j^{n-1} = \hat{u}_j^n - \frac{1}{2}\lambda\left(\hat{u}_{j+1}^{n-1} - \hat{u}_{j-1}^{n-1} + \hat{u}_{j+1}^n - \hat{u}_{j-1}^n\right) \tag{19}$$

is a second order approximation of (1) which can be solved for the $\hat{u}_j^{n-1}$, and in fact, the mesh refinement program already solves such equations when setting up the initial data for a calculation. From analytic considerations, it is fairly easy to argue that with the first technique, the use of linear interpolation in time, combined with the leap–frog scheme, will lead to a breakdown in the Richardson expansion. The situation is not so clear when cubic interpolation, which I have also tried, is used.

Figure 2 shows some results from a numerical experiment which illustrate some of these points. The experiment involves the integration of (17) with $\epsilon = 0.5$ on a single mesh, with periodic regridding of the solution to a 2:1 refinement which becomes the new grid for the calculation. The quantity $\|\tau_{est.}\|$ is the l_2 norm (r.m.s. value) of the truncation error estimate analagous to the leading term of (11), but with a factor of Δx^3 removed. The vertical dotted lines indicate regridding times, while the horizontal dashed line is the norm of the leading term of the true truncation error (again, scaled by Δx^{-3}) which is calculated analytically. At each level, 50 time steps are taken. In the top plot, the retarded data at regridding times comes from the solution of (19), in the other case it is obtained from cubic interpolation. Clearly, with the former technique, there seems to be no problem in going to finer and finer levels, but as suggested above, the interpolation method seems to introduce spurious modes at the $O(h^2)$ level, which eventually swamp the truncation error calculation. These results are preliminary, however, and need to be analyzed further.

Finally, it should be stressed that the issue of truncation error estimation is most

Figure 2. Results of the numerical experiment described in the text for investigating the l.t.e. estimation procedure with a leap-frog scheme. The top plot is from a run where the retarded data is determined using differenced equations of motion at regridding times; in the bottom case, the data was computed using cubic interpolation (in space and time) of coarse grid values.

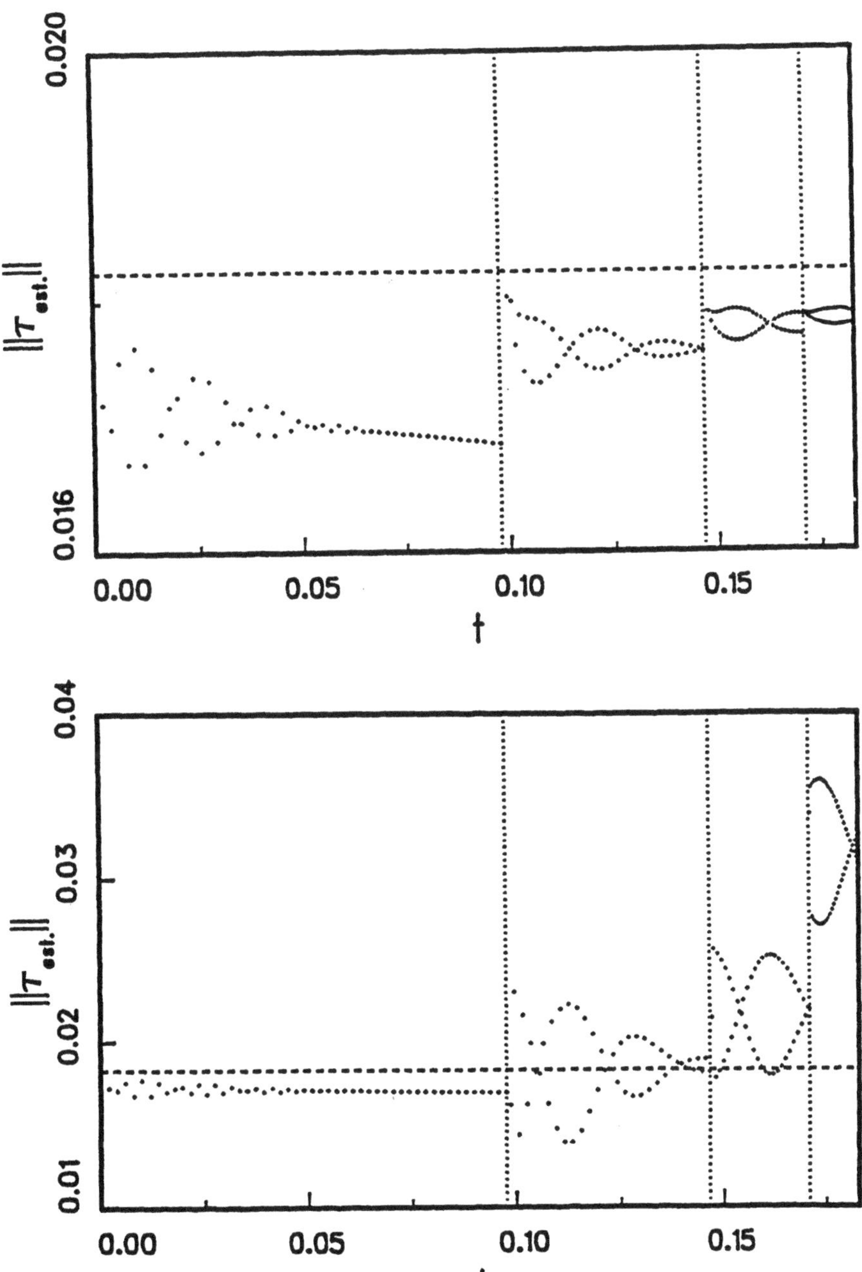

relevant to the *efficiency* of the computation, not the accuracy. When the Richardson expansion breaks down, the performance of the algorithm can degrade significantly, since a refinement by a factor of ρ will lead to an estimate of τ which is reduced by some factor *less* than ρ^3 (for a second–order scheme). Clearly, the algorithm will still introduce fine meshes in those regions where they are needed. It is also likely to introduce them in areas where they are *not* needed. Some of the results shown in the next section (particularly the strong–field calculations) are good examples of this phenomenon.

SOME MESH REFINEMENT RESULTS

The results in this section come from a mesh refinement treatment of the spherically–symmetric equations for a massless scalar field, ϕ, minimally coupled to the gravitational field. This problem has also been solved numerically by Goldwirth & Piran (1987, this proceedings 1988), who, following analytic work by Christodoulou (1986 a,b), used a characteristic method to study the possibility of cosmic censorship violations in the model. I use a 3+1 approach, but have chosen a coordinate system which produces a particularly simple set of equations in order to minimize the overall complexity of the refinement algorithm. Specifically, by combining polar slicing with a radial (areal) coordinate, r, (Unruh 1976, Bardeen & Piran 1983, Choptuik 1986, Shapiro & Teukolsky 1986) and adopting a fully constrained approach (Choptuik 1986), the 4–metric is

$$ds^2 = -\alpha^2 dt^2 + a^2 dr^2 + r^2 d\Omega^2, \tag{20}$$

and the equations

$$\Phi_t = \left(\frac{\alpha}{a}\Pi\right)_r \tag{21}$$

$$\Pi_t = \frac{1}{r^2}\left(r^2\frac{\alpha}{a}\Phi\right)_r \tag{22}$$

$$\frac{d\alpha}{dr} = \left(\frac{a^2-1}{r} + \frac{1}{a}\frac{da}{dr}\right)\alpha \tag{23}$$

$$\frac{da}{dr} = \frac{1}{2}\left(\frac{a-a^3}{r}\right) + 2\pi r\left(\Pi^2+\Phi^2\right)a \tag{24}$$

are sufficient to determine the evolution of the system. Here, $\Phi \equiv \phi_r$ and $\Pi \equiv a\phi_t/\alpha$ are auxiliary variables which allow the wave equation to be written in 3+1 form. All functions depend on r and t alone. Another quantity of interest is the mass aspect

$$m \equiv \frac{1}{2}r\left(1-a^{-2}\right) \tag{25}$$

which in a vacuum region, measures the total mass within the origin–centred sphere of radius r. I will not discuss the incorporation of the slicing and Hamiltonian constraint equations (23) and (24), which are not hyperbolic, into the

Figure 3. Typical mesh patterns produced by weak– (left) and strong–field runs. Each rectangle represents a grid which (except for the base grid) has a temporal extent of $4\Delta t_l$. Additional levels of refinement were used in the strong–field case but are not displayed here.

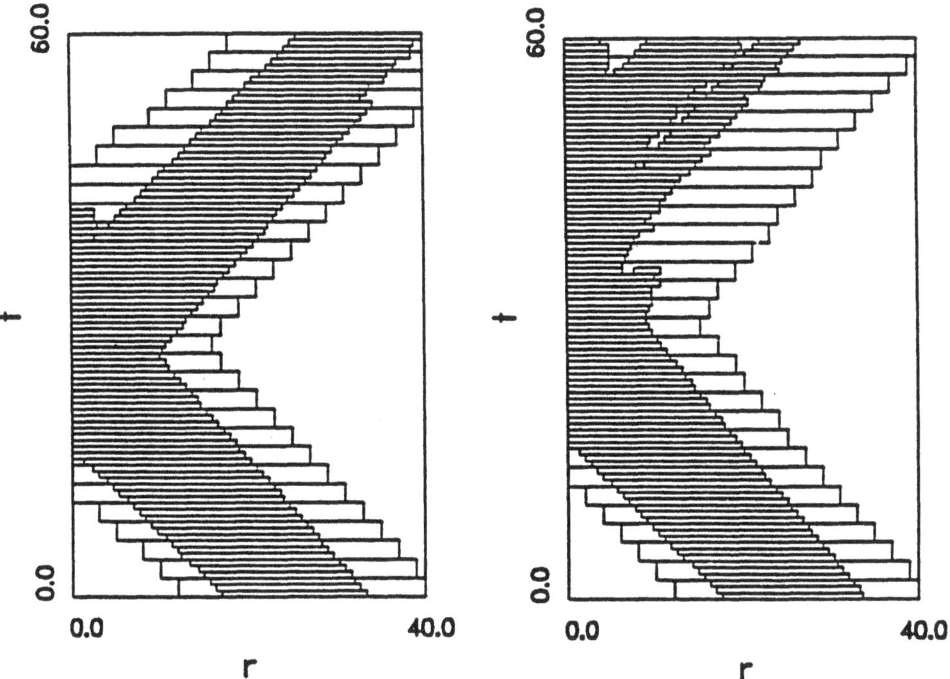

mesh refinement program. However, it should be stressed that this is one aspect of the method which demands special care and further examination, since the Berger/Oliger algorithm in its original incarnation asssumes locality of the differential system.

Initial data for the model is set up by specifying $\phi(r,0)$ in the form

$$\phi(r,0) = Ar^3 \exp\left(-\left(\Delta^{-1}\left(r-r_0\right)\right)^p\right) \tag{26}$$

where A, r_0, Δ and p are parameters, and requiring that the pulse be ingoing. This fixes Φ and Π; α and a are then computed from differenced versions of (23) and (24). For all of the results presented here, $p = 2$.

Figure 3 shows representations of the grid structure produced by the refinement algorithm in integrating a very weakly self–gravitating pulse and a pulse which forms a black hole. It is clear that the algorithm does a good job of automatically tracking features in the solution.

Figure 4 displays plots of the mass aspect for two distinct initial configurations at

Figure 4. Late–time plots of the mass aspect function for two runs with slightly different values of the initial field amplitude A. The base grid had $\Delta x_0 = 1.0$, $\lambda = 0.5$ and up to 8 additional levels of grid were allowed with a constant refinement factor of 3. A black hole forms only in the second case.

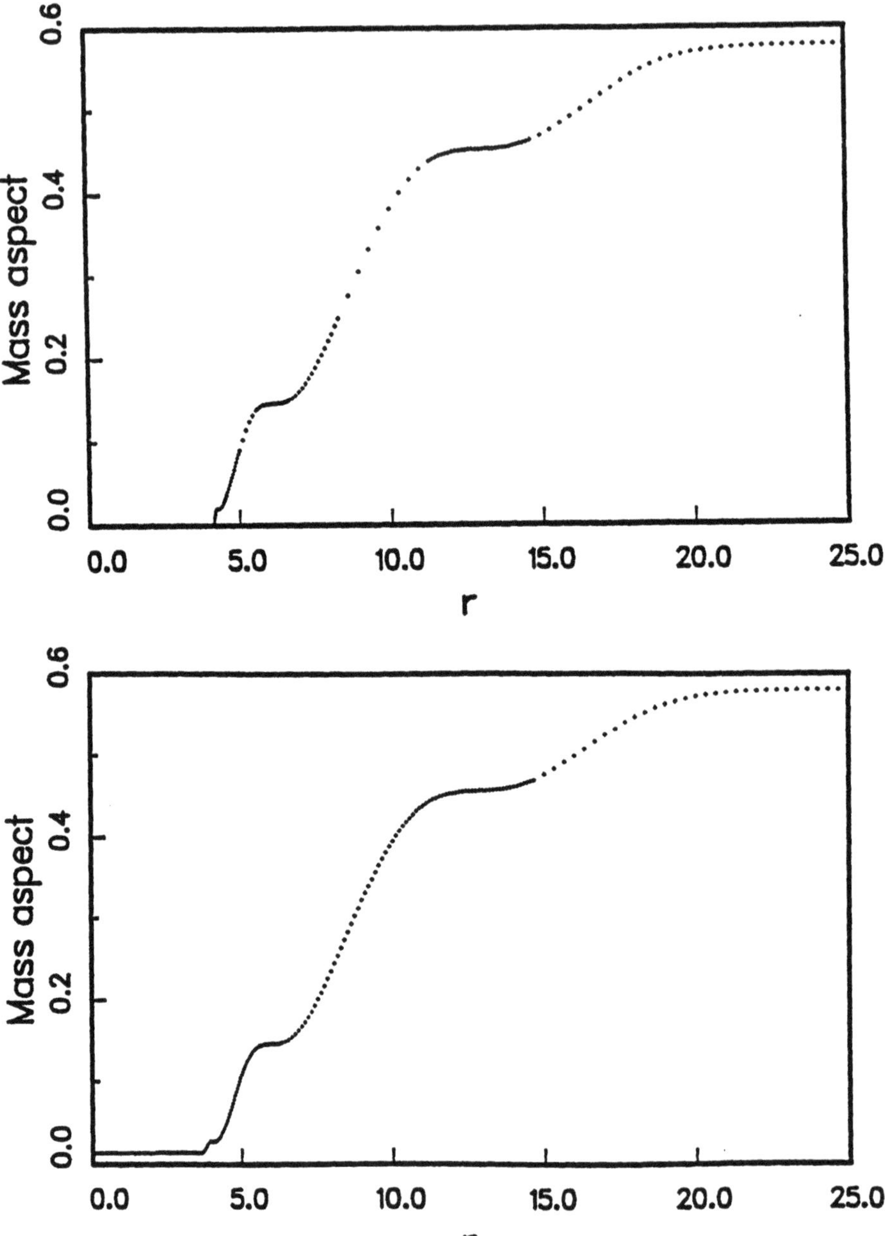

Figure 5. Detailed views of m and Φ from the black-hole run. The dashed line is $r = 2m$. Features in both plots near the "horizon" may not be trustworthy due to the combination of dissipation in the difference scheme and the fact that the algorithm would like to introduce new levels of refinement there.

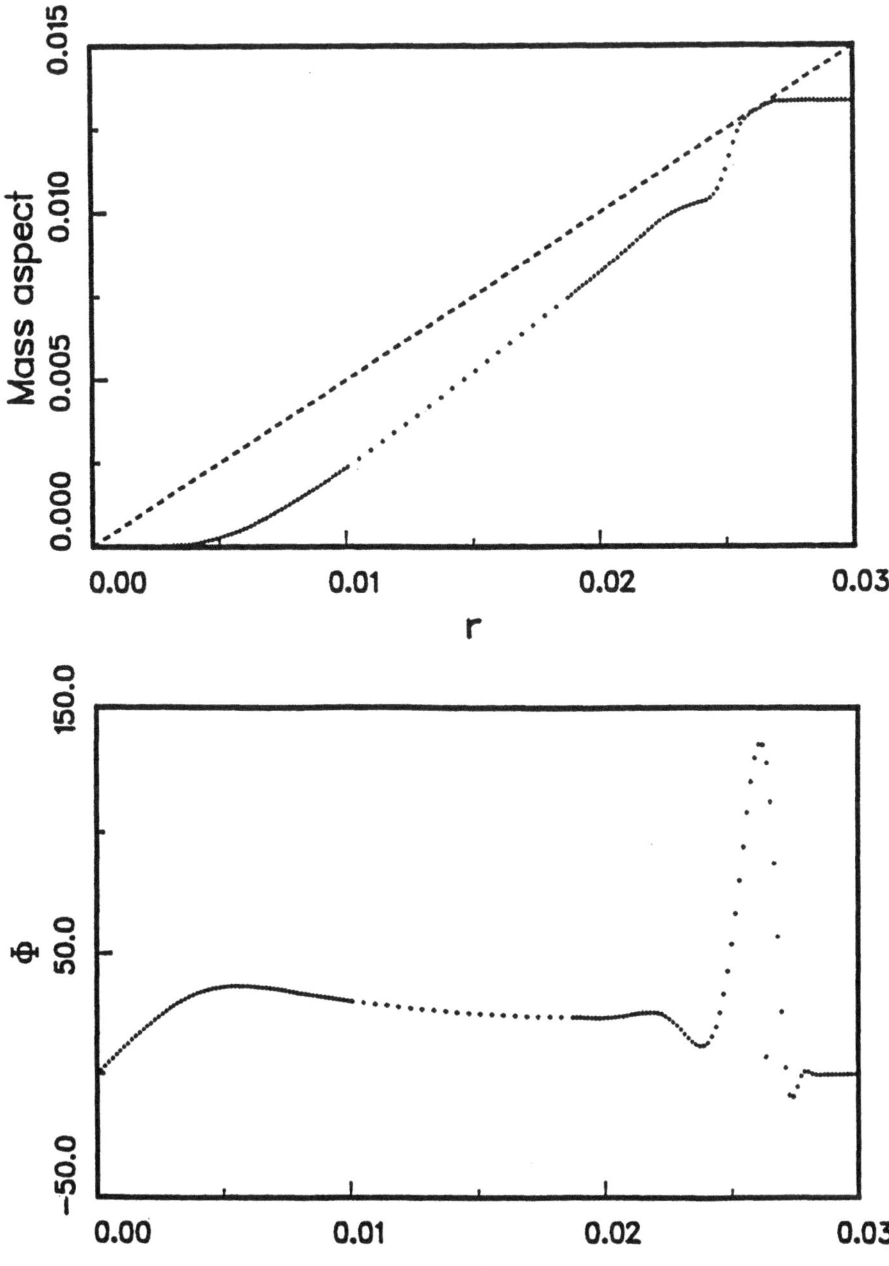

Figure 6. Results of Richardson–extrapolating the outer value of the mass aspect using two mesh–refinement runs which have identical grid patterns, but a 2:1 ratio between mesh spacings on all corresponding grids. Note that, particularly at early times, the fluctuations in the unextrapolated quantities are much smaller than the actual error in the values.

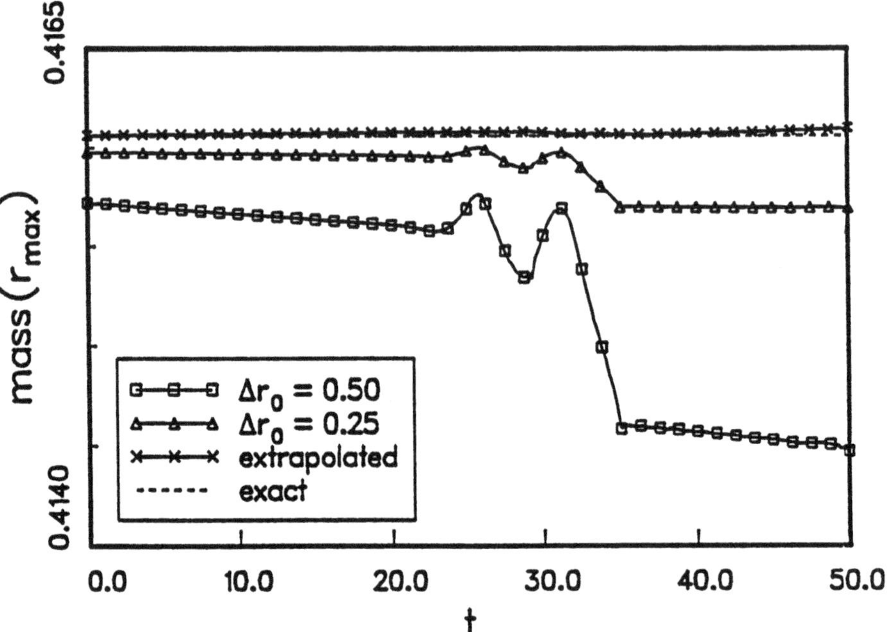

a time when the pulses have "self–reflected" through $r = 0$ and are propagating outwards. Both runs used $r_0 = 25.0$ and $\Delta = 3.0$. The only difference was in the parameter A which is 2.08369×10^{-5} in the first case and 2.08372×10^{-5} in the second. In the second run, a black hole containing about 2% of the total mass of the pulse appears, whereas there is no black hole in the first calculation. The sensitivity of the calculation to the initial conditions in the regime where a black hole just begins to form is quite remarkable and seems to be a generic feature of the model. Figure 5 shows the structure of m and Φ in the vicinity of the "horizon" of the hole (polar/radial coordinates can not actually penetrate the event horizon). The ability of the refinement algorithm to automatically resolve features on a wide range of scales is apparent.

Finally, Figure 6 shows the results of Richardson–extrapolating the value of the mass aspect at the outer edge of the grid using the results of two calculations which differed only in the scale of each component grid. The calculation at the coarse scale was done with the automatic regridding enabled and a record of the gridding structure was maintained. This record was then used as a "script" to generate an identical pattern for the second calculation. The outer value of the mass should stay constant during this calculation (except for the very small losses due to curvature backscattering of the scalar field), and the success of the extrapolation provides good empirical evidence that at least some quantities remain fully second order accurate, despite all the regridding which takes place.

To conclude briefly, adaptive methods such as the Berger/Oliger algorithm have great potential in advancing the field of numerical relativity. Not only do they extend the class of problems which can be solved with a given amount of computer resources, they represent an important step in making the numerical solution of PDE's less of a "black-art".

It is pleasure to thank D.F. Chernoff, L.S. Finn, S.L. Shapiro, A. Summers and S.A. Teukolsky for useful discussions and encouragement. This work was supported by the Natural Sciences and Engineering Research Council of Canada and NSF PHY 86-03284. Computations were performed with the resources of the Cornell National Supercomputing Facility.

REFERENCES

Babuska, I. & Bieteman, M. (1985) *J. Comp. Phys.*, 63, 33-66.
Baines, M.J. & Wathen, A.J. (1986) *Appl. Num. Math*, 2, 495–514.
Bardeen, J.M. & Piran, T. (1983). *Phys. Rep.*, 96, 205–250.
Berger, M.J. & Oliger, J. (1984). *J. Comp. Phys.*, 53, 484–512.
Berger, M.J. (1985). *Math. Comp.*, 45, 310–318.
Berger, M.J. & Jameson, A. (1985). *AIAA J.*, 23, 561–568.
Berger, M.J. (1986). *SIAM. J. Sci. Stat. Comput.*, 7, 904–916.
Choptuik, M.W. (1986). University of British Columbia Ph.D. Thesis (unpublished).
Christoudolou, D. (1986 a) *Commun. Math. Phys*, 105, 337.
Christoudolou, D. (1986 b) *Commun. Math. Phys*, 106, 587.
Goldwirth, D.S. & Piran, T. (1987) *Phys. Rev. D*, 36, 3575–3581.
Kopriva, D.A. (1986) *Appl. Num Math*, 2, 221–241.
Kreiss, H.–O. & Oliger, J. (1973). Methods for the Approximate Solution of Time Dependent Problems. Global Atmospheric Research Programme Publications Series No. 10.
Lõhner, R. (1987). *Intl. J. Num. Meth. Eng.*, 24, 1741–1756.
Richardson, L.F. (1910). *Phil. Trans. Roy. Soc.*, 210, 307–357.
Richtmeyer, R.D. & Morton, K.W. (1957). Difference Methods for Initial–Value Problems. New York: Wiley.
Shapiro, S.L. & Teukolsky S.A. (1986) *Astrophys. J.*, 298, 33–57.
Unruh, W.G. (1976) *Phys. Rev. D*, 14, 870-892.

The Multigrid Technique

Gregory B. Cook
Institute of Field Physics
Department of Physics and Astronomy
University of North Carolina at Chapel Hill
Chapel Hill, NC 27599, USA

Abstract. It is quite common in the field of Numerical Relativity to be faced with the task of solving numerically an elliptic boundary value problem on a very large grid. There are many methods available for solving the large numbers of algebraic equations produced by finite differencing or finite element techniques. For very large grids, it is often too time consuming to attempt a solution by simple relaxation techniques and the memory requirements are too great for direct methods. One approach to solving such large grids, which has received a great deal of attention recently, is the multigrid method. There are several ways to interpret the basic multigrid algorithm, all of which are useful. In this article, I will attempt to describe these differing points of view and to describe an improvement in the algorithm derived from a combination of these points of view.

Introduction.

Anyone familiar with the numerical solution of difference equations knows the standard iterative technique of simultaneous over-relaxation . The method begins with an initial guess to the solution and this initial guess is iteratively corrected until the solution is know to some accuracy. Making a good guess can aid greatly in decreasing the amount of time needed to converge to a solution. One method for obtaining an initial estimate of the solution is to use the solution from a coarser grid, the solution on that grid being found more easily because there are fewer grid points on the coarser grid. This is a simple example of using multiple grids to decrease the amount of work needed to obtain the solution to a set of difference equations. Multigrid (MG) techniques employ a much more sophisticated interplay between grids which can greatly speed up the convergence to a solution. There are many variations of the MG algorithm. In this article we shall deal only with those variations which are applicable to both linear and non-linear problems. Many of the details necessary for implementing an algorithm will be omitted below. Many excellent articles on implementing MG algorithms are listed in the references.

The Full Approximation Storage (FAS) Algorithm.

Probably the most widely used MG algorithm for non-linear problems is the FAS schema proposed by Brandt (1977). Let the differential equation which we are solving be denoted by

$$\mathfrak{I}u = f, \tag{1}$$

and let P^k be an operator which projects a function onto a grid labeled by k. The representation of the source function f on the grid G^k is thus $P^k f = f^k$. The finite difference approximation, on G^k, to the differential equation will be denoted

$$\mathfrak{I}^k u^k = f^k. \tag{2}$$

$\mathfrak{I}^k$ is a set of possibly non-linear algebraic equations and $u^k \neq P^k u$ since u^k is the set of values which must satisfy the difference equations. These values will differ from $P^k u$ due to the truncation error in the finite difference approximation to $\mathfrak{I}$. We can trivially rewrite the projection of (1) onto G^k as

$$\begin{aligned} \mathfrak{I}^k P^k u &= f^k + \mathfrak{I}^k P^k u - P^k \mathfrak{I} u \\ &= f^k + \tau^k . \end{aligned} \tag{3}$$

τ^k is the *local truncation error* of the finite difference operator $\mathfrak{I}^k$ and is defined as

$$\tau^k \equiv \mathfrak{I}^k P^k u - P^k \mathfrak{I} u . \tag{4}$$

If τ^k could somehow be found without knowing the exact solution u, then the solution to (3), the corrected finite difference equation, would be the projection of the exact solution of (1), the differential equation.

Consider now a two level system consisting of a fine grid G^k and a coarser grid G^{k-1}. Each grid has its respective finite difference approximation to $\mathfrak{I}$ and we have two operators which smoothly take values from one grid to the other. The prolongation operator p^k_{k-1} takes values form the coarse grid to the finer grid and the restriction operator r^{k-1}_k takes values from the fine grid to the coarser grid. If we have a solution on the fine grid satisfying (2) then we can define, in a manner similar to (4), the *relative local truncation error* τ^{k-1}_k

$$\tau^{k-1}_k \equiv \mathfrak{I}^{k-1} r^{k-1}_k u^k - r^{k-1}_k \mathfrak{I}^k u^k . \tag{5}$$

The relative local truncation error can be used as an estimate to the local truncation error of the operator on G^{k-1} relative to the operator on G^k. If we now solve

$$\mathfrak{I}^{k-1} \tilde{u}^{k-1} = f^{k-1} + \tau^{k-1}_k , \tag{6}$$

$\tilde{u}^{k-1}$ will be a better approximation to $P^{k-1}u$ since its truncation error will be on the order of that for G^k. We can now view (6) as a new set of finite difference equations on G^{k-1} which have been corrected to decrease the truncation error. This point of view, however, has not aided in decreasing the work needed to find the solution to a given truncation error. In order to find $\tilde{u}^{k-1}$ to the truncation error of level G^k, we must have already solved for u^k.

If we have an estimate of the solution on G^k, $\grave{u}^k$, then we can estimate the local truncation error

$$\grave{\tau}_k^{k-1} \equiv \mathfrak{I}^{k-1} r_k^{k-1} \grave{u}^k - r_k^{k-1} \mathfrak{I}^k \grave{u}^k . \tag{7}$$

Using (7) in (6) gives a coarse grid equation which approximates the fine grid equations. The equations can be solved more easily on the coarser grid since there are fewer equations there. Once the solution $\tilde{u}^{k-1}$ is know, then subtracting $r_k^{k-1}\grave{u}^k$ gives the amount by which the estimated solution $r_k^{k-1}\grave{u}^k$ has been corrected. This quantity is the *coarse grid correction*

$$\tilde{u}^{k-1} - r_k^{k-1} \grave{u}^k . \tag{8}$$

Projecting this correction to the fine grid updates the estimated solution to

$$\grave{u}^k = \grave{u}^k + p_{k-1}^k (\tilde{u}^{k-1} - r_k^{k-1} \grave{u}^k) \tag{9}$$

and gives a better solution to (2) without solving the equations on the fine grid by conventional techniques.

At this point, the general philosophy of MG can be seen. The fine grid solution is found by taking an estimate of the solution on the fine grid and using this to generate a correction to the finite difference equations on the next coarser grid. The solution on the coarser grid is then used to correct the solution on the fine grid. This requires the solution of the equations on the coarser grid. If there are still too many equations on the coarser grid to solve them efficiently, then a MG step can be made to an even coarser grid. This promotion of the problem to coarser and coarser grids is continued until the problem is on a grid with few enough points to be solved efficiently.

The General Multigrid Derivation.

Another derivation of the MG approach due to Hackbusch (1985) gives a more general form for the coarse grid correction and another interpretation to the MG approach. Consider again a two level system with fine and coarse grids labeled G^k and G^{k-1} respectively. We seek the solution on level k to equation (2). We take an approximate solution $\grave{u}^k$ which satisfies

$$\mathfrak{I}^k \grave{u}^k - f^k = d^k , \tag{10}$$

where d^k is the defect in the solution which is assumed to be smooth. The difference between the solution we seek, u^k, and the approximate solution, $\grave{u}^k$, is the fine grid correction v^k.

$$v^k = u^k - \grave{u}^k = (\mathfrak{J}^k)^{-1}(f^k) - (\mathfrak{J}^k)^{-1}(d^k + f^k)\,, \tag{11}$$

where $(\mathfrak{J}^k)^{-1}$ denotes the inverse operation to $\mathfrak{J}^k$. If we expand in a Taylor series about f^k, we get

$$\begin{aligned} v^k &= (\mathfrak{J}^k)^{-1}(f^k) - (\mathfrak{J}^k)^{-1}(f^k) - ((\mathfrak{J}^k)^{-1})'(f^k)d^k + O(d^{k^2}) \\ &\approx - ((\mathfrak{J}^k)^{-1})'(f^k)d^k\,, \end{aligned} \tag{12}$$

where $((\mathfrak{J}^k)^{-1})'(f^k)$ is the Jacobian of $(\mathfrak{J}^k)^{-1}$ evaluated at f^k. (12) is no easier to solve on the fine grid than the original equation so we wish to use a coarser grid to approximate v^k. Let u^{k-1} be a solution on G^{k-1} of the equation

$$\mathfrak{J}^{k-1}u^{k-1} = \phi^{k-1}\,. \tag{13}$$

Note that the source function in (13) is not the physical source. We will define a coarse grid defect

$$d^{k-1} \equiv \phi^{k-1} - s(r_k^{k-1}d^k) \tag{14}$$

and a coarse grid function $\tilde{u}^{k-1}$ which satisfies

$$\begin{aligned} \tilde{u}^{k-1} &\equiv (\mathfrak{J}^{k-1})^{-1}(d^{k-1}) \\ &= (\mathfrak{J}^{k-1})^{-1}(\phi^{k-1} - sr_k^{k-1}d^k)\,. \end{aligned} \tag{15}$$

The parameter s is simply a coefficient in the linear combination. If we now expand (15) in a Taylor series about ϕ^{k-1} we get

$$\begin{aligned} \tilde{u}^{k-1} &= (\mathfrak{J}^{k-1})^{-1}(\phi^{k-1}) - s((\mathfrak{J}^{k-1})^{-1})'(\phi^{k-1})(r_k^{k-1}d^k) + O((sr_k^{k-1}d^k)^2) \\ &\approx u^{k-1} - s((\mathfrak{J}^{k-1})^{-1})'(\phi^{k-1})(r_k^{k-1}d^k)\,. \end{aligned} \tag{16}$$

We now define the coarse grid correction as

$$v^{k-1} \equiv \frac{1}{s}(\tilde{u}^{k-1} - u^{k-1})\,, \tag{17}$$

which, when projected onto G^k gives the coarse grid approximation to (12)

$$\underline{v}^k = p^k_{k-1}\frac{1}{s}(\tilde{u}^{k-1} - u^{k-1}) = -p^k_{k-1}((\mathfrak{I}^{k-1})^{-1})'(\phi^{k-1})(r^{k-1}_k d^k)\,. \tag{18}$$

In this approach to MG, we correct the fine grid estimate of u^k by solving for $\tilde{u}^{k-1}$ on the coarse grid. From (15) or (16) we can view $\tilde{u}^{k-1}$ as resulting from a perturbation of the known solution satisfying (13). The perturbation is forced by the defect in the approximate solution on G^k and is scaled by the parameter s. In theory, any non-zero value of s which maintains the validity of the approximation in (16) can be used. Practically, the choice of s will effect the rate of convergence when solving for $\tilde{u}^{k-1}$ since the magnitude of s determines by how much $\tilde{u}^{k-1}$ differs from u^{k-1}, the initial guess. If not for computational roundoff error, s could be made arbitrarily small resulting in high rates of convergence.

The two MG approaches can be made to look more similar. If we take the known solution on G^{k-1} to be $u^{k-1} = r^{k-1}_k \grave{u}^k$ and if (13) defines ϕ^{k-1}, we see from (15) and (10) that $\tilde{u}^{k-1}$ satisfies

$$\mathfrak{I}^{k-1}\tilde{u}^{k-1} = \phi^{k-1} - s r^{k-1}_k d^k = \mathfrak{I}^{k-1} r^{k-1}_k \grave{u}^k - s r^{k-1}_k(\mathfrak{I}^k \grave{u}^k - f^k)$$

$$= s f^{k-1} + \mathfrak{I}^{k-1} r^{k-1}_k \grave{u}^k - s r^{k-1}_k \mathfrak{I}^k \grave{u}^k\,. \tag{19}$$

When s =1, the last two terms on the right in (19) are the estimated relative local truncation error and (19) reverts to (6). Finally, the updated approximate solution becomes

$$\grave{u}^k = \grave{u}^k + p^k_{k-1}\frac{1}{s}(\tilde{u}^{k-1} - r^{k-1}_k \grave{u}^k) \tag{20}$$

which is equivalent to (9) when s = 1. We see that, formally, the FAS algorithm is a special case of this more general MG algorithm. However, the philosophy of the two methods is quite different.

Current Research.

There are many advantages to the point of view taken by the FAS schema. The estimated relative local truncation error which can be obtained between the finest and next coarser levels provides a natural means for judging the accuracy of the solution on the finest level. Furthermore, the interpretation of the solution on the coarser level as being a coarse grid approximation of the solution on the finer level is very intuitive. However, the interpretation of (7) as the relative local truncation error is limited to the step between the finest and next coarser level.

The more general MG schema defined by (19) and (20) offers an extra freedom not available in the FAS schema. The freedom to choose s can in theory allow one to optimize the rate of convergence of the equations on the coarser grids. One possible avenue to utilizing the s parameter would be to choose it so that the perturbation added to the source in (14) is small. If we take ε_s to be a parameter gauging the size of the perturbation then we can choose s so that

$$\| s(r_k^{k-1} d^k) \| = \varepsilon_s \| \phi^{k-1} \| , \tag{21}$$

or

$$s = \frac{\varepsilon_s \| \phi^{k-1} \|}{\| r_k^{k-1} d^k \|} , \tag{22}$$

where $\| \bullet \|$ is some appropriate vector norm. I denote this method for choosing s as "adaptive scaling" where ε_s is the adaptive scaling parameter.

Tests of this approach were made with an FAS code modified to allow adaptive scaling. The routine allows adaptive scaling between all levels except the finest level and its coarse grid. On this lowest level, standard FAS is used, enabling the relative local truncation error estimate to be used as a natural gauge of convergence. However, since (5) cannot be interpreted as a relative local truncation error on the coarser levels there is nothing lost in not calculating it as an intermediate step. Listed below are two tables corresponding to sets of tests on both linear and non-linear cases of the Hamiltonian constraint equation. The model problem of Bowen and York (1980) was used, for which there exists an analytic solution of the full nonlinear problem.

Both tables represent the work done in solving the equations on level 4, where level 0 is the coarsest level. The first column in each table indicates the current level on which a coarse grid correction will be made. Each column to the right indicates the value of 's' chosen for that correction. Each column is labeled to indicate whether the FAS schema was used or if adaptive scaling was used. If adaptive scaling was used, the value of ε_s heads the column. Below each column is the time used to obtain the solution and the L_1 and L_∞ norms of the difference between the solution on level 4 and the exact solution.

Table 1: Linear case

level	FAS	$\varepsilon_s = 1$	$\varepsilon_s = 10^{-3}$	$\varepsilon_s = 10^{-6}$	$\varepsilon_s = 10^{-9}$
4	s = 1	1	1	1	1
3	s = 1	11.0954	0.0110954	1.10954e-05	1.10451e-08
2	s = 1	6.12952	5.99314	5.99346	5.98666
1	s = 1	3.67508	3.65255	3.65275	3.65469
time	30:28.7	30:24.5	30:24.6	29:28.3	29:24.3
L_1	0.000615978	0.000615978	0.000615978	0.000615959	0.000607174
L_∞	0.000319898	0.000319898	0.000319898	0.000319895	0.000318293

Table 2: Non-linear case

level	FAS	$\varepsilon_s = 1$	$\varepsilon_s = 10^{-3}$	$\varepsilon_s = 10^{-6}$	$\varepsilon_s = 10^{-9}$
4	s = 1	1	1	1	1 1
3	s = 1	2.42077	0.00242479	2.42477e-06	2.42531e-09
2	s = 1	3.84437	3.67861	3.67854	3.67376
1	s = 1	3.21175	3.09449	3.09443	3.09544
1	s = 1	8.2754	5.43329	5.43022	5.43396
3	s = 1	7.87781	0.00784082	7.84073e-06	7.81537e-09
2	s = 1	2.75552	3.14434	3.14465	3.14369
1	s = 1	2.89549	3.08104	3.08116	3.09415
1	s = 1	3.88691	4.96438	4.96738	4.99193
time	43:10.8	44:25.8	42:58.8	40:50.7	39:51.8
L_1	0.00432647	0.00434234	0.00432676	0.00432679	0.00433726
L_∞	0.000865437	0.000882332	0.000865583	0.000865607	0.000864812

For the linear case, adaptive scaling produces no significant effect. The same sequence of grid changes is followed, the time to solution is nearly the same for all cases and the error in the solution is virtually identical. For the non-linear case, adaptive scaling does produce a small systematic effect. While the sequence of grid changes is again identical and the error in the solution does not change, the time to solution systematically decreases as the value of ε_s decreases.

The cause of this behavior is straightforward. For non-linear cases, each smoothing step and the exact solution on the coarsest level require the solution of non-linear algebraic equations. This is accomplished by Newton iterations which converge to the solution in a number of iterations depending on their initial distance from the solution. Using a smaller ε_s keeps the initial distance from the solution smaller. For the linear case, no Newton iterations are needed resulting in no net effect for adaptive scaling.

Conclusions.

Implementing an optimal FAS algorithm for a given equation is a difficult task requiring a careful choice of restriction and prolongation operators, and smoothing functions. While adaptive scaling shows little improvement over a well coded FAS algorithm, it is useful to note that when small errors were in the code, adaptive scaling and FAS did not follow the same path and the time to solution for adaptive scaling was as much as 25% better for the same accuracy. Thus, adaptive scaling can be used as a debugging tool which is very sensitive to subtle programming errors while requiring very little additional work.

On a more speculative note, while adaptive scaling does not provide a better method for solving this particular equation, it is possible that it will prove more useful for ill-behaved non-linear problems. If the desired solution is surrounded by local minima, then using adaptive scaling to keep the initial guess to solutions on coarser grids close to the correct minima could also prove useful.

Acknowledgments

I would like to thank J. Rauber, B. Whiting and J. York for numerous helpful discussions.

References

Bowen,J.M. & York, J.W. Jr. (1980). Time-asymmetric Initial Data for Black Holes and Black-hole collisions. Physical Review D, 21, no. 8, 2047-2056.

Brandt, A. (1977). Multi-Level Adaptive Solutions to Boundary-Value Problems. Mathematics of Computation, 31, no. 138, 333-390.

Brandt, A. (1982). Guide to Multigrid Development. In Multigrid Methods, ed. W. Hackbusch & U. Trottenberg, pp. 220-312. Berlin: Springer-Verlag.

Choptuik, M.W. (1982). A Study of Numerical Techniques For the Initial Value Problem of General Relativity. University of British Columbia: Masters thesis.

Hackbusch, W. (1985). Multi-Grid Methods and Applications. Berlin: Springer-Verlag.

Rauber, J.D. (1985). Initial Data For Black Hole Collisions. University of North Carolina -Chapel Hill: Ph.D. dissertation.

Finite Element Methods in Numerical Relativity

P.J. Mann,
Department of Astronomy, University of Western Ontario,
London, Ontario, Canada.

Abstract. The finite element method is very successful in Newtonian fluid simulations, and can be extended to relativistic fluid flows. This paper describes the general method, and then outlines some preliminary results for spherically symmetric geometries. The mixed finite element-finite difference scheme is introduced, and used for the description of spherically symmetric collapse. Baker's (Newtonian) shock modelling method and Miller's moving finite element method are also mentioned. Collapse in double-null coordinates requires non-constant time slicing, so the full finite element method in space and time is described next. The paper concludes with a brief comparison with finite differences, and some suggestions for future research.

1 *INTRODUCTION*

The Finite Element Method (FEM) has become popular in Newtonian fluid flow problems, but it has been more or less ignored in Relativity. The FEM has its own advantages (and disadvantages), and it is reasonable to apply it to problems in numerical relativity. This paper will be a summary of the work I have done with this application.

In general it seems that many relativists are not familiar with the FEM, so I will include a short introduction to the method. Then I will outline the basic first applications to spherical symmetry, and describe some preliminary attempts at non-constant time slicing in double null coordinates.

2 *THE FINITE ELEMENT METHOD*

There are many text books describing the FEM. A comprehensive introduction and comparison with finite differences and other methods can be found in Lapidus and Pinder (1982), while a detailed study of FEMs in fluids is available in Baker (1983). The details of the following overview can be found in either of these excellent texts.

The basic technique is based on the Rayleigh-Ritz variational method. An action:

$$S = \int L(u,u',x)\,dx \quad (u' \equiv \partial u/\partial x)$$

where u is the function to be varied, is assumed to exist, and an approximation to u is required. Usually the approximation is taken to be a point in a linear subspace of the space of allowed trial functions, in which case we have

$$u \cong \sum_{i=1}^{n} u_i N_i(x)$$

where the $N_i(x)$ are known basis functions which span the linear subspace. The u_i are unknown parameters which must be calculated from the variational principle. This is carried out by inserting the approximation u in S, performing the integration (the basis functions give the x dependence), and extremizing the resulting algebraic expression with respect to the u_i.

This method can be extremely powerful, and may be important in relativity, but in practice it is difficult to find actions. I used Schutz' variational principle (1970) for perfect fluids, but this involves the use of velocity potentials which require higher order approximations for the higher order derivatives; a waste of computational effort. Instead different methods of finding a best approximation can be used.

The most general FEM is the weighted residual method. Here any differential equation L(u)=f can be discretized. Just insert the approximation and minimize the residual r=L(u)-f with the weighted average:

$$\int (L(u)-f)W_i(x)dx = 0$$

for n independent weight functions $W_i(x)$. As in the variational method the integrals can be evaluated so a set of algebraic equations results. In general the W_i are chosen to be the basis functions $N_i(x)$, in which case the method is termed a Galerkin method.

It is evident that the choice of basis functions is all-important. The method was first used by civil engineers who used as basis functions the beams which were to be used in the construction of a building or bridge. These beams are defined locally with respect to their bending and twisting properties (the finite element), and globally with respect to their connections with the adjoining beams. In a general FEM the $N_i(x)$ are therefore given as simple functions on each element, with some suitable degree of continuity across inter-element boundaries. The archetype is a spline approximation for u(x), in which case the $N_i(x)$ form a spline basis. The "hat" basis functions for linear splines are extensively used, and have the property that the u_i can be chosen to be nodal values. With this property the approximation in an element depends on surrounding nodal values, giving a localized approximation. This is extremely useful and almost all FEM's implement it. Note that globally dependent approximations can be used, in which case a spectral method or boundary integral method results, but that the FEM uses the localized system.

The weight function is also localized, with local support. Then the discretization extends over this local support, which may contain many adjoining elements, and links the nodal values to form an equation similar to a difference approximation.

The method is therefore grid based, with the solution defined by parameters which are usually given as function values on nodes. Note that the grid is made up of individual elements, and is almost invariably not uniform. The resulting discretizations are similar to, but not the same as, those resulting from the finite difference method.

The method bridges the gap between finite differences (very simple "basis" functions but many elements) and boundary element methods (very complicated basis functions but few elements). The FEM shares with boundary element methods the advantage of coping with complicated geometries. In the case of the FEM any asymmetries in the element geometry are hidden implicitly in the integration, which, in most cases, is performed with a numerical quadrature.

This advantage is also a disadvantage when compared to finite differences. A relatively large amount of work must be done in deriving any discretization because the whole machinery is implicitly included. However, some improvement in computational efficiency can be expected, and for complicated problems (Relativity!) a completely different discretization results which can be compared to finite difference and other methods. This point is important, and I believe completely justifies the investigation of as many other methods as possible.

3 *MIXED METHODS*

In the ADM formulation of relativity the time coordinate is numerically very simple: constant t slicing. All the difficult slicing conditions are included analytically in the choice of lapse and shift. Therefore it seems reasonable to use finite differences for the time direction, but to split off the spatial slices and use the FEM. This is, of course, consistent with the ADM philosophy, as well as being useful numerically.

The "mixed" method proceeds as follows. Let the approximation be

$$u(t,x) = \sum u_i(t)N_i(x)$$

where the u_i are now unknown functions of t. The weighted residual method is applied to a differential equation $L(\partial u/\partial t, \partial u/\partial x) - f = 0$ only in x, to give

$$\int\left[L\left(\frac{\partial u}{\partial t}, \frac{\partial u}{\partial x}, x\right) - f\right]W_i dx = 0.$$

The integration is carried out, leaving a set of ordinary differential equations in t for the $u_j(t)$. These can be solved with a suitable differencing scheme.

As an example consider the conservation equation $\dot{u} + (V\cdot\nabla)u = 0$. The weighted residual method gives

$$\int_{e_i} \left\{ \dot{u}_j N_j(\underline{x}) + u_j(\underline{V}\cdot\nabla)N_j(\underline{x}) \right\} W_i \, d^3x = 0$$

where repeated indices are summed over the basis functions. The integration can be performed, giving an equation

$$\sum_{j=1}^{n} \left[a_{ji}\dot{u}_j(t) + b_{ji}u_j(t) \right] = 0$$

or

$$\mathbf{A}\underline{\dot{u}} + \mathbf{B}\underline{u} = 0$$

where **A** and **B** are matrices and $\underline{u}$ is the vector of parameters. With local support **A** is banded and the system solved for $\underline{\dot{u}}$. However, this can be time-consuming and the technique of condensation, in which the **A** matrix is rather artificially diagonalized, is extensively used. The method lowers the order of accuracy, but can be advantageous when carefully implemented.

The basis functions are usually defined with a grid, so they also depend on nodal positions. These nodal positions can be time dependent, and including this in the integral results in "natural" grid velocity terms. The moving finite element method (Miller and Miller, 1981) considers the node positions to be just another set of unknown parameters, and produces another set of weighted residual equations for their evolution. This has been very successful in moving nodes to important regions (ie, near shocks) but is expensive (many more equations to solve) and I have not implemented the method in relativity. Instead the grid velocities are given with some simple formula.

4 *SIMPLE FIRST APPLICATIONS*

The method has been applied in one dimension to the solution of the TOV equations for relativistic spherical polytropes. This proved the general method, and also resulted in some extremely stable and robust (implicit) ordinary differential equation solvers. (Mann, 1983)

Next the FEM was applied to the velocity potential variational principle of Schutz (1970) for the case of spherical collapse. As mentioned above the potentials required unnecessary computation, but also shocks were poorly modelled due to the isentropic equation of state used. (Mann, 1985)

Instead of continuing with this code a weighted residual mixed method was applied directly to a standard set of spherical collapse equations (Mann, 1987). The mass variable was intrinsically a cubic in radius, and required a cubic Hermite spline approximation, but the hydrodynamic variables could be adequately approximated with simple linear splines.

This code showed various trouble spots. The boundaries (centre and surface) were inclined to be unstable, but a careful choice of weights was able to overcome the inclination. For instance, a weight of $r^2 N_i(r)$ removed instabilities from the (m/r^2) terms in Euler's equation. Secondly the time differencing had to be robust. Runge-Kutta methods were very unstable, and in general methods of the predictor-corrector type were to be preferred.

In this code the nodes were more or less co-moving with the fluid so an artificial bulk viscosity was added to the pressure to smooth shocks. The viscosity was similar in form to the standard methods, but some care was required to ensure consistent continuity across element boundaries.

This code was successful in that it duplicated the results from other codes (finite differences) with about the same computational work. Since I did not expect the FEM to be as efficient in these one dimensional situations as finite differences the results were encouraging.

5 *FURTHER INVESTIGATIONS*

I have been concentrating on two further FEM investigations. The first involves improved shock modelling, as the artificial viscosity method is not completely satisfactory. It was obvious that internal energy and density were not being adequately smoothed. Baker (1983) has introduced a smoothing technique for Newtonian fluid flows which is simple and effective, and can be generalized to Relativity. The method is as follows.

Take as an example the standard form $\dot{u} + (V \cdot \nabla)u = f$. Add a smoothing term of the form

$$\underline{\beta} \circ \nabla(\dot{u} + (\underline{V} \cdot \nabla)u)$$

to the left side of the equation, where $\underline{\beta}$ is a constant vector. Now apply the standard weighted FEM to the modified equation.

This extra term is a measure of the slope of the time derivative across characteristics, and is a smoothing influence. An analysis of diffusion and dispersion (for $f=0$) gives an optimum value:

$$\underline{\beta} = \frac{\Delta x}{\sqrt{15}} \left[1,1,1\right]^T$$

where Δx is a typical element size. This technique is applied to Euler's equation, the conservation equation, and to the internal energy equation, and has given excellent results in the Riemann shock tube problem, and in various simple 2d shock simulations (see Baker 1983). In practice the coupling between the equations is important and Baker has found that a form:

$$\underline{\beta}^t \circ \nabla \dot{u} + \underline{\beta}^x \circ \nabla (\underline{V} \cdot \nabla) u$$

is to be preferred, with $\underline{\beta}^t$ and $\underline{\beta}^x$ chosen experimentally (different values are allowed for each of the three equations, giving six parameters to choose).

This method can easily be included in relativistic fluid flow, as all the hydrodynamic equations can be cast in the $\dot{u}+v\circ\nabla u$ form. It can be argued that some co-variant form should be used, with a slope calculated perpendicular to the characteristics, but in practice these seem to have little effect, and the simple Baker smoothing is just as good.

6 *FINITE ELEMENTS IN SPACE AND TIME*

Recent advances in the study of naked singularities suggest that such objects may exist in Einstein's theory, and it seemed reasonable to investigate perfect fluid collapse. It is evident that some single-null or double-null coordinate system is required, but this means the time-slicing should not be constant t. In this case the co-ordinate choice has been chosen to fit the problem, and not, as in ADM, to fit the solution method. Therefore a method for non-constant t slicing must be introduced.

The FEM in space and time is such a method. The hyberbolic nature of some of the equations requires a special choice of weight and element geometry, but otherwise there is little change to the overall weighted residual method. For a first try spherical symmetry is again used, making the problem two dimensional, with two coordinates, t and x.

I have chosen to use quadrilateral elements with bi-linear approximations for those quantities which are evolved. This element geometry is best implemented with the iso-parametric technique, in which the element is transformed to a rectangle, and then the approximation defined in the transformed region (see Baker for instance). All the integration is performed in the transformed coordinates, which results in a considerable simplification.

The weight function is chosen to be non-zero over only two elements:

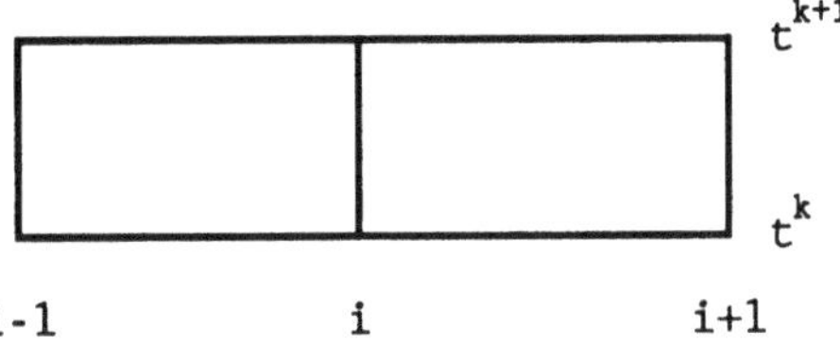

The integral becomes:

$$\iint dx\,dt \left\{ \dot{u} + \underline{V}\circ\nabla u + \underline{\beta}\circ\nabla\left[\dot{u} + \underline{V}\circ\nabla u \right] - f \right\} W_i^k(t,x) = 0$$

where the integration region extends over the two elements and u (and any other required variables) is approximated with the form:

$$u \cong \sum_{k=1}^{nt} \sum_{i=1}^{nx} u_i^k N_i^k(t,x).$$

The u_i^k are chosen to be nodal values, so it is evident that the integral is an algebraic equation in the six nodal values appearing in the two elements. This is just an evolution system, with the three values on the future time slice depending on the three values on the past time slice.

The Baker smoothing term is included as usual in the equation, and a special scheme is not required. It is advantageous to use integration by parts to remove the second derivative.

Note that the system is implicit because f is integrated over the whole region. It is possible to use a delta function weight, but I have found it easiest to use the evolution equation as a corrector in a predictor-corrector evolution scheme. This method also gives an interpretation to the weight function W(t,x). A centre-weighted (in time) scheme is similar to a Crank-Nicholson finite difference scheme, and is marginally stable. Future weighting gives more stable schemes, but some accuracy is lost.

Spatially the weight is chosen to ensure some continuity across element boundaries. A modified Galerkin weight of the form:

$$W_i^{k+1}(t,x) = N_i^{k+1}(t,x) + N_i^k(t,x)$$

(which is time-centred) has been successfully implemented for the relativistic shock tube problem.

In the collapse case constraint equations must also be solved. The evolution scheme is not a good choice, and a different method must be used. I have found that a single triangle is quite adequate for the first degree differential equations of spherical symmetry ($dm/dx=4\pi r^2\rho\ldots$). The weight function is non-zero only on a region

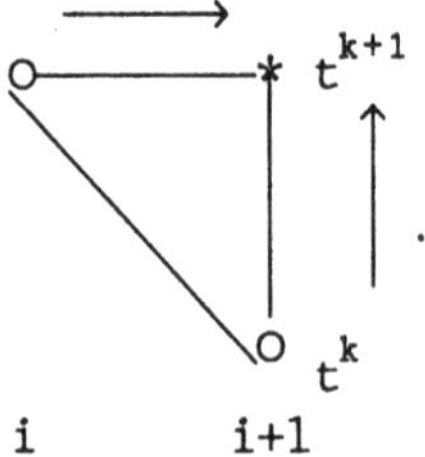

The o indicates known values, and values are required at *. The standard weighted residual method is applied to give a relation between these points, which is then solved for the unknown at *. Some

variables are intrinsically cubic in x (the mass for instance) and for these variables a cubic approximation in the triangle has been useful. A complete cubic has ten parameters, which are usually taken to be three nodal function values, three nodal t-derivative values, three nodal x-derivative values, and one centroidal function value. The centroidal value is a great nuisance because a complete new vector of coordinate values must be calculated and stored, so a restricted cubic with nine parameters has been used (see Lapidus and Pinder, 1982).

The nodal positions are not nailed down, and can be moved as required by the behaviour of the model. Any nodal movement is contained in the integration, and no changes are required in the program. A grid velocity is therefore automatically included, but in addition the time position can be changed ensuring that non-constant time slicing can be used.

The method has been tested for the relativistic shock tube, with acceptable results. The models are not as good as those produced with specialized differencing schemes (see Hawley et al 1984 for instance), but there is room for improvement, and the freedom in element geometry is still included.

A double-null collapse code has been written, and has successfully passed some preliminary tests. It is evident that the grid can indeed be moved in space and time, but that the movement algorithm has to be carefully chosen. Further investigations in this area are in progress.

7 *CONCLUSIONS*

I think that the finite element method has proved itself to be a reasonable tool in attacking relativistic fluid flow problems. It has both advantages and disadvantages, but I think that modern programs using finite differences or finite elements are generally similar in performance, and in preparation. The important aspect is that the discretizations produced are quite different, especially when non-linearities are present (relativity!), so an excellent check on results is available.

There are many avenues open for further FEM research. I am continuing to explore double-null spherical collapse. This is a fascinating numerical problem as well as being extremely interesting to relativists. The classical mixed FEM-FDM should be extended to non-spherical configurations, where the real power of the FEM will become more evident. The moving FEM shows great promise in the treatment of shocks, and I am sure could be implemented in relativity.

The FEM shows lots of promise, and I hope others will begin using this fine numerical method.

REFERENCES

Baker, A.J. (1983). Finite Element Computational Fluid Mechanics, Washington: Hemisphere Publishing Corporation.

Hawley, J.F., Smarr, L.L. & Wilson, J.R. (1984). Astrophys. J. Suppl. 51, p.211.
Lapidus, L. & Pinder, G.F. (1982). Numerical Solution of Partial Differential Equations in Science and Engineering. New York: John Wiley and Sons.
Mann, P.J. (1983). Comput. Phys. Commun. 30, p. 127.
Mann, P.J. (1985). J. Comput. Phys. 58, p. 127.
Mann, P.J. (1987). J. Comput. Phys. 72, p. 467.
Miller. K. & Miller, R.N. (1981). SIAM J. Numer. Anal. 18, p.1019.
Schutz, B.F. (1970). Phys. Rev. D 2, p. 2762.

PSEUDO–SPECTRAL METHODS APPLIED TO GRAVITATIONAL COLLAPSE

Silvano Bonazzola
and
Jean-Alain Marck

Groupe d'Astrophysique Relativiste
Observatoire de Paris, section de Meudon
F–91195 Meudon Principal Cedex, France

Abstract : We present codes for solving Newtonian gravitational collapse in spherical coordinates for the spherical, axial and true 3–D cases. The pseudo–spectral techniques are used. Each quantities are expanded in Chebychev or Legendre polynomials or Fourier series for the periodic parts. The codes are able to handle in a rigorous way the pseudo–singularities $r = 0$ and $\theta = 0, \pi$ typical of the spherical coordinates. Illustrative results of each one of the three cases are shown.

1. Introduction

The present "*super–computers*" generation and - a *fortiori* the next one – allow to think on realistic numerical experiments on gravitational collapse. However, even if the storage troubles are disappearing with the present generation of computers, the problem of computational time limits the spatial resolution achievable. Let us discuss shortly this point. Consider a 3–D simulation. The computational time is at least proportional to N^4, where N is the number of grid points in each direction (the extra factor N comes out of the necessity of describing with an accuracy independent on N the high frequency modes). Now, discarding parallel computers, it is easy to see that clock's cycle time cannot be improved by a factor of say 10^4 (because of the limit imposed by the atomic time of say $10^{-14}s$), and, consequently, the number of degrees of freedom cannot be improved by a factor greater than 10. Of course, this situation will be drastically changed when high parallelised computers will be available. However, it is difficult to forecast when such a machine will be born. It is then clear that the number of degrees of freedom cannot be arbitrarily large and, consequently, that numerical techniques giving rise to hightly accurate and reliable results have to be worked out.

A class of applications needs a high spatial resolution. Consider for instance the computation of gravitational radiation due to a 3–D gravitational collapse. The strongest power emitted is expected to arise at the time as a shock formation. Numerical codes cannot handle shock without (explicit or intrinsic) viscosity. We want to point out that viscosity may have dramatic effects on such a computation : if the motion is viscous dominated, the gravitational power is proportional to $1/\nu^6$. Consequently, brute force must be prohibited, and more sophisticated techniques are necessary.

This paper is organized in the following way. We present in section 2 the spherically symmetric code and we discuss how to give a correct treatment of the shock. We show in section 3 how initial physical conditions can be constructed, and we describe In sections 4 and 5 the axisymetric and the 3–D code respectively.

2. Spherically symmetric collapse

Newtonian spherical collapse using pseudo-spectral method has already been studied and described in a previous paper (Bonazzola and Marck 1986, quoted thereinafter BM86). In this paper, density and velocity field were expanded in Chebychev polynomials series and shock was handled by adding a "natural" viscosity constant in space and time. This preliminary work did not take into account the even (odd) parity of the density (velocity) with respect to the origin $r = 0$. Therefore, the sampling grid presents the concentration of points near the origin and near the boundary of the grids, typical of Chebychev expansion.

As shown by Bonazzola and Marck (1988) and in the following sections, one can reap advantages of the regularity properties of physical quantities (even parity of the density and gravitational potential and odd parity of the radial component of the velocity field). Using ρ and v as unknown quantities, and introducing the dimensionless coordinate $x = r/\big(R\alpha(t)\big)$, the evolution equations for the isoenthropic case read :

$$\frac{\partial v}{\partial t} = -(v - R\dot{\alpha}x)\frac{1}{R\alpha}\frac{\partial v}{\partial x} - \frac{1}{R\alpha\rho}\frac{\partial P}{\partial x} + \frac{1}{R\alpha}\frac{\partial \Phi}{\partial x} + \nu\Delta v$$

$$\frac{\partial \rho}{\partial t} = -(v - R\dot{\alpha}x)\frac{1}{R\alpha}\frac{\partial \rho}{\partial x} - \frac{2\rho v}{R\alpha x} - \frac{\rho}{R\alpha}\frac{\partial v}{\partial x}$$

$$\frac{\partial^2 \Phi}{\partial x^2} + \frac{2}{x}\frac{\partial \Phi}{\partial x} = 4\pi G R^2 \alpha^2 \rho$$

$$\alpha = \alpha(t) \tag{2.1}$$

where ν is the viscocity (constant in space and time).

The moving grid has been introduced in order to take into account the large scale change of the star during the collapse (see BM86). The equations written in this form are solved by giving a boundary condition on v only of the form :

$$a(t)v + b(t)\frac{\partial v}{\partial r} = c(t) \tag{2.2}$$

It is to be noticed that the viscous coefficient can be of order $1/N^2$ where N is the number of degrees of freedom, but that the code is not able to handle the case $\nu = 0$, because, when $\nu = 0$, there is no dissipation and, therefore, the wave generated by any arbitrary initial condition steeps until a discontinuity forms.

During the collapse of the star a shock forms. Spectral methods can handle a shock if viscosity is high enough to ensure that the shock lies on at least three points of the grid. If we want to reduce the viscocity, one possibility is to take advantages of the concentration of sampling points near the boundaries by introducing two grids on the intervals $[0, r_0]$ and $[r_0, R]$ where r_0 is the location of the shock. The problem now consists of solving two sets of equations and to match them on the common boundary $r = r_0$. Figure 1 shows how the method works. Note that a more sophisticated scheme would consist to split the interval $[0, R]$ in three regions $[0, r_0 - \delta]$, $[r_0 - \delta, r_0 + \delta]$ and $[r_0 + \delta, R]$ where 2δ is the thickness of the shock. However, there is a more elegant and exact way to treat a shock without viscosity and to verify exactly the Rankine-Hugoniot conditions.

Introducing the Riemann "Invariants"

$$I^{\pm} = v \pm \int \frac{\mathrm{d}P}{\rho c}, \tag{2.3}$$

where c is the local speed of sound, the evolution equations take the simple form :

$$\frac{\partial I^{\pm}}{\partial t} = -(v \pm c)\frac{\partial I^{\pm}}{\partial r} \mp 2\frac{vc}{r} \pm \frac{\partial \Phi}{r}. \tag{2.4}$$

Note that in this case, the regularity properties of ρ and v can no more be used, but, when using $I^{\pm}$ as unknown quantities, the problem of treating the boundary conditions disappears. More precisely, boundary condition is imposed at a boundary of the interval $[0, R]$ either on I^+ or I^-, depending if the corresponding characteristic is ingoing or outgoing.

It is then possible, when a shock appears, to split the numerical grid into two moving grids at the location of the shock. Rankine-Hugoniot conditions and the fact that v has to vanishe at the origin, gives a complete set of informations to solve

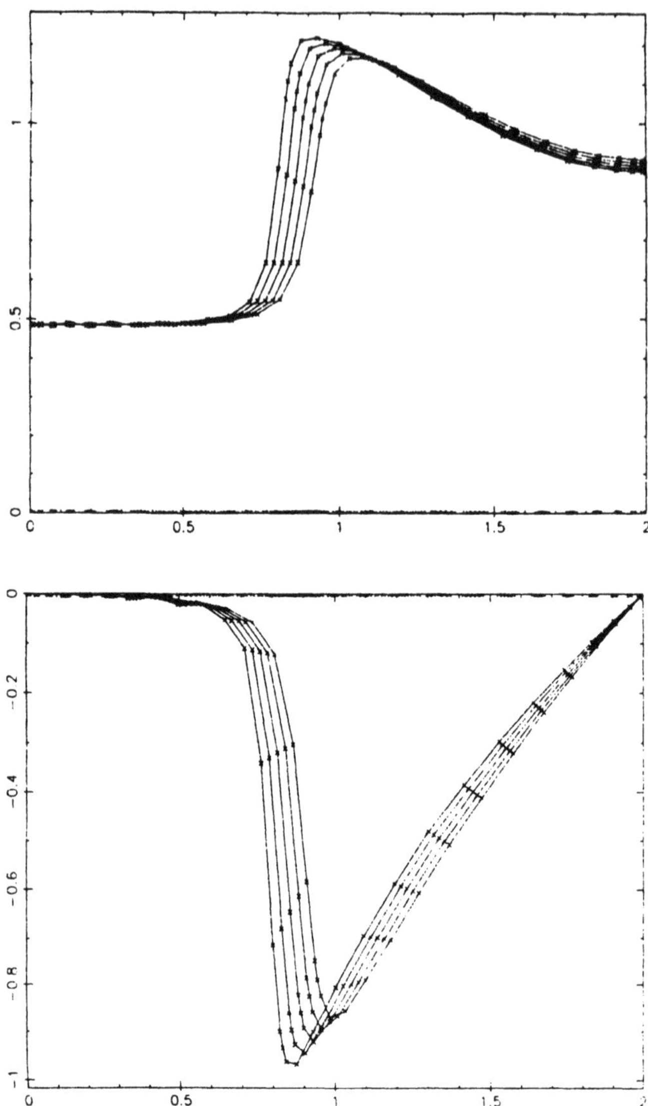

Figure 1 : *Density profiles (up) and radial velocity fields (down) at successive time-steps for an hydrodynamic calculation performed with 2 (moving) numerical grids. The grids match at the shock location. The density and the velocity are expanded in Chebychev polynomials on both the interval* $[0, r_0]$ *and* $[r_0, r_b]$. *Therefore, the sampling points are concentrated near the shock which allows to give a fine description of it.*

the matching problem and to know the grid's (shock's) velocity (see e.g. Dang et *al.* 1986 for a description of these technics in the case of pure hydrodynamics in flat geometry). This work is in progress.

3. Initial conditions problem

The simulation of a gravitational collapse needs realistic initial conditions. The computation of initial conditions in spherical symetry is reconducted to the resolution of an ordinary differential equation. However, for the multi-dimentional cases, the problem becomes more general and more complicated. We will describe in this section the method we use to solve this problem.

Let us consider the 2 dimension case. The axisymetric cases of astrophysical interest are : isolated rotating star and one object of a corrotating binary system. We then have to find a stationary solution of the Navier-Stockes equation. In the case of an isotropic isolated rotating star, this consists of finding a solution for n of

$$\frac{{v_\varphi}^2}{\rho} - \frac{\partial P}{\partial \rho}\frac{1}{n} = \frac{\partial U}{\rho}$$

$$\frac{\partial P}{\partial z} = \frac{\partial U}{\partial z} \tag{3.1}$$

where ρ is the radial cylindrical coordinate, P is the pressure, n is the density and U is the self-gravitating potential. The solution being a stationary solution, v_φ has to be a function of ρ only. Writing

$$\frac{{v_\varphi}^2}{\rho} = \frac{\partial \Omega}{\partial \rho} \tag{3.2}$$

we are lead to solve

$$\Omega - \mu - U = C^{ste}$$

$$\Delta U = -4\pi G n$$

$$\mu = \mu(n) \tag{3.3}$$

μ being the heat function.

The solution is obtained by iteration : given an arbitrary density profile, the gravitational potential U is computed and a new density profile is obtained from (3.3) and so on until convergence is obtained. This method rapidly converges (typically 10 iterations are needed). The gravitational potential is obtained by inverting the Laplacian operator in Chebychev-Legendre representation as described in Bonazzola and Marck (1988).

In the second case, v_φ has to vanish in order to preserve the axisymetry and an external (tidal) potential has to be added to the Navier-Stockes equations. This method has be used in the General Relativistic case by Bonazzola and Schneider (1974).

In the 3–D case, the method described above cannot be used for general purposes. However, quasi-stationary solutions can be computed by using the 3–D hydrodynamic code with important bulk viscocity in order to relax until a quasi-stationary divergence-free solution is obtained.

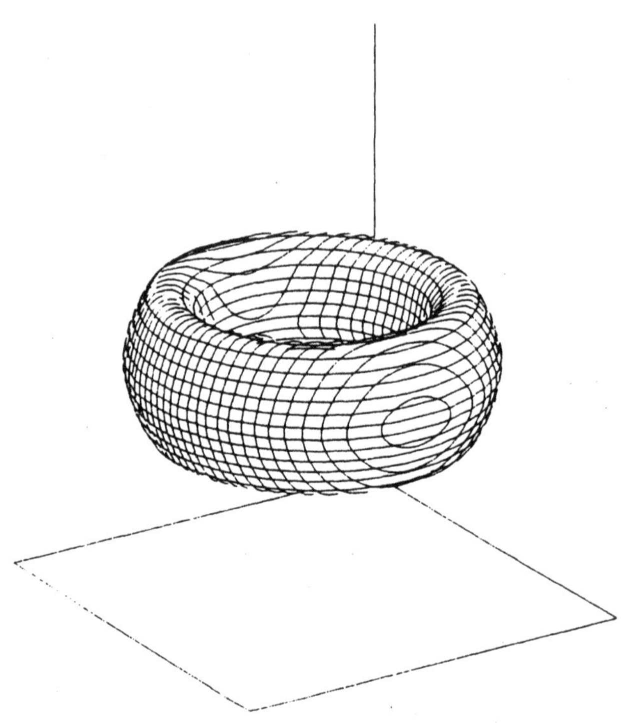

Figure 2 : *plot of isodensity surface of an axisymetric equilibrium configuration obtained for a star imbedded in the "exotic" external potential* $U_{ext} = .5(r - r^6)\rho^2\Omega^2$.

We present in figure 2 an axisymetric solution obtained for the following external potential :

$$U_{ext} = .5(r - r^6)\rho^2\Omega^2. \tag{3.4}$$

4. Axisymetric case

In the case of axisymetric calculations (with or without rotation around the axis of symmetry) we choose to describe the scalar quantities by means of the spherical coordinates system (r, θ, φ) and to expand the vector field with respect to the canonical orthonormal triad associated to the cylindrical coordinates system. This choice allows to take into account the symmetries of the problem (non-dependance on φ of all the quantities) while the numerical grid is such that its boundary is a sphere.

The equations to be solved read

$$\frac{\partial v_\rho}{\partial t} = -(v_\rho \nabla_\rho v_\rho + v_z \nabla_z v_\rho) - \frac{v_\varphi^2}{r \sin\theta} - \frac{1}{\rho}\nabla_\rho P - \nabla_\rho \Phi + \nu \Delta_1 v_\rho,$$

$$\frac{\partial v_\varphi}{\partial t} = -(v_\rho \nabla_\rho v_\varphi + v_z \nabla_z v_\varphi) - \frac{v_\rho v_\varphi}{r \sin\theta} + \nu \Delta_1 v_\varphi,$$

$$\frac{\partial v_z}{\partial t} = -(v_\rho \nabla_\rho v_z + v_z \nabla_z v_z) - \frac{1}{\rho}\nabla_z P + \nu \Delta_0 v_z,$$

$$\frac{\partial \rho}{\partial t} = -(\nabla_\rho(\rho v_\rho) + \frac{\rho v_\rho}{r \sin\theta} + \nabla_z(\rho v_z)\ ),$$

$$\Phi = -4\pi G \Delta_0^{-1} \rho,$$

$$P = P(\rho), \tag{4.1}$$

where we have introduced the operators ∇_ρ, ∇_z and Δ_m defined as

$$\nabla_\rho = \sin\theta \frac{\partial}{\partial r} + \cos\theta \frac{1}{r}\frac{\partial}{\partial \theta},$$

$$\nabla_z = \cos\theta \frac{\partial}{\partial r} - \frac{\sin\theta}{r}\frac{\partial}{\partial \theta},$$

and

$$\Delta_m = \frac{\partial^2}{\partial r^2} + \frac{2}{r}\frac{\partial}{\partial r} + \frac{1}{r^2}(\frac{\partial^2}{\partial \theta^2} + \frac{\cos\theta}{\sin\theta}\frac{\partial}{\partial\theta} - \frac{m^2}{\sin^2\theta}). \tag{4.2}$$

The "scalar" quantities ρ, P, Φ and v_z are expanded in the form of the truncated series :

$$S(t, r, \theta) = \sum_i \sum_l a_{il}(t) T_{2i+[l,2]}(r) \cos l\theta, \tag{4.3}$$

where $T_{2i+[l,2]}(r)$ are even (respectively odd) Chebychev polynomials for even (repectively odd) values of l.

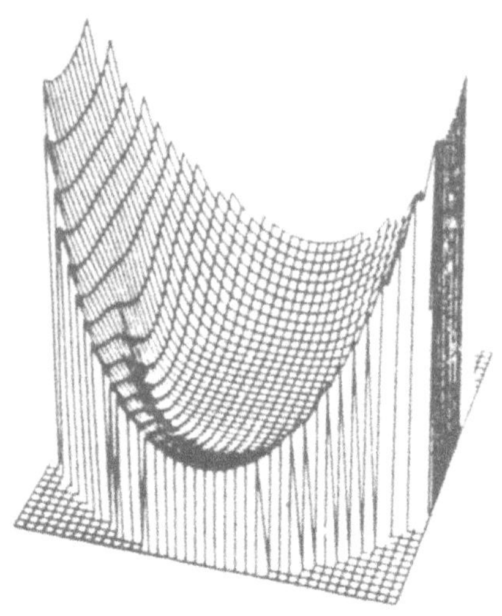

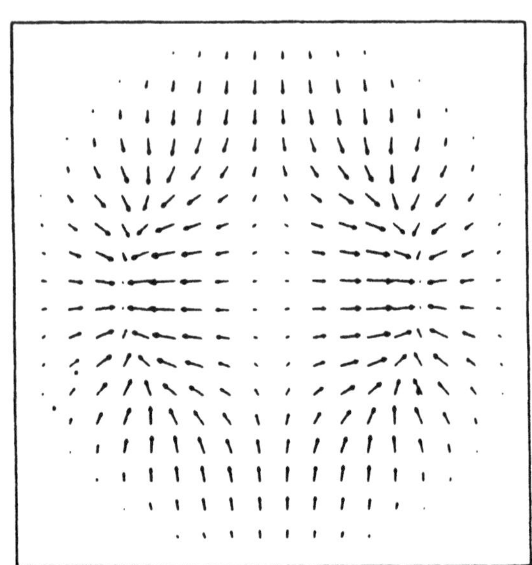

Figure 3a : *Density as a function of x, y and projection of the velocity field onto the $x - z$ plane after the matter was reflected at the boundary of the grid. It can be seen that a shock begins to form which goes to the center of the sphere.*

On another hand, v_ρ and v_φ being the components of $\mathbf{v}$ with respect to a singular triad must be expanded in the form :

$$V(t, r, \theta) = \sum_i \sum_l a_{il}(t) T_{2i+[l,2]}(r) \sin l\theta. \tag{4.4}$$

Let us consider now the finite difference temporal scheme. In order to ensure numerical stability in the case of an explicit scheme, when using Chebychev polynomials expansions, the time-step dt has to satisfy the strong condition :

$$dt \propto \mathrm{Min}\left(\frac{1}{|v_b|N^2}, \frac{1}{|v_{max}|N}\right), \tag{4.5}$$

where N is the number of degrees of freedom in the r–direction, v_b is the radial component of the velocity at the boundary and v_{max} its maximum value (Gottlieb and Orzag 1977). In a 2–D calculation, where the number of grid points can be large, this condition becomes too prohibitive. We have then used a semi-implicit scheme which ensure numerical stability with $dt \propto \frac{1}{|v_{max}|N}$.

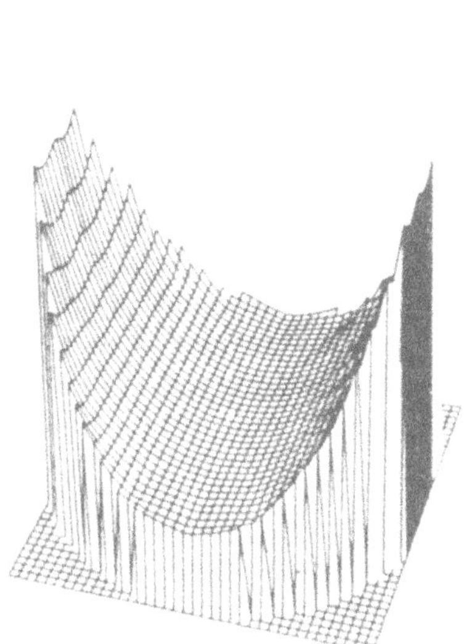

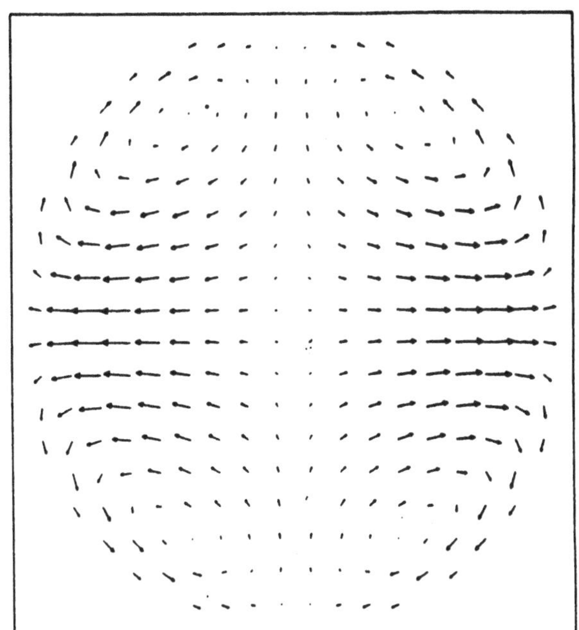

Figure 3b : *Density as a function of x, y and projection of the velocity field onto the $x-z$ plane for the quasi-stationnary solution obtained when the solution becomes relaxed.*

This scheme reads for the evolution equation of v_ρ as :

$$\begin{aligned}
v_\rho^{j+1} = v_\rho^j + dt \times \Big(\\
&- \Big[\big(v_\rho \sin\theta + v_z \cos\theta - \alpha_0(t,\theta) - \alpha_1(t,\theta) r - \alpha_2(t,\theta) r^2\big) \frac{\partial v_\rho}{\partial r}\Big]^{j+1/2} \\
&- \frac{1}{2}\Big[\alpha_0(t,\theta) + \alpha_1(t,\theta) r + \alpha_2(t,\theta) r^2\big) \frac{\partial v_\rho}{\partial r}\Big]^{j+1} \\
&- \Big[v_\rho \frac{\cos\theta}{r} \frac{\partial v_\rho}{\partial r}\Big]^{j+1/2} + \Big[\frac{v_\varphi^2}{r \sin\theta}\Big]^{j+1/2} \qquad (4.6) \\
&+ \Big[\big(v_z - \beta(t)\big) \frac{\sin\theta}{r} \frac{\partial v_\rho}{\partial \theta}\Big]^{j+1/2} \\
&+ \frac{1}{2}\Big[\beta(t) \frac{\sin\theta}{r} \frac{\partial v_\rho}{\partial \theta}\Big]^{j+1} + \frac{1}{2}\Big[\beta(t) \frac{\sin\theta}{r} \frac{\partial v_\rho}{\partial \theta}\Big]^{j} \\
&- \frac{1}{2}\Big(\frac{\sin\theta}{\rho} \frac{\partial}{\partial r} + \frac{\cos\theta}{r\rho} \frac{\partial}{\partial \theta}\Big)(P^{j+1} + P^j) + \frac{\nu}{2} \Delta_1 (v_\rho^{j+1} + v_\rho^j) \Big)
\end{aligned}$$

where the quantities written at the time $j+1$ are computed in an implicit way, and where the coefficients $\alpha_i(t,\theta)$ are chosen such that the advective terms vanish at the boundary $r = 1$, the coefficient $\beta(t)$ being determined by $\beta = v_z|_{r=0}$.

A numerical simulation without gravitational field, that was performed on the VAX–8600 of the Observatoire de Meudon using a 33×33 numerical grid points, is presented in figures 3. The initial conditions were :

$$\rho(t_0, r, \theta) = C^{\underline{ste}},$$

$$v_\rho(t_0, r, \theta) = v_z(t_0, r, \theta) = 0,$$

$$v_\varphi(t_0, r, \theta) = \omega r, \tag{4.7}$$

with boundary conditions : $v_\rho(t, 1, \theta) = v_z(t, 1, \theta) = 0$, and equation of state $P = k\rho^\gamma$. The parameters of this simulation, expressed in dimensionless units, are $\gamma = 5/3$, the sound velocity $c_s = \sqrt{5/3}$, the acoustic crossing time $\tau_c = D/c_s = 1.6$, where D is the diameter of the box, the rotational time at $t = 0$ $\tau_r = 1.8$ and the diffusion time $\tau_D = D^2/\nu = 40$.

The matter is initially going to the boundary of the box, then reflected on the boundary and a shock begins to form. After a lot of crossing time, because of the dissipation, the solution tends to a stationary one and give rise to a circulation typical of the axisymetric equilibrium configurations.

5. 3–D case

For 3–D calculations, we choose to describe the scalar quantities by means of the spherical coordinates system. The reasons of this choice are the following. The boundary surface associated to these coordinates is a smooth and regular one (at the opposite of the boundary associated to cartesian coordinates), using spherical coordinates allows a very simple matching of internal solutions with external analytical ones (especially for the gravitational potential) and, moreover, these coordinates are the most natural ones for the problem in which we are interested.

The vector fields (and more generally any tensor fields) can be described by means of their components with respect to an arbitrary basis. The choice of the basis of decomposition is a crucial one. If the basis of decomposition is singular (this is the case for the canonical basis associated to the spherical coordinates system), then the components of any vector field have to be singular. It is possible to treat these singularities, however a simple way consists to expand the vector fields on a regular basis. We choose to express the vector fields by means of their components with respect to the canonical cartesian basis $(\partial/\partial x, \partial/\partial y, \partial/\partial z)$.

All the quantities, that, in the present case, are regular single-valued functions satisfying the same regularity rules near the axis $\sin\theta = 0$ and the origin $r = 0$, are expanded as :

$$Q(t,r,\theta,\varphi) = \sum_i \sum_l \sum_m a_{ilm}(t) T_{2i+[l,2]}(r) T_l^{\dagger[m,2]}(\theta) e^{im\varphi}, \tag{5.1}$$

where we use the notation $[j,2] = 0$ (resp. 1) for j even (rep. odd). The Chebychev polynomials used in the previous expansion are defined as :

$$T_j(\psi_r) = \cos(j\psi_r), \qquad \text{with} \qquad r = \cos\psi_r, \qquad \psi_r \in [\frac{\pi}{2}, \pi],$$

$$T_l^{\dagger 0}(\theta) = \cos l\theta,$$

$$T_l^{\dagger 1}(\theta) = \sin l\theta, \tag{5.2}$$

θ lying in the range $[0,\pi]$.

This expansion is used to compute all the spatial operators except in the case of the Laplacian operator inversion where we use spherical harmonics expansion for the angular parts. The coefficients of the spherical harmonics expansion are obtained from the coefficients of the Chebychev/Fourier expansion by means of a matrix multiplication.

The equations (written in a fixed grid) take the simple form :

$$\frac{\partial v_i}{\partial t} = -v_x \frac{\partial v_i}{\partial x} - v_y \frac{\partial v_i}{\partial y} - v_z \frac{\partial v_i}{\partial z} - \frac{1}{\rho}\partial_i P - \partial_i \Phi - \partial_i \Omega_{ext} + \nu \Delta v_i,$$

$$\frac{\partial \rho}{\partial t} = -\frac{\partial(\rho v_x)}{\partial x} - \frac{\partial(\rho v_y)}{\partial y} - \frac{\partial(\rho v_z)}{\partial z}$$

$$\Delta\Phi = -4\pi G\rho, \tag{5.3}$$

where $i = x, y, z$ and $P = P(\rho)$ is the pressure and G the gravitational constant and where Ω_{ext} is an external potential if any.

The formalism we use to expand the quantities allows to treat in an exact way the pseudo-singularities $r = 0$ and $\sin\theta = 0$ typical of the spherical coordinates system which appear in the evaluation of the elementary operators $\partial/\partial x$, $\partial/\partial y$ and $\partial/\partial z$. The code is written in such a way that, for any quantity Q, the coefficients of $\partial_i Q$ are computed *globally* by means of the coefficients of Q (see Bonazzola and Marck 1988 for details).

The finite difference temporal scheme is a second order scheme, the source terms and the advective terms being computed explicitely, while the dissipative terms are computed implicitely. This scheme reads :

$$\rho^{j+1} = \rho^j - dt(\frac{3}{2}\mathrm{div}\rho\mathbf{v}^j - \frac{1}{2}\mathrm{div}\rho\mathbf{v}^{j-1}),$$

$$\rho^{j+1/2} = \frac{1}{2}(\rho^{j+1} + \rho^{j}),$$

$$\Phi^{j+1/2} = -4\pi G \Delta^{-1} \rho^{j+1/2},$$

$$P^{j+1/2} = P(\rho^{j+1/2}),$$

$$\begin{aligned} v_i^{j+1} = v_i^j + dt \times \Big(& -\frac{3}{2}\mathbf{v}\cdot\mathbf{grad}v_i{}^j + \frac{1}{2}\mathbf{v}\cdot\mathbf{grad}v_i{}^{j-1} \\ & - \frac{1}{\rho^{j+1/2}}\partial_i P^{j+1/2} - \partial_i \Phi^{j+1/2} \\ & + \frac{1}{2}\nu(\Delta v_i^{j+1} + \Delta v_i^j)\Big) \end{aligned} \tag{5.4}$$

The boundary conditions, which are imposed on the cartesian components of the velocity by means of the τ–approximation (Lanczos 1956), can be of the form :

$$\alpha(t)\frac{\partial v_i}{\partial r} + \beta(t)v_i = \gamma(t,\theta,\varphi), \tag{5.5}$$

where $i = x, y, z$ and α, β, γ are arbitrary functions.

We present in figures 4 a preliminary 3–D collapsing star simulation. The initial model is a spherical $1.4M_\odot$ star at equilibrium with adiabatic index $\gamma = 1.34$. The collapse is started by changing the equation of state to

$$P(\rho) = \begin{cases} k_1\rho^{1.29} & \text{for } \rho \le \rho_n, \\ k_2\rho^2 & \text{for } \rho \ge \rho_n, \end{cases} \tag{5.6}$$

where ρ_n is the nuclear density (Bonche and Vautherin 1982), and by making the star orbit on a closed circular orbit around a $1M_\odot$ companion located initially in the x–direction. We used in this calculation a moving grid determined such that the matter is everywhere outgoing (or at rest) at the boundary of the grid. The boundary conditions were $\partial v_x/\partial r = \partial v_y/\partial r = \partial v_z/\partial r = 0$.

As it can be seen on figure 4a, the star is initially drawn out along the x–axis while it is collapsing in the other direction because of the combination of the tidal forces and the gravitational forces. The collapse grows while the star is orbiting, which results in the apparition of a differential rotation in the star, as can be seen in figure 4b. This preliminary calculation was performed on a VAX–8600 using a $17 \times 9 \times 8$ grid points, which is obviously not enough to give a good description of the bounce.

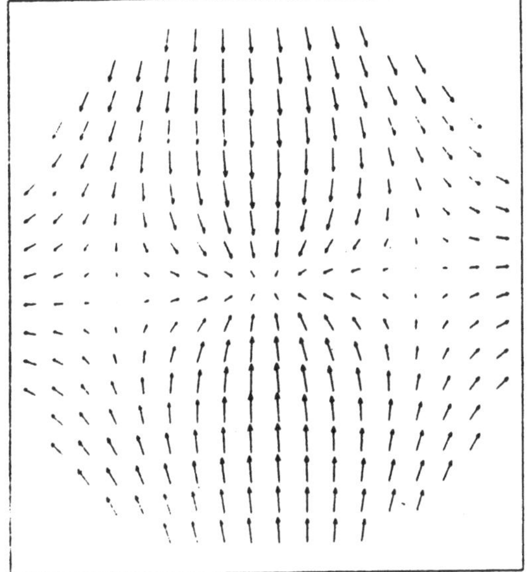

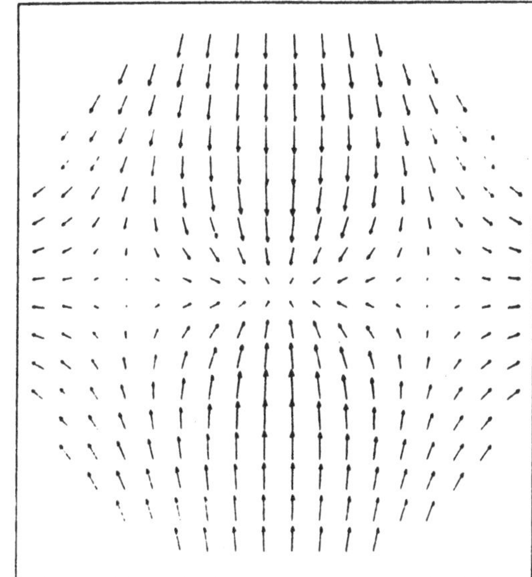

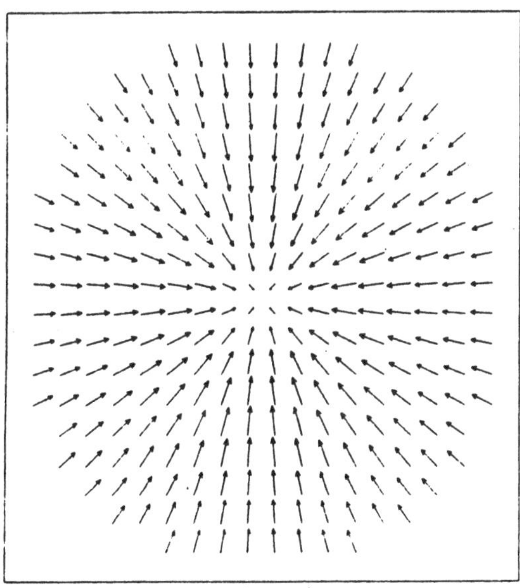

Figure 4a : *Plots of the 2-dimentional vector fields whose componants are* $(v^x|_{z=0}, v^y|_{z=0})$ *(up left),* $(v^x|_{y=0}, v^z|_{y=0})$ *(up right) and* $(v^z|_{x=0}, v^y|_{x=0})$ *(down) when the collapse begins.*

6. Conclusion

We have described a numerical method for solving self-gravitating 2–D and 3–D hydrodynamics equations in spherical coordinates system. The method, which takes into account the analytical properties of the physical quantities allows to give a rigourous treatment of the pseudo-singularities $\sin\theta = 0$ and $r = 0$. The quantities are expanded in Fourier series for the longitudinal part, in first kind or second kind Chebychev polynomials series for the azymuthal part and in even or odd Chebychev polynomials series for the radial part. These codes will be used for the study of gravitational waves emission during the collapse of a star in the Newtonian approximation by means of the quadrupole formula.

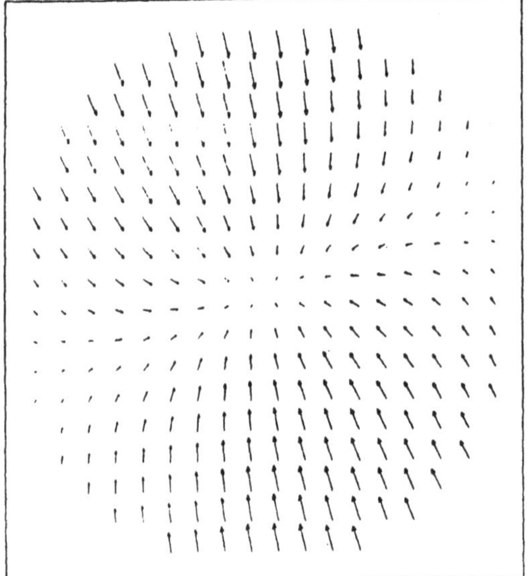

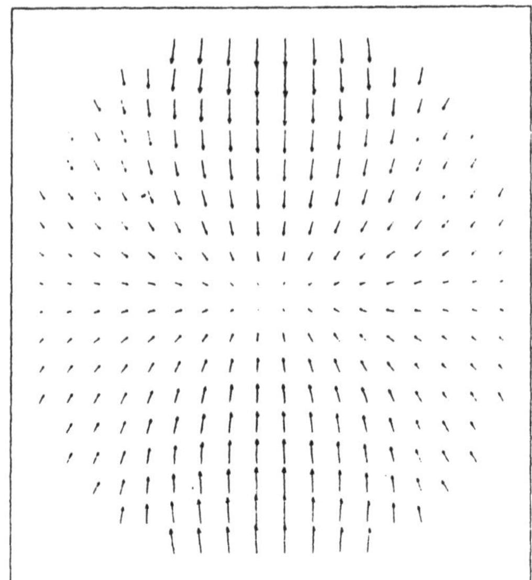

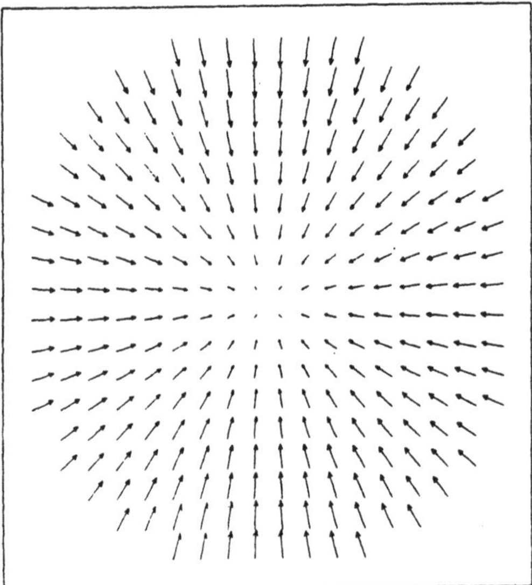

Figure 4b : *Plots of the same vector fields as in figure 4a after several time-steps. One can see on figure up-left the direction of the companion which is no more the x-axis because the star is (rapidly) orbiting.*

These codes are developped and used on the VAX-8600 of the Observatoire de Paris, section de Meudon, and on the VP-200 of the Centre Interrégional de Calcul Électronique (Orsay). We gratefully acknowledge financial support of the M.P.B. Department of C.N.R.S.

7. References

Bonazzola S., Marck J.A., 1986, *Astr. Astroph.*, **164**,300.
Bonazzola S., Marck J.A., 1988, *submitted to J. Comp. Phys.*

Bonazzola S., Schneider J., 1974, *Ap. J.*, **191**,273.
Bonche P., Vautherin D., 1982, *Astr. Astroph.*, **112**,268.
Dang K., Maday Y., Pernaud-Thomas B., Vandeven H., 1986, *La Recherche Aérospatiale*, **5**,315.
Gottlieb D., Orzag S., 1977, *Numerical Analysis of Spectral Methods: Theory and Application, Regional Conference Series in Applied Mathematics.*
Lanczos C., 1956, *Applied Analysis, Prentice–Hall, Englewood Cliffs, NJ.*

Methods in 3D Numerical Relativity

Takashi Nakamura

Department of Physics, Kyoto University, Kyoto 606, Japan

Ken-ichi Oohara

National Laboratory for High Energy Physics
Oho, Tsukuba-shi, Ibaraki-ken, 305, Japan

Abstract

Two methods in 3D numerical relativity are presented. In the first method all the metric quantities are expanded by spherical harmonics Y_{lm}. As a first test of this method, time evolution of pure gravitational waves is shown. In the other method, we use (x, y, z) coordinates as grids in which no coordinate singularities exist. One of the most powerful super computers such as VP400 of FACOM enables us to use 200Mbytes core memories with a practical speed of 400MFLOPS for the present numerical calculation. One can use $80 \times 80 \times 80$ grids in numerical relativity, which is marginally sufficient. As a first step of 3D code in (x, y, z) coordinate, we show the time evolution of pure gravitational waves. We have also been developing a 3D hydrodynamics code using the Leblanc's transport method for advection terms and the tensor artificial viscosity. This hydrodynamics code passed through various tests such as advection tests, shock tube problems, Sedov solutions and collapse of a homogeneous ellipsoid. As a demonstration of the applicability of this code we present coalescence of a binary neutron star and the estimate of the amount of gravitational waves using a quadrupole formula. We used two methods for the third time derivatives of the quadrupole moment. We performed 7 simulations with different total angular momentum, the initial radius and the seperation of two neutron stars. The energy radiated ranges 10^{-3} to 10^{-2} of the rest mass at the first $\sim$ 1ms. As the luminosity at the final stage of the numerical calculation is still large, simulaions including the back reacyion is urgent. We also try to include the radiation reaction terms to know the final destiny of two neutron stars.

1. Coordinate Singularities

One of the major difficulties in constructing axially symmetric codes in numerical relativity is how to avoid coordinate singularities in cylindrical and spherical polar coordinates. Let us consider a point on the z-axis in the cylindrical coordinate. This point is clearly a single point in the real space but is assigned to many points in the coordinate space, that is, to arbitrary values of ϕ. In order to guarantee that this is a single point in the real space, one can easily show that certain relations called regularity conditions should be satisfied by components of metric tensors such as

$$\gamma_{RR} = \gamma_{\phi\phi}/R^2. \tag{1.1}$$

However in a naive finite difference method it is very difficult to guarantee the above relation numerically due to the truncation errors. The violation of the regularity usually yields numerical instabilities. Physically the violation of the regularity on the axis means that there is a true physical singularity. This situation is essentially the same in the Newtonian gravitational collapse problem. However in numerical relativity the difficulty becomes greater due to the second spatial derivative terms in the evolution equations of the extrinsic curvatures. One of the methods of avoiding this difficulty is to use the regularized variable defined by

$$g = (\gamma_{RR} - \gamma_{\phi\phi}/R^2)/R^2. \tag{1.2}$$

If one use g and γ_{RR} instead of γ_{RR} and $\gamma_{\phi\phi}$, the regularity is automatically satisfied.

Now in non-axially symmetric cases, what are the regularity conditions ? If one uses (x, y, z) coordinate, each coordinate point has one to one relation to a point in the space-time and there are no problems concerning the regularity. However if one uses the spherical polar coordinate, for example, the transformation becomes as

$$\gamma_{rr} = \gamma_{xx}\sin^2\theta\cos^2\phi + \gamma_{xy}\sin^2\theta\sin 2\phi + \gamma_{yy}\sin^2\theta\sin^2\phi$$

$$+\gamma_{yz}\sin 2\theta\sin\phi + \gamma_{zz}\cos^2\theta + \gamma_{zx}\sin 2\theta\cos\phi,$$

$$\gamma_{\theta\theta}/r^2 = \gamma_{xx}\cos^2\theta\cos^2\phi + \gamma_{xy}\cos^2\theta\sin 2\phi + \gamma_{yy}\cos^2\theta\sin^2\phi$$

$$-\gamma_{yz}\sin 2\theta\sin\phi + \gamma_{zz}\sin^2\theta - \gamma_{zx}\sin 2\theta\cos\phi,$$

$$\gamma_{\phi\phi}/r^2/\sin^2\theta = \gamma_{xx}\sin^2\phi - \gamma_{xy}\sin 2\phi + \gamma_{yy}\cos^2\theta\cos^2\phi,$$

$$\gamma_{r\theta}/r = 0.5(\gamma_{xx}\sin 2\theta\cos^2\phi + \gamma_{xy}\sin 2\theta\sin 2\phi + \gamma_{yy}\sin 2\theta\sin^2\phi)$$

$$+\gamma_{yz}\cos 2\theta\sin\phi - 0.5\times\gamma_{zz}\sin 2\theta + \gamma_{zx}\cos 2\theta\cos\phi,$$

$$\gamma_{r\phi}/r/\sin\theta = -0.5(\gamma_{xx}\sin 2\phi + \gamma_{xy}\cos 2\phi + \gamma_{yy}\sin 2\phi)$$

$$-\gamma_{yz}\cos\theta\sin\phi - \gamma_{zx}\cos\theta\cos\phi,$$

and

$$\gamma_{\theta\phi}/r = 0.5\cos\theta(-\gamma_{xx}\sin^2\phi + 2\gamma_{xy}\cos 2\phi + \gamma_{yy}\sin 2\phi)$$

$$-\gamma_{yz}\sin\theta\cos\phi + \gamma_{zx}\sin\theta\sin\phi.$$

Let us consider the origin r=0 for example. As $\gamma_{xx}, \gamma_{yy}, \gamma_{zz}, \gamma_{xy}, \gamma_{yz}$ and γ_{zx} have definite independent values even at the origin, the above relations tell us that metric tensor in the spherical polar coordinate should depend on θ and ϕ even at the origin to guarantee the regularities. So it will be very difficult to find the regularized variables like Eq. (1.2) . The only variables which behave well at r=0 and on the axis seem to be $\gamma_{xx}, \gamma_{yy}, \gamma_{zz}, \gamma_{xy}, \gamma_{yz}$ and γ_{zx} although there is no proof for this statement. (See also R.Stark this volume for this line of constructing 3D codes.)

2. Spectral Method

The above consideration suggests us that we had better use $\gamma_{xx}, \gamma_{yy}, \gamma_{zz}, \gamma_{xy}$, γ_{yz} and γ_{zx} as basic variables. Then there are no merits to write down the Einstein equations in (r, θ, ϕ) coordinate. However unless one has a super computer with memories greater than 200Mbytes or so, it is impossible to use (x, y, z) coordinates as grids. Thus we encounter dilemma. We must use (r, θ, ϕ) as grids with γ_{xx}, $\gamma_{yy}, \gamma_{zz}, \gamma_{xy}$, γ_{yz} and γ_{zx} as basic variables. A naive method to overcome this difficulty is to convert $\partial/\partial x, \partial/\partial y, \partial/\partial z$ to finite difference version of $\partial/\partial r, \partial/\partial\theta, \partial/\partial\phi$. In reality one of the present authors(T.N) tried this method but encountered serious numerical instabilities on the axis due to zero-divided-zero type terms on the axis. So we need a more sophisticated method.

Now let us assume every quantity Q in (x, y, z) coordinates has a Taylor expansion as

$$Q(x, y, z, t) = \sum_{a,b,c=0}^{\infty} A_{abc}(t)\frac{x^p y^q z^r}{a!b!c!}. \tag{2.1}$$

We try to express Eq.(2.1) using spherical harmonics Y_{lm}. If we notice that $\sin^{p+q}\theta\cos^p\phi\sin^q\phi\cos^r\theta$ can be expressed as a sum of spherical harmonics of the form as $Y_{p+q-r-2n,m}(\theta,\phi)$, $n = 0,1,2,\cdots$, we can show that Q can be reexpressed as

$$Q = \sum_{l,m} r^l Q_{lm}(r^2,t) Y_{lm}(\theta,\phi). \tag{2.2}$$

Here note that Q_{lm} has a Taylor expansion with respect to r^2 (not r). We need the first and the second derivatives of Q in order to calculate Ricci tensor and other quantities needed in numerical codes. We first need the first spatial derivatives of Q. For example let us consider $\partial Q/\partial z$. Since $\partial Q/\partial z$ should have the Taylor expansion like Eq.(2.1) ,it should have also the spherical harmonics expansion as

$$\partial Q/\partial z = \sum_{l,m} r^l Q^z_{lm}(r^2,t) Y_{lm}(\theta,\phi), \tag{2.3}$$

where $Q^z_{lm}(r^2,t)$ has a relation to $Q_{lm}(r^2,t)$ as

$$Q^z_{lm}(r^2,t) = C_1(l,m)\frac{\partial}{\partial w}Q_{l-1,m}$$

$$+C_2(l,m)(2w\frac{\partial}{\partial w}Q_{l+1,m} + (2l+3)Q_{l+1,m}), \tag{2.4}$$

with $C_1(l,m)$ and $C_2(l,m)$ being Clebsh-Gordon like coefficients and $w = r^2$. We have similar formula for $\partial Q/\partial x$ and $\partial Q/\partial y$. Thus once $Q_{lm}(r^2,t)$ is given the first derivative is determined without zero-divided-zero type terms. For the second derivatives we perform the above procedure twice. Since formulae for the second derivatives are too complicated, we use REDUCE to avoid mistakes and we obtain FORTRAN programs directly from the output of the REDUCE code.

Now in our spectral method we adopt $Q_{lm}(r^2,t)$ as the basic variables of the Einstein equations. (See also J.Marck this volume for another spectral method.) We express γ_{ij} and K_{ij} as

$$\gamma_{ij} = \sum r^l \gamma^{lm}_{ij}(r^2,t) Y_{lm}(\theta,\phi) \tag{2.5}$$

and

$$K_{ij} = \sum r^l K^{lm}_{ij}(r^2,t) Y_{lm}(\theta,\phi). \tag{2.6}$$

The basic equations for the time evolution of pure gravitational waves are

$$\frac{\partial \gamma_{ij}}{\partial t} = -2K_{ij}, \tag{2.7}$$

$$\frac{\partial K_{ij}}{\partial t} = R_{ij} + KK_{ij} - 2K^m_i K_{mj} = S_{ij}, \tag{2.8}$$

where we adopt $\alpha = 1$ and $\beta^i = 0$ coordinate conditions for simplicity. Constraint equations become

$$R + K^2 = K_{ij}K^{ij}, \tag{2.9}$$

$$K^j_{i|j} = K_{|i}. \tag{2.10}$$

One of the present authors has already shown a method of solving constraint equations at t=0 for general pure gravitational waves.[1] As for the time evolution, we have already shown the details of the method[2] Here we show only the essential points. Eq.(2.7) is rewritten in our basic variables as

$$\frac{\partial \gamma_{ij}^{lm}}{\partial t} = -2K_{ij}^{lm}. \tag{2.11}$$

From the trace of Eq.(2.8) we have

$$\frac{\partial K}{\partial t} = R + K^2. \tag{2.12}$$

Using the Hamiltonian constraint equation, we rewrite the r.h.s as

$$\frac{\partial K}{\partial t} = K_{ij}K^{ij}. \tag{2.13}$$

The determinant of $\gamma_{ij} (= \gamma)$ obeys

$$\frac{\partial \gamma}{\partial t} = -2K\gamma \tag{2.14}$$

From Eq.(2.13) for large r, it is clear that the K is the second order quantity although K_{ij} is the first order. Thus if we use Eqs.(2.13) and (2.14) to determine K and γ, $\gamma - 1$ at large distance from the source of the gravitational waves is guaranteed to be the second order of the amplitude, that is, traceless even numerically.

Next we try to guarantee numerically the transverse nature of the wave. For this purpose we must rewrite the Ricci tensor. We first rewrite the momentum constraint equations as

$$\frac{\partial K_{ij}}{\partial x^j} = -\tilde{\gamma}^{jl}\frac{\partial K_{ij}}{\partial x^l} + \gamma^{lj}(\Gamma^m_{jl}K_{im} + \Gamma^m_{il}K_{mj}) + \frac{\partial K}{\partial x^i} \equiv F_i, \tag{2.15}$$

where

$$\tilde{\gamma}^{jl} = \gamma^{jl} - \eta^{jl}.$$

Since the Christoffel symbols and $\tilde{\gamma}$ are the first order quantity, it is numerically guaranteed that F_i is the second order quantity. This means that the transversality is established as

$$\frac{\partial G_i}{\partial t} = -2F_i, \tag{2.16}$$

where

$$G_i \equiv \frac{\partial \gamma_{ij}}{\partial x^j}$$

Therefore the transverse-traceless nature of the wave is guaranteed under the use of Eqs.(2.13) to (2.16) .

Ricci tensor is the most complicated source term for K_{ij}. We first define R^{lin}_{ikjm} and R^{nonlin}_{ikjm} as

$$R^{lin}_{ikjm} = 0.5\left(\frac{\partial^2 \gamma_{im}}{\partial x^k \partial x^j} + \frac{\partial^2 \gamma_{kj}}{\partial x^i \partial x^m} - \frac{\partial^2 \gamma_{ij}}{\partial x^k \partial x^m} - \frac{\partial^2 \gamma_{km}}{\partial x^i \partial x^j}\right),$$

and

$$R^{nonlin}_{ikjm} = 0.5(\Gamma_{n,kj}\Gamma^n_{im} - \Gamma_{n,km}\Gamma^n_{ij}).$$

It is clear that R^{lin}_{ikjm} is the first order quantity while R^{nonlin}_{ikjm} is the second order one. Using R^{lin}_{ikjm} and R^{nonlin}_{ikjm}, we can express Ricci tensor as

$$R_{ij} = -0.5\Delta\tilde{\gamma}_{ij} + \frac{\partial G_i}{\partial x^j} + \frac{\partial G_j}{\partial x^i} + \tilde{\gamma}^{km} R^{lin}_{ikjm} + \gamma^{km} R^{nonlin}_{ikjm}$$

$$-0.5\frac{\partial^2(\tilde{\gamma}_{11} + \tilde{\gamma}_{22} + \tilde{\gamma}_{33})}{\partial x^i \partial x^j}. \tag{2.17}$$

$\gamma - 1$ is explicitly expressed as

$$\Gamma \equiv \gamma - 1 = (1 + \tilde{\gamma}_{11})(1 + \tilde{\gamma}_{22})(1 + \tilde{\gamma}_{33}) + 2\tilde{\gamma}_{12}\tilde{\gamma}_{23}\tilde{\gamma}_{31}$$

$$-(1 + \tilde{\gamma}_{11})\tilde{\gamma}_{23}^2 - (1 + \tilde{\gamma}_{22})\tilde{\gamma}_{13}^2 - (1 + \tilde{\gamma}_{33})\tilde{\gamma}_{12}^2 - 1$$

$$\equiv \tilde{\gamma}_{11} + \tilde{\gamma}_{22} + \tilde{\gamma}_{33} + det2. \tag{2.18}$$

Since Γ is the second order, Eq.(2.18) clearly tells us that the last term in Eq.(2.17) is the second order. As a whole except the first term the r.h.s of the Ricci tensor is the second order even numerically. In the limit of zero amplitude the equations for K_{ij} become the wave equations in the usual senses definitely.

Now the basic equations we use are

$$\frac{\partial \tilde{\gamma}_{ij}}{\partial t} = -2K_{ij}, \tag{2.19}$$

$$\frac{\partial \Gamma}{\partial t} = -2K(1+\Gamma), \tag{2.20}$$

$$\frac{\partial K}{\partial t} = K_{ij}K^{ij}, \tag{2.21}$$

$$\frac{\partial K_{ij}}{\partial t} = R_{ij} + KK_{ij} - 2K_i^m K_{mj}, \tag{2.22}$$

$$\frac{\partial G_i}{\partial t} = -2F_i. \tag{2.23}$$

We expand $\tilde{\gamma}_{ij}, K_{ij}, \Gamma, K$ and G_i as

$$\tilde{\gamma}_{ij} = \sum r^l \tilde{\gamma}_{ij}^{lm}(r^2, t) Y_{lm}(\theta, \phi), \tag{2.24}$$

$$K_{ij} = \sum r^l K_{ij}^{lm}(r^2, t) Y_{lm}(\theta, \phi), \tag{2.25}$$

$$\Gamma = \sum r^l \Gamma^{lm}(r^2, t) Y_{lm}(\theta, \phi), \tag{2.26}$$

$$K = \sum r^l K^{lm}(r^2, t) Y_{lm}(\theta, \phi), \tag{2.27}$$

$$G_i = \sum r^l G_i^{lm}(r^2, t) Y_{lm}(\theta, \phi). \tag{2.28}$$

Inserting these expression to Eqs.(2.19) to (2.23) , we have

$$\frac{\partial \tilde{\gamma}_{ij}^{lm}}{\partial t} = -2K_{ij}^{lm}, \tag{2.29}$$

$$\frac{\partial \Gamma^{lm}}{\partial t} = -2K^{lm} - \int 2K\Gamma Y_{lm}^* d\Omega / r^l, \tag{2.30}$$

$$\frac{\partial K^{lm}}{\partial t} = \int K_{ij}K^{ij} Y_{lm}^* d\Omega / r^l, \tag{2.31}$$

FIGURE 1

Time evolution of $\gamma_{xx}-1$ for $A=10^{-2}$

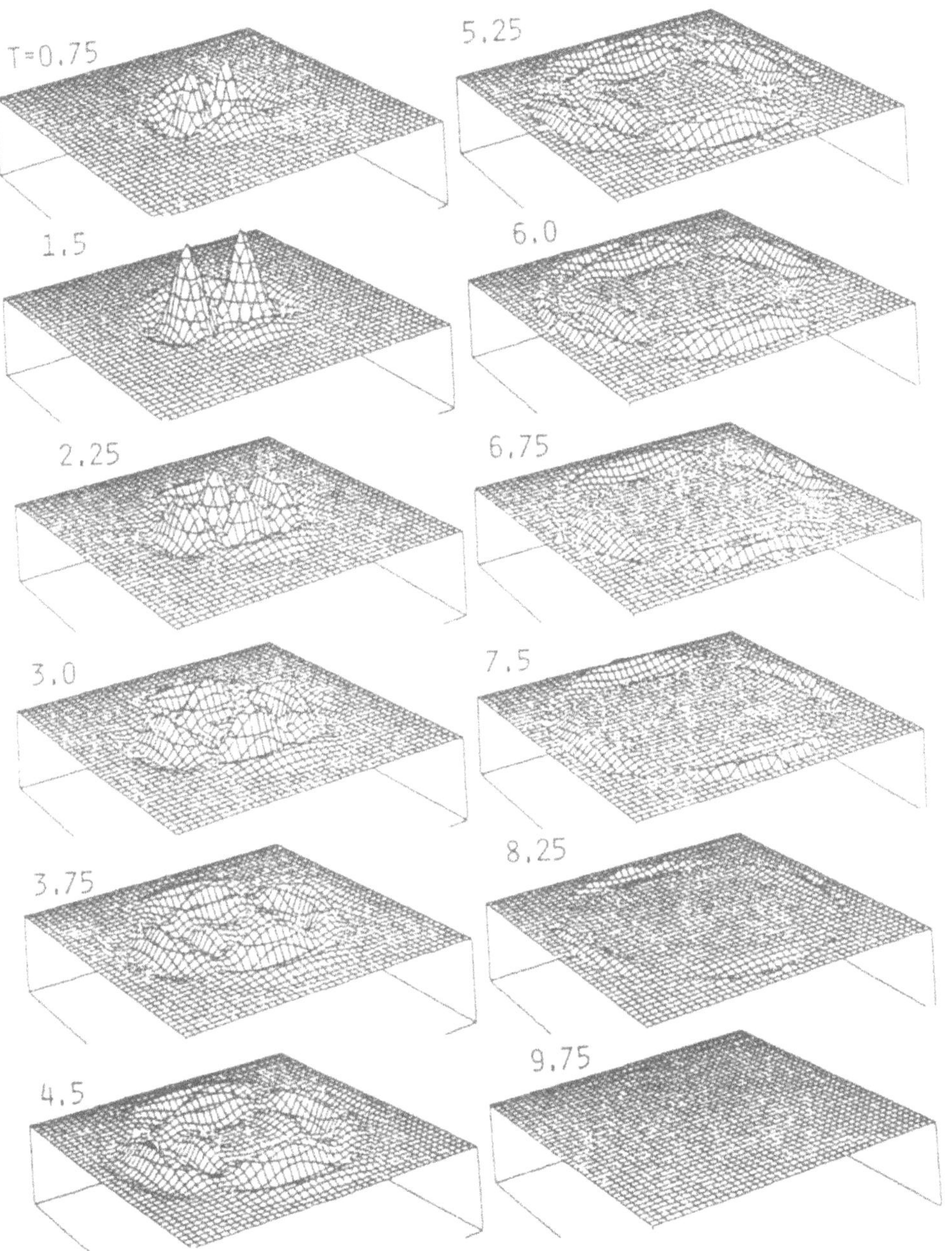

FIGURE 2

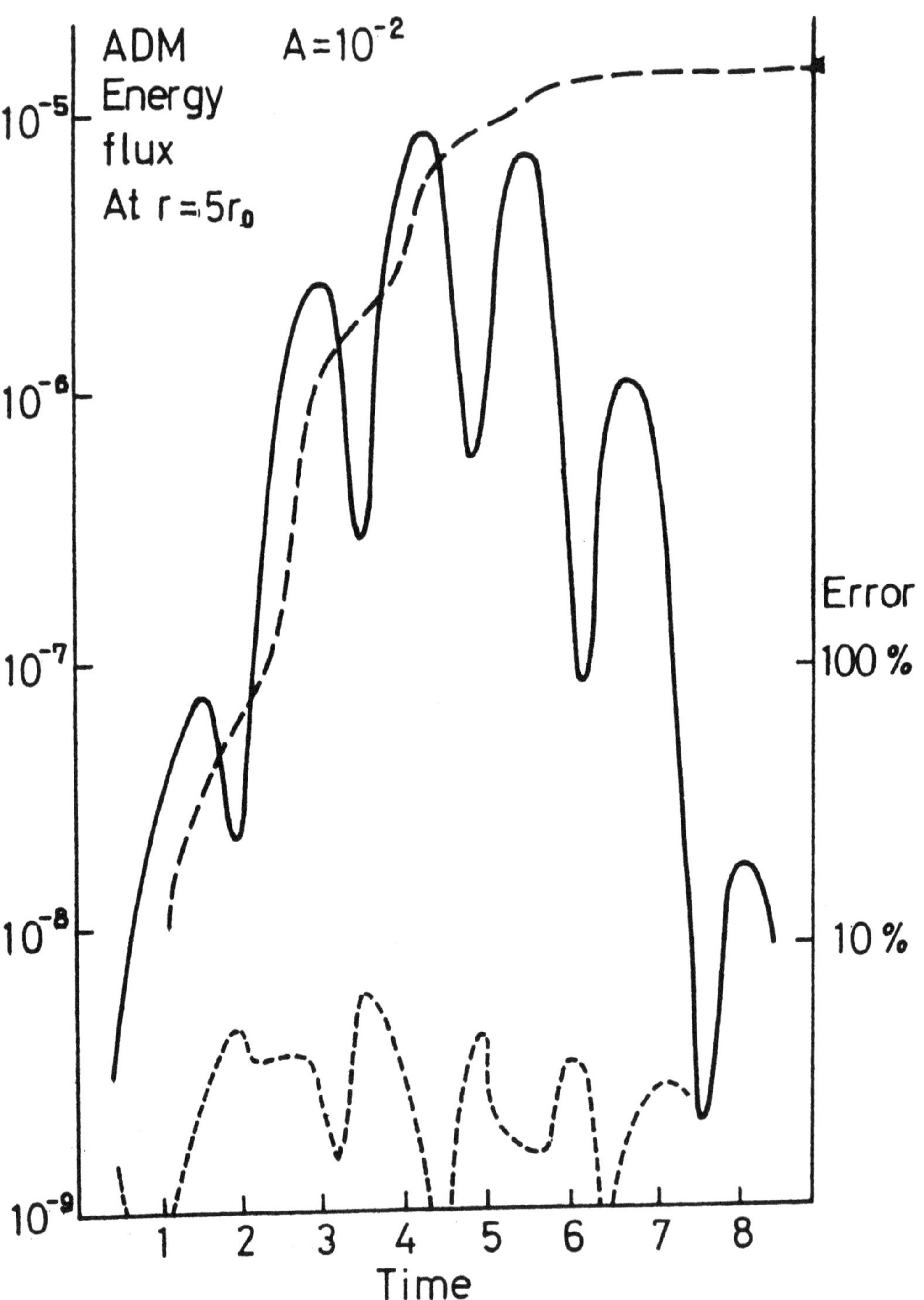
ADM
Energy
flux
At r = 5r0
A=10^-2
10^-5
10^-6
10^-7
10^-8
10^-9
Error
100 %
10 %
1
2
3
4
5
6
7
8
Time

$$\frac{\partial K'^{lm}_{ij}}{\partial t} = \int S_{ij} Y^*_{lm} d\Omega / r^l, \tag{2.32}$$

$$\frac{\partial G^{lm}_i}{\partial t} = \int F_i Y^*_{lm} d\Omega / r^l. \tag{2.33}$$

We solved the even parity mode of Teukolsky wave with l=2 and m=2. Initially we put the localized wave packet with an exponential amplitude distribution. The amplitude of the wave is 10^{-2} in the sense of h. In Fig.1 we show the time evolution of $(\gamma_{xx} - 1)r$ in the equatorial plane. In Fig.1-(a) there is a quadrupole wave in the center. This wave propagates outward and finally passes through the numerical outer boundary without artificially reflected waves. In this calculation the number of grids is (100,41,16) in (r, θ, ϕ) with 650 time steps. A method of the finite difference is essentially a leap frog method. At the outer boundary of numerical grids we impose the outgoing wave condition as

$$\left(\frac{\partial}{\partial r} - \frac{\partial}{\partial t}\right)(r^{l+1} Q^{lm}) = 0. \tag{2.34}$$

The computing time by VP200 of FACOM is about three hours. In Fig.2 we show the ADM energy flux estimated at the radius five times as far as the initial size of the wave packet by a solid line. A dashed line shows a relative deference between a linearized solution and a numerical solution. As the amplitude of the wave is 1%, a few % difference is natural and Fig.2 shows that in this method we can trace propagation of gravitational waves quite well.

3. 3D Pure Gravitational Waves with (x, y, z) Meshes

One of the main purposes of 3D numerical relativity is to attack highly non-axisymmetric phenomena such as coalescence of a binary system due to the emission of the gravitational waves and anisotropic inhomogeneous early universe. In such a phenomena there is no reason to use (r, θ, ϕ) coordinates because there is no special place which should be identified as r=0. The reason for the use of (r, θ, ϕ) has been from the restriction of the number of memories one can use. However the situation has changed recently. For example one can use 200Mbytes in VP400 by FACOM. What is the meaning of this number for 3D numerical relativity? In 3D in general one need $\sim$ 60 variables including metric quantities and matter quantities. Then for the double precision calculation (64bit), 200Mbytes memories enable us to use 80^3 grids. If we remember that 2D numerical relativity began with $\sim 40^2$ grids, we can say that we are now standing in front of the door to the kingdom of 3D numerical relativity.

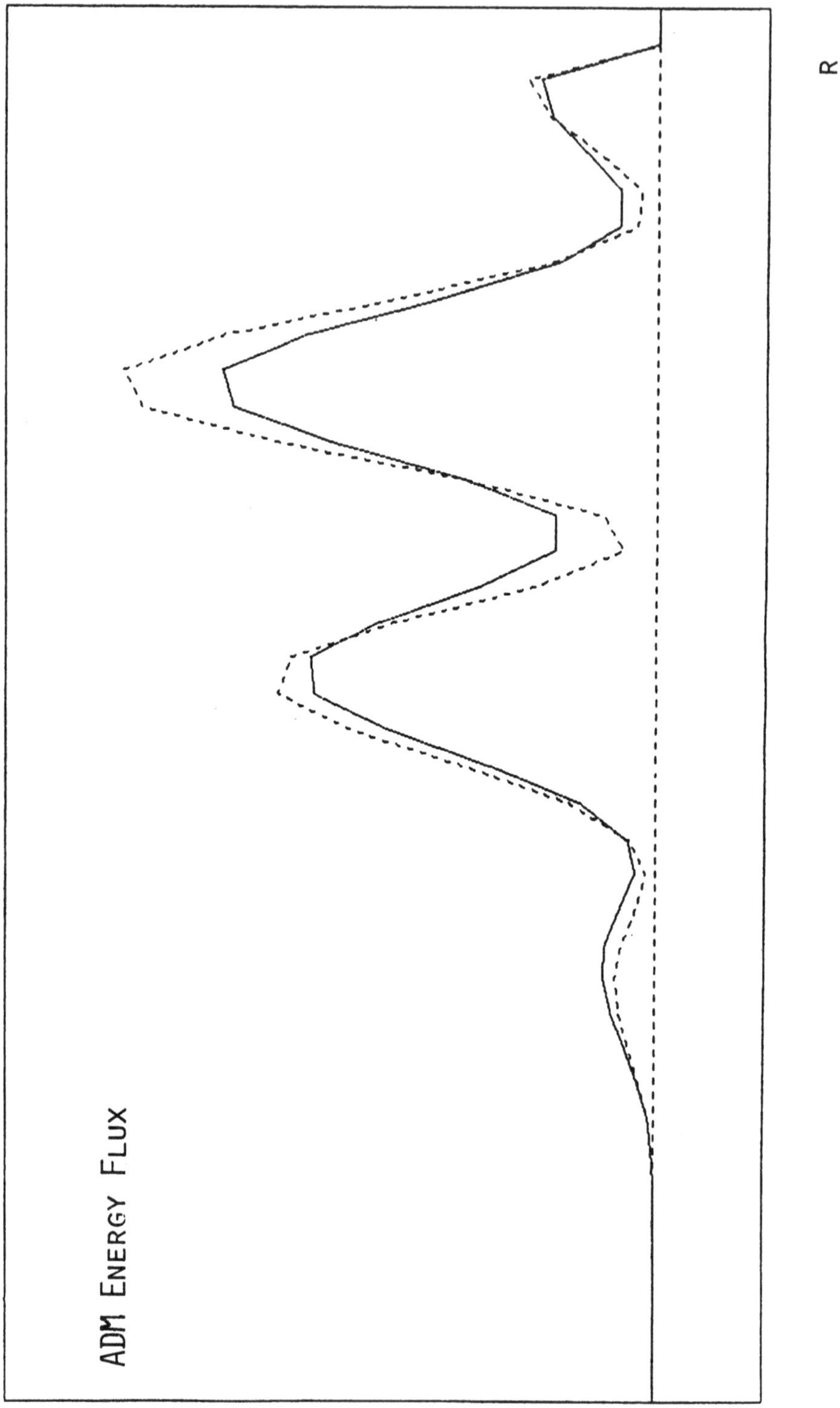

FIGURE 3

ADM ENERGY FLUX

R

As a first step of 3D numerical relativity we here show the time evolution of the Teukolsky wave with 80^3 grids. The basic equations we use are Eqs.(2.19) to (2.23) . How to solve initial value equations is shown by Oohara and Nakamura in this volume. As for the method of finite difference we use a leap frog method and at present no viscosity terms are introduced. Since each numerical boundary in this case is defined by such as x=const, θ and ϕ at the numerical boundary are not constants. Therefore we can not use the numerical boundary condition as Eq.(2.34) in principle. However we use here a simple boundary condition similar to Eq.(2.34) as

$$\frac{\partial(rQ)}{\partial x} = \frac{x}{r}\frac{\partial(rQ)}{\partial t}. \tag{3.1}$$

The above outgoing wave condition is derived assuming the dependence of the quantity Q on θ and ϕ is weak.

We computed the same problem as the previous chapter in (x,y,z) coordinate using 80^3 grids. The CPU time for whole the evolution which has typically 3000 time steps is about three hours. In Fig.3 we show the time evolution of energy flux. In each figure, a solid line shows computed ADM energy luminosity as a function of the distance from the center. Here ADM energy luminosity means the surface integral of the ADM energy flux at the constant radius. A dashed line shows the ADM energy luminosity estimated from the analytic solutions. Fig.3 shows globally solid lines agree with dashed lines. This means that the propagation of the gravitational waves are successfully simulated by our 3D metric code including all the non-linear terms of the Einstein equations. Fig.3 also shows there is no artificially reflected wave at the numerical boundary, which means the numerical outer boundary condition (3.1) has practically no problem.

Fig.3 tells us that the error of the numerical calculation is typically 10% . If we have a machine with 2Gbytes, we can double the number of grid points in each direction. Since our finite difference has second order accuracy, we will easily obtain a few % error code in near future when we have a 2Gbytes-2GFLOPS machine.

4. 3D Hydrodynamical Code

To construct a 3D numerical relativity code in (x,y,z) coordinate we need an accurate hydrodynamics code in (x,y,z) coordinate. In the hydrodynamics code using grids , one of the major problems is how to treat the advection terms. Leblanc has recently found a simple second order method for advection. If one uses his method, details of which will be explained by Clincy et al. in this volume, both a box type and Gaussian type density distribution is successfully transported. Therefore we adopt Leblanc's method for the advection terms.[3]

To express shock waves we use the tensor artificial viscosity terms as

$$Q_{ij} = \rho l^2 div\mathbf{v}\left(\frac{\partial v^i}{\partial x^j} + \frac{\partial v^j}{\partial x^i} - \frac{2}{3}\delta_{ij} div\mathbf{v}\right) \quad if \quad div\mathbf{v} \leq 0. \tag{4.1}$$

We use $P\delta_{ij} + Q_{ij}$ as pressure tensor where P is the gas pressure. The above artificial viscosity terms with Leblanc's transport term expresses a one dimensional shock tube problem quite accurately. In 3D problem, we have checked a point explosion. We have performed a numerical experiment of a hydrogen bomb explosion with energy $8 \times 10^{20} ergs$ in the air (the adiabatic index is 1.4). Fig.4 shows the density distribution in the equatorial plane (xy plane). In the other planes (yz and zx) also, figures are essentially the same. Thus the explosion is spherically symmetric, which is not a trivial result in the numerical code with (x,y,z) coordinate. The number of grids in this simulation is 100^3. In Fig.5 we show the density distribution as a function of a distance from the center. A solid line shows the numerical result and a dashed line shows the Sedov solution. Since the density enhancement at the shock is 6 in the air, we need more meshes to express this large contrast. However we can say our code can trace the Sedov solution quite well.

5. Coalescence of a Binary Neutron Star

We add the Poisson equation to the above hydrodynamics code to take into account the gravity. Then the basic equations we solve are

$$\frac{\partial \rho}{\partial t} + div(\rho\mathbf{v}) = 0, \tag{5.1}$$

$$\frac{\partial \rho v^i}{\partial t} + \frac{\partial \rho v^i v^j}{\partial x^j} = -\frac{\partial P_{ij}}{\partial x^j} - \rho\frac{\partial \psi}{\partial x^i}, \tag{5.2}$$

$$\frac{\partial \rho\epsilon}{\partial t} + \frac{\partial \rho\epsilon v^j}{\partial x^j} = -P_{ij}e^{ij}, \tag{5.3}$$

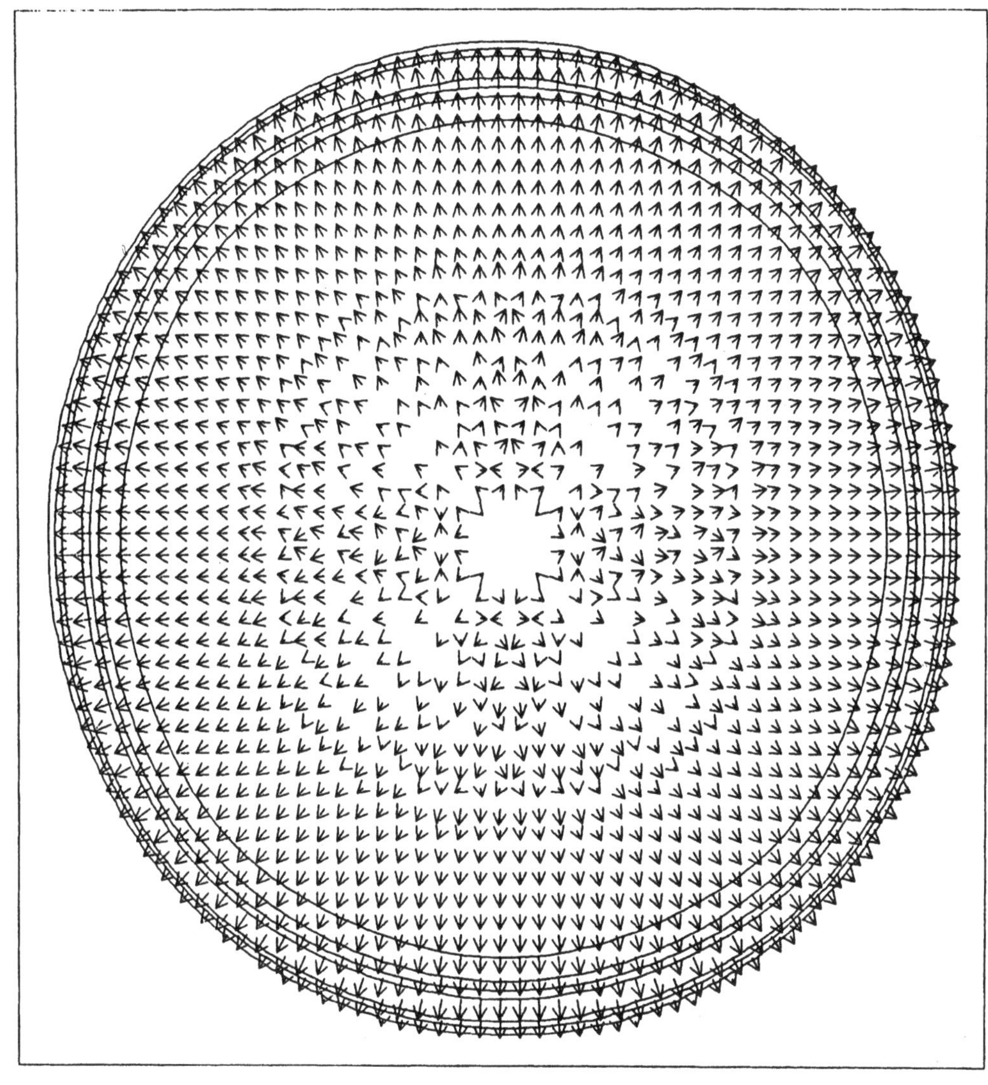

FIGURE 4

FIGURE 5

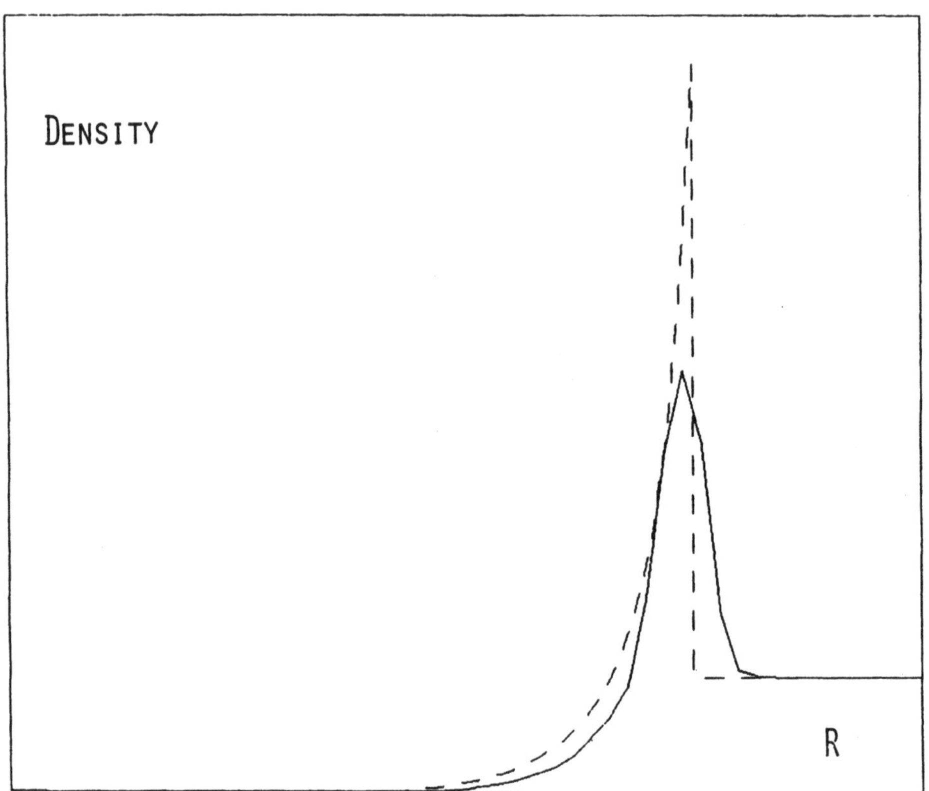

$$P_{ij} = P\delta_{ij} + Q_{ij},$$

$$2e^{ij} = \frac{\partial v^i}{\partial x^j} + \frac{\partial v^j}{\partial x^i},$$

$$P = (\gamma - 1)\rho\epsilon,$$

and

$$\Delta\psi = 4\pi G\rho. \tag{5.4}$$

We solve Poisson equation using ICCG method described by Oohara and Nakamura in this volume. As a check of the code we solved the equilibrium of a polytropic star and confirmed the stability. We also solved collapses of homogeneous ellipsoid with P=0. We compared the results with those obtained by the integral of ordinary differential equations for the length of each axis and confirmed the agreement.

Now let us use the following units as

$$M = M_\odot, \;\; L = \frac{GM_\odot}{c^2} = 1.5km, \;\; T = \frac{GM_\odot}{c^3} = 5 \times 10^{-6} sec.$$

Although the mass of the system in this problem can be freely specified, we use the above units for concreetness. To express the equationof state of high density matter appropriately for neutron stars, we use γ=2. We put two neutron stars of mass M_0=0.7$M_\odot$ and radius r_0 at $y = \pm r_0$ on the y-axis so that two neutron stars just contact each other. As for the initial velocity we assume that

$$v_x = -y\Omega, \;\; v_y = x\Omega \; and \; \Omega = 0.5\sqrt{\frac{GM_0}{r_0^3}}q,$$

where q is a parameter to specify the total angular momentum of the system. In the above initial condition the system is rigidly rotating with respect to the origin x=y=z=0. We solved four models with 70^3 grids. They are specified by

Model A: $r_0 = 6$ and q=0.25

Model B: $r_0 = 6$ and q=0.5

Model C: $r_0 = 6$ and q=0.8

Model D: $r_0 = 6$ and q=0

The CPU time needed was 3 hours for each model with 5000 time steps up to t=200 (1ms). As an example of the simulation we show several snapshots of Model B in Fig.6.

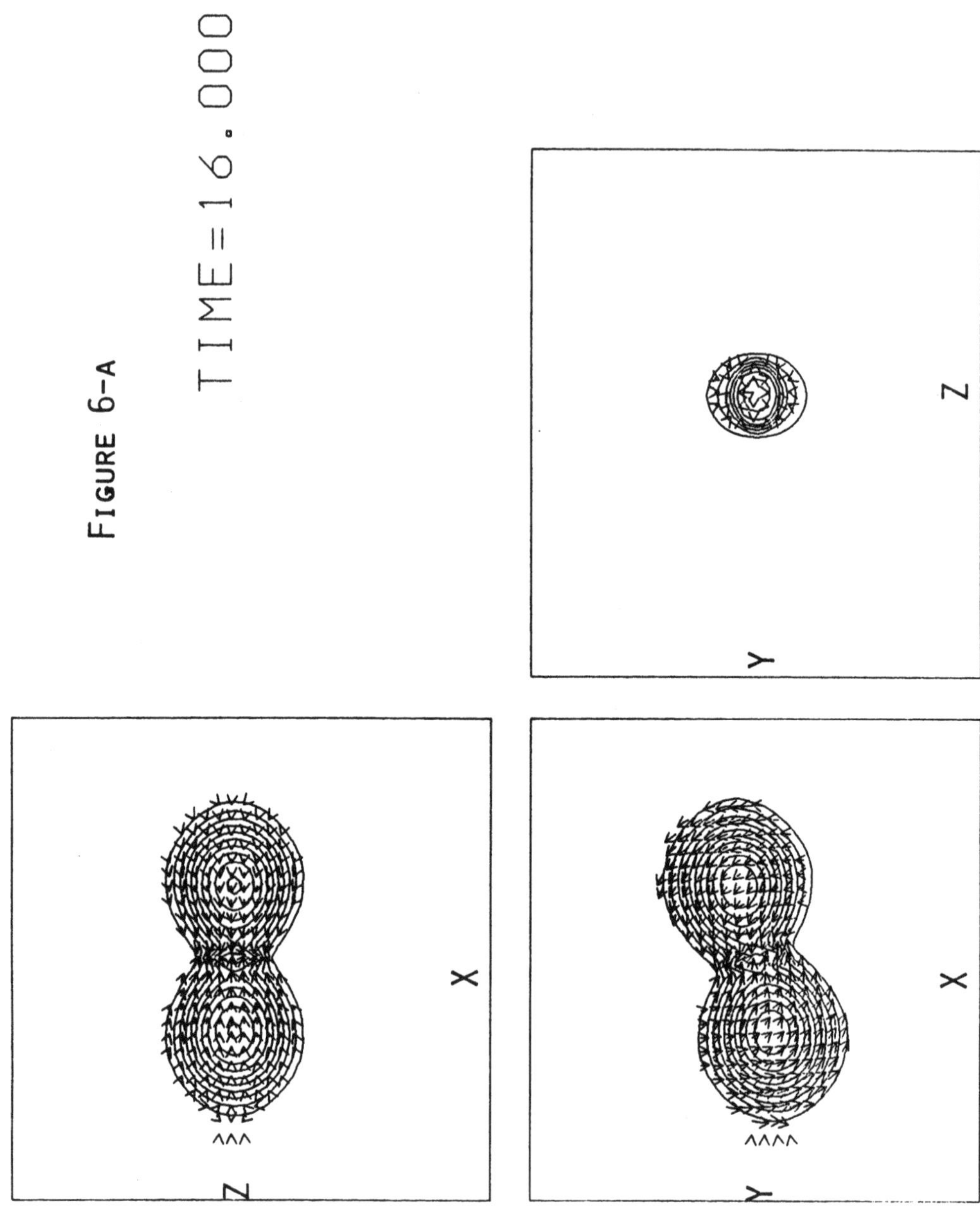
FIGURE 6-A
TIME=16.000
Y
Z
Z
X
Y
X

FIGURE 6-B

TIME=23.999

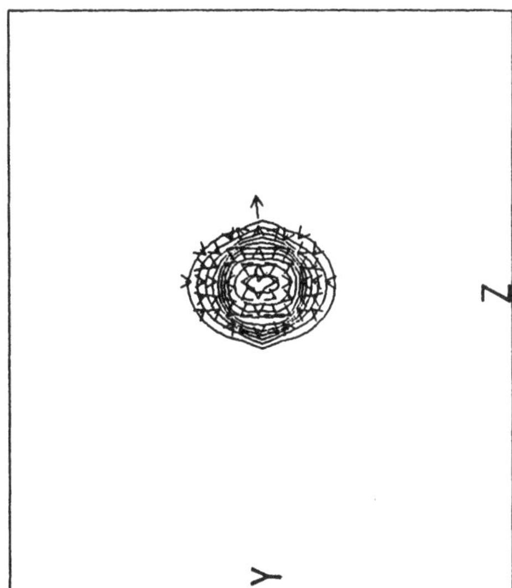

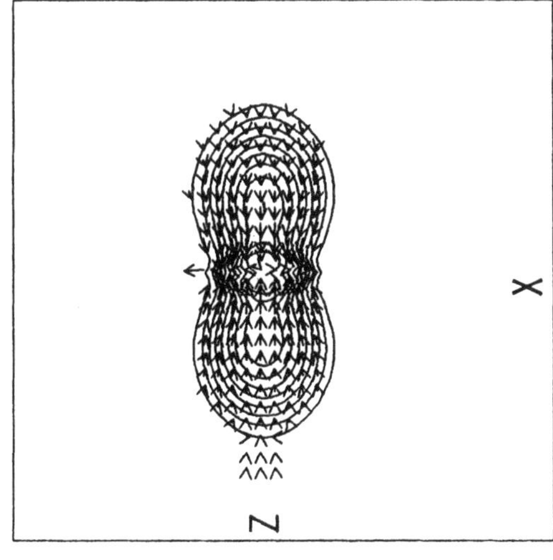

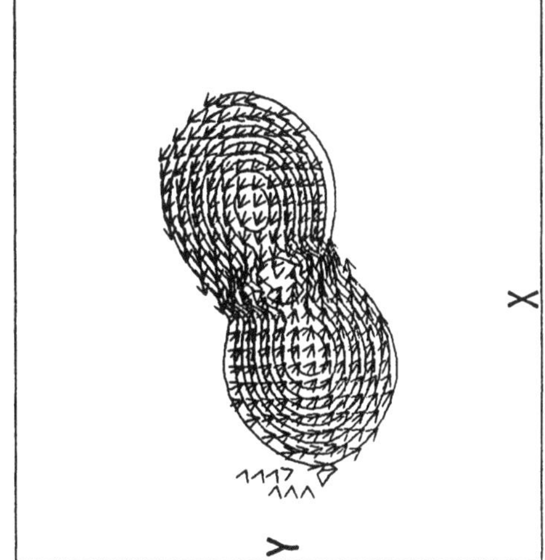

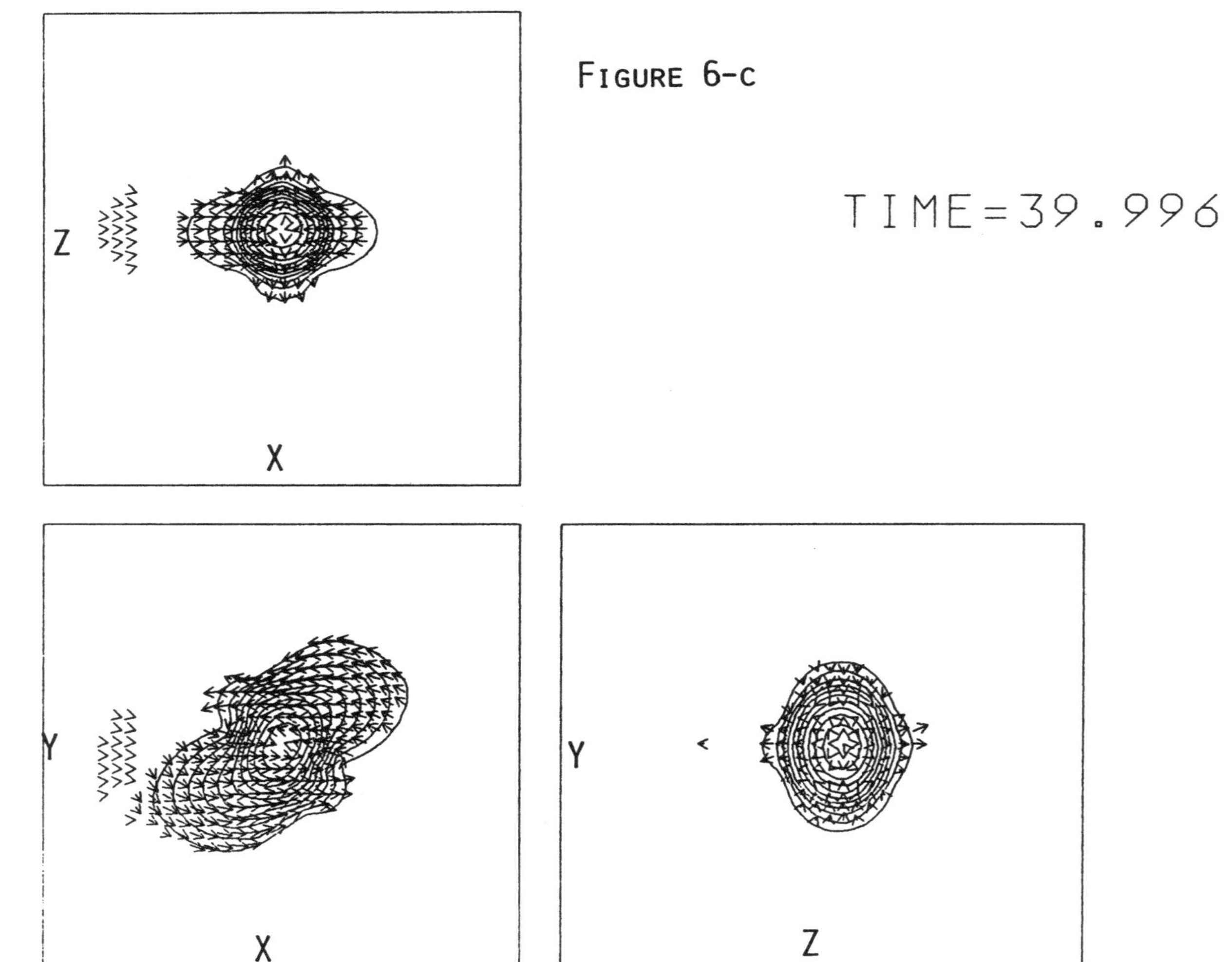

FIGURE 6-C

FIGURE 6-D

TIME=55.993

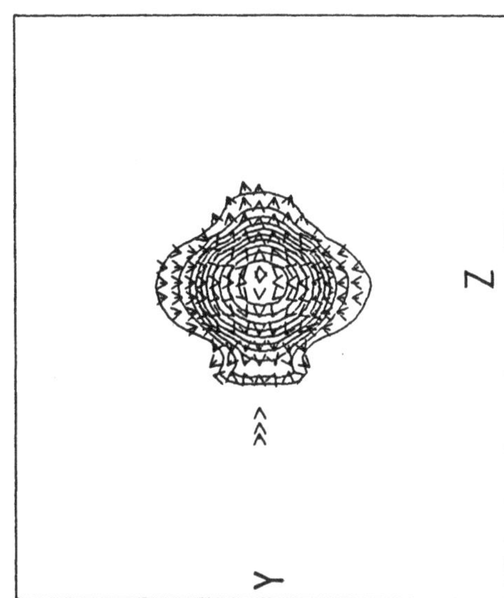

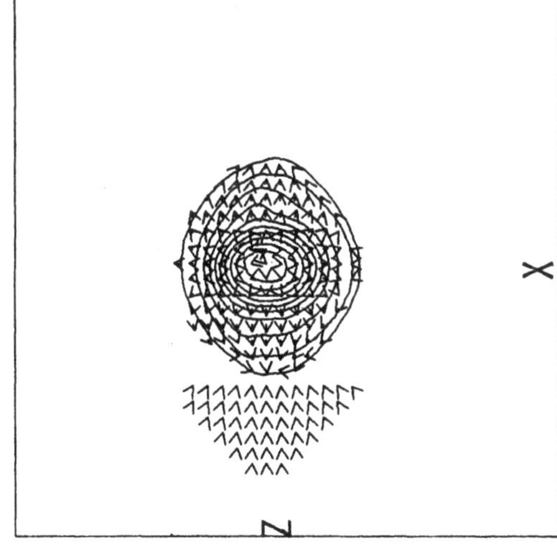

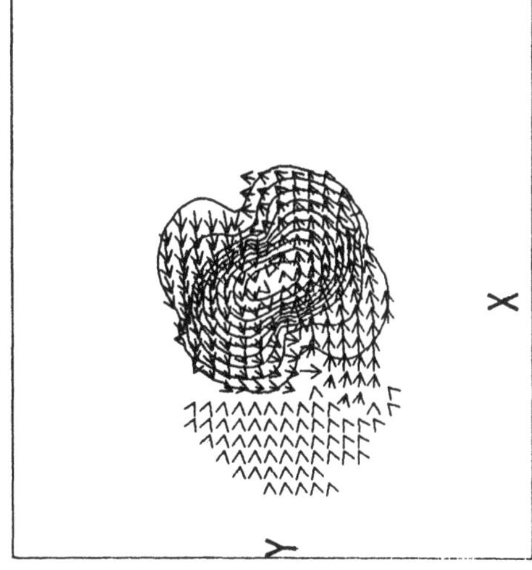

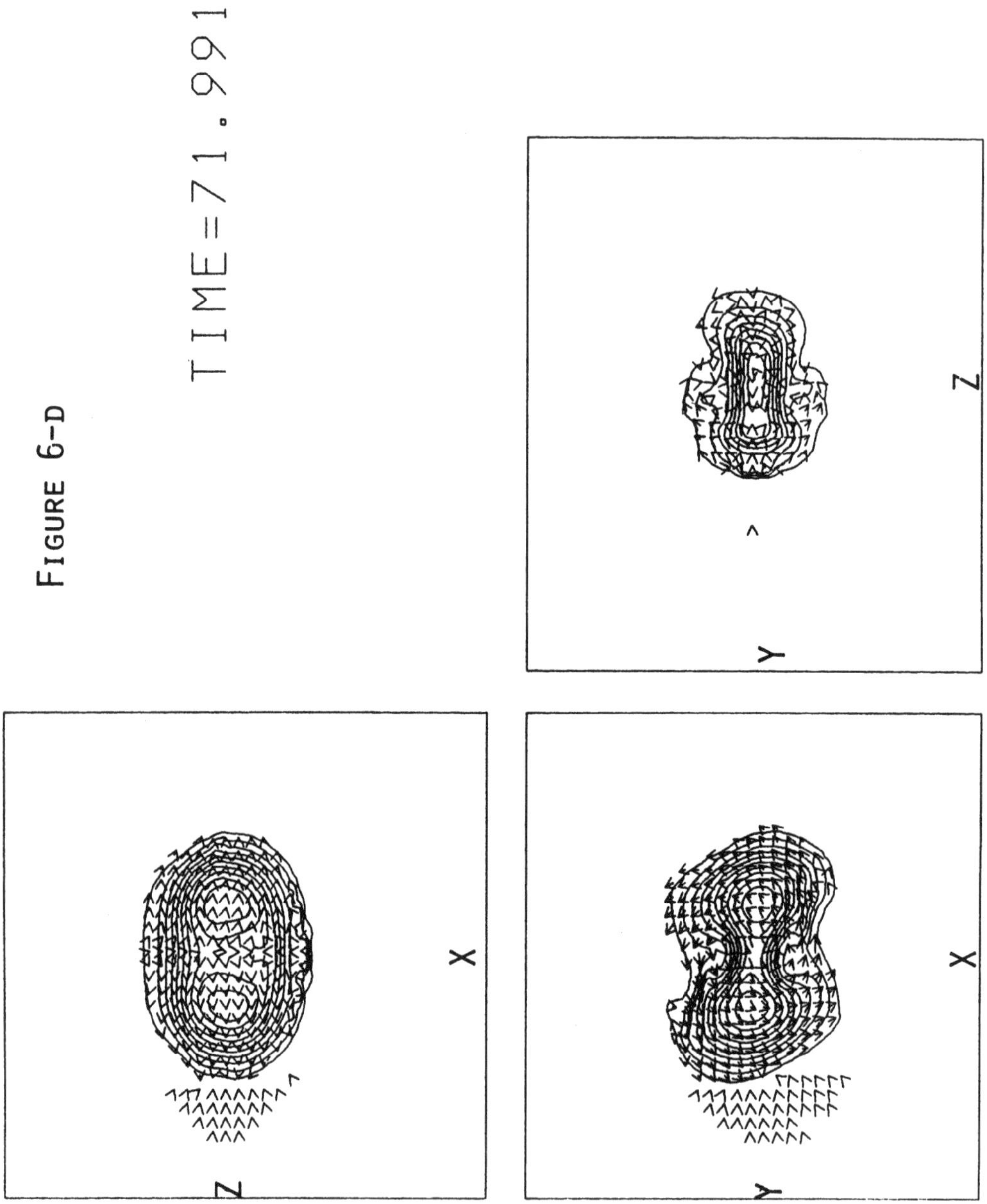

FIGURE 6-D

TIME=71.991

Z X

Y X

Y Z

FIGURE 6-E

TIME=87.988

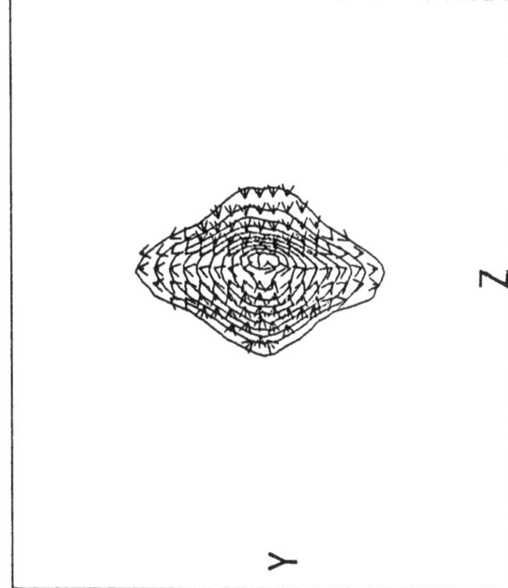

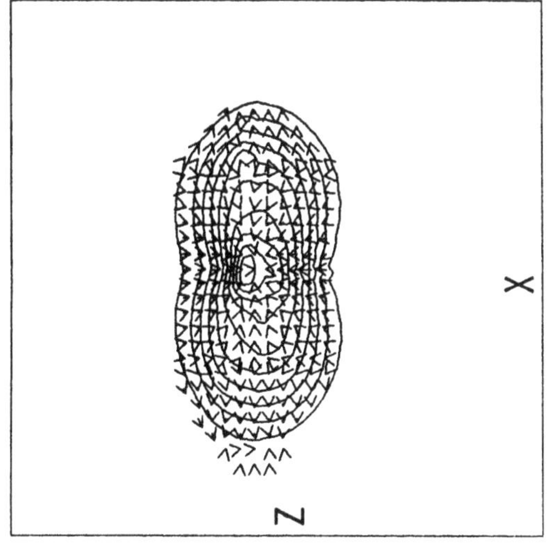

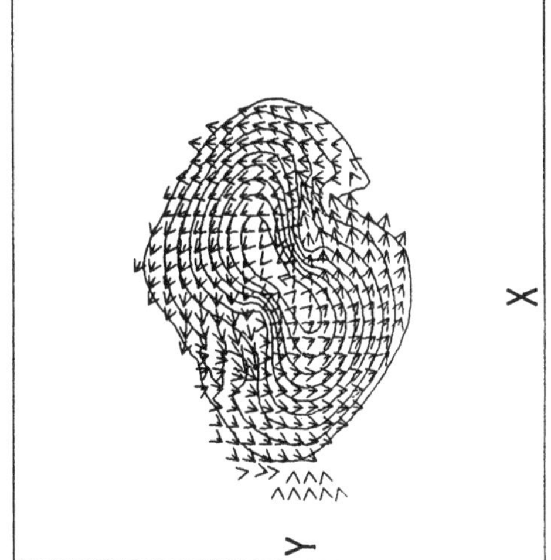

We also estimated the amount of gravitational radiation emitted using the quadrupole formula. For this purpose we need to carry out the third time derivatives of quadrupole. Finn in this volume has shown a good method to estimate the third time derivative. Following his method we use two methods. Using the continuity equation we can show the first time derivative of the quadrupole moment can be reduced to

$$\dot{D}_{ij} = \int \rho(x^i v^j + x^j v^i)dV. \tag{5.5}$$

To obtain the third time derivative we perform the second time derivative of the above quantity numerically. We call the energy flux obtained by this method FLUX1.

Using the equations of motion we can show the second time derivative of the quadrupole moment can be written as

$$\ddot{D}_{ij} = \int (2\rho v^i v^j + 2P\delta_{ij} - \rho(\frac{\partial\psi}{\partial x^i}x^j + \frac{\partial\psi}{\partial x^j}x^i))dV. \tag{5.6}$$

Although the integrand in the above expression contains ψ, we do not need to integrate up to infinity because ρ is multiplied to the potential terms. To obtain the third derivatives we perform the first time derivative of the above quantity numerically. We call the energy flux obtained by this method FLUX2. Note that for the luminosity we use the traceless part of the third time derivatives of D_{ij} in reality. We checked both methods for the collapse of homogeneous ellipsoid and compared results to the semi analytic ones from the integration of the ordinary differential equations. We found the results are satisfactory.

In Fig.7 we show the luminosity of the gravitational waves by both methods for Model B. One can see that FLUX1 is noisy due to the higher numerical time derivative while FLUX2 is very smooth. A very interesting point is FLUX2 seems to be a very mean of FLUX1. The total energy radiated up to the end of numerical simulations is 1.42×10^{-3} in our unit, that is, about 0.1% of the rest mass of the system. For Models A and C the total energy radiated are 0.96×10^{-3} and 1.05×10^{-3}, respectively, while for non-rotating Model D it is only 0.22×10^{-3}. We have also simulated the coalescences of more compact binary systems up to t=100. They are

Model E: $r_0 = 3$ and q=0.25

Model F: $r_0 = 3$ and q=0.5

Model G: $r_0 = 3$ and q=0.8

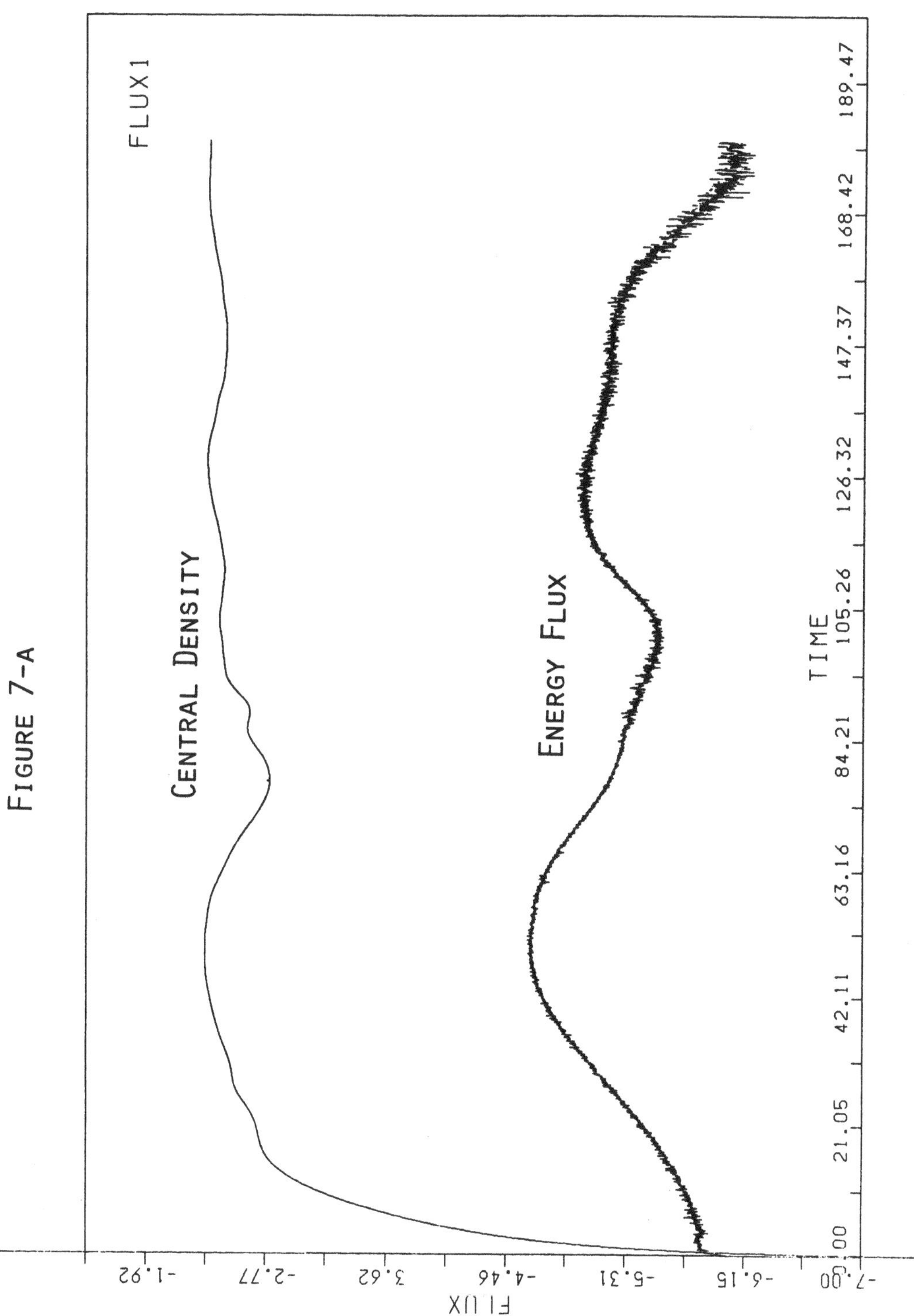
FIGURE 7-A
CENTRAL DENSITY
ENERGY FLUX
FLUX1
TIME
FLUX
0.00
21.05
42.11
63.16
84.21
105.26
126.32
147.37
168.42
189.47
-1.92
-2.77
3.62
-4.46
-5.31
-6.15
-7.00

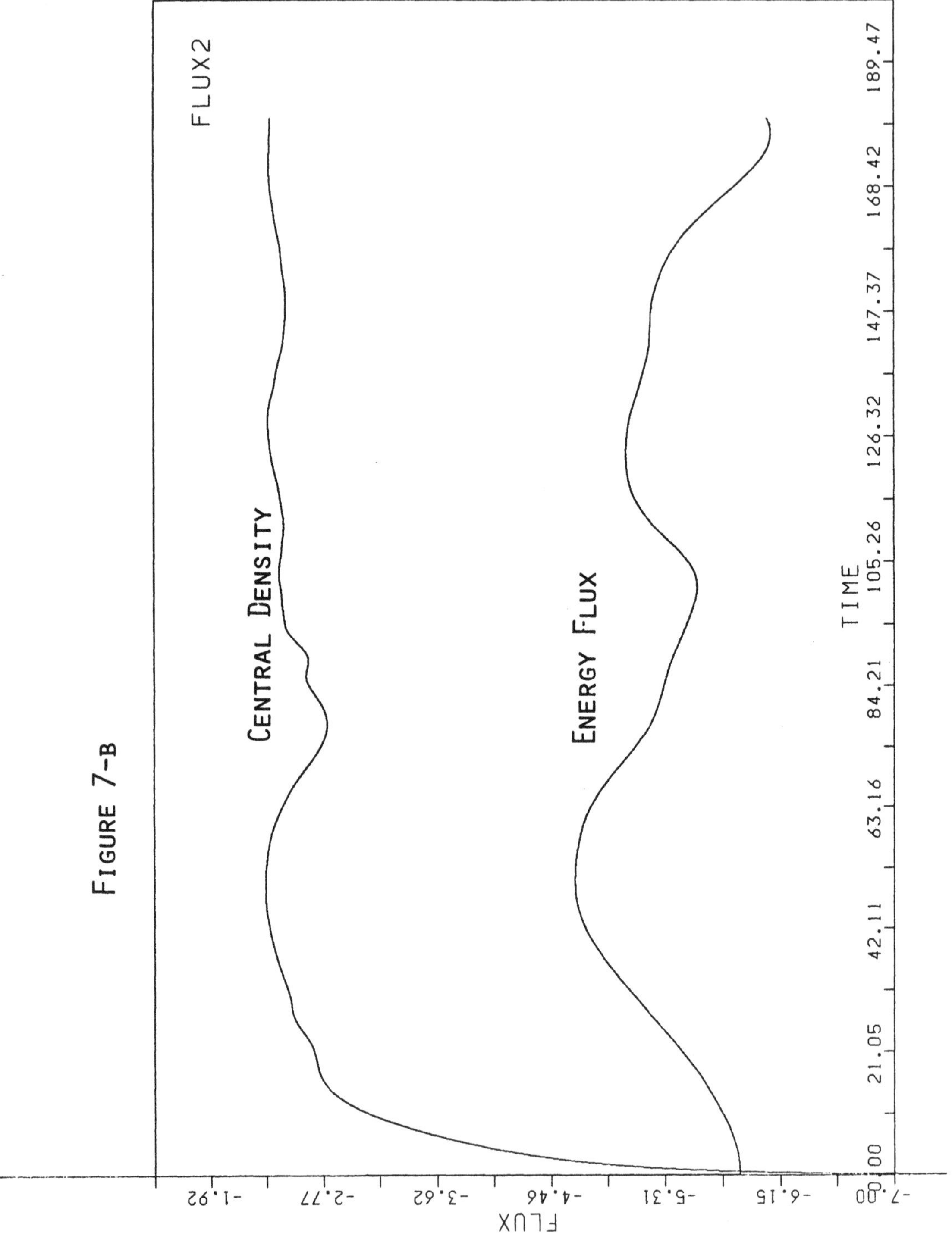
FIGURE 7-B
FLUX2
CENTRAL DENSITY
ENERGY FLUX
TIME
0.00
21.05
42.11
63.16
84.21
105.26
126.32
147.37
168.42
189.47
FLUX
-1.92
-2.77
-3.62
-4.46
-5.31
-6.15
-7.00

For Models E, F and G the total energy liberated up to t=100 are 1.4×10^{-2}, 1.7×10^{-2} and 1.3×10^{-2}, respectively. This is because the increase of the energy by an order of magnitude is due to the increase of the initial Ω by a factor of 3. The increase of Ω overcomes the decrease of the moment of inertia.

In reality Models E, F and G are unrealistically compact systems. But if we consider the radiation reaction in Models A, B and C, the angular momentum will be lost eventually since the luminosity at the final stage of Models A, B and C are rather large as is seen from Fig.7. We can expect the increase of the luminosity due to the loss of the angular momentum for the same reason as in Models E, F and G. One may consider Models E, F and G as the future of Models A, B and C qualitatively. In this sense, to take into account the radiation reaction in our code is an urgent matter.

In the framework of the quadrupole formula, the radiation reaction potential is expressed by the fifth time derivatives of the quadrupole moment. A simple method to take into account the radiation reaction is to estimate this fifth derivatives and put them to the equations of motion although there are theoretical questions on applying simple fifth derivatives to our systems such as those on the gauge conditions and the existense of many time constants.[4] Here we show, however, it is possible to estimate the fifth time derivatives if one solves two more Poisson equations.

Let us consider the time derivative of Eq.(5.6) . If we use the equation of motion, the continuity and the energy equation , all the time derivatives except for $\dot{\psi}$ can be converted to the spatial derivatives. So if we can estimate $\dot{\psi}$ accurately, we will have a smooth third derivatives of D_{ij}. We show that this is possible. $\dot{\psi}$ obeys

$$\Delta\dot{\psi} = 4\pi G\dot{\rho}. \tag{5.7}$$

Using the continuity equation we have

$$\Delta\dot{\psi} = -4\pi G div(\rho \mathbf{v}). \tag{5.8}$$

The source term of this equation has no time derivative. Therefore if we solve this Poisson equation we can expect the smooth solution.

For the fourth derivatives we repeat the above procedure. Again the source term for $\ddot{\psi}$ has no time derivatives if one uses the equation of motion. In reality we coded this method to our hydrodynamics code. We checked the fifth time derivatives of the quadrupole moment for the collapse of the homogeneous ellipsoid. We compared the results with those obtained from the solution of ordinary differential equations and confirmed the results are satisfactory. Then we performed preliminary computations with back reaction potentials for Models E, F and G with 50^3 grids. No numerical instabilities appeared. Thus simulations using finer grids are urgent. However needless to say we need much more CPU time since we solve three Poisson equations for each time step.

We will publish the details of the results presented in this chapter in a separate paper.

6. Conclusion

In this article we showed the metric code and hydrodynamics code in (x,y,z) coordinate. We showed many tests for both codes and at present there are no serious difficulties. We are now ready to combine two codes to construct a *general* general relativistic code in numerical relativity. We believe that we can open the door to the kingdom of 3D numerical relativity

This work was partly supported by a Grant-in-Aid for Scientific Research of Ministry of Education, Science and Culture (62540188).

REFERENCES

1. T.Nakamura, Prog.Theor.Phys.**72** (1984) 746.
2. T.Nakamura, K.Oohara and Y.Kojima, Prog.Theor.Phys.supple. **90**(1987)1-218.
3. Leblanc, private communication.
4. Anderson, private communication in this conference

NON-AXISYMMETRIC ROTATING GRAVITATIONAL COLLAPSE AND GRAVITATIONAL RADIATION

Richard F. Stark
Institute for Theoretical Physics,
University of California, Santa Barbara,CA, 93106, U.S.A.

ABSTRACT: We describe a preliminary approach taken to solve numerically the full non-axisymmetric coupled Einstein/hydrodynamic equations, and in particular to obtain directly the gravitational wave emission from non-axisymmetric rotating gravitational collapse. The complete equations solved are given, and the approach employed summarised.

I Introduction:

In this talk we give the equations solved and summarise the formalism and methods that have been used to construct a three (spatial) dimensional non-axisymmetric code for the solution of the complete coupled Einstein and hydrodynamic equations. The main aim of this work is to evolve non-axisymmetric rotating gravitational collapse (in particular to a black hole) and to study *directly* the gravitational wave emission from such an evolution. (The problem of the non-axisymmetric collision of two neutron stars and their resulting gravitational emission may also be studied). Thus we are interested in the transverse traceless asymptotic waveforms as a function of (retarded) time for the even and odd polarization states, as well as the radiation efficiency or fraction of mass converted to gravitational radiation. Although it has been completed, the code we discuss here has not yet been extensively tested, so that this talk should be regarded only as a preliminary description of work still in progress. It remains to be seen whether the approach described here will prove adequate.

The present work is a generalisation of an axisymmetric (rotating) code already completed and used to study in particular gravitational wave emission from the collapse of rotating polytropes to black holes. (For details of the code see Stark & Piran, 1987; for collapse results see Stark & Piran, 1985, 1986). That work has shown that the gravitational emission from axisymmetric (rotating) gravitational collapse to a black hole: (i) has a waveform shape with a characteristic signature corresponding to the excitation of the lowest normal modes of the black hole formed (see *Fig.1*a) (ii) has a rather low efficiency: less than 7×10^{-4} of the mass of the star is converted to gravitational radiation. To what extent can we expect similar behaviour from non-axisymmetric collapse to a black hole?

Fig.1 (a) Waveforms computed from the axisymmetric collapse of rotating polytropes (with dimensionless angular momenta $a \equiv J/M^2$ as shown) to a black hole. (Stark & Piran, 1985, 1986). (b) Perturbation waveform from the infall of a rotating dust ring (mass μ) onto a Schwarzschild black hole (mass $M \gg \mu$). (Nakamura, Oohara & Kojima, 1987). (See Stark & Piran, 1987 for sign convention used).

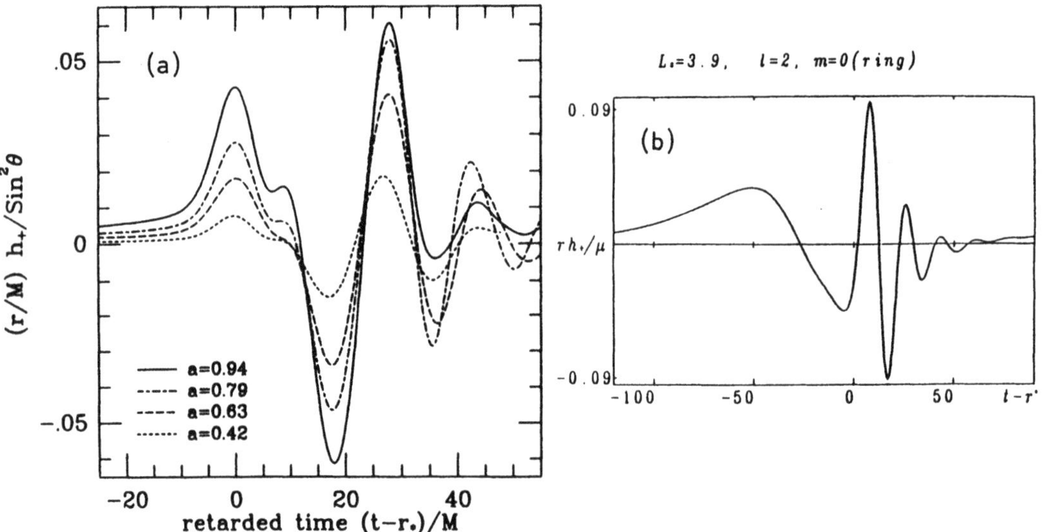

The characteristic waveform of the axisymmetric collapse results because the collapsing star radiates most of its emission at a retarded time when it is less than $3M$ in size (with M the stellar mass). The star is then already smaller in size than the peak in the potential which determines the black hole normal mode structure, resulting in their characteristic excitation. The lowest axisymmetric black hole normal mode frequencies are very nearly independent of the black hole angular momentum (Detweiler, 1980), and this is reflected in the computed collapse waveforms having a characteristic shape that does not depend much on the rotation of the collapsing star (see *Fig.1*a; the amplitude of the waveform does, of course, depend on the rotation). In non-axisymmetric collapse, we may well expect a similar process except that (a) the frequencies of the non-axisymmetric modes may become significantly split at high rotations; (b) there may be significant radiation prior to the actual black hole formation because of a bar mode instability.

With regard to the radiation efficiency, the scaling up of approximate perturbation studies suggests that non-axisymmetric collapse may be at least an order of magnitude more efficient. Previous experience suggests that such scaling arguments work remarkably well. *Fig.1a* shows the waveform obtained from the full scale calculation of the axisymmetric collapse to a black hole of initially pressure reduced rotating polytropes (Stark & Piran, 1985, 1986). In comparison, *Fig.1b* shows the waveform from an approximate perturbation study (Nakamura, Oohara & Kojima, 1987) of a rotating dust ring (of mass μ) falling in from infinity onto a Schwarzschild black hole (mass $M \gg \mu$) with a specific angular momentum comparable to that of the collapsing stars. The collapsing dust ring mimics the flattening in the equatorial plane of the collapsing polytrope. The agreement in the waveform shape is remarkable. One also finds very near agreement in the amplitudes on scaling up μ to the reduced mass. (Both efficiencies then also agree at a few $\times 10^{-4}$).

In Section II of this talk we discuss the formalism we have used and the choice of variables to be computed. In Section III we give a complete listing of the final form of the coupled Einstein/hydrodynamic equations which are solved. Section IV discusses questions to do with regularity and expansions near the origin and polar axis. Section V briefly summarises the main differences which are numerically encountered in the non-axisymmetric code as compared to the axisymmetric one; and we end with a brief outlook in Section VI.

II Formalism and Approach:

We extend to three spatial dimensions the approach used in the (rotating) axisymmetric calculations (Stark & Piran, 1985, 1986, 1987). Thus we use the '3+1' ADM formalism (Arnowitt, Deser & Misner, 1962) and evolve the spatial metric, extrinsic curvature and hydrodynamic variables from one selected spatial hypersurface to another. We employ the radial gauge (first suggested in this context by Eardley in unpublished work and fully developed for the axisymmetric case by Bardeen & Piran (1983), herafter denoted as BP); we use spherical polar coordinates (r, θ and ϕ) and (following BP) a mixed hypersurface condition consisting of a linear combination of maximal and polar slicing. As shown by BP, such a combination is necessary in order to preserve the origin boundary conditions. Many of the equations in this gauge turn out to be parabolic (rather than elliptic), and we are able to read off the outgoing transverse traceless waveform directly from the evolved metric variables without being swamped by additional gauge dependent contributions.

As far as possible we have tried to directly generalise the numerical methods we have employed in the axisymmetric code (Stark & Piran, 1987) and have simply included the azimuthal ϕ dependence for all the geometrical and hydrodynamic variables. As would be expected some differences and complications do arise, mainly due to the normalization of the variables; the periodicity requirement in ϕ (in contrast to θ); the mixed (θ,ϕ) derivatives; the origin and polar axis boundary conditions and the regularity of the variables there. These questions are discussed further below.

We will limit ourselves to non-axisymmetric configurations which:
(i) Possess a stable rotation axis where $U^\theta = 0$ (where U^α is the matter 4-velocity).
(ii) Possess only even azimuthal matter distributions
(iii)Possess an (equatorial) reflection symmetry plane.

We choose our coordinates so that the polar axis ($\theta = 0$) corresponds to the rotation axis; the reflection symmetry plane corresponds to the equator; and the intersection of these corresponds to the origin ($r = 0$). Conditions (i),(ii) and (iii) ensure that there is no recoil of the star and means that we do not need to introduce a (spatially) constant global time dependent shift vector to keep track of the star. This leads to some simplifications in the origin boundary conditions. Condition (i) also allows us to maintain some (but not all) of the variable normalizations (i.e., division by factors of $\sin\theta$) which we used in the axisymmetric case. Condition (iii) simplifies the boundary conditions and allows us to concentrate our attention on one hemisphere only. In principle these conditions can be easily lifted if necessary.

Fig.2 Part of the additional non-axisymmetric terms contained in the lapse equation as directly generated by MACSYMA. In this figure the labels (x,y,z) are used (for typographical convenience) for (r,θ,ϕ) and are not to be confused with 'cartesian' coordinates.

$$
- \frac{2\, bc^2\, nc\, (xic_z)^2}{x^2 \sin^2(y)} - \frac{2\, bc^2\, nc_z\, xic\, xic_z}{x^2 \sin^2(y)} - \frac{4\, etc_z\, nc\, xic\, xic_z}{x^2 \sin^2(y)} - \frac{4\, ac_z\, bc^2\, nc\, xic\, xic_z}{ac\, x^2 \sin^2(y)} - \frac{bc^2\, nc_{zz}\, xic^2}{x^2 \sin^2(y)} - \frac{etc_z\, nc_z\, xic^2}{x^2 \sin^2(y)}
$$

$$
- \frac{etc_{zz}\, nc\, xic^2}{x^2 \sin^2(y)} - \frac{2\, ac_z\, etc_z\, nc\, xic^2}{ac\, x^2 \sin^2(y)} - \frac{2\, ac_{zz}\, bc^2\, nc\, xic^2}{ac\, x^2 \sin^2(y)} - \frac{nc_{zz}}{bc^2\, x^2 \sin^2(y)} + \frac{etc_z\, nc_z}{bc^4\, x^2 \sin^2(y)} - \frac{ac_z\, nc_z}{ac\, bc^2\, x^2 \sin^2(y)}
$$

$$
+ \frac{2\, ac_z\, etc_z\, nc}{ac\, bc^4\, x^2 \sin^2(y)} - \frac{2\, ac_{zz}\, nc}{ac\, bc^2\, x^2 \sin^2(y)} - \frac{bc^2 \frac{dnc}{dcos(y)}\, xic_z}{x^2} - \frac{3 \frac{detc}{dcos(y)}\, nc\, xic_z}{x^2} - \frac{2 \frac{dac}{dcos(y)}\, bc^2\, nc\, xic_z}{ac\, x^2} - \frac{2\, bc^2\, nc \frac{d^2 xic}{dcos(y)\, dz}}{x^2}
$$

$$
- \frac{etc_z\, nc \frac{dxic}{dcos(y)}}{x^2} - \frac{2\, ac_z\, bc^2\, nc \frac{dxic}{dcos(y)}}{ac\, x^2} - \frac{\frac{detc}{dcos(y)}\, nc\, xic_z}{x^2} - \frac{\frac{dac}{dcos(y)}\, bc^2\, nc_z\, xic}{ac\, x^2} - \frac{2\, bc^2 \frac{d^2 nc}{dcos(y)\, dz}\, xic}{x^2} - \frac{etc_z \frac{dnc}{dcos(y)}\, xic}{x^2}
$$

$$
- \frac{ac_z\, bc^2 \frac{dnc}{dcos(y)}\, xic}{ac\, x^2} - \frac{2 \frac{dac}{dcos(y)}\, etc_z\, nc\, xic}{ac\, x^2} - \frac{2 \frac{d^2 etc}{dcos(y)\, dz}\, nc\, xic}{x^2} + \frac{etc_{zz}\, nc}{bc^4\, x^2 \sin^2(y)} - \frac{2\, (etc_z)^2\, nc}{bc^6\, x^2 \sin^2(y)} - \frac{bc^2\, nc_z \frac{dxic}{dcos(y)}}{x^2}
$$

The axisymmetric equations derived by BP were derived by hand. In extending these to the non-axisymmetric case, we were not brave enough to attempt this by hand, and instead relied on symbolic computer algebra (MACSYMA) to obtain these. As a side result we were able to verify that the axisymmetric results given by BP and used in the axisymmetric code were indeed correct and without error! MACSYMA is used to obtain simplified algebraic expressions for the 3-D equations (e.g., *Fig.2* shows some of the additional non-axisymmetric terms for the lapse). MACSYMA allows for the output of terms directly in FORTRAN ready for inclusion in the code, but we have not used this feature. This is because it is most important to group terms carefully in a manner best fitted to the grid structure and finite differencing used. This is particularly important in getting angular factors arranged optimally with respect to the derivatives. In principle this could be automated, but would require more effort than is needed to do this by hand. MACSYMA is used however to check afterwards that no errors have been made in this procedure.

The 3-dimensional spatial metric in the radial gauge (BP) is determined by the three functions h_{rr}, η, and ξ, each functions of (r, θ, ϕ). The off-diagonal (r,θ) (r,ϕ) components of the metric are zero, leading to a considerable simplification of the equations, while the angular spatial determinant is fixed to its euclidean value so that $r^2 \sin\theta d\theta d\phi$ measures the element of proper area. The 4-dimensional metric has the form:

$$ds^2 = -(N^2 - N^i N_i)dt^2 - 2N_i dx^i dt + h_{rr}dr^2 + r^2 B^{-2} d\theta^2 + r^2 B^2 (\sin\theta d\phi + \xi d\theta)^2$$

with $B^2 = 1 + \eta$, and where η and ξ are respectively the even and odd dynamical degrees of freedom in the sense that they tend at large radii to the even and odd transverse traceless amplitudes. The time coordinate t labels the hypersurfaces, N is the lapse, and N^i the shift vector. The foliation consists of a combination of polar and maximal slicing: we set $trace(K_i{}^j) = (1 - C(r))K_r{}^r$ where $0 \leq C(r) \leq 1$ is a chosen smooth function of radius such that $C(0) = 1$ and $C(r > r_0) = 0$ (corresponding to maximal slicing at the origin and polar slicing for radii larger than a chosen transition radius r_0 - see below for the actual form used).

The actual variables computed during the evolution are chosen carefully to be suitable normalized, and to have appropiate numerical behaviour. In the axisymmetric case it proved possible to normalise many of the variables by factors of $\sin\theta$. In the non-axisymmetric case, we must allow for radiation emission along the polar axis and this restricts the possibility of such normalization to only some of the variables. Our assumptions (i)-(iii) above of a stable rotation axis without recoil enable us to normalise two of the extrinsic curvature variables ($K_{(1)(2)}$, $K_{(1)(3)}$ - see below); the shift vector; the matter velocity; as well as the matter momentum density.

Our final choice for computed variables are:

Spatial metric: η; ξ; $A = \sqrt{h_{rr}}$.

Extrinsic curvature: K_1; K_2; K_3; K_+; $K_\times$.

Lapse and shift: N; $\beta^r = N^r/r$; $G = N^\theta/\sin\theta$; N^ϕ.

Hydrodynamic variables: $D = NAU^0 n$; $H = (NAU^0)^\Gamma \epsilon$;

$$S_r = NAU^0(n+\epsilon+p)U_r; \qquad S_x = NAU^0(n+\epsilon+p)U_\theta/\sin\theta;$$

$$S_\phi = NAU^0(n+\epsilon+p)U_\phi/\sin^2\theta;$$

$$V^r = U^r/U^0; \qquad V^x = U^\theta/(U^0\sin\theta); \qquad V^\phi = U^\phi/U^0$$

where $x = \cos\theta$; n, ϵ, p are respectively the co-moving number density, energy density and pressure of the matter, and U^α its 4-velocity. D is the coordinate density; H the coordinate energy density (multiplied by $(NAU^0)^{\Gamma-1}$); S_r, S_θ, S_ϕ the radial and angular momentum densities; and V^r, V^x, V^ϕ the radial and angular matter velocites. The assumed polytropic equation of state is:

$$p \propto n^\Gamma \quad (p = (\Gamma-1)\epsilon) \qquad (\Gamma = \text{adiabatic index})$$

The components of the extrinsic curvature we use are based on orthonormal components of K_{ij}:

$$K_1 = K_{(1)(1)}; \qquad K_2 = K_{(1)(2)}/\sin\theta; \qquad K_3 = K_{(1)(3)}/\sin\theta;$$

$$K_+ = \frac{1}{2}(K_{(3)(3)} - K_{(2)(2)}); \qquad K_\times = K_{(2)(3)}$$

where the orthonormal components $K_{(i)(j)}$ $(i,j = 1,2,3)$ are the components of K_{ij} projected onto the orthonormal triad of basis vectors:

$$e^i_{(1)} = [A^{-1},0,0]; \quad e^i_{(2)} = [0, B/r, -\xi B/(r\sin\theta)]; \quad e^i_{(3)} = [0,0,1/(Br\sin\theta)].$$

The scheme we employ is partially constrained, in that the dynamic metric variables (η, ξ) and the five independent extrinsic variables are all time evolved, while the radial metric function, A, is obtained from the Hamiltonian constraint. (In the axisymmetric code we were able to use $\xi_{,r}$ rather than ξ in order to eliminate gauge freedom arising from asymptotic rotation, but this does not appear to be possible in the non-axisymmetric case).

III The form of the equations solved:

All together we need to solve 17 coupled PDE in the four variables (t, r, $x = \cos\theta$, ϕ). It is perhaps worth remarking that only about 10% of the effort goes into solving the hydrodynamic equations, so that this code should probably be considered a 'geometrical' code. The equations we solve are:

(i) Field evolution equations: $Y_{,t} + r\beta^r Y_{,r} - G(1-x^2)Y_{,x} + N^\phi Y_{,\phi} = XY + S$

where:

Y	X	S
A	$-(r\beta^r)_{,r} - NK_1$	0
η	$-((1-x^2)G_{,x} + 2NK_+)$ $-(N^\phi{}_{,\phi} + 2\xi G_{,\phi})$	$-((1-x^2)G_{,x} + 2NK_+)$ $-(N^\phi{}_{,\phi} + 2\xi G_{,\phi})$
ξ	$(1-x^2)G_{,x} + N^\phi{}_{,\phi}$	$(1-x^2)N^\phi{}_{,x} - 2NK_\times B^{-2}$ $+(\xi^2 - B^{-4})G_{,\phi}$
K_1	0	$2N(1-x^2)(K_2^2 + K_3^2)$ $-2AB(1-x^2)K_2\beta^r{}_{,x}$ $-A^{-1}r^{-2}[(1-x^2)NB^2A_{,x}]_{,x}$ $-A^{-1}[A^{-1}N_{,r}]_{,r} + 2Nr^{-1}A^{-3}A_{,r} + NKK_1$ $-\frac{1}{2}NA^{-2}[B^{-4}\eta_{,r}^2 + B^4\xi_{,r}^2]$ $+2AB^{-1}\beta^r{}_{,\phi}(K_3 - B^2\xi K_2)$ $-\{[A_{,\phi}N(B^2\xi^2 + B^{-2})]_{,\phi}/(1-x^2)$ $+(A_{,x}B^2N\xi)_{,\phi} + (A_{,\phi}B^2N\xi)_{,x}\}/(Ar^2)$ $+4\pi N\{[(1-x^2)r^{-2}(S_{(2)}^2 + S_{(3)}^2) -$ $S_{(1)}^2]/T + p - E\}$
K_2	$-xG + NK$ $+N[\frac{1}{2}CK_1 + K_1 + K_+]$	$2NK_3K_\times + Br^{-1}[A^{-1}N_{,r}]_{,x}$ $+r(AB)^{-1}G_{,r}[\frac{1}{2}CK_1 + K_1 + K_+]$ $-BA^{-2}r^{-2}[NA]_{,x}$ $+\frac{1}{2}\{[A^{-1}N\eta_{,r}]_{,x} -$ $2xNA^{-1}\eta_{,r}/(1-x^2)\}/(Br)$ $+\{AN[(B^2\xi)_{,r\phi} - \eta_{,\phi}\xi_{,r} + \eta_{,r}\xi_{,\phi}]$ $+(B^2\xi)_{,r}(N_{,\phi}A - NA_{,\phi})$ $+2B^2\xi[AN_{,r\phi} - A_\phi N_r -$ $(AN)_{,\phi}/r]\}/(2A^2Br(1-x^2))$ $+AB^{-1}K_\times\beta^r{}_{,\phi}/(1-x^2) + K_3G_{,\phi}/B^2$ $-8\pi NS_{(1)}S_{(2)}r^{-1}/T$

Y	X	S
K_3	$-xG + NK$ $+N[-K_1 - \frac{1}{2}CK_1 + K_+]$	$-2NK_2K_\times - r(AB)^{-1}G_{,r}K_\times$ $-\{[NB^4A^{-1}\xi_{,r}]_{,x} -$ $2xNB^4A^{-1}\xi_{,r}/(1-x^2)\}/(2Br)$ $+\{A^2N^r{}_{,\phi}(2K_+ + K - 3K_1)$ $-\frac{1}{2}A[B^4NA^{-1}(\xi^2)_{,r}]_{,\phi}$ $+AB^2[NA^{-1}B^{-4}\eta_{,r}]_{,\phi} - B^4N\xi_{,r}\xi_{,\phi}$ $-2A[N_{,r}A^{-1}]_{,\phi} +$ $2A^{-1}(NA)_{,\phi}/r\}/(2ABr(1-x^2))$ $-K_2G_{,\phi}/B^2 - 8\pi NS_{(1)}S_{(3)}r^{-1}/T$
K_+	NK	$-N[2K_\times^2 + (1-x^2)(K_2^2 + K_3^2)]$ $-r(AB)^{-1}G_{,r}(1-x^2)K_2$ $-[r^2NA^{-1}\eta_{,r}]_{,r}/(2AB^2r^2)$ $+\frac{1}{2}B^2r^{-2}(1-x^2)[N_{,xx} + NA^{-1}A_{,xx}]$ $+\frac{1}{2}NA^{-2}[B^{-4}\eta_{,r}^2 + B^4\xi_{,r}^2]$ $+\{(AN_{,\phi\phi} + NA_{,\phi\phi})(B^2\xi^2 - B^{-2})$ $+B^2(1-x^2)^2[(AN)_{,\phi}(\xi/(1-x^2))_{,x}$ $-(AN)_{,x}(\xi/(1-x^2))_{,\phi}$ $+2(\xi/(1-x^2))(AN_{,x\phi} + NA_{,x\phi})]\}$ $/(2Ar^2(1-x^2))$ $-2K_\times G_{,\phi}B^{-2} - AK_3N^r{}_{,\phi}/(Br)$ $-4\pi N(1-x^2)(S_{(3)}^2 - S_{(2)}^2)r^{-2}/T$
$K_\times$	$NK + 2NK_+$	$-\frac{1}{2}A^{-1}(Br)^{-2}[r^2NB^4A^{-1}\xi_{,r}]_{,r}$ $+rG_{,r}(1-x^2)K_3/(AB)$ $+\{-(1-x^2)[B^2(AN)_{,\phi}]_{,x} + 2B^2x(AN)_{,\phi}$ $+(1-x^2)[(B^2(AN)_{,x})_{,\phi} +$ $2B^2(AN_{,x\phi} + NA_{,x\phi})]$ $+2B^2\xi[AN_{,\phi\phi} + NA_{,\phi\phi}]\}/(2AB^2r^2(1-x^2))$ $+2K_+G_{,\phi}/B^2 - AK_2N^r{}_{,\phi}/(Br)$ $-8\pi N(1-x^2)S_{(2)}S_{(3)}r^{-2}/T$

where:

$S_{(1)} = A^{-1}S_r$; $S_{(2)} = B(S_x - \xi S_\phi)$; $S_{(3)} = B^{-1}S_\phi$ and $T = (NAU^0)^2(n + \epsilon + p)$

(ii) The shift vector β^r, G, N^ϕ:

$$G_{,r} = \frac{1}{2} r^{-1}(AB)^2 \lambda^{-1}\{[(1-x^2)G]_{,xx} + D(N^\phi) + S\}$$

where:

$$D(N^\phi) = -[N^\phi{}_{,x\phi} + 2B^2 r\xi A^{-2} N^\phi{}_{,r}]$$

$$S = -[-C(NK_1)_{,x} + 4N(AB)^{-1}K_2 + 4A^{-1}BN\xi K_3]$$

$$(\lambda = 1 + B^4\xi^2)$$

$$N^\phi{}_{,r} = \frac{1}{2} r^{-1} A^2 B^{-2} N^\phi{}_{,\phi\phi}/(1-x^2) + D'(G) + S'$$

where:

$$D'(G) = -\frac{1}{2} A^2 B^{-2} r^{-1}[(1-x^2)G]_{,x\phi}/(1-x^2) - \xi G_{,r}$$

$$S' = -2NK_3 A r^{-1} B^{-1} - \frac{1}{2} A^2 B^{-2} C r^{-1}[NK_1]_{,\phi}/(1-x^2)$$

$$\beta^r = \frac{1}{2}[(1-x^2)G]_{,x} + \frac{1}{2} NCK_1 - \frac{1}{2} N^\phi{}_{,\phi}$$

(iii) The lapse N:

$$CA^{-1}(A^{-1}N_{,r})_{,r} + 2A^{-2}r^{-1}N_{,r} + CA^{-1}r^{-2}[(1-x^2)AB^2N_{,x}]_{,x}$$
$$+ (1-C)r^{-2}[(1-x^2)B^2N_{,x}]_{,x} = S + \Lambda N + D''(N)$$

with:

$$S = rC_{,r}[\beta^r - (r_0)_{,t}/r_0]K_1 + 2(1-C)AB(1-x^2)\beta^r{}_{,x}K_2$$
$$+ 2(1-C)AB^{-1}\beta^r{}_{,\phi}(\xi B^2 K_2 - K_3)$$

$$\Lambda = (1-C)\{r^{-2}(1-A^{-2}) - \frac{1}{2} r^{-2}[(1-x^2)\eta]_{,xx}$$
$$+ \frac{1}{4} A^{-2}[B^{-4}\eta_{,r}^2 + B^4\xi_{,r}^2]\} + \cdots$$

$$\cdots(1+C)(K_+^2+K_\times^2)$$
$$+(3C-1)(K_2^2+K_3^2)(1-x^2)+C(1-\frac{1}{4}C)(1+C)K_1^2$$
$$-\frac{1}{2}(1-C)\{[(1-x^2)(B^2\xi)_{,\phi}]_{,x}$$
$$+(1-x^2)(B^2\xi_{,x\phi}+2\eta_{,x}\xi_{,\phi}+\xi\eta_{,x\phi})$$
$$+[B^{-2}]_{,\phi\phi}+[B^2\xi^2]_{,\phi\phi}\}/(r^2(1-x^2))$$
$$+4\pi\{CE+(2+C)p+[C(1-x^2)r^{-2}(S_{(2)}^2+S_{(3)}^2)$$
$$+(2-C)S_{(1)}^2]/T\}$$
$$D''(N)=-A^{-1}r^{-2}[(N_{,\phi}AB^2\xi)_{,x}+(N_{,x}AB^2\xi)_{,\phi}$$
$$-(1-C)B^2\xi(A_{,x}N_{,\phi}+A_{,\phi}N_{,x})]$$
$$-A^{-1}r^{-2}\{[AN_{,\phi}(B^{-2}+B^2\xi^2)]_{,\phi}$$
$$-(1-C)A_{,\phi}N_{,\phi}(B^{-2}+B^2\xi^2)\}/(1-x^2)$$

with:

$$E=(n+\epsilon)(NU^0)^2+p[(NU^0)^2-1]$$

($C(r)=[1-(r/r_0)^2]^n$ in the present work with $r_0=r_0(t)$, n chosen values).

(iv) The metric function A (Hamiltonian constraint):

$$A_{,r}=\frac{1}{2}A^2r^{-1}\{[(1-x^2)B^2A_{,x}]_{,x}+[(B^2\xi^2+B^{-2})A_{,\phi}]_{,\phi}/(1-x^2)$$
$$+[B^2\xi A_{,\phi}]_{,x}+[B^2\xi A_{,x}]_{,\phi}+A(S_1-1)+A^{-1}(S_2+1)\}$$

with:

$$S_1=r^2\{8\pi E+(K_+^2+K_\times^2)+(K_2^2+K_3^2)(1-x^2)$$
$$+\frac{3}{4}C^2K_1^2+CKK_1\}+\frac{1}{2}[(1-x^2)\eta]_{,xx}$$
$$+\frac{1}{2}\{[B^2\xi^2+B^{-2}]_{,\phi\phi}/(1-x^2)+[(1-x^2)B^2\xi]_{,x\phi}/(1-x^2)$$
$$+[\xi(B^2)_{,x}]_{,\phi}+[B^2\xi_{,\phi}]_{,x}\}$$

$$S_2=\frac{1}{4}r^2[B^{-4}\eta_{,r}^2+B^4\xi_{,r}^2]$$

(v) The hydrodynamic equations:

$$D_{,t} + r^{-2}[r^2 D V^r]_{,r} - [(1-x^2)DV^x]_{,x} + [DV^\phi]_{,\phi} = 0$$

$$H_{,t} + r^{-2}[r^2 H V^r]_{,r} - [(1-x^2)HV^x]_{,x} + [HV^\phi]_{,\phi} =$$
$$-(\Gamma-1)H\{r^{-2}[r^2 V^r]_{,r} - [(1-x^2)V^x]_{,x} + [V^\phi]_{,\phi}\}$$

$$(S_\phi)_{,t} + r^{-2}[r^2 S_\phi V^r]_{,r} - [(1-x^2)^2 S_\phi V^x]_{,x}/(1-x^2) + [S_\phi V^\phi]_{,\phi} =$$
$$-(ANp_{,\phi} + \mathrm{Source}_\phi)/(1-x^2)$$

$$(S_r)_{,t} + r^{-2}[r^2 S_r V^r]_{,r} - [(1-x^2)S_r V^x]_{,x} + [S_r V^\phi]_{,\phi} = -ANp_{,r} - \mathrm{Source}_r$$

$$(S_x)_{,t} + r^{-2}[r^2 S_x V^r]_{,r} - [(1-x^2)^{\frac{3}{2}} S_x V^x]_{,x}/(1-x^2)^{\frac{1}{2}} + [S_x V^\phi]_{,\phi} =$$
$$ANp_{,x} + \mathrm{Source}_x$$

where: $(L = r, x, \phi)$

$$\mathrm{Source}_L = S^0 N N_{,L} + S_r (r\beta^r)_{,L} + (1-x^2)^{\frac{1}{2}} S_x[(1-x^2)^{\frac{1}{2}} G]_{,L}$$
$$+(1-x^2)S_\phi N^\phi{}_{,L} + \frac{1}{2} S_0^{-1}\{S_r^2 A^{-2}{}_{,L} + (1-x^2)S_x^2[B^2 r^{-2}]_{,L}$$
$$+(1-x^2)^2 S_\phi^2[r^{-2}(1-x^2)^{-1}(B^{-2} + B^2\xi^2)]_{,L}$$
$$-2(1-x^2)^{\frac{3}{2}} S_x S_\phi[(1-x^2)^{-\frac{1}{2}} B^2 \xi r^{-2}]_{,L}\} \qquad (L = r, x, \phi)$$

$$(NS^0)^2 = (nNAU^0)^2 + A^{-2}S_r^2 + r^{-2}(1-x^2)B^2 S_x^2$$
$$+r^{-2}(1-x^2)[B^{-2} + B^2\xi^2]S_\phi^2$$
$$-2(1-x^2)r^{-2}B^2\xi S_\phi S_x$$

(this condition being the normalization of the 4-velocity $U^\alpha U_\alpha = -1$)

$$(NU^0)^2 = (NS^0)/[(n+\epsilon+p)A]$$

$$V^r = r\beta^r + A^{-2}S_r/S^0$$

$$V^x = G + \{r^{-2}B^2 S_x - r^{-2}B^2\xi S_\phi\}/S^0$$

$$V^\phi = N^\phi + (1-x^2)^{\frac{1}{2}}\{r^{-2}B^2 S_\phi - r^{-2}B^2\xi S_x\}/S^0$$

and where the comoving quantities n, p, ϵ are:

$$n = D/(NAU^0); \qquad p = (\Gamma-1)H/(NAU^0)^\Gamma; \qquad \epsilon = p/(\Gamma-1)$$

with the assumed polytropic equation of state $p \propto n^\Gamma$.

IV Regularity and expansions:

An important problem associated with using spherical coordinates for a non-axisymmetric code is ensuring that regularity conditions are maintained near the polar axis and the origin. In our previous axisymmetric code, this was automatically accomplished by a suitable normalization of variables. As we have seen, such normalization is not generally possible for the non-axisymmetric case, making it necessary to impose these regularity conditions directly on the variables. An example of the need for this can be seen by considering some of the additional non-axisymmetric contributions to the source terms; e.g., that for K_2 which, near the polar axis, contains the contribution proportional to:

$$[(B^2\xi)_{,r\phi} - B^2_{\phi}\xi_{,r} + B^2{}_{,r}\,\xi_{,\phi} - 2B^2{}_{,r}]/\sin^2\theta$$

Near the polar axis numerator and denominator both tend to zero. η and ξ however are constrained by regularity to possess a particular ϕ dependence near the polar axis which ensures that this source contribution limits correctly. To see this, transform the spatial metric to 'cartesian' coordinates $x = r\sin\theta\cos\phi$, $y = r\sin\theta\sin\phi$, $z = r\cos\theta$. For $\theta = 0$, the angular part of the spatial metric contributes:

$$dx^2[B^{-2}\cos^2\phi + B^2(\sin^2\phi + \xi^2\cos^2\phi - 2\xi\sin\phi\cos\phi)]$$

$$+dy^2[B^{-2}\sin^2\phi + B^2(\cos^2\phi + \xi^2\sin^2\phi + 2\xi\sin\phi\cos\phi)]$$

$$+2dxdy[B^{-2}\sin\phi\cos\phi + B^2(-\sin\phi\cos\phi + \xi(\cos^2\phi - \sin^2\phi) + \xi^2\sin\phi\cos\phi)]$$

These 'cartesian' components must, by uniqueness, be independent of ϕ, which allows us to deduce the following ϕ dependence on the polar axis:

$$B^2 = B^2_{\phi=0}\cos^2\phi + [B^{-2} + B^2\xi^2]_{\phi=0}\sin^2\phi - [B^2\xi]_{\phi=0}\sin 2\phi$$

$$2\xi B^2 = [2\xi B^2]_{\phi=0}\cos 2\phi - [B^{-2} + B^2\xi^2 - B^2]_{\phi=0}\sin 2\phi$$

Similar arguments may be used for the behaviour of the extrinsic curvature.

To obtain expansions near the origin, we follow the regularity condition suggested by BP, namely that the components of a quantity in the 'cartesian' (x,y,z) system of coordinates should be expandable in non-negative powers of (x,y,z). Consider first the spatial metric. We express the spherical components of the spatial metric in terms of the cartesian ones. Next we use the conditions $h_{r\theta} = 0$; $h_{r\phi} = 0$ and the angular determinant condition $(h_{\theta\theta}h_{\phi\phi} - h^2_{\theta\phi}) = r^4\sin^2\theta$ to express the spherical components in terms of h_{xx}, h_{yy}, h_{xy} (say) only (these being even functions of $\cos\theta$ with our assumption of equatorial reflection symmetry). Expanding h_{xx}, h_{yy}, h_{xy} and requiring all angular factors in the denominator be cancelled, one deduces the following origin expansions:

$$B^2 = 1 + r^2\{\sin^2\theta f(\phi) + \cos^2\theta(b\cos 2\phi - b'\sin 2\phi)\} + \cdots$$

$$\xi = r^2 \cos\theta(b \sin 2\phi + b' \cos 2\phi) + \cdots$$

$$A^2 = 1 + r^2\{a + (1 + \cos^2\theta)f(\phi) - \cos^2\theta(b\cos 2\phi - b' \sin 2\phi)\} + \cdots$$

with:

$$f(\phi) = c\cos 2\phi + d \sin 2\phi + e$$

(a,b,b',c,d,e depending on t only). For unique polar axis values of A, $c = b/2$ and $d = -b'/2$.

For the extrinsic curvature, at the origin the 'cartesian' components K_{xz}, K_{yz} are zero (for equatorial reflection symmetry) and the slicing condition $K = 0$ results in only three independent origin 'cartesian' components. Transforming our orthonormal extrinsic curvature components (listed above) to 'cartesian' components, we find the following extrinsic curvature origin values:

$$K_+(r=0) = \frac{3}{2}k_+ \sin^2\theta + \frac{1}{2}(1+\cos^2\theta)[-k_\times \sin 2\phi + k_- \cos 2\phi]$$

$$K_\times(r=0) = \cos\theta[k_- \sin 2\phi + k_\times \cos 2\phi]$$

$$K_1(r=0) = k_+(1 - 3\cos^2\theta) + \sin^2\theta[-k_- \cos 2\phi + k_\times \sin 2\phi]$$

$$K_2(r=0) = \cos\theta[3k_+ - k_- \cos 2\phi + k_\times \sin 2\phi]$$

$$K_3(r=0) = [k_- \sin 2\phi + k_\times \cos 2\phi]$$

where:

$$k_+ = \frac{1}{2}(K_{xx} + K_{yy})|_{r=0}; \quad k_- = \frac{1}{2}(K_{yy} - K_{xx})|_{r=0}; \quad k_\times = K_{xy}|_{r=0}$$

(k_+, k_-, $k_\times$ depending on t only).

The origin shift vector (spherical) components are found again from expanding the 'cartesian' components, using the equatorial reflection symmetry, and requiring zero 'cartesian' components at the origin (by our assumption of no recoil). Substituting these into the shift vector equations and using the above extrinsic curvature origin values, we can express the origin shift in terms of k_+, k_- and $k_\times$:

$$N^r/r(r=0) = N_0\{k_+(3\cos^2\theta - 1) + (k_- \cos 2\phi - k_\times \sin 2\phi)\sin^2\theta\}$$

$$G(r=0) = N_0\{-3k_+ - k_\times \sin 2\phi + k_- \cos 2\phi\}\cos\theta$$

$$N^\phi(r=0) = N_0\{k_\times - k_- \sin 2\phi - k_\times \cos 2\phi\}$$

These origin values also provide starting values for the outward radial parabolic integration.

V Numerical methods:

We will not attempt to give here a complete description of the numerical methods used, but will only point out some of the differences encountered between the non-axisymmetic and axisymmetric cases. The equations are finite differenced maintaining second order accuracy wherever possible both spatially and in time. As in the axisymmetric code, the variables are spatially staggered on four interlocking grids in the (r, $x = \cos\theta$) plane and in time and also in the azimuthal (ϕ) direction. Unequal spacing and a grid velocity is included both in the r and x grids. (A full account of the axisymmetric numerical methods is given in Stark & Piran, 1987).

The field evolution equations and the hydrodynamic equations are straight forwardly generalised to three dimensions and we simply refer to Stark & Piran (1987) for the numerical methods used. It is in the solution of the parabolic equations that significant differences are encountered.

The parabolic equations for A, G, N^ϕ and the outer lapse (i.e., for $r > r_0$) are solved using iterated Gaussian elimination with forced periodicity fixing in ϕ. To illustrate the method consider the Hamiltonian constraint for A. This has the form:

$$A_{,r} = \frac{1}{2}A^2r^{-1}\{[(1-x^2)B^2A_{,x}]_{,x} + [(B^2\xi^2 + B^{-2})A_{,\phi}]_{,\phi}/(1-x^2)$$
$$+[B^2\xi A_{,\phi}]_{,x} + [B^2\xi A_{,x}]_{,\phi}$$
$$+A(S_1 - 1) + A^{-1}(S_2 + 1)\}$$

On the RHS we have double x, double ϕ and mixed (x,ϕ) angular derivatives. We can treat the finite differenced form of these equations as tridiagonal, in x and ϕ, of the form:

$$b_1A_1 + c_1A_2 = d_1$$
$$a_kA_{k-1} + b_kA_k + c_kA_{k+1} = d_k \quad (k = 2, ..., n-1)$$
$$a_nA_{n-1} + b_nA_n = d_n$$

by employing the following steps:
1) Use the previous timestep values as the zeroth solution.
2) Treat the mixed (x,ϕ) derivatives as part of the source. The $\sin^2\theta$ factor in the double x derivative then results in tridiagonal equations in x.
3) Sweep over (r,x) (Stark & Piran, 1987) treating the ϕ and mixed derivatives as part of the source.
4) Update the coefficients a,b,c which, because of non-linearity, depend on A.
5) Sweep over (r,ϕ) treating the x and mixed (x,ϕ) derivatives as part of the source. Enforce the ϕ periodicity and obtain tridiagonal equations by employing the previous values of A to write:

$$b_1A_1 + c_1A_2 = d_1 - a_1A_0(old)$$

$$a_n A_{n-1} + b_n A_n = d_n - c_n A_{n+1}(old)$$

with (by the ϕ periodicity): $A_0 = A_n$; $A_{n+1} = A_1$
6) Update the coefficients a,b,c which, because of non-linearity, depend on A. Return to 2) and iterate to convergence. (Generally it is found that the iterations are quite rapid, so that the solution is not too time consuming).

The parabolic part of the lapse (i.e., for $r > r_0$) is solved similarly, although this is somewhat easier in that it is linear in the lapse. The angular shift components G and N^ϕ have coupled parabolic equations. These have the form:

$$G_{,r} = \frac{1}{2} r^{-1} (AB)^2 \lambda^{-1} \{[(1-x^2)G]_{,xx} + D(N^\phi) + S\}$$

$$N^\phi{}_{,r} = \frac{1}{2} r^{-1} A^2 B^{-2} N^\phi{}_{,\phi\phi} / (1-x^2) + D'(G) + S'$$

with:

$$D(N^\phi) = -[N^\phi{}_{,x\phi} + 2B^2 r \xi A^{-2} N^\phi{}_{,r}]$$

$$D'(G) = -\frac{1}{2} A^2 B^{-2} r^{-1} [(1-x^2)G]_{,x\phi} / (1-x^2) - \xi G_{,r}$$

This is very different from the axisymmetric case where G is parabolic in r, x independent of N^ϕ, and N^ϕ requires only one straight forward radial integration. To solve the non-axisymmetric coupled case we simultaneously iterate for G and N^ϕ using similar methods as described above for A:
1) Use the previous timestep values of G and N^ϕ as the zeroth solution.
2) Sweep over (r,x) for G treating $D(N^\phi)$ as part of the source.
3) Sweep over (r,ϕ) for N^ϕ treating $D'(G)$ as part of the source.
4) Return to 2) and iterate to convergence.

VI Outlook:

The code at present is undergoing testing, and it remains to be seen whether the approach used here is adequate. The run time per timestep using an 80 by 12 by 16 grid is of the order of 1 to 2 seconds on a Cray XMP-48. It is expected that the initial data for the non-axisymmetric collapse studies will be the same as that used in the axisymmetric code studies: a pressure reduced TOV polytrope with rotation added (Stark & Piran, 1985, 1986, 1987). (The constraint equations for this initial data being easily solved). With sufficient rotation (but still sufficiently low for collapse to a black hole), numerical round-off will be expected to break the axisymmetry and lead to the growth of the unstable non-axisymmetric modes.

Acknowledgment:

This work was supported by NSF grant PHY82-17853 to the Institute for Theoretical Physics.

References:

Arnowitt, R., Deser, S. & Misner, C. (1962). In *Gravitation: Introduction to Current Research*, ed. L.Witten, Wiley, New York.

Bardeen, J.M. & Piran, T. (1983). Physics Reports,**96**,205. (Denoted here as BP).

Detweiler, S. (1980). Ap.J.,**239**,292.

Nakamura, T., Oohara, K. and Kojima, Y. (1987). Prog.Th.Phys.Suppl.,**No.90**.

Stark, R.F. & Piran, T. (1985). Phys.Rev.Lett.,**55**,891.

Stark, R.F. & Piran, T. (1986). In *Proceedings of the fourth Marcel Grossmann Meeting on general relativity*, Ed. R.Ruffini, North Holland, Amsterdam.

Stark, R.F. & Piran, T. (1987). Computer Physics Reports, **5**, 221.

Nonaxisymmetric Neutron Star Collisions: Initial Results Using Smooth Particle Hydrodynamics

Christopher S. Kochanek
Charles R. Evans
Theoretical Astrophysics 130-33
California Institute of Technology
Pasadena, CA 91125, U.S.A.

Abstract. We compare the gravitational radiation generated in the axisymmetric, Newtonian collision of $\Gamma = 2$ polytropes using both finite difference and smooth particle hydrodynamics. The agreement between the two techniques is remarkably good. We then calculate the energy radiated in gravitational waves as a function of pericentric radius R_p for nonaxisymmetric parabolic orbits. The parabolic orbit with the highest radiative efficiency has $R_p \sim R/2$ where R is the radius of the polytrope. This orbit emits five times more radiation than in the axisymmetric case, with the total energy radiated being $\simeq 3.4M(M/R)^{7/2}$.

1: *INTRODUCTION*

Smooth particle hydrodynamics (SPH) is a relatively recent addition to the numerical hydrodynamicists toolbox. Here we wish to apply this technique to an extremely limited problem – approximately computing the quadrupole moments and gravitational radiation waveforms generated by the collision of two neutron stars. We follow the stars through the collision, the passage of the recoil shock through the stars, and the reexpansion of the system. In §2 we briefly describe the SPH method for non-relativistic, self-gravitating hydrodynamics. In §3 we try to convince the reader that SPH can achieve certain limited goals by comparing the results for the head-on collision of two $\Gamma = 2$ polytropes, which can be done using both SPH and a two dimensional, finite difference (FD) method. In §4 we describe some preliminary, low resolution results for non-axisymmetric collisions, and finally in §5 we discuss the future of the study.

2: SPH: A BRIEF INTRODUCTION

SPH is a method which attempts to do a free Lagrangian hydrodynamics calculation by associating the bulk properties of a fluid with individual particles (Gingold and Monaghan, 1977, 1983; Wood, 1981; Monaghan and Lattanzio, 1985; Benz and Hills 1988). The particles carry a mass m_i, an entropy K_i, a position $\vec{x}_i$, and a velocity $\vec{v}_i$. The first problem is to construct a density ρ_i from the masses and positions of the particles. The pressure can be computed given the density and the entropy through the equation of state, in this case a simple adiabatic equation of state, $P = K\rho^\Gamma$ with $\Gamma = 2$. The process of generating a density field depends on a *smoothing length* h which controls the number of particles over which we average. If we assume that within a volume $\sim h^3$ there are a large number of particles, then each particle represents a density spike characterized by its mass and a length scale much smaller than h. Then

$$\rho_i = \int \rho(\vec{x}) W(\vec{x} - \vec{x}_i; h) d^3x \simeq \sum_{j=1}^{N} m_j W(\vec{x}_j - \vec{x}_i; h) \tag{2.1}$$

gives the average density at point x_i. The *kernel*, $W(\vec{x} - \vec{x}_i; h)$, is a function that has compact support over a region with size $\sim h$ and a unit norm $\int W(\vec{x} - \vec{x}_i; h) d^3x = 1$. The choice of the kernel is fairly arbitrary provided it is compact, smooth, and correctly normalized. Among the kernels in use are exponentials, Gaussians, and cubic splines. We have been using a cubic spline of the type described in Monaghan and Lattanzio (1985) [although there are errors in their equations (19) and (21) for the spline].

The hydrodynamic equations are generally broken into advection terms and acceleration terms. The advection terms are automatically satisfied in SPH by keeping the masses of the particles fixed and by moving the particles with the local velocity. The acceleration terms require gradients of the bulk quantities – this is achieved through the use of the kernel. If we want to compute the gradient of the density at point i, we differentiate equation (2.1), and the gradient of the density becomes a gradient of the kernel, so that

$$\vec{\nabla}\rho(x_i) \simeq \sum_{j=1}^{N} m_j \vec{\nabla} W(\vec{x} - \vec{x}_j; h) \tag{2.2}$$

The acceleration terms are comprised of the usual menagerie: pressure acceleration, gravitational acceleration, and an "artificial" viscous acceleration to satisfy shock jump conditions. The pressure terms have been implemented in the manner of Gingold and Monaghan (1982). The gravitational accelerations are computed using an N-body tree code; descriptions of various N-body tree codes can be found in Press (1986) and Barnes (1986). The main advantage of a particle based gravity scheme is that it eliminates the need for a grid on which to solve Poisson's equation, thus leading to a completely gridless calculation. A tree code is used because it allows an order $N \log N$ calculation of the gravitational forces, and because the tree which is constructed to compute the gravitational acceleration also allows rapid identification of the particles needed to compute the pressure and viscous accelerations. We have also included a variable smoothing length and individual particle time steps to increase the dynamic range and efficiency of the program for highly inhomogeneous systems. Tree based SPH codes have also been implemented by Benz *et al.* (1988) and Hernquist and Katz (1988). We have found that the linear viscosity law advocated by Gingold and Monaghan (1988) leads to problems with anomalous entropy generation in regions of nearly smooth flow. Therefore, we implemented the artificial viscosity in our SPH code with a simple generalization of the standard quadratic viscosity law from FD computations.

3: A DIRECT COMPARISON BETWEEN SPH AND FD RESULTS

The reader might well be concerned about the ability of a method such as SPH to correctly calculate the quadrupole moments and gravitational radiation waveforms in the collision of a pair of neutron stars. Fortunately there is an excellent test problem at hand – the head-on collision of two neutron stars. This calculation was done using FD techniques in two independent efforts by Gilden and Shapiro (1984) and Evans (1987). Replicating the highly resolved, symmetry constrained, FD results with a lower resolution, unconstrained, three dimensional calculation during a violent collision surely represents a rigorous test of our ability to calculate bulk quantities. This is not the only test problem which we have used. Other tests include advecting polytropes, radial oscillations of polytropes, the collapse of underpressured polytropes, and the explosion of over pressured polytropes. Note that while the collapse and explosion problems are simple, one dimensional problems for a spherical symmetry FD code, the SPH code knows nothing about spherical coordinates. SPH correctly replicates the bulk properties of the flow even with extremely low numbers of particles, while the fine details of the flow, naturally, require larger particle numbers. The calculations shown in this paper have been done with either 500 or 2000 particles per neutron star. This is, without question, a low resolution study of the problem. However, as we will emphasize later, these studies can be done at significantly higher resolutions.

We have made a direct comparison of SPH and FD waveforms for the Newtonian, axisymmetric, parabolic collision of two $\Gamma = 2$ polytropes. In terms of bulk quantities, such as the center of mass positions and quadrupole moments, the agreement between the two techniques is nearly exact – in fact the difference between the SPH and Evans' (1987) FD waveforms is less than the difference between the two published FD waveforms (see Figure 1abc)! This is true even though we used only 500 particles per cluster in our preliminary survey. Why is this true? The basic answer is that the bulk properties of the fluid in the parabolic collision are "kinematic" structures – they result from correctly transferring energy between kinetic, potential, and internal forms. If the bulk energy transfer is handled correctly, the bulk properties of the flow will follow. At late times the quadrupole moments begin to diverge, but this is almost entirely due to mass flow off the FD grid. The differences remain insignificant if the quadrupole moments for the SPH simulations are computed with the same outer boundary as that used in the FD simulations.

This does not mean that SPH can reproduce the fine detail found in the FD calculations without a significant increase in the number of particles. As can be seen in the contour plots of the pressure computed using the two methods, the contours agree fairly well in the regions where most of the mass lies (see Figure 2). Shocks are largely washed out in the SPH calculation as a consequence of the lower resolution. After the collision there is an outflow of low density material several radii from the collision. In this region SPH (with these low particle numbers) cannot compete with FD. The integrated mass in the outflow is the same (*as are its quadrupole moments!*), but SPH does not have the dynamic range of the FD method, and eventually the discreteness of the SPH solution becomes apparent. However, for computing the gravitational waveforms, the lack of smoothness in the outflowing regions is unimportant provided the amount and distribution of mass in the outflow is approximately correct.

The bottom line is that SPH is well suited for computing the gravitational waves emitted by colliding neutron stars. It is not well suited (for small numbers of particles) for reproducing the fine detail of the FD calculations, because of problems in both resolution and dynamic range. These preliminary studies have been made with many fewer particles than will eventually be possible, and our future studies will be done with significantly higher spatial resolution. However, if you maintain an attitude of healthy skepticism, and consider only the bulk features of the flow as being well represented in these low particle number results, the quadrupole formula results for the waveforms and luminosities are believable at the 10 percent level.

4: NONAXISYMMETRIC NEUTRON STAR COLLISIONS

Now that we have cautioned the reader as to the limitations of the present calculations, we will briefly examine a series of non-axisymmetric, parabolic collisions. Again, we emphasize that these are preliminary results from fairly low resolution studies, but we believe that the waveforms and luminosities we calculate are fairly robust based on the results of the axisymmetric case. We can already reproduce the FD results for orbits with a pericentric radius of $R_p = 0$ and we know the results for pericentric radii of $R_p \gg R$ which is the point mass limit. We now attempt to find the smooth curve connecting these two limiting cases. In each case the polytropes (which are of radius R, mass M, and $R/M = 17$) start with a separation of $4R$ along the z-axis and are released along a parabolic orbit with pericentric radius R_p. The time scale for the collision is the dynamical time

Figure 1. A comparison of the SPH and FD quadrupole moment (a), waveform (b), and luminosity (c) for an axisymmetric, parabolic collision of two $\Gamma = 2$ polytropes. In each case the solid line shows the SPH result and the dashed line shows the FD result. In (a) the bold, solid line shows the difference between the two calculated moments, and the open squares show the correction due to mass loss from the FD grid.

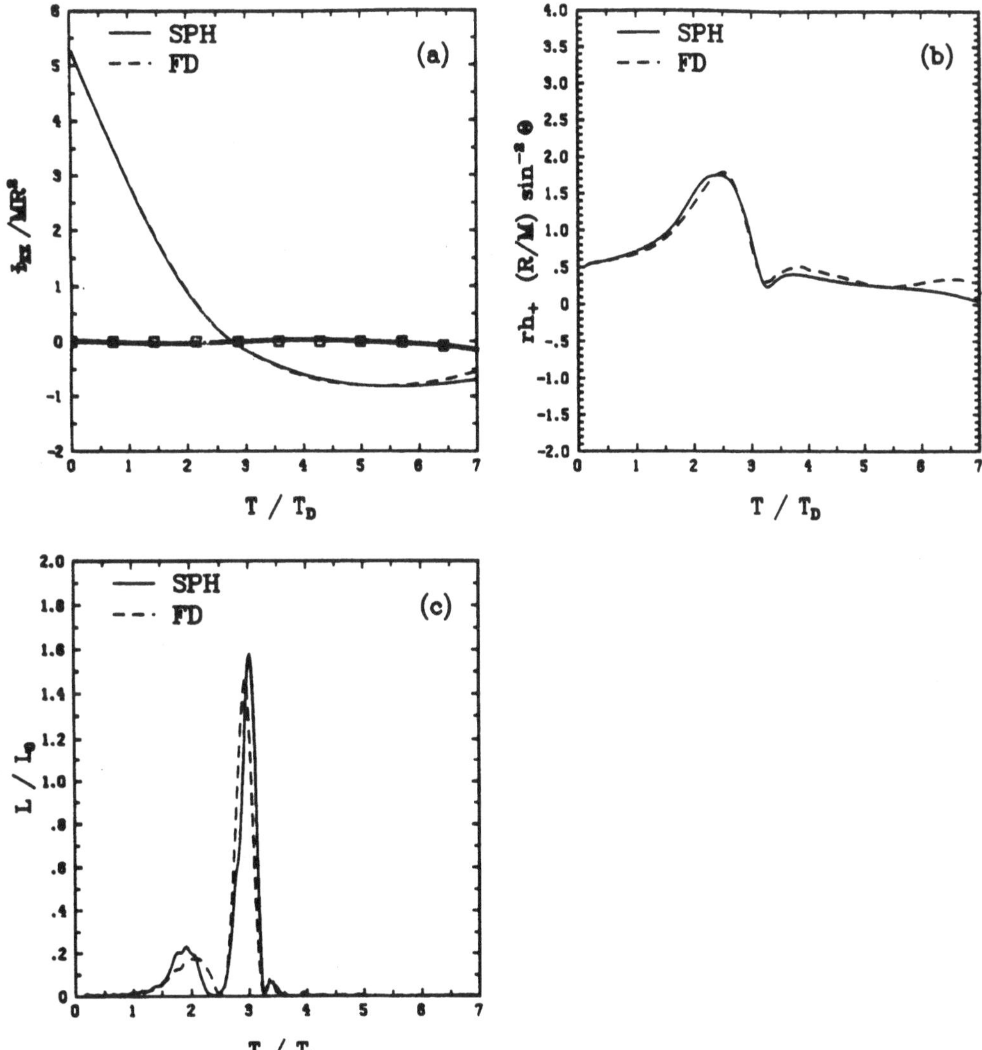

$t_d = M(R/M)^{3/2}$, the scale of the gravitational wave luminosity is $L_0 = (M/R)^5$ and the scale of the emitted energy is $L_0 t_d = M(M/R)^{7/2}$.

We find that the radiated energy increases dramatically once the col-

Figure 2. Pressure contours for the $\Gamma = 2$ axisymmetric collision for the FD (left) and SPH (right) calculations shortly before the point of maximal compression. The contours are equally spaced in the logarithm of the pressure.

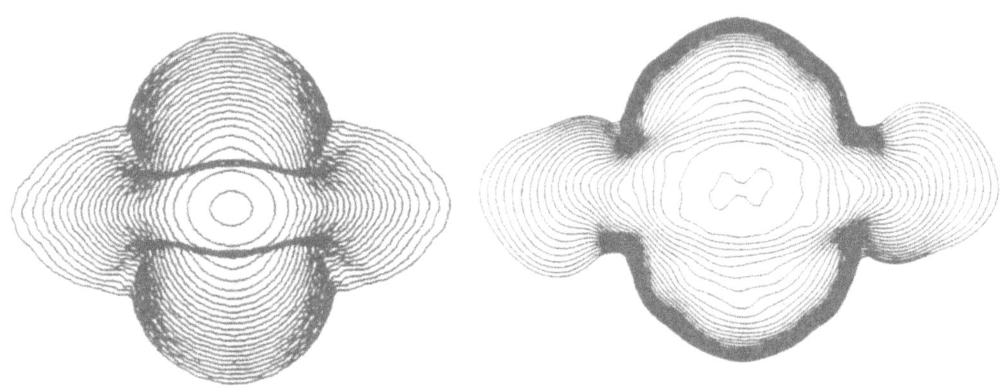

lisions cease to be axisymmetric (see Figure 3). This can be attributed to two causes. The first is that axial symmetry "artificially" limits the number of modes into which energy can be radiated. The axisymmetric case has two discrete luminosity spikes (Fig 1c), while the nonaxisymmetric cases have broader, smoother, luminosity curves as the different modes fill in the gap between the spikes. The axisymmetric case radiates only the + polarization in a $m = 0$ tensor spherical harmonic, while the non-axisymmetric collisions radiate both polarizations into $m = 0$, ± 1 and ± 2 tensor spherical harmonics. The second cause is what we might call the "coherence" of the collision. The energy radiated is highly sensitive to whether the material accelerates coherently or incoherently. This is most clearly seen in the fact that a single point mass of mass M will radiate twice as much energy as two point masses each of mass $M/2$ which follow the same orbit but with a time lag between them greater than the width of the individual luminosity peaks (see also Haugan *et al.*, 1982 for a general relativistic example). In the case of extended neutron stars, this can be seen by decomposing the quadrupole moments of the system into a part due to the center of mass motion of the system (determined by the quadrupole moments found by moving point masses with the mass of the neutron stars along the actual center of mass trajectories of the stars) and into a part due to changes in the structure of the neutron stars. The result of the exercise is that while each subcomponent of the waveform exhibits a very large, sharp peak, they interfere with each other to cause the much smaller, broader peak which characterizes the collision (see Figure 4). The radiative luminosity is reduced because the neutron stars do not decelerate coherently with the center of mass.

The peak orbital velocities decrease as the pericentric radius is increased, eventually leading to a decline in the radiated energy. The peak radiative

Figure 3. The energy radiated as a function of pericentric radius for parabolic collisions of $\Gamma = 2$ polytropes. The solid line shows the energy radiated by two point masses on a parabolic orbit. The filled squares show the results of the SPH simulations. The open circle shows the energy radiated in Evans' (1987) FD simulation.

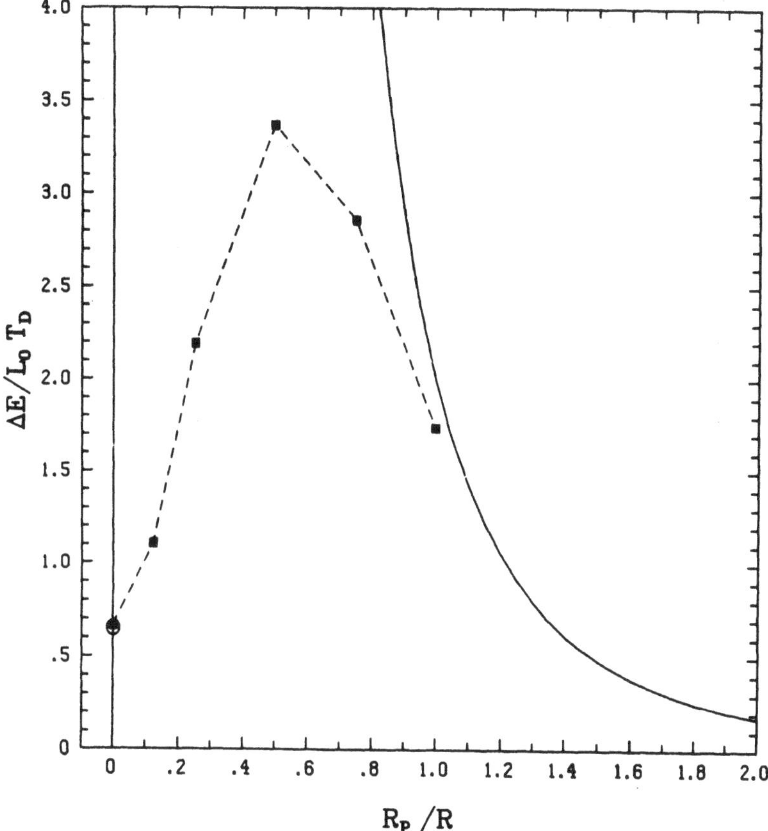

efficiency is about five times the axisymmetric value, and it corresponds to pericentric radii near $r_p \simeq R/2$ where the cores of the stars would be grazing the opposing star's surface. In this case, the dynamical behavior is mostly confined to stripping off the envelopes of each star. The shocked envelope then expands outward into a shell surrounding the two cores. Most of the gas in the shell is still bound to the system and does not escape.

Note that because we have followed the neutron stars only through the initial collision in this preliminary study, we miss the later recollapse and coalescence. The radiation from this secondary phase is unlikely to be as *intense* as in the first phase. The velocities are lower (as the system is now bound) and the luminosity is a strong function of the peak velocities. However, the secondary phases may make a significant contribution to the radiated energy if a rotating,

Figure 4. The waveform of the axisymmetric collision divided into a center of mass component, and an internal structure component (dashed lines) compared to the total (solid line). Notice the larger, sharper peaks of the subcomponents compared to the broader, lower peak of the linear combination.

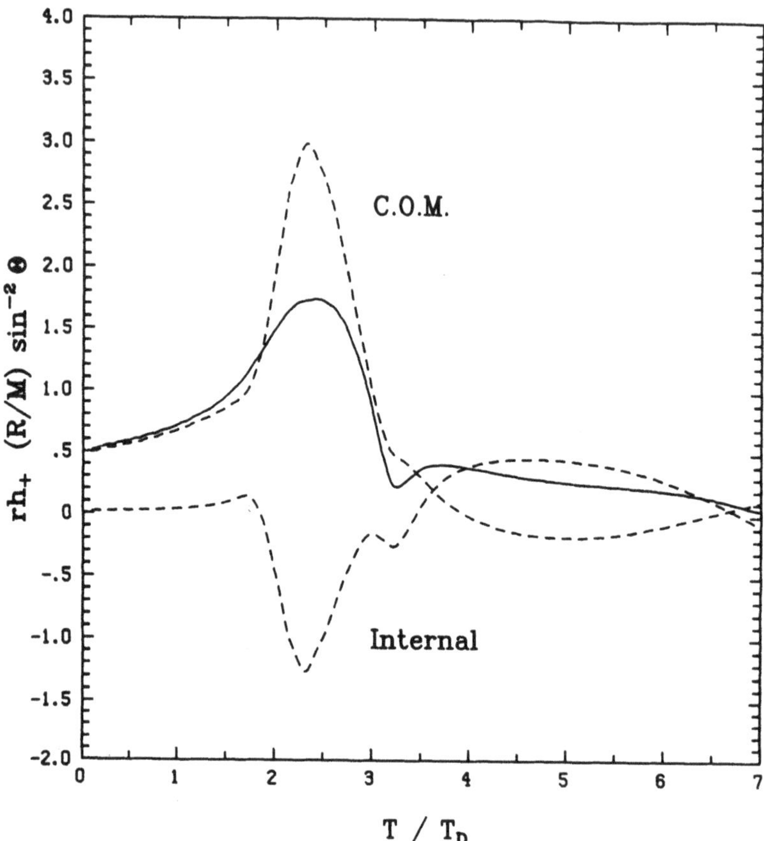

non-axisymmetric structure forms and persists for many dynamical times (Ipser *et al.*, 1984). We hope to follow these later phases in our future studies.

5: WHERE DO WE GO FROM HERE?

The next step is to redo the calculations with a significantly larger numbers of particles – we hope to eventually reach the equivalent of 10^4-10^5 particles per neutron star compared to 500 per neutron star in some of these crude calculations. This represents an improvement in spatial resolution by a factor of 3 to 6, and in mass resolution by a factor of 40 to 400. Note that a FD grid containing 10^4 grid zones in the neutron star (and using the symmetries of the problem) has 21 zones across the star – however, because the FD method is not Lagrangian, the requirement for a region outside the star of at least the same size renders the problem nearly intractable on currently accessible supercomputers. Moreover, mass outflow is clearly a substantial perturbation to the quadrupole moments at late times even when the grid boundary is *several* stellar radii from the point of collision. The main improvement from using more particles will not be in the waveforms and luminosities – it seems clear that the low resolution simulations are adequately computing these – but rather in better resolving the details of the flow, particularly the outflow from the collision. It will then be of interest to examine the problem of the decay of circular orbits and the coalescence of two neutron stars in the Newtonian limit.

The SPH method shows great promise for doing calculations involving the dynamics of compact objects in which most of a finite difference grid would be wasted in non-dynamical regions of the calculation. SPH does not immediately appear to be generalizable to true relativistic problems without sacrificing its main advantage – the ability to simulate the problem without the use of a grid.

ACKNOWLEDGEMENTS

The authors would like to thank Sam Finn and James LeBlanc for their comments on aspects of this work. This work was supported by a graduate fellowship from the AT&T Foundation and NSF grants AST86-15325 and AST85-14911.

REFERENCES

Barnes, J.E., 1986, in *The Use of Supercomputers in Stellar Dynamics*, P. Hut and S. McMillan, *ed.*, (Berlin: Springer-Verlag), p.175

Benz, W., and Hills, J.G., 1987, *Astrophys. J.,* **323**, 614.

Benz, W., Bowers, R.L., Cameron, A.G.W., and Press, W.H., 1988, *CFA preprint.*

Evans, C.R., 1987, in *Proceedings of the 13th Texas Symposium on Relativistic Astrophysics*, M.P. Ulmer, *ed.*, (Singapore: World Scientific), p.152.

Gilden, D.L., and Shapiro, S.L., 1984, *Astrophys. J.,* **287**, 728.

Gingold, R.A., and Monaghan, J.J., 1977, *Mon. Not. R. astr. Soc.,* **181**, 375.

Gingold, R.A., and Monaghan, J.J., 1983, *Mon. Not. R. astr. Soc.,* **204**, 715.

Haugan, M.P., Shapiro, S.L., and Wasserman, I., 1982, *Astrophys. J.,* **257**, 283.

Hernquist, L, and Katz, N., 1988, *IAS preprint.*

Ipser, J.R., and Managan, R.A., 1984, *Astrophys. J.,* **282**, 287.

Misner, C.W., Thorne, K.S., and Wheeler, J.A., 1973, *Gravitation* (San Francisco: Freeman)

Monaghan, J.J., and Lattanzio, J.C., 1985, *Astr. Astrophys.,* **149**, 135.

Press, W.H., 1986, in *The Use of Supercomputers in Stellar Dynamics*, P. Hut and S. McMillan, *ed.*, (Berlin: Springer-Verlag), p.184

Wood, D., 1981, *Mon. Not. R. astr. Soc.,* **194**, 201.

Relativistic Hydrodynamics

James R. Wilson and Grant J. Mathews
Lawrence Livermore National Laboratory

Abstract

An algorithm for hydrodynamics suitable for high velocity shock waves is presented along with a comparison of experiment and calculations of heavy ion collisions. An outline of a method of calculating 3D radiation free hydrodynamics for the solution of the problem of binary neutron star coalescence is given.

I. Nuclear Collision Hydrodynamics Model

The work in the following section was done with C. Alonzo, T. McAbee, J. Wilson, and J. Zingman. We wish to model heavy ion collisions, in particular the collisional experiments performed at CERN which have a relativistic gamma factor of 10.5 in the equal velocity target and projectile frame. Past experience with numerical methods for Eulerian codes gave good accuracy only for gamma factors below 2 to 4. We started from a 3 + 1 dimensional hydrodynamics program that was developed for binary star systems. The program we started with was quite similar to that described in Hawley, Smarr and Wilson 1985. We will describe just those changed parts of the program that are relevant to shock wave modeling.

We use a cartesian coordinate system (X, Y, Z). Density and internal energy are zone centered. Velocities and momentum densities are face centered.

State variables:

$$D = W\rho$$
$$E = W\rho\varepsilon$$
$$S_i = (D + E + WP)U_i,$$

where ρ is proper particle density, ε is specific internal energy, and W is the special relativistic gamma factor. To calculate velocities we proceed as follows:

Let,

$$U_i = 2S_i \,/\, (D_+ + E_+ + P_+W_+ + D_- + E_- + P_-W_-) \quad ,$$

where + and - refer to zones ahead and behind the zone face on which S_i resides.

$$W(\text{face } i) = 1 + U^2{}_i + \bar{U}_j{}^2 + \bar{U}_k{}^2,$$

where $\bar{U}_j$, $\bar{U}_k$ are averages on the face i of the zone face U's in the other two directions.

$$V_i = U_i \,/\, W\ (\text{face } i).$$

W (zonal) = Average of W (face) over all 6 faces.

Compressional heating of material is evaluated by using

$$\dot{E} = P(\nabla \cdot U + \dot{W}),$$

and the U and W calculated as described above.

The method used for calculating the artificial viscosity stress is based on a method of R. Christensen developed for a Lagrangian computer code. Christensen's method produced shock profiles narrow in zone number with good accuracy. Since the shock is narrow in number of zones corrections must be made in going to an Eulerian frame for the advection of momentum in the shock region, which causes a loss of kinetic energy. Three methods were tried. In the first method the total energy (T^{tt}) was calculated before the advection calculations. Then after advection (T^{tt} was also advected) the discrepancy in T^{tt} (before) and T^{tt} (after) was placed into E. This method worked well but it adds a lot of numerical operations to the program. The second method is based on an estimate of the numerical error induced in the kinetic energy by momentum advection and places that energy into the internal energy, E. Again the program is complicated. The last method which works almost as well as the first two was adapted because it is computer efficient. It is based on a formula for the loss in kinetic energy assuming the momentum advection is done to first order in space. We now describe the latter method. The artificial stress is calculated in each direction separately. We will describe the computational process in one dimension.

Step 1 - We use a monotonicity algorithm to determine the slope of the four-velocity, U, at the zone faces and with the values of U on the faces and the slopes the velocities are extrapolated to the zone center and a velocity jump ΔU is evaluated at the zone center.

Step 2 - Quadratic and linear vicosity stresses are determined by:

$$Q_q = \frac{(\alpha + 1)}{2} (D + \gamma E) (\Delta U)^2 \qquad \Delta U < 0$$

$$Q_\ell = - (D + \gamma E)\ C\ (\Delta U) \quad \text{all } \Delta U,$$

where C is the matter sound speed and γ is the gas adiabatic index.

Step 3 - A corrector for momentum advection diffusion is taken as:

$$Q_d = \frac{1}{2} (D + \gamma E) (\Delta U) \left(\frac{U + U+}{2} \right) / W^2 \qquad \Delta U < 0 \ .$$

Step 4 - Momentum and energy are advanced by

$$S_i^{new} = S_i^{old} - \Delta t \cdot (Q_{q+} + Q_{\ell+} - Q_{q-} - Q_{\ell-}) / (x_+ - x_-)$$

$$E^{new} = E^{old} - \Delta t\ (Q_q + Q_\ell + Q_d)\ (U_+ - U_-) / (x_+ - x_-)$$

For three dimensional calculations the Q_q for the three directions are added to form a "scaler" Q which is used as if it were a pressure.

In Table I results for two methods applied to wall shocks are presented for a gas with $\gamma = 5/3$ and $E = 0$ initially.

Table I

method 2			method 3		
W	η	%Error	W	η	%Error
1.01	4.07	1.	2	6.49	-1.
1.5	5.26	.5	4	11.6	1.
2.0	6.45	-1.	8	22.3	2.5
10.0	11.5	-1.	16	42.5	3.0

W is U^t for the matter initially, η is the compression ratio, and the error is the error in compression. For material initially at rest accelerated up to W=2, a compression ratio of 6.51 was obtained by the code. For higher values of the adiabatic index the errors are less. For smaller adiabatic index errors are larger. For $\gamma = 4/3$ the error at W = 10 is 5%.

Figure 1 shows the comparison of a code calculated shock tube expansion versus theory. Figure 2 shows a typical shock structure. The shock is thin but it has an overshoot.

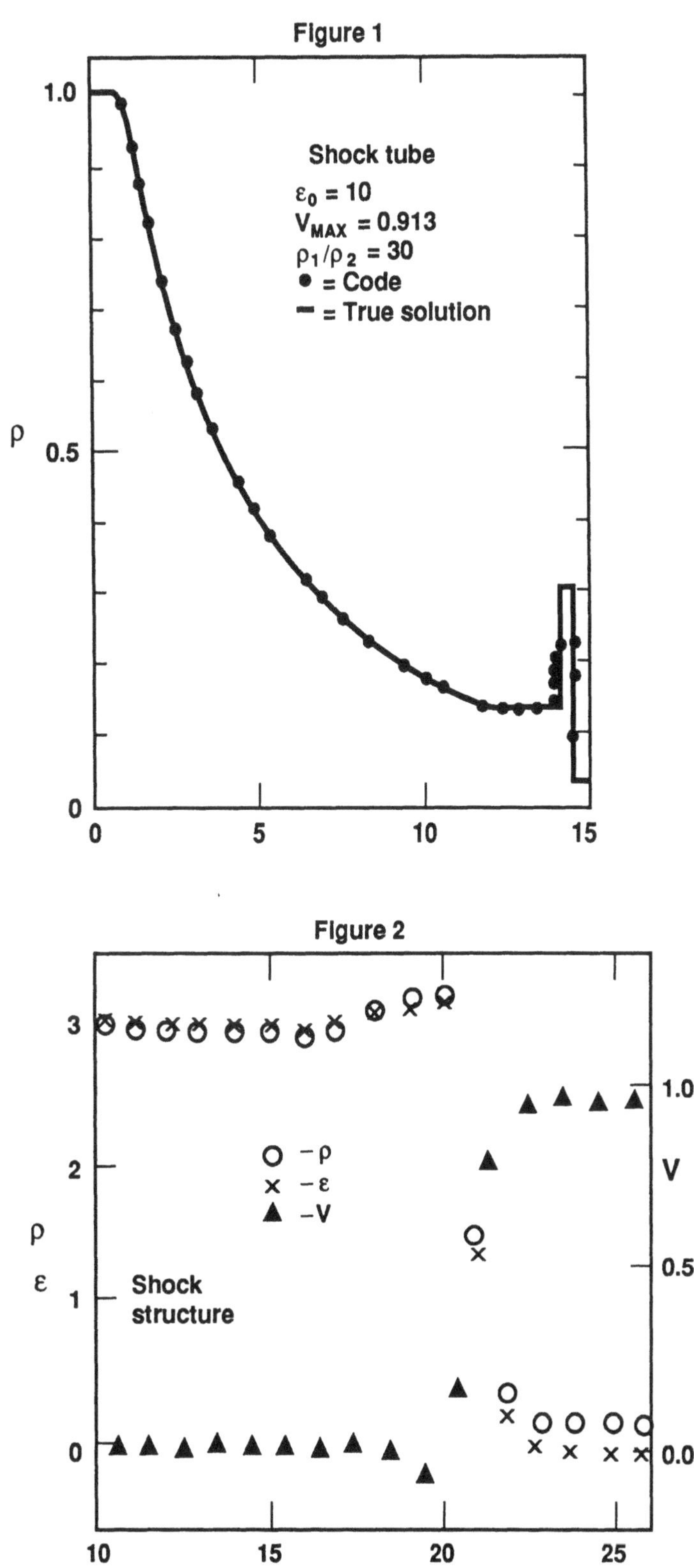
Figure 1
Shock tube
ε0 = 10
VMAX = 0.913
ρ1/ρ2 = 30
● = Code
— = True solution
1.0
ρ
0.5
0
0
5
10
15
Figure 2
3
2
ρ
ε
1
0
○ –ρ
× –ε
▲ –V
Shock
structure
1.0
V
0.5
0.0
10
15
20
25
Zone number

We now present the results of a calculation of the collision of a 100 Gev/nucleon oxygen nucleus on a lead nucleus, and compare the calculation with experimental results from CERN. Calculations were made for an on axis collisions only.

Figure 3 shows the number of baryons per rapidity interval for a calculation with an adiabatic index of 4/3. Included in figure 3 is the experimental distribution versus rapidity of baryons (dashed line). Previous calculations by another group gave a rapidity shift of two so that there was virtually no overlap with experiment. Figure 4 shows the rapidity plots for three different adiabatic indicies. Figure 5 shows the distribution of energy in the perpendicular direction with rapidity. Experiment peaks at a rapidity of about -2 with an energy of 60$\pm$ 10 Gev. The total perpendicular energy experimentally is 160$\pm$ 40 Gev. For γ = 4/3 we calculate 120 Gev for E and with γ = 5/3 E = 155 Gev. We, thus, are in moderately good agreement with experiment with a γ = 4/3 equation of state. Agreement with experiment may be fortuitous. Particle theorists say nuclei should interpenetrate and invalidate this type of hydrodynamics model. We will press on with the hydrodynamic model.

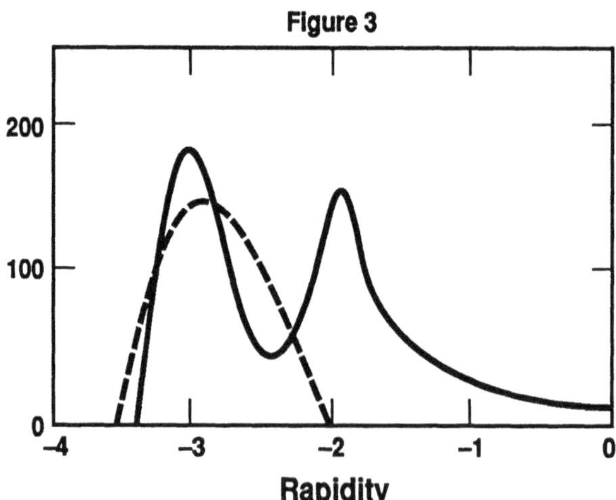

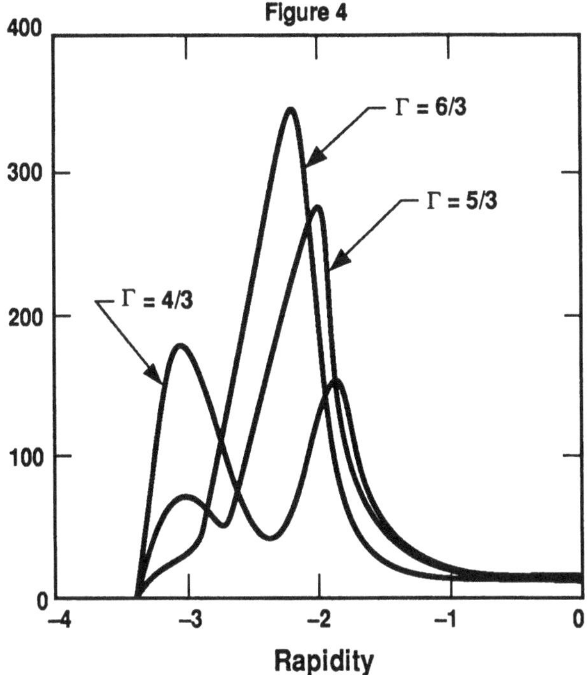
Figure 4
400
300
200
100
0
Γ = 6/3
Γ = 5/3
Γ = 4/3
–4
–3
–2
–1
0
Rapidity

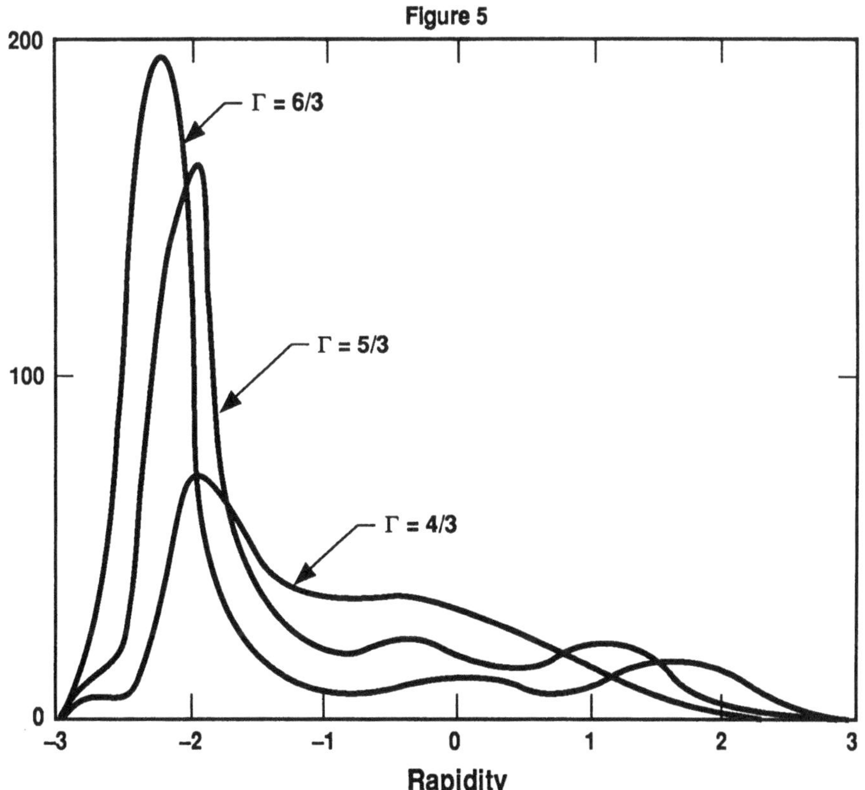
Figure 5
200
100
0
Γ = 6/3
Γ = 5/3
Γ = 4/3
–3
–2
–1
0
1
2
3
Rapidity

II. Stellar Collision Computer Code.

This work here presented was done by Mathews, Evans, Detweiler, and Wilson. The coalescense of neutron binaries is thought to be a likely source of detectible gravitational radiation hence we started a program to calculate the late stages of neutron star binary life. Most recent calculations have shown little gravitational radiation is emitted in stellar systems. To simplify the calculations we have assumed that the gravitational radiation energy is almost negligible. We start with the metric defined by:

$$ds^2 = -(\alpha^2 - \beta_i \beta^i)\, dt^2 + 2\beta_i \cdot dt\, dx^i + g(dX^2 + dY^2 + dZ^2).$$

The effects of radiation reaction will be applied as a perturbation using a radiation reaction potential evaluated from the quadrapole moments. We take maximal slicing and use that condition to find α. To preserve the flat space three metric we take the extrinsic curvature as given by:

$$2\alpha K_{ij} = D_i \beta_j + D_j \beta_i - \frac{2}{3} g \delta_{ij} D_i \beta^i .$$

This formula is also clearly consistent with $K^i_i = 0$. Placing this K^i_j in the momentum constraints formula yields elliptic equations for the β_i. Finally g is evaluated by solving the Hamiltonian constraint.

The method of solution for these elliptic equations now follows. We follow York's principles to put equations in flat space. Let $A = \alpha g^{1/6}$ then from the trK equation and the Hamiltonian equation we arrive at the equation for A

$$\Delta A = 4\pi A \left[K^i_j K^i_j + D\left(3W^2 - 2\right) E \left(3\gamma\left(W^2 + 1\right) - 5.\right)\right] /W,$$

Where D, E, γ, and W are defined in section I. The operator Δ is the flat space Laplancian. Since Δ is a flat space Laplancian we can use a moment expansion of to evaluate A at the boundary. We use spherical harmonics up to l = 4, m = 4 and can thus have our boundaries close to the region of activity. Next let $B = g^{1/6}$ with the Hamiltonian constraint yielding the equation.

$$\Delta B = 4\pi B \left[\frac{DW}{g^{1/3}} + \frac{E(\gamma(W^2 - 1) + 1)}{\alpha W g} + \frac{K^i_j K^i_j}{16\pi} \right] .$$

The momentum constraint equations are less tractible since α occurs inside the divergence operator. We take the derivative of α terms to the other side of the equation in order to arrive at flat space elliptic equations. We take the vector elliptic equation for β and arrive at a scaler elliptic equation by letting:

$$\beta_i = X_i - \frac{1}{4} \Delta_i C .$$

Then

$$\Delta C = \nabla_i X^i$$

$$\Delta X^i = Q^i$$

$$Q^i = 16\pi \alpha g S^i + \frac{\Delta_i H}{H} A^{ij}$$

$$H = \alpha / g^{3/2}$$

$$A^{ij} = \nabla^i \beta^j + \nabla^j \beta^i - \frac{2}{3} \delta^{ij} \nabla_k \beta^k ,$$

where Δ and ∇_i are flat space operators.

All the elliptic equations are solved by a conjugate gradient method using fourth order moments of the sources to establish the boundary conditions. Since the elliptic equations are not independent we iterate between them as necessary.

We will need to follow the binary system for several orbits which potentially could use much computer time. To save time we plan to use a gravity wave loss multiplier early on to bring the stars close together. We also use a pressure, gravitational force multiplier which is valid when the system is almost stationary. This latter process allows the Courant sound speed stability condition for explicit hydro to be ignored. With an early version of the program that did not have the shift vector solver in yet one rotation of a binary star system, 360°, required 5 minutes of Cray time. To achieve an almost stationary configuration and hence a large time step, we use the boundary conditions on the shift vector such that the lines of centers between two stars does not move in our coordinates. For equal mass stars we can use a reflection and inversion boundary condition thru the center of mass so as to only require the calculation of one star. As the stars begin to merge neutrino losses will become important. We have introduced a field representing the energy density of all the neutrinos. This field obeys a simple diffusion equation and is coupled to the matter. Our goal is to develop a solution method that is accurate and can do a complete binary star collapse calculation in less than a few hours of Cray computer time.

References for CERN experiments:

H. Schmidt et al. (WA80 Collaboration), in "Proceedings of the 27th International Summer School for Theoretical Physics", Zakopane, Poland, 1986, to be published.

A. Bamberger et al. (NA35 Collaboration), Phys. Lett. B184, 271 (1987).

R. Albrecht et al. (WA80 Collaboration), Phys. Lett B199, 297 (1987).

COMPUTATIONAL DYNAMICS OF $U(1)$ GAUGE STRINGS: PROBABILITY OF RECONNECTION OF COSMIC STRINGS

Richard A. Matzner
Center for Relativity and Department of Physics,
The University of Texas at Austin, Austin, Texas 78712-1081

ABSTRACT A series of large scale (64^3) computations have been carried out to investigate the evolution of interactions of $U(1)$ gauge cosmic strings, both nonconducting and superconducting. The interactions of such strings are important in various scenarios of structure formation in the early universe. In all cases investigated (for a range of approach speeds and crossing angles), reconnection occurs. This means that the "head" of one string reconnects after the collision with the "tail" of the other. The conclusion is that cosmic string loops will be easily shaken off long strings; string loops can act as massive condensation centers for the formation of structure in the universe. In the case of interactions occurring for superconducting strings, the reconnection can result in very substantial electromagnetic radiation as the currents in the reconnected strings attempt to equalize. This may provide a dominant mechanism for void production in a superconducting string populated universe.

I. $U(1)$ GAUGE COSMIC STRINGS

I begin by describing $U(1)$ cosmic strings. These can arise in any field theory that passes through a period in which the matter density is dominated by a complex scalar whose effective potential admits a high temperature, symmetric false vacuum together with low temperature symmetry breaking. Such a scalar field would be described by a Lagrangian:

$$\mathcal{L} = \tfrac{1}{2}\partial_\mu\phi^*\partial_\nu\phi\eta^{\mu\nu} + V\big(|\phi|\big) \ . \tag{1}$$

(I use the Minkowski metric with signature $-+++$.)

In the cosmological setting I consider, the transition to the symmetry broken, low temperature, state takes place with a coherence length no longer than the size of the cosmic horizon at the time of the transition. Hence in different regions of the universe, the scalar field ends up in the true vacuum with a particular value of its complex phase. The phase is a gauge variable with no local significance, but the wrapping of the phase is a topological invariant that cannot be removed by any gauge transformation. The scalar field ϕ must then have a representation (in cylindrical coordinates ρ, φ, z) like

$$\phi = \phi_s(\rho)e^{in\varphi} \tag{2}$$

where ϕ_s is real and positive and where n is an integer for singlevaluedness at general ρ. Regularity around the axis $\rho = 0$ then requires $\phi_s(\rho) \to 0$ as $\rho \to 0$ (otherwise ϕ is multivalued as $\rho \to 0$). This means that there must be a line along which the scalar field ϕ is at its false vacuum value, i.e., $|\phi| = 0$, and this high energy, high density line is the cosmic string. For string formation, one requires the central maximum in the effective potential, but the shape of the potential is not very important to the remaining string structure. For the computations described here, I always take

$$V(|\phi|) = \lambda_\phi(|\phi|^2 - \eta^2)^2 \ . \tag{3}$$

For these simple string structures, the fact that the gradient of the phase is nonzero at infinity (i.e., far away from the string) means that there is a nonzero energy density far from the string and the mass per unit length diverges (logarithmically, as the integration cylinder is taken to infinity). The string-string forces fall off very slowly ($\sim 1/\rho$).

However, it is natural to consider *gauged* scalar fields, in which case the Lagrangian becomes:

$$\mathcal{L} = \tfrac{1}{4}\eta^{\alpha\rho}\eta^{\beta\gamma}F_{\alpha\beta}F_{\rho\gamma} + \tfrac{1}{2}\eta^{\mu\nu}(\partial_\mu - ieA_\mu)\phi^*(\partial_\nu + ieA_\nu)\phi + V(|\phi|) \ . \tag{4}$$

Here A_μ are the components of a gauge vector field, and $F_{\mu\nu} = \partial_\mu A_\nu - \partial_\nu A_\mu$ is its curl. The coupling to the gauge field is denoted e. It can be seen that if $|\phi| = \eta$ and $\phi = \sigma e^{in\varphi}$, then $eA_\mu\, dx^\mu = -n\, d\phi$ will cancel the gradient of phase contribution at infinity, because of the way the derivative and A enter the covariant derivative, $D_\mu = \partial_\mu + ieA_\mu$. There will be no contribution to the energy density at infinity from just variations in phase, and the strings are localized objects with finite energy per unit length and with exponential fall-off of the forces between them.

The major portion of this paper will be devoted to these $U(1)$ gauge strings. In §VIII et seq. below, however, I will discuss *superconducting* gauge strings, which differ from the nonsuperconducting strings in the addition of another complex scalar coupled to another (the electromagnetic) gauge vector. For now we concentrate on strings described by the Lagrangian (4).

II. *ASTROPHYSICS OF COSMIC STRINGS*

The development of cosmic strings is not required in every field theory. Furthermore, except for extremely fine-tuned scenarios, strings will have no astrophysical effect if they form before a period of inflation. So, we need a period of ordinary cosmic expansion during which time the effective potential (3) is accurate and during which time the string field ϕ cools into the minimum channel of $V(|\phi|)$. Although the minimum energy string configuration is then long and straight, in fact the existence of horizons means that the string will be "kinky" on horizon size scales. As the horizon expands, it exposes more and more of the kinked string, which now begins to flop and, in general, to self-intersect (Kibble and Turok 1982; Turok 1983a; Albrecht and Hindmarsh 1987). The self-intersection presents an opportunity to break off loops, and such loops—or their daughter loops—can act as gravitational centers to seed structures (either galaxies or clusters of galaxies, depending on the parameters of the strings and the size of the loops).

If one *postulates* that crossing strings always reconnect, then it is possible to simulate the large-scale evolution of their distribution by using the Nambu action (which describes strings as arbitrarily thin, 1-spatial dimensional objects (Nambu 1970), a valid approximation away from self-intersections) and switching partners at every intersection. In this case, computations by Turok (1983b, 1984, 1985), Turok & Brandenberger (1986), Vilenkin (1981), Zel'dovich (1980), Bennet & Bouchet (1987), and Albrecht & Turok (1985) have shown that a large population of string loops forms with spatial distribution and correlation that is apparently consistent with observed galaxy-galaxy and cluster-cluster correlations.

The crucial question with regard to the astrophysical implications of cosmic strings is, then, what is the probability that reconnection occurs? More correctly, in which range of the parameter space (2-dimensional, depending only on the speed and the angle of approach) does reconnection occur, in which range is it excluded? It is this question these computations were meant to address.

III. *SINGLE ISOLATED STRINGS*

If the equations resulting from Lagrangian (4) are specialized to static axial symmetry, different solutions describe different states: in true vacuum, the phase of the scalar field is constant and the magnitude $|\phi|$ is at the minimum channel true vacuum value; in states describing cosmic strings with $n = 1, 2, \ldots$; cf. Eq. (2). (The computations here will concentrate on $n = 1$ cosmic strings; higher n can be obtained by superposing $n = 1$ strings.) Solution of the second order, coupled radial differential equations describing strings is not, except for particular choices of the shape of the potential, analytically feasible [It is not feasible, for instance, when using the "ϕ^4" potential, Eq. (3).], but is straightforward by computational techniques. One can analytically derive—for potential (3), for instance—that the asymptotic forms of the fields

$$\phi = \phi_s(\rho)e^{i\varphi} \qquad \text{and} \qquad A_\mu\, dx^\mu = \left[P(\rho) - 1\right] d\varphi \tag{5}$$

are

$$\left.\begin{array}{ll} \phi_s & \approx a\rho \\ P & \approx 1 - c\rho^2 \end{array}\right\} \quad \text{near } \rho = 0, \text{ and}$$

$$\left.\begin{array}{ll} \phi_s & \approx \eta(1 - be^{-\sqrt{8\lambda\phi\eta^2}\rho}) \\ P & \approx de^{-e\eta r} \end{array}\right\} \quad \text{near infinity} \tag{6}$$

With these asymptotic forms in hand, one can consider parameterized analytic trial functions and numerically minimize the energy of the configuration as an alternative approach to constructing the strings. This is the computational approach taken here.

IV. *MULTISTRING SOLUTIONS AND INITIAL CONDITIONS*

The process of obtaining a single string is straightforward, at least if the gauge of the A_μ field is chosen appropriately: if z is the direction along the string, then $A_z = 0$, as is A_t, the timelike component. For the case of a single static string in this gauge, it is a simple matter to verify also that the field A_μ satisfies the Lorentz gauge:

$$A^\mu{}_{,\mu} = 0\,. \tag{7}$$

This gauge is linear and Lorentz invariant, so the Lorentz boosted string also satisfies the Lorentz gauge condition; the whole string structure can be given a velocity perpendicular to its length by such a boost. Conceptually, then, it is easy to consider an initial string offset from the center of the cube, rotated at some angle and boosted toward the center. Such a configuration provides a partial data set for the colliding string problem. Because the equations solved are second-order field equations (see §VI below), the initial data set must also include momenta, i.e., time derivatives, of the field. Here these are again straightforward. Simply evaluate the boosted string at a later time Δt, take the difference of the field configurations divided by Δt, and thus numerically calculate the momenta.

So far, however, I have only shown how to put one string in the cube. Such a problem was in fact computationally evolved as a test of the stability of the code. However, the important questions involve collisions of two strings. This requires the superposition of two differently moving cosmic string solutions.

The superposition of the gauge vectors is straightforward (Müller-Hartmann 1966; Abrikosov 1957) because the fields go to zero far from the strings and because the equations describing A_μ are linear: add the components ${}^R A_\mu$ and ${}^L A_\mu$ of the *right* and *left* string gauge vectors. But the ϕ fields are definitely nonlinear. They can be superposed, however, by noting the additivity of A_μ and its already-noted relationship to cancelling the gradients of the phase far from the strings. Hence, add the phases. The rules for superposing right and left strings are:

$$A_\mu = {}^R A_\mu + {}^L A_\mu \tag{8}$$

$$\phi = {}^R\phi {}^L\phi / \eta \tag{9}$$

where the last division by η makes the ϕ composition simply an addition of phase. The momenta can be computed by recomputing this configuration at a later time Δt. This formulation is correct for widely separated strings. For closer strings, it is still a valid data set, but its interpretation as superposed strings is less clear.

V. BOUNDARY CONDITIONS

The initial data setting as described in §IV. above clearly sets the initial boundary conditions. The problem of boundary condition maintenance at later times is solved by repeating the initial data construction on the boundary at each time step. This means that the strings at the boundary behave as if the interactions at the center never occurred. Hence the parts of the strings away from the center continue in motion as if they were attached to infinitely long strings that extend out of the computational box. This accurately reflects the physics of the interaction. Disturbances from the center of the box travel outward at the speed of light, and the strings themselves are moving at about the speed of light. Hence we expect that the boundary conditions will be good only until signals have had time to propagate out from the center to the boundaries, but by this time the strings will have almost evolved out of the box anyway. Because the string fields are strongly nonlinear, it is difficult to see how one can improve this boundary situation short of finding a complete analytic solution to the intersection problem.

VI. NUMERICAL SIMULATION OF COSMIC STRINGS

The equations derived from the Lagrangian (4) can be written:

$$\partial_t \pi_{\phi_X} = \Delta\phi_X - e^2 A^2 \phi_X - 2e\varepsilon_{XY}(A^t \pi_{\phi_Y} + A^i \partial_i \phi_Y) - \frac{\partial V}{\partial \phi_X} \tag{10}$$

$$\partial_t \pi_i = \Delta A_i - e\varepsilon_{XY}\phi_X \partial_i \phi_Y - e^2 A_i \phi^2 \tag{11}$$

where the labels X and Y run over 1 and 2;

$$\phi^2 = |\phi|^2 \tag{12}$$
$$A^2 = A_i A_i \tag{13}$$
$$\pi_{\phi_X} = \partial_t \phi_X \tag{14}$$

and

$$\pi_{A_i} = \partial_t A_i \ . \tag{15}$$

Additionally, ϕ has been split into real and imaginary parts

$$\phi = \phi_1 + i\phi_2 \ , \tag{16}$$

and in Eqs. (10)–(11), ε_{XY} is the antisymmetric unit tensor with

$$\varepsilon_{12} = 1 \ . \tag{17}$$

The set is closed by the Lorentz condition:

$$\partial_t A_t = \pi_{A_t} \tag{18}$$
$$\pi_{A_t} = \partial_i A_i \ . \tag{19}$$

Differencing these equations is carried out in a staggered leap-frog scheme. The quantities ϕ_X, A_i, and A_t are all defined on the spatially even and temporally even nodes, their gradients on spatially odd, temporally even steps. The momenta are defined on the spatially even, temporally odd steps, with their spatial gradients on the spatially odd, temporally odd nodes. This leads to a natural evolution scheme with no interpolation or extrapolation, except for two terms involving A_t. To define π_t on the spatially even zones requires spatial averaging from its natural location on spatially odd, temporally even nodes; to define it on the odd temporal steps requires an extrapolation to those zones. Furthermore, Eq. (10) contains the term $A^t \pi_{\phi_Y}$. For consistency, this should be defined on spatially even, temporally even nodes, which requires a further extrapolation of π_{ϕ_Y}. In fact, both these problems seem to be associated with the gauge covariant derivative $\partial_\mu + ieA_\mu$ in this formulation. Meyers and Rebbi (1988) have given a numerical string-collision formalism and implemented a code based on lattice gauge theory concepts. One product of such an approach is that the gauge covariant derivatives are correctly handled (and gauge invariance is explicitly guaranteed) in the discrete version of the theory. Leblanc (1988) has also raised the question of accuracy in the differencing of covariant derivatives. The approach I take is to accept the inaccuracies of averaging A_t to the appropriate node and of computing with an off-time centered term in $\partial_t \pi_Y$. An attempt is being made to find a more natural differencing that will alleviate these problems.

VII. *RESULTS FOR NONSUPERCONDUCTING COSMIC STRINGS*

A number of simulations of cosmic string collisions at various incident speeds and crossing angles have been carried out. In the flat space cases considered here, one can always set $\eta = 1$, and I do so. These simulations used $\lambda = 0.01$, $e = 0.2$. I considered parallel string, string (repulsive for the parameters considered here) and antiparallel string, string (i.e., string-antistring) encounters for initial velocity $v = 0$ and for initial velocity $v = 0.75c$. The string-antistring cases led to annihilation; the $v = 0$ string-string case had an initial repulsive acceleration followed by continued drift. The $v = 0.75$ string-string case led to the interesting result [predicted by Ruback (1988) for strings that have in our notation $e^2/\lambda = 8$ and which approach slowly] that the final parallel string configuration lies in the plane perpendicular to the plane containing the initial configuration. High speed ($v/c = 0.9$ and $v/c = 0.95$) string-antistring collisions surprisingly do not lead to annihilation. There is a substantial energy loss by radiation from the collision, but the two strings re-emerge. The strings re-emerge with some translational velocity out of the initial plane but generally in the direction opposite that from which they entered. That is, they do not pass through one another but instead "bounce back." This result was apparently first found by Meyers and Rebbi (1988).

More interesting, perhaps, are the string, string collisions at various crossing angles: $\theta = \pi/2$ with $v/c = 0.1, 0.5, 0.75, 0.85, 0.9$; $\theta = \pi/4$ and $\theta = 3\pi/4$ with $v/c = 0.75$, and $\theta = 7\pi/8$ with $v/c = 0.9$. In all these cases there is cross-connection—the string ends switch partners.

I present here figures from two of the simulations. Plotted here is the 40% contour of the deviation of the scalar field from the true vacuum where 100% corresponds to the value at the core of the strings. In all cases, a plot of the (gauge invariant) vector field ($\nabla \times A$) shows identical structure to those plotted here.

The first sequence shows the $v/c = 0.75$, right angle case, the paradigm for all these simulations. Figure 1c shows slice 40 in this simulation with the strings just touching. The reconnection can be clearly seen, proceeding with the production of a bubble of radiation (Fig. 1b, slice 60). The released radiation expands out to contact the now-separated strings (Fig. 1c, slice 80), and in Fig. 1d, slice 100, we see the final structure in which the string is straight at the boundaries, then has a kink, a straight "join" segment, another kink, then the final straight segment leaving the cube. The kinks move out into the undisturbed string at the speed of light. This straight-kink-straight-kink-straight structure can be obtained from an analysis of the Nambu equations

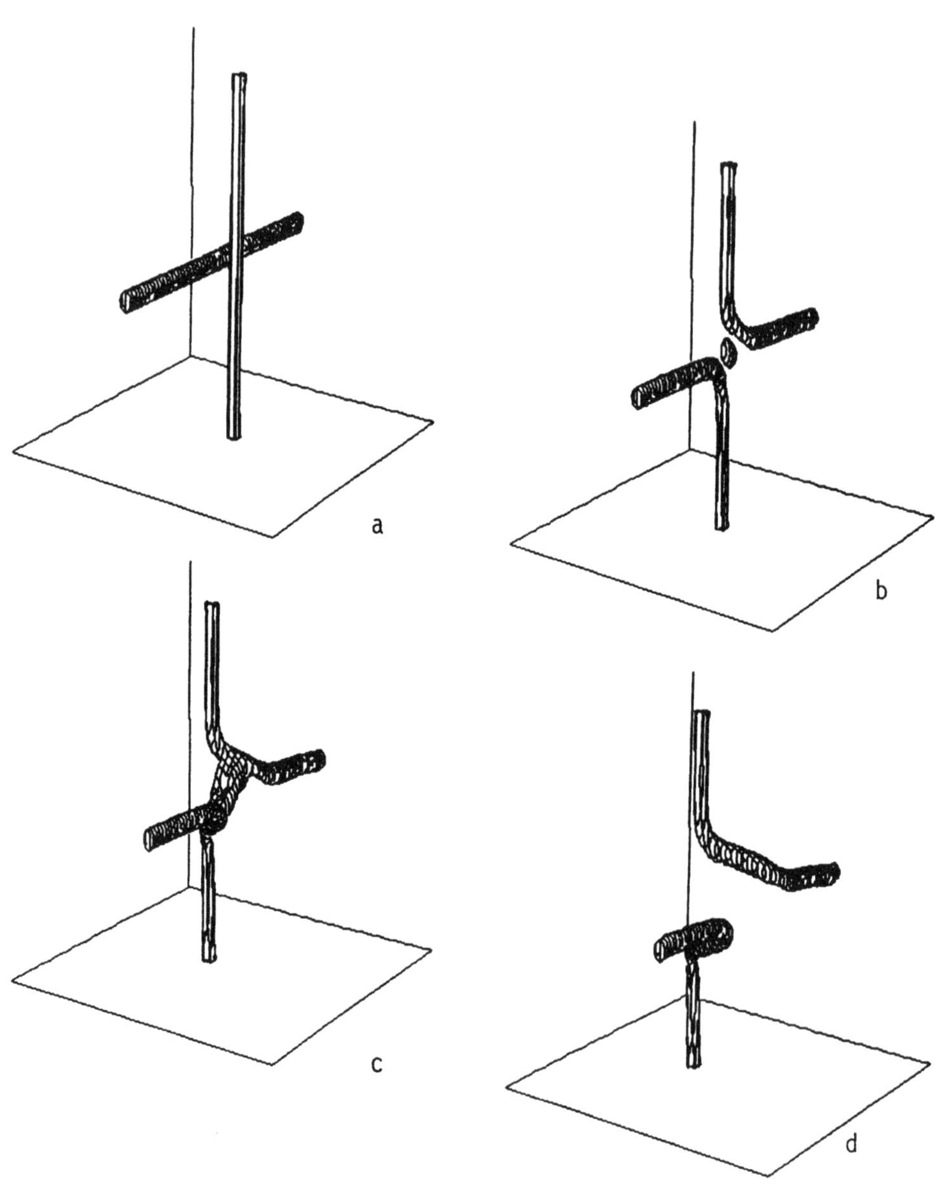

FIGURE 1

beginning immediately after the reconnection.

The second sequence I show involves the $v/c = 0.9$ collision of strings oriented at $7\pi/8$. There is thus a central region that is closely antialigned but where the initial behavior is somewhat like the high speed "bounceback" of antiparallel strings discussed above. Slice 40 (Fig. 2a) shows the strings just touching as they approach. In Fig. 2b, slice 60, they have reconnected, and a bubble of apparent radiation forms in the central region. By slice 100, Fig. 2c, it has expanded into a sheet of radiation between the two, well separated strings. Slice 120 (Fig. 2d) shows that this radiation rearranges itself into a string loop, which persists, oscillating and shrinking (Figs. 2e, 2f, slices 140, 160), and eventually disappears. The regions of strings farther away from the crossing point reconnect, even as the central regions bounce back. The result is a (microscopic) string loop, which, because it is under tension, shrinks and disappears.

All cases examined so far show reconnection of the string ends.

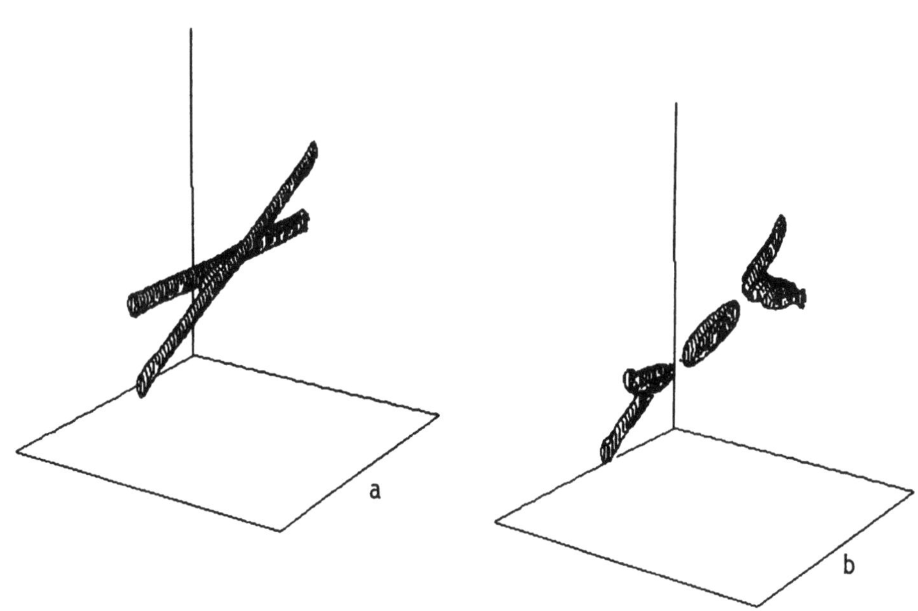

FIGURE 2

FIGURE 2 (Continued)

VIII. SUPERCONDUCTING COSMIC STRINGS

This work was done in collaboration with Dr. P. Laguna.

In 1985 Witten pointed out that cosmic strings can support superconductivity, both from bosonic and fermionic charge carriers. We here consider bosonic superconducting strings.

These are described by an additional charged (complex) scalar field σ coupled via a charge g to a gauge vector field C_α. This charge is meant to be literally electromagnetism, and C_α is the electromagnetic vector potential. In order for this to be an accurate description, corresponding to our commonsense (low energy) ideas of electromagnetism, it is necessary that the scalar field σ vanish far from the string but be nonzero at the center of the string. (This will become clear shortly.)

The Lagrangian describing these superconducting cosmic strings is a straightforward extension of (4):

$$\mathcal{L} = \tfrac{1}{4} F_{\alpha\beta}F^{\alpha\beta} + \tfrac{1}{4}\tilde{F}_{\alpha\beta}\tilde{F}^{\alpha\beta} + \tfrac{1}{2} D_\alpha\phi(D^\alpha\phi)^+ + \tfrac{1}{2}\tilde{D}_\alpha\sigma(\tilde{D}^\alpha\sigma)^+ + V(\,|\phi|\,,\,|\sigma|\,) \quad (20)$$

where $\tilde{F}_{\alpha\beta} = \partial_\alpha C_\beta - \partial_\beta C_\alpha$ and $\tilde{D}_\alpha\sigma = (\partial_\alpha + igC_\alpha)\sigma$. The potential for these coupled fields is now taken as

$$V(\,|\phi|\,,\,|\sigma|\,) = \lambda_\phi(\,|\phi|^2 - \eta^2)^2 + \tfrac{1}{4}\lambda_\sigma|\sigma|^4 + \tfrac{1}{2} f|\sigma|^2\,|\phi|^2 - \tfrac{1}{2}m^2\,|\sigma|^2\,. \quad (21)$$

The new parameters, involving σ and its coupling to ϕ, can be chosen so that the global minimum of the potential is at $\sigma = 0$, $|\phi| = \eta$. This is the behavior we want at infinity. Then the potential reduces to that already seen in Eq. (3) and it imposes a restriction on f:

$$f\eta^2 \geq m^2\,. \quad (22)$$

The equations derived for ϕ and A_μ from the Lagrangian (17) look identical to those already given in Eqs. (10)–(19), except that $V(\,|\phi|\,)$ there is replaced by $V(\,|\phi|\,,\,|\sigma|\,)$ here. The equations for σ and C_μ are obtained by taking $e \leftrightarrow g$, $\sigma \leftrightarrow \phi$, and $A_\mu \leftrightarrow C_\mu$ and similarly for the momenta in Eqs. (10)–(19).

The requirement $\sigma \rightarrow 0$ at infinity can be understood by considering the σ, C_μ version of Eq. (11). The C_μ field is supposed to be electromagnetism with a line source. Hence, far away from the string, C_μ should obey the wave equation, which it does if $\sigma \rightarrow 0$. On the other hand, for there to be a current flow one needs the term $\varepsilon_{XY}\sigma_X\partial_i\sigma_Y$ nonzero. Hence σ should not vanish within the string where $|\phi| \sim 0$. For this to be energetically favorable requires

$$4\lambda_\sigma\lambda_\phi > m^4/\eta^4\,, \quad (23)$$

giving a second constraint on the parameters of the new potential. Because of the finite value of σ near the center, its phase *cannot* wrap the string but rather increases along the string. Hence the superconducting strings are like long wires carrying current. In particular, the magnetic fields associated with the current wraps around the string, and its strength falls off only as $1/\rho$ away from the string. Notice that for small σ the last term in the potential (18) reduces the total energy when σ is nonzero, so a supercurrent is formed. On the other hand, the current itself contributes positively to the electromagnetic $\tilde{F}^2$ energy; so for a very large current, it is energetically favorable to be in the completely nonsuperconducting state, $\sigma = 0$ everywhere. Hence there is a maximal supercurrent.

We (P. Laguna and I) chose parameters in the potential so that currents on the strings near their critical values can be achieved, while all fields have comparable asymptotic fall-off radii, and so that the superconducting phase, without current, is favored by reducing $\sim$ 20% of the nonsuperconducting energy density. Parameters used in our numerical simulations that satisfy these conditions and the constraints are: $\eta = 1$ (which is a convention allowed by the scale freedom in this quantity), $\lambda_\sigma = 60$, $\lambda_\phi = 1/8$, $m^2 = 5$, $f = 5.5$, $e^2 = 1/2$, and $g^2 = 4\pi/137$.

To construct the initial data for our simulations, we looked for static solutions for infinitely long strings, where $\phi = \phi_s(\rho)e^{i\varphi}$, $eA_\alpha = [P(r) - 1]\nabla_\alpha\varphi$, $\sigma = \sigma_s(\rho)e^{i\chi(z)}$ and $qC_\alpha = K(\rho)\nabla_\sigma z$. Thus the string "magnetic field" $\nabla \times A$ runs along the string, while $\nabla \times C$, the electromagnetic magnetic field due to the current, wraps around the string. When there is no current, one takes $\chi(z) \equiv 0$; then $C_\alpha = 0$ is a consistent solution. There is still then an interaction between the string scalar field ϕ and the electrically charged scalar σ.

IX. *ASTROPHYSICS OF SUPERCONDUCTING COSMIC STRINGS*

From even small primordial magnetic fields, a moving segment of superconducting cosmic string can pick up large currents. In addition, there have been simulations carried out (Spergel, Press, and Scherrer 1988) suggesting a self-dynamo process may be possible, where the string pumps up its electric current at the expense of its mechanical motion. Since there is no correlation between the direction defined by $\nabla \times A$ (the curl of the string gauge vector) and the current, one expects that typically half the strings will have supercurrent aligned with the $U(1)$ string "magnetic" field and half oppositely. A distribution of currents from zero to the critical current is expected.

Superconducting loops can have a substantial effect on cosmic evolution. For instance, the nonsuperconducting loops can support the formation of cusps, spacetime events at which part of the string is moving at the speed of light. In nonconducting strings, this can lead to substantial gravitational radiation. In the charged case, much stronger electromagnetic radiation could presumably be produced. This, or the radiation from oscillating or spinning superconducting loops, can play a large role in the formation of voids and of other structure (Ostriker, Thompson and Witten 1986). Here the direct gravitational effects of the string are unimportant compared to the forces produced in its electromagnetic interactions.

In the simulations here, I shall show that string cross-connection still occurs (at least for the cases we have investigated), even when there is large current flowing oppositely (with respect to the $\nabla \times A$ directions on the two strings) on the two strings. The result is abrupt cancellation of the current in the central region and explosive electromagnetic radiation. Here is another mechanism that may contribute to void/structure formation.

X. *INITIAL DATA FOR INTERACTION SUPERCONDUCTING STRINGS*

When there is a superconducting bosonic field σ present, the asymptotic behavior given in (6) is extended:

$$\left.\begin{array}{ll} \phi_s & \approx a\rho \\ P & \approx 1 - c\rho^2 \\ \sigma_s & \approx \sigma_0(1 - t^2\rho^2/4) \\ K & \approx K_0(1 + g^2\sigma_0^2\rho^2/4) \end{array}\right\} \quad \text{near } \rho = 0, \text{ and}$$

$$\left.\begin{array}{ll} \phi_s & \approx \eta(1 - be^{-m_\phi \rho}) \\ P & \approx de^{-m_A \rho} \\ \sigma_s & \approx \sigma_\infty e^{-jr} \\ K & \approx K_\infty \ell n(k\rho) \end{array}\right\} \quad \text{near infinity} \tag{24}$$

where $K_0, K_\infty, a, b, c, d, t, j, k, \sigma_0$, and σ_∞ are constants adjustable in our fits, and $m_\phi^2 = 8\lambda_\phi\eta^2$ and $m_A^2 = e^2\eta^2$.

Numerical minimization, using these asymptotic forms, can then produce, as before, the initial static single string. We impose the Lorentz gauge on the electromagnetic vector potential C_μ. As before, we add the vector potentials and add the phases of each scalar field to superpose the (appropriately boosted and rotated) fields of the two strings. Since $|\phi| \rightarrow \eta$ far from the string, we still can take $\phi^{\text{TOT}} = ({}^R\phi {}^L\phi)/\eta$. For the σ fields, because they go to zero

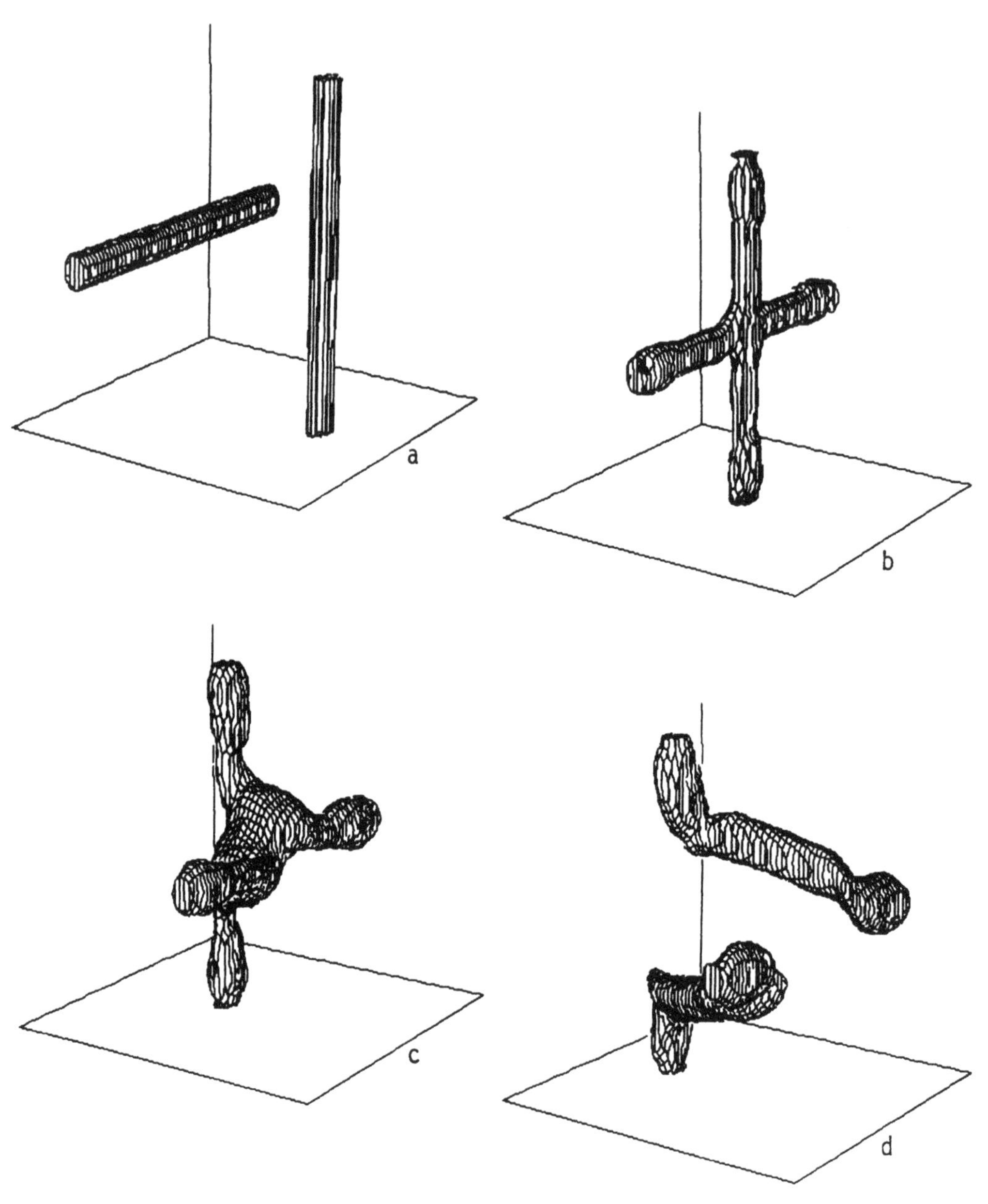

FIGURE 3

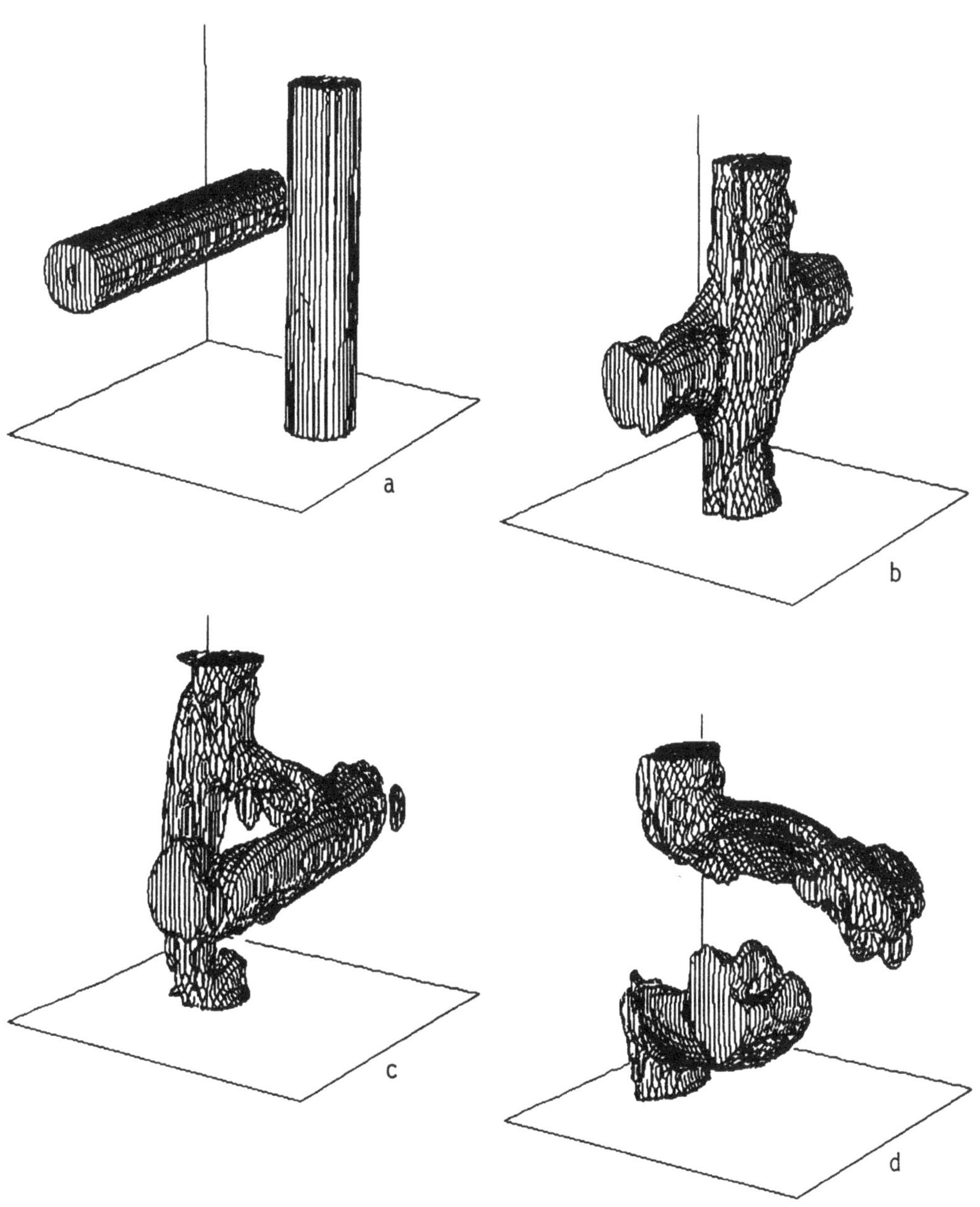

FIGURE 4

far from the strings, we can simply add the complex scalars also. Formation of the initial momenta, by considering the fields at Δt later, is carried out as in the nonsuperconducting core. Evolution of the differential equations and boundary handling are done by a natural extension of the process used in the nonsuperconducting case.

XI. NUMERICAL RESULTS: INTERSECTION OF SUPERCONDUCTING COSMIC STRINGS

We have now done a number of simulations of the interaction of superconducting strings. All the simulations done so far have been carried out at a 90° crossing angle and at $v = 0.75c$ approach speed. In all cases the strings were superconducting with the parameters given above. In all cases considered, reconnection of the cosmic strings with the direction expected from the $U(1)$ "magnetic" field occurs. The cases considered so far are:

a) zero current in both strings,
b) parallel equal currents (85% of critical),
c) opposite equal currents (85% of critical), and
d) nonequal parallel currents (85%/35% of critical).

In runs b)–d), "parallel" refers to current in each string parallel to the direction of the $U(1)$ "magnetic" direction, and "opposite" indicates that the current in *one* string is opposite the direction of the $U(1)$ "magnetic" direction. The importance of this difference is that when reconnection occurs in the direction specified by the $U(1)$ field, the parallel case allows smooth reconnection of its current also; in the opposite case, there is an abrupt break of the "circuit" and very substantial electromagnetic radiation is produced. In case d) above, there is still a discontinuity at the join because the currents in the two strings are not equal. (This, of course, is the situation expected in nature.) Substantial electromagnetic radiation occurs here also.

I discuss here cases b) and c) in which two strings carrying 85% of the maximal current cross and reconnect. The structures arising in the string scalar quantities are quite similar to those for the nonsuperconducting string (Figs. 1), so I concentrate here on the superconducting scalar and the (electromagnetic) magnetic field arising from the superconducting current.

Figures 3 and 4 show the parallel case. Figures 3 show the current-carrying scalar σ. Figure 3a shows the initial configuration, slice 0. Notice that we start the strings farther apart in the box than for the nonsuperconducting case in Figs. 1 to ensure that the magnetic fields, which fall off only as $1/\rho$, are well separated. By slice 105, Fig. 3b, the two strings are just touching. Slice 165, Fig. 3c, shows that the two strings have broken, though there is a

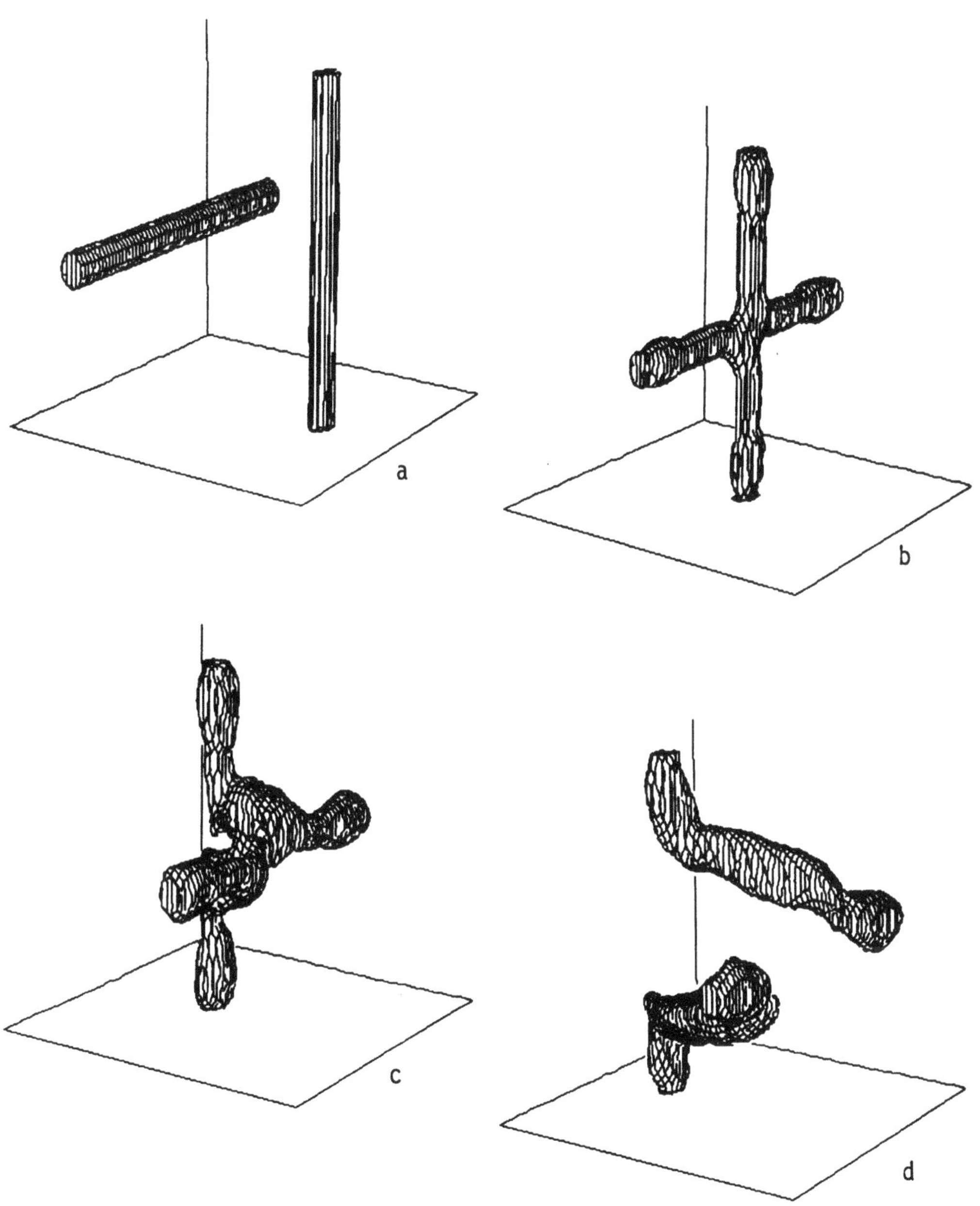

FIGURE 5

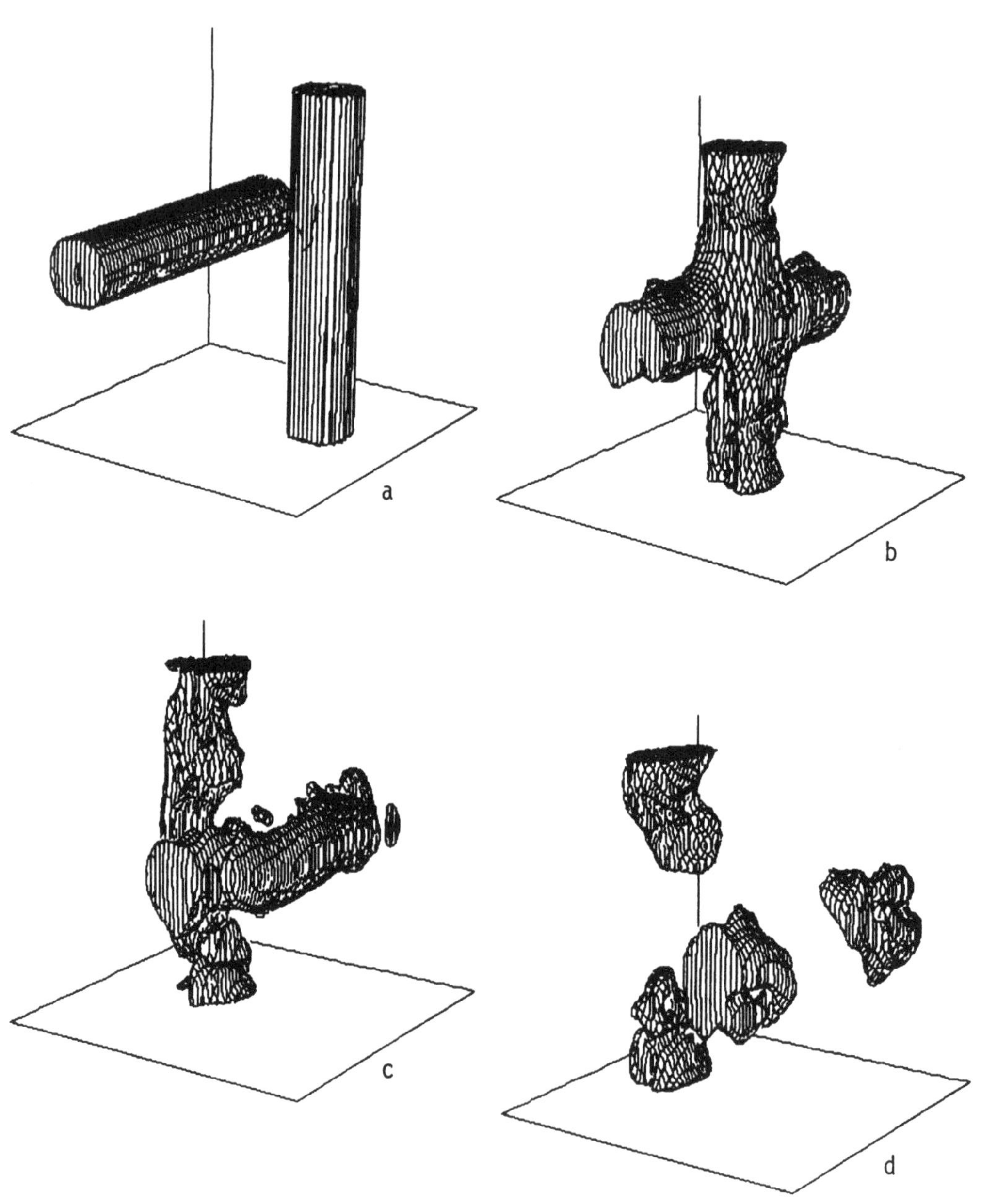

FIGURE 6

substantial region of excited scalar field σ between the two strings. Slice 210, Fig. 3d, shows that the strings have broken and assumed the expected final form. These figures plot the 50% contour of $|\sigma|$, normalized by σ_0, the value of $|\sigma|$ at the core of the initial strings.

Figures 4 show the evolution of the (electromagnetic) magnetic field associated with the two strings. The time slices correspond to those in Fig. 3. We see that the magnetic field, which wraps around the string, reconnects in the same way as does the scalar field σ; compare Fig. 4d with Fig. 3d. The ring-like structure in Fig. 4c and the appearance of Fig. 4d that there is a spherical "void" in the center of the simulation arise from electromagnetic radiation coming from the collision at the cube center. The forward propagating "bar" in Fig. 4c can be viewed as a kind of "virtual photon" phenomenon associated with the collision. The contouring here is at 50% of the maximum value of the magnetic fields in the initial configuration.

Figures 5 show the scalar field σ for the opposite current case. There is some difference in detail, but the general features are like those of the scalar field in the parallel case, Figs. 3. Figures 6 show the magnetic field for this case. Here the currents in the central part of the reconnected strings cancel, and so there is no magnetic field there after the collision, Fig. 6d. We again see the "virtual photon" in Fig. 6c and the spherical "void" in Fig. 6d.

CONCLUSION

I have been able to show that reconnection occurs in a wide range of string-string collisions. Hence the putative role of ordinary string loops in producing large scale structure passes this test; loops have a maximal probability of budding off. In the superconducting case we find that in all cases considered, reconnection is allowed, even if it means cancellation of almost maximal currents in the collision. The astrophysical implications of this can be substantial. If string collisions are frequent, a substantial amount of the string energy can be radiated electromagnetically in an explosive manner. Superconducting strings would then play a very dynamic role in driving substantial large scale structure through the interactions of the radiation with the background plasma.

A videomovie of the nonsuperconducting string interactions has been prepared and can be obtained by sending a blank videocassette (VHS, 8 mm, or 3/4 inch) with return postage.

ACKNOWLEDGEMENTS

I thank Dr. P. Laguna for his assistance on this paper, and also

R. Jacob and J. McCracken for very valuable help in computation, graphics, and copy proofing. This work was supported by NSF grants PHY-8404931 and PHY-8806567.

REFERENCES

Abrikosov, A. A. (1957). Zh. Eksp. Teor. Fiz. 32, 1442 [Sov. Phys.-JETP 5, 1174].
Albrecht, A. & Hindmarsh, M. (1987). Private communication.
Albrecht, A. & Turok, N. (1985). Phys. Rev. Lett. 54, 1868.
Bennett, D. P. & Bouchet, F. R. (1988). Preprint.
Kibble, T. W. B. & Turok, N. (1982). Phys. Lett. B 116, 141.
Leblanc, J. (1988). Private communication.
Meyers, E. & Rebbi, C. (1988). Preprint.
Müller-Hartmann E. (1966). Phys. Lett. 23, 521.
Nambu, Y. (1970). In Symmetries and Quark Models, ed. R. Chand, New York: Gordon and Breach.
Ostriker, J.P., Thompson, C., & Witten E. (1986). Phys. Lett. B 180, 231.
Ruback, P. (1988). Nucl. Phys. B 296, 699.
Spergel, D. N., Press, W. H., & Scherrer, R. J. (1988). In The Proceedings of the Yale Workshop on Cosmic Strings (to be published).
Turok, N. (1983a). Phys. Lett. B 123, 387.
Turok, N. (1983b). Phys. Lett. B 126, 437.
Turok, N. (1984). Nucl. Phys. B 242, 520.
Turok, N. (1985). Phys. Rev. Lett. 55, 1801.
Turok, N. & Brandenberger, R. (1986). Phys. Rev. D 33, 2175.
Vilenkin, A. (1981). Phys. Rev. Lett. 46, 1169.
Witten, E. (1985). Nucl. Phys. B 249, 557.
Zel'dovich, Ya. B. (1980). Mon. Not. R. Astron. Soc. 192, 663.

DYNAMICALLY INHOMOGENEOUS COSMIC NUCLEOSYNTHESIS

Hannu Kurki-Suonio
Drexel University, Philadelphia, PA 19104

Abstract. The effects of large-amplitude horizon-scale inhomogeneity in the energy density and flow on the cosmic nucleosynthesis are investigated with a fully general relativistic inhomogeneous nucleosynthesis code. Several mechanisms affecting the ^{4}He-abundance in such inhomogeneous models are identified.

1. Introduction

We present here some preliminary results of our ongoing work (Kurki-Suonio, Centrella, Matzner, Rothman, & Wilson 1988b) on inhomogeneous cosmic nucleosynthesis. The Centrella-Wilson (1983, 1984) plane-symmetric cosmology code provides the evolution of an inhomogeneous spacetime. The code is fully general relativistic allowing us to study the effects of nonperturbative horizon-scale inhomogeneities in the energy density and spacetime. These inhomogeneities affect each other and we refer to them as dynamical inhomogeneities, in contrast to mere baryon inhomogeneity, which also affects nucleosynthesis, but has practically no effect on the spacetime or on the hydrodynamical flow of energy in the early universe.

A study of inhomogeneous nucleosynthesis has a twofold motivation. First, we want to see how standard nucleosynthesis results are modified in the presence of inhomogeneity. Second, nucleosynthesis is one of the few probes we have on the conditions in the early universe; we would like to use it to obtain some information on the possible inhomogeneity present then. Inhomogeneity, of course, implies a large number of degrees of freedom; one cannot pretend, that doing a few computer runs would provide definite answers to these questions. On the other hand these questions must not be ignored.

This work is a continuation of Centrella *et al.* (1986). We have added ^{7}Li to the code, updated the nuclear reaction rates, and done some other improvements. Because of improved supercomputer access, a more extensive study has become possible. We now give results as a function of the present baryon density.

2. About the code

The inhomogeneous nucleosynthesis code (Centrella *et al.* 1986) is a combination of the Texas nucleosynthesis code developed by Rothman & Matzner (1984) and Kurki-Suonio & Matzner (1985) and the plane-symmetric inhomogeneous cosmology and hydrodynamics code by Centrella and Wilson (1983, 1984). The smaller reaction network (up to ^{4}He) used in Centrella *et al.* (1986) has now been replaced with a larger network with the 30 strong reaction rates listed by Fowler *et al.* (1967, 1975) and Harris *et al.* (1983) that involve nuclei with mass numbers $A \leq 7$ only, and their inverse reactions. Thus we can calculate all the cosmologically significant isotopes ^{2}H, ^{3}He, ^{4}He, and ^{7}Li. Because in an inhomogeneous calculation the time it takes to do a nucleosynthesis computation is multiplied by the number of grid zones, computer time is at a premium. We save time in not going beyond ^{7}Li.

Because of the smallness of the baryon to entropy ratio in the universe, the nuclear matter has a negligible contribution to the total energy density at the epoch of nucleosynthesis. Thus the nucleosynthesis part of the code does not affect the cosmology part. The cosmology part provides a fully general relativistic evolution of inhomogeneous spacetime and (radiation dominated) matter. The small baryon contamination flows with the bulk of the matter and the nucleosynthesis part of the code runs the nucleosynthesis within it.

The cosmology is plane-symmetric, *i.e.*, we use an xyz-coordinate system and the xy-planes are assumed homogeneous. The inhomogeneous z-direction is divided into 50 zones (in a typical run) and we impose periodic boundary conditions. The code uses $trK = const.$ slicing and the shift vector is used to keep the 3-metric diagonal, $\gamma = diag(A^2, A^2h^2, A^2)$. The evolution of the Einstein equations is fully constrainted: the Hamiltonian constraint is solved for h, and the momentum constraint is solved for K^z_z. For more details on the code see Centrella & Wilson (1983, 1984). The latter reference presents tests of the cosmology part of the code.

In connecting the cosmology part to the nucleosynthesis part, the temperature of a zone is obtained from the proper energy density using the equation of state. We use 3 flavors of neutrinos in all runs presented here. The proper time of the flowing matter is computed to obtain the correct time dilation effects for the nuclear reaction rates. We use a neutron mean life-time $\tau_n = 926$ s. The code reproduces the correct standard homogeneous nucleosynthesis results.

The neutron-to-proton ratio begins to deviate significantly from equilibrium at around $T_9 \sim 20$ (T_9 = temperature in 10^9 K). A nucleosynthesis run should thus be started above that temperature. For numerical reasons and to have better control over the inhomogeneity close to the main nucleosynthesis action ($T_9 \sim 1$) the cosmology is initialized at $T_9 \sim 10$. The code, without the nucleosynthesis part, is then run backward in time to $T_9 \sim 20$, where the nucleosynthesis is initialized, and then the code is run forward, with nucleosynthesis, to $T_9 \sim 0.03$, at which time the primordial abundances have been determined. In the runs presented here, the length of the grid (which is comoving) at initialization ($T_9 \sim 10$) is two horizon radii, or 4.33 light-seconds.

3. Results

There are many different kinds of inhomogeneity to consider. Here we study two kinds of initial density inhomogeneities. In both cases we set an inhomogeneity in the energy density and flow velocity at the initial slice ($T_9 \sim 10$). In an *adiabatic* case we then set the baryon density so that n_b/n_γ is constant in the turn-around slice ($T_9 \sim 20$). In a *non-adiabatic* case we simply set the baryon density to be constant in the turn-around slice. This leads to an inhomogeneous baryon-to-photon ratio. The form of the inhomogeneity used here is the same as in Centrella *et al.* (1986): the initial data for energy density and momentum density are

$$E = E_0 \left(1 + \sum_n E_n cos(2n\pi z)\right)$$

$$S_z = E_0 \sum_n v_{zn} cos(2n\pi z)$$

with $E_n = 0.6, 0.3, 0.15, .075, 0.0375$ and $v_{zn} = 0.8, 0.4, 0.2, 0.1, 0.05, 0.025$. This defines (at $T_9 \sim 10$) a narrower high-density region and a wider low-density region with a density contrast of about 4:1. In the runs, the energy density 'sloshes' back and forth and the highest instantaneous density constrast reached is over 10:1. We did both kinds of runs with 6 different values of average baryon density, corresponding to present-day ($T_\gamma = 2.7$ K) values $\rho_b = 0.04, 0.1, 0.2, 0.4, 1.0$ and 2.0 in units of 10^{-30}g cm^{-3}. The results are shown in Figure 1. (In Centrella *et al.* (1986), in the non-adiabatic runs, the actual average baryon densities were not the same as in the corresponding adiabatic and homogeneous runs. We have corrected this problem and the ρ_b values given here are the actual present-day values to which each model will evolve).

Comparing the results to homogeneous runs shows that the adiabatic inhomogeneity has raised ^{4}He by 0.4–1.0 % (absolute), with little effect on the other isotopes, whereas the nonadiabatic inhomogeneity has had less effect on ^{4}He, but a stronger effect on the other isotopes.

The observed behaviour of ^{2}H, ^{3}He, and ^{7}Li is what one would expect. Their final abundances are mainly determined by the n_b/n_γ-ratio during (the most active period of) nucleosynthesis. In the adiabatic case n_b/n_γ has remained fairly constant, and the results are close to what a homogeneous model with the same n_b/n_γ-ratio would give, even though because of the energy density inhomogeneity nucleosynthesis happens at different times in different regions. In the non-adiabatic case we then see essentially a baryon inhomogeneity effect.

The situation with ^{4}He is different. ^{4}He is mainly determined by the neutron fraction at the onset of nucleosynthesis, which depends on the dynamical history preceding nucleosynthesis. Let us first consider the adiabatic case, so that the n_b/n_γ-effect does not intrude, and nucleosynthesis begins everywhere at the same temperature. We note that ^{4}He has been raised (compared to the homogeneous case). This means that more neutrons have survived to nucleosynthesis temperature. Why? Review the history of the neutron fraction in standard nucleosynthesis: At high temperature neutrons follow their equilibrium abundance. As temperature falls, the rates of the weak $n \leftrightarrow p$ reactions drop very rapidly and the n fraction is left behind (above) the decreasing equilibrium abundance, and freezes at a certain value. As the time scale grows, the n fraction again begins to go down, now due to the free decay of neutrons; but then the nucleosynthesis begins.

The free decay of neutrons depends on the proper time experienced by them. In our runs the matter is flowing at relativistic speeds. Time dilation thus causes neutrons to decay slower, and thus more neutrons survive till nucleosynthesis time. We ran one of the adiabatic runs again, but with time dilation disabled in the code. It turned out that in this particular run time dilation accounted for one-third of the total increase in ^{4}He. To see the cause of the remaining two-thirds we examined the run more closely: Just before neutron freeze-out, one side of the grid has a higher density and thus a higher temperature, keeping the neutron equi-

Figure 1. The results of the inhomogeneous adiabatic (○) and nonadiabatic (+) runs compared to homogeneous runs (•). We plot the mass fractions of ^{2}H and ^{7}Li (effects on ^{3}He were small) as a function of the ^{4}He-fraction produced. The corresponding present baryon densities for the six runs of each type are (from left to right) 0.04, 0.1, 0.2, 0.4, 1.0, and 2.0 in units of 10^{-30}g cm^{-3}. We also show the adiabatic $\rho_b = 0.2 \times 10^{-30}$g cm^{-3} run with time dilation disabled (◇), and the nonadiabatic $\rho_b = 0.2 \times 10^{-30}$g cm^{-3} run with shifted baryon inhomogeneity (×).

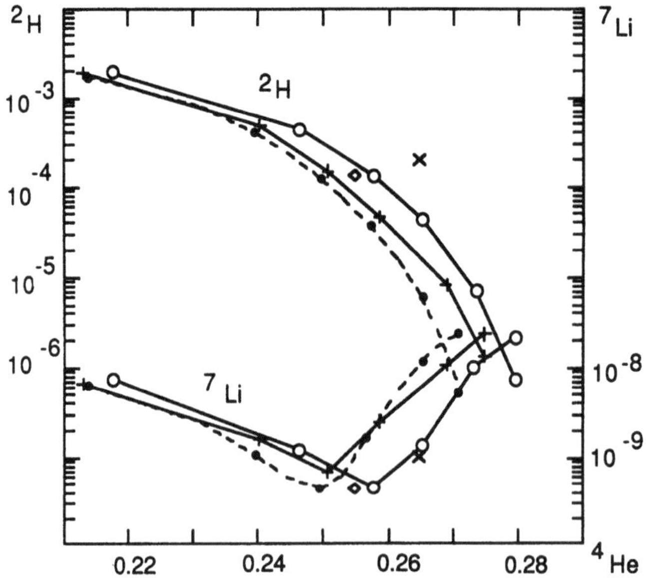

librium abundance fairly high. Then energy flows out, causing a sudden drop in temperature which leaves the neutron fraction frozen at a high value. In the other half the opposite happens; the inflow of energy slows down the cooling, delaying the freeze-out, and bringing the neutron fraction to a lower value. Because most of the nucleons were on the first half of the grid, the total ^{4}He production goes up.

In the nonadiabatic run one would expect ^{4}He to be increased even more, since baryon inhomogeneity alone raises ^{4}He. However, the opposite happens in our runs. To see how the n_b/n_γ-inhomogeneity managed to lower ^{4}He, we again had to examine the runs in detail. The neutron fraction before nucleosynthesis is independent of baryon density, and depends only on the temperature history (*vs.* proper time) along the flow world lines. The previoulsly mentioned effects cause the n fraction to be higher (compared to homogeneous runs) on one side of the grid and lower on the other at the onset of nucleosynthesis. In our nonadiabatic runs the n_b/n_γ-inhomogeneity happened to have such a distribution that most of the nucleons were on the low n fraction side. We ran one more run, where the baryon inhomogeneity had been shifted (we set it constant at a later time slice), so that baryon density now was enhanced on the side with the high n fraction at nucleosynthesis. This caused the ^{4}He-production to rise well above that of the corresponding adiabatic run.

4. About diffusion

A new development in inhomogeneous nucleosynthesis is the neutron diffusion scenario of Applegate, Hogan, & Scherrer (1987). In this scenario neutrons diffuse out from regions of very high baryon density just before nucleosynthesis.

In the work reported on in this paper we have ignored diffusion. We study horizon-scale inhomogeneities, and the neutron and proton mean free paths are much too short compared to the horizon for diffusion to have any effect in the time available. Any baryon inhomogeneity will just follow the flow of the radiation fluid. For diffusion to become important, the baryon inhomogeneity scale must be many orders of magnitude shorter than the horizon scale at nucleosynthesis. Only after research on the cosmic quark-hadron phase transition revealed that a scale of such magnitude could emerge from the transition, and possibly leave behind a strong inhomogeneity in baryon number, was this new scenario conceived. Nucleosynthesis is very directly affected by the neutron-to-proton ratio and thus this new scenario allows very dramatic changes to the standard picture of nucleosynthesis.

As reported elsewhere (Kurki-Suonio *et al.* 1988a) diffusion was added to our code to study this new effect (without dynamical inhomogeneities). We found that the effect on final abundances was much less than what had been thought at first. In our runs neutrons diffused back into the high-density regions once nucleosynthesis began there, and thus most of the effect was eliminated. Because of this back-diffusion the neutron-proton segregation has a significant effect only in a very narrow range of baryon inhomogeneity distance scale. Since our code is plane-symmetric, we were restricted to a slab type inhomogeneity. It has been argued that a different geometry might lead to different results. We believe that if the high density regions cover a significant fraction of total volume, the results should not be sensitive to the shapes of these regions; but if the high density regions are small and far apart the geometry could be important. We are presently investigating this question.

5. Conclusion

We have identified several mechanisms which affect the nucleosynthesis results in an inhomogeneous model. ^{4}He, because of its dependence on the pre-nucleosynthesis dynamics, proved to be the most interesting. Its production was seen to depend in part on a purely relativistic effect, time dilation. In the case of an adiabatic inhomogeneity, we only found effects that tend to raise ^{4}He. If the adiabaticity requirement is removed, there is a much larger freedom in setting the initial data, which allows the ^{4}He-production to be 'manipulated'.

In these runs most of the inhomogeneity had smoothed out by the onset of nucleosynthesis; this is part of the reason why the effects on other isotopes than ^{4}He were small. These abundances are determined in a fairly short (measured in logarithmic time) period, and mainly depend on n_b/n_γ at that time. Inhomogeneity effects on the isotopes would presumably be magnified if the amplitude of the inhomogeneity were increased. In a nucleosynthesis computation these inhomogeneities need to be carried over a very long time period and the present code is not robust enough to handle much stronger dynamical inhomogeneities without difficulty.

Acknowledgements

The computations were done at the National Center for Supercomputing Applications. This research was supported in part by National Science Foundation Grants Nos. PHY84-04931, PHY84-51737, and PHY87-06315.

References

Applegate, J.H., Hogan, C.G., & Scherrer, R.J. (1987). Cosmological baryon diffusion and nucleosynthesis. Phys. Rev. **D35**, 1151–60.

Centrella, J. & Wilson, J.R. (1983). Planar numerical cosmology. I. The differential equations. Astrophys. J. **273**, 428–35.

Centrella, J. & Wilson, J.R. (1984). Planar numerical cosmology. II. The difference equations and numerical tests. Astrophys. J. Supp. Ser. **54**, 229–49.

Centrella, J., Matzner, R.A., Rothman, T., & Wilson, J.R. (1986). Cosmic nucleosynthesis and nonlinear inhomogeneities. Nucl. Phys. **B266**, 171–227.

Fowler, W.A., Caughlan, G.R., & Zimmermann, B.A. (1967). Thermonuclear reaction rates. Ann. Rev. Astron. Astrophys. **5**, 525–70.

Fowler, W.A., Caughlan, G.R., & Zimmermann, B.A. (1975). Thermonuclear reaction rates, II. Ann. Rev. Astron. Astrophys. **13**, 69–112.

Harris, M.J., Fowler, W.A., Caughlan, G.R., & Zimmermann, B.A. (1983). Thermonuclear reaction rates, III. Ann. Rev. Astron. Astrophys. **21**, 165–76.

Kurki-Suonio, H. & Matzner, R. (1985). Anisotropy and cosmic nucleosynthesis of light isotopes including ^{7}Li. Phys. Rev. **D31**, 1811–4.

Kurki-Suonio, H., Matzner, R.A., Centrella, J.M., Rothman, T., & Wilson, J.R. (1988a). Inhomogeneous nucleosynthesis with neutron diffusion. Phys. Rev. **D38**, no. 4.

Kurki-Suonio, H., Centrella, J.M., Matzner, R.A, Rothman, T., & Wilson, J.R. (1988b). Unpublished.

Matzner, R.A., Centrella, J.M., Rothman, T., & Wilson, J.R. (1988). Primordial nucleosynthesis in a universe with nonlinear inhomogeneities. *In* Origin and distribution of the elements, ed. G.J. Mathews, pp. 124–59. Singapore: World Scientific.

Rothman, T. & Matzner, R. (1984). Nucleosynthesis in anisotropic cosmologies revisited. Phys. Rev. **D30**, 1649–68.

INITIAL VALUE SOLUTIONS IN PLANAR COSMOLOGIES

Peter Anninos and Joan Centrella
Drexel University, Philaladelphia, Pa 19104

Richard Matzner
The University of Texas at Austin, Austin, Texas 78712

Abstract. We derive the Hamiltonian and momentum constraint equations for a vacuum plane symmetric cosmology using the York decomposition. We also discuss a numerical code developed to solve these equations for standing wave and pulse-like initial data. As a check on our code, we obtain exact and perturbative analytic solutions. We present graphic results of both weak and strong perturbations.

INTRODUCTION

Because of the complexity of the Einstein equations, few analytic solutions exist which describe gravity exactly. It is our purpose to develop a 1-D numerical code describing an inhomogeneous cosmology. Although the restriction to 1-D is severe, it still provides a framework within which to study a number of interesting fully nonlinear phenomena. In particular, we will explore some of the nonlinear properties of wave propogation. How do large amplitude waves interact with themselves, other such waves and/or the background curvature? How do these waves distort themselves and spacetime? Is the generation of harmonics of any consequence? Do waves steepen to form shocks? Do they disperse? Belinski and Zakharov (1980), Carr and Verdaguer (1983), and Ibanez and Verdaguer (1983, 1985) have recently derived exact "soliton" solutions to the Einstein equations possessing two Killing vectors. What is the nature of the interaction between nonlinear and dispersive properties of these spacetimes? These are some of the questions we hope to address with our code.

The importance of such a code has been justified by several authors. Centrella and Wilson (1983, 1984) have developed a 1-D planar cosmological code with hydrodynamics which has been used to study a number of different phenomena, in particular, nonlinear gravitational waves (Centrella 1986), cosmic nucleosynthesis (Centrella *et al.* 1986), and inflationary cosmology (Kurki-Suonio *et al.* 1987). Our code differs from that of Centrella and Wilson in the way we set up and solve the initial value equations. We use the York formalism to freely specify the metric and momentum variables up to a conformal transformation. This important difference allows us to explore the parameter space of initial data more easily. The York procedure is described in more detail in the next section.

In this paper, we concentrate on solutions to the initial value equations only; the evolution equations will be discussed in future papers. We present standing wave as well as pulse-like initial data for the freely-specifiable variables. We also present analytical solutions as tests of our code when available. These analytic solutions are in the form of perturbations of spatially homogeneous Kasner spacetimes. In some cases, the simplicity of the momentum constraint equations allows us to find exact nonperturbative solutions for the momentum variables.

FORMALISM

We write our 1-D metric in the usual 3+1 splitting of spacetime:

$$ds^2 = -(\alpha^2 - \beta_i \beta^i)\, dt^2 + 2\beta_i\, dx^i dt + \phi^4 (dx^2 + h^2 dy^2 + dz^2)\ , \tag{1}$$

where all variables are functions only of z and t. ϕ is the conformal factor and the variable h describes anisotropic shear . The lapse function α determines the orthogonal proper time interval between slices of spacetime and the shift vector β shifts the spatial coordinates. Since we will here only deal with the initial value problem, α and β will not be used. We impose periodicity in the spatial coordinate z on all variables to produce spacelike slices which are topologically three-tori and characteristic of Gowdy T^3 models (Gowdy 1971, Berger 1974).

In the 3+1 split for vacuum, we write the Hamiltonian constraint

$$R + (\mathrm{tr}\,K)^2 - K_{ij} K^{ij} = 0 \tag{2}$$

and the momentum constraint

$$D_j (K^{ij} - \gamma^{ij}\, \mathrm{tr}\,K) = 0\ . \tag{3}$$

For further details see York (1979). Our convention uses Latin indices for spatial quantities: i,j = 1,2,3. γ_{ij} is the 3-metric of the spacelike slice, K_{ij} is the extrinsic curvature tensor and $\mathrm{tr}\,K = \gamma^{ij} K_{ij}$. D_j is the covariant spatial derivative and R is the Ricci scalar formed from the 3-metric.

We briefly review the York formalism used to solve Eqs. (2) and (3). This procedure begins by defining a conformal transformation of the metric

$$\gamma_{ij} = \phi^4 \hat{\gamma}_{ij}\ , \tag{4}$$

where the conformal factor ϕ is a positive scalar function. Next, define the trace-free part of the extrinsic curvature,

$$A_{ij} = K_{ij} - \frac{1}{3}\gamma_{ij}\, \mathrm{tr}\,K\ \ , \tag{5}$$

and then conformally transform A_{ij} as

$$\hat{A}^{ij} = \phi^{10} A^{ij}\ . \tag{6}$$

The quantity tr K will be treated as a given scalar function and will not be conformally transformed. The transformed traceless momentum tensor is further split into a transverse-traceless part and a longitudinal part:

$$\hat{A}^{ij} = \hat{A}_*^{ij} + (\hat{l}w)^{ij} \quad . \tag{7}$$

The longitudinal part (**w**) vanishes in our case since our vacuum model has no z Killing vectors (Evans 1986). Now we may write the conformally transformed Hamiltonian constraint as

$$\hat{\Delta}\phi = \frac{1}{8}\phi\left[\hat{R} - \hat{A}_{ij}\hat{A}^{ij}\phi^{-8} + \frac{2}{3}\phi^4(\mathrm{tr}\,K)^2\right] \tag{8}$$

and the momentum constraint as

$$\hat{D}_j\hat{A}_*^{ij} = 0 \,, \tag{9}$$

where

$$\hat{\Delta} \equiv \hat{\gamma}^{ij}\hat{D}_i\hat{D}_j \; .$$

Since we require that K_{ij} be diagonal, then A_{ij} is also diagonal. Eqs. (5) and (9) then imply that there is one free component of A_{ij}. We choose this to be

$$\eta = A_x^x - A_y^y \quad . \tag{10}$$

We also define

$$\hat{\eta} = \hat{A}_x^x - \hat{A}_y^y \qquad \text{and} \qquad \hat{\eta}_* = \hat{A}_{*x}^x - \hat{A}_{*y}^y \tag{11a,b}$$

and, since **w** = 0, Eq. (7) results in the identities

$$\hat{A}_{*j}^i = \hat{A}_j^i \qquad \text{and} \qquad \hat{\eta}_* = \hat{\eta} \; . \tag{12a,b}$$

Using Eqs. (12) and the metric variables of Eq. (1), we rewrite the Hamiltonian constraint (8) as

$$h^{-1}(\phi' h)' = \frac{\phi}{8}\left[-2\frac{h''}{h} + \frac{2}{3}(\mathrm{tr}\,K)^2\phi^4 - \frac{1}{2}\left(\hat{\eta}^2 + 3(\hat{A}_z^z)^2\right)\phi^{-8}\right] \tag{13}$$

and the momentum constraint (9) as

$$(\hat{A}_{*z}^z h^{3/2})' = -\frac{1}{2}h^{1/2}h'\hat{\eta}_* \; . \tag{14}$$

We now solve the constraint equations in their final form of (13) and (14) by first specifying the free data

$$(\hat{\gamma}_{ij},\ \mathrm{tr}\,K,\ \hat{\eta})$$

such that they satisfy the periodicity condition in the spatial coordinate z. Then, following the prescription laid out by York, we recompose the transformed quantities to obtain the data set (γ_{ij}, K_{ij}) satisfying fully the constraint Eqs. (2) and (3). Specifically, our conformal metric is

$$\hat{\gamma}_{ij} = \mathrm{diag}\,(1, h^2, 1)$$

so that h is the only freely specifiable metric component. We also choose constant mean curvature slices which implies that tr K is a function only of time. In the following sections, we present initial data for the free variables in the form of standing wave and pulse-like data and explore the solution space with a number of parameter runs.

PERTURBATION ANALYSIS

In another paper (Anninos *et al.* 1988), hereafter referred to as Paper 1, we presented a perturbation expansion for our metric and momentum variables within the context of homogeneous Kasner solutions possessing the metric

$$ds^2 = -dt^2 + t^{2p_1}dx^2 + t^{2p_2}dy^2 + t^{2p_3}dz^2 \ . \tag{15}$$

We considered two axisymmetric cases ($p_1 = p_3$):

case (C) :	$p_1 = 2/3$ $p_2 = -1/3$	axisymmetric vacuum expanding cosmology
case (F) :	$p_1 = 0$ $p_2 = 1$	flat space in expanding Milne-like coordinates

and presented the perturbation expansion as:

case (C):	case (F):	
$\phi = t^{1/3}(1+\phi_1+\phi_2)$	$\phi = 1+\phi_1+\phi_2$	(16a)
$h = t^{-1}(1+h_1+h_2)$	$h = t(1+h_1+h_2)$	(16b)
$\mathrm{tr}\,K = -t^{-1}$	$\mathrm{tr}\,K = -t^{-1}$	(16c)
$\hat{A} = -\frac{1}{3}t(1+\hat{A}_1+\hat{A}_2)$	$\hat{A} = \frac{1}{3}t^{-1}(1+\hat{A}_1+\hat{A}_2)$	(16d)
$\hat{\eta} = -t(1+\hat{\eta}_1+\hat{\eta}_2)$	$\hat{\eta} = t^{-1}(1+\hat{\eta}_1+\hat{\eta}_2)$.	(16e)

Our notation of Eqs. (16) uses subscripts to indicate the order of smallness. For example, $h_1 \sim O(\varepsilon)$, where $\varepsilon \ll 1$. Also, to simplify the appearance of the momentum variables, we made the following notational change

$$\hat{A} \equiv \hat{A}_z^z = \hat{A}_{*z}^z \qquad \text{and} \qquad \hat{\eta} = \hat{\eta}_* \ . \tag{17}$$

In this paper, we present two types of freely specifiable data : (1) standing wave perturbations of an axisymmetric Kasner cosmology and (2) pulse-like perturbations of flat space.

First we consider the standing wave data. In Paper 1, we presented this type of perturbation against flat space. Because of the sign reflection symmetries in the momentum variables, the perturbed constraint equations for a Kasner background are identical to those derived in Paper 1 for flat space if we set the initial slice at time (t=1). The perturbation equations to first order are obtained by substituting Eqs. (16) into (13) and (14) to yield:

$$\hat{A}_1' + 3\,h_1' = 0 \tag{18}$$

$$\phi_1'' = -\frac{1}{4}h_1'' + f(t)\left[\phi_1 - \frac{1}{8}\hat{\eta}_1 - \frac{1}{24}\hat{A}_1\right] \ , \tag{19}$$

where $f(t) = t^{-2/3}$ for case (C) and $f(t) = t^{-2}$ for case (F).

Following Paper 1, we assume periodic standing wave forms for our free variables. Since we are completely free to specify these variables, we put the waves only in the "first order" terms and set the "higher order" terms to zero. Utilizing the case (C) perturbation expansion of Eqs. (16), the free variables take the following exact form:

$$h = 1 + h_1 = 1 + \sum_m h_{1(m)} \sin\left(2\pi m \frac{z}{L} + \delta_{1(m)}\right) \tag{20a}$$

$$\hat{\eta} = -\,(1 + \hat{\eta}_1) = -\left(1 + \sum_m \hat{\eta}_{1(m)} \sin\left(2\pi m \frac{z}{L} + \delta_{1(m)}\right)\right) \ , \tag{20b}$$

where L is the length of the grid. The quantity $\delta_{1(m)}$, a function of m, appears in both Eqs. (20) to guarantee linearization stability (Brill 1982). To determine the relative importance of the momentum and metric variables, we analyze initial data of different modes. The following choices are the simplest which both meet the requirement of linearization stability and allow an exact analytical solution (see Paper 1):

$$h_1 \equiv a \sin\left(2\pi \frac{z}{L}\right) \qquad \text{and} \qquad \hat{\eta}_1 \equiv b \sin\left(6\pi \frac{z}{L}\right) \ . \tag{21a,b}$$

Thus,

$$\lambda_h = 3\,\lambda_{\hat{\eta}} = L \ .$$

Because of their complexity, we will not write out the perturbative and exact solutions to the constraints. For some of these solutions we will refer the reader to Paper 1. However, plots of these solutions are given in Figs. (2a), (2b) and (2c), with (a, b, c) being the direction of increasing perurbation amplitudes. More detailed explanations of these figures are given in the next section.

As a further test of our code, we set up pulse-like initial data. This type of data will eventually be used to describe traveling waves in an effort to address the questions raised in the beginning of this paper. We construct this pulse in the following manner :

$$\begin{aligned} I &= 1 & & 0 \le z \le z_0 - \frac{1}{4} \\ I &= \frac{1}{1 - \frac{\varepsilon}{1+\varepsilon}\sin^{2m}(2\pi(z - z_0 + \frac{1}{4}))} & & z_0 - \frac{1}{4} \le z \le z_0 + \frac{1}{4} \\ I &= 1 & & z_0 + \frac{1}{4} \le z \le 1 \, , \end{aligned} \tag{22}$$

where I represents the free data, ε is the amplitude of the pulse against a flat background (see Eq. (24) below), z_0 is the location of the peak, and m defines the half-width at half maximum by the relation

$$z_{max} - z_{max/2} = \frac{1}{4} - \frac{1}{2\pi}\sin^{-1}\left[\frac{1+\varepsilon}{2+\varepsilon}\right]^{1/2m} . \tag{23}$$

This form was chosen mainly because it provides an easy means of obtaining a first order solution to the Hamiltonian constraint (Eq. 19) in the limit where $\varepsilon << 1$.To find ϕ_1, we expand I, which we identify as our free metric and momentum variables, to first order as

$$I = h = \hat{\eta} = 1 + \varepsilon \sin^{2m}(2\pi(z - z_0 + \frac{1}{4})) + O(\varepsilon^2) \, . \tag{24}$$

Because the metric and momentum variables are identical (Eq. 24), linearization stability is guaranteed (see Paper 1). Making the specific choice of m = 2 and using the flat space perturbation expansion, case (F) of Eqs. (16), we identify :

$$h_1 = \hat{\eta}_1 = \frac{\varepsilon}{8}\left[\cos(8\pi(z - z_0 + \frac{1}{4})) - 4\cos(4\pi(z - z_0 + \frac{1}{4})) + 3\right] . \tag{25}$$

To solve the Hamiltonian constraint (19) to first order, it is necessary to identify three separate regions across the z-axis. This is a result of the construction of the free data by means of Eq. (22).The differential equations satisfied by ϕ_1 across the entire grid are

$$\phi''_{1a} - \phi_{1a} = 0 \qquad 0 \le z \le z_0 - \frac{1}{4} \tag{26a}$$

$$\phi''_{1b} - \phi_{1b} = -\frac{h''_1}{4} \qquad z_0 - \frac{1}{4} \le z \le z_0 + \frac{1}{4} \tag{26b}$$

$$\phi''_{1c} - \phi_{1c} = 0 \qquad\qquad z_0 + \frac{1}{4} \leq z \leq 1 \, . \tag{26c}$$

We obtain the following general solution to the above equations by matching ϕ_1 and its first derivatives at $z = 0$, $z_0 - 1/4$ and $z_0 + 1/4$:

$$\phi_{1a} = \frac{\sqrt[4]{e}\, d}{(1 + \sqrt{e}\,)} \cosh\left(z - z_0 + \frac{1}{2}\right) \tag{27a}$$

$$\begin{aligned} \phi_{1b} = &- \frac{\sqrt[4]{e}\, d}{(1 + \sqrt{e}\,)} \cosh(z - z_0) \\ &- \frac{2\pi^2 \varepsilon}{1 + (8\pi)^2} \cos\left(8\pi\left(z - z_0 + \frac{1}{4}\right)\right) \\ &+ \frac{2\pi^2 \varepsilon}{1 + (4\pi)^2} \cos\left(4\pi\left(z - z_0 + \frac{1}{4}\right)\right) \end{aligned} \tag{27b}$$

$$\phi_{1c} = \frac{\sqrt[4]{e}\, d}{(1 + \sqrt{e}\,)} \cosh\left(z - z_0 - \frac{1}{2}\right) , \tag{27c}$$

where we define the coefficient d as

$$d = 2\pi^2 \varepsilon \left[\frac{1}{1 + (4\pi)^2} - \frac{1}{1 + (8\pi)^2} \right] . \tag{28}$$

An exact analytical solution of the momentum constraint may also be easily derived since we selected identical metric and momentum free data. We write this solution as

$$\hat{A} = \frac{8}{15} h^{-3/2} - \frac{h}{5} \, . \tag{29}$$

We derive the first order perturbative momentum solution by setting $h = 1 + h_1$ and expanding Eq. (29) in powers of ε. The result

$$\hat{A} = \frac{1}{3}(1 + \hat{A}_1) = \frac{1}{3}(1 - 3\, h_1) \tag{30}$$

is consistent with Eq. (18).

We present plots of this momentum solution in Figs. (3a), (3b) and (3c) along with the numerically computed results. We also show graphs of the perturbative and numerical solutions to ϕ. These results are presented with increasing amplitude of perturbation with the uppermost plots (a) being the smallest and (c) the largest. These results are explained in more detail in the next section.

NUMERICS

The complete initial value problem for a metric of the form of Eq. (1)

reduces to solving just two equations, (13) and (14). In this section, we discuss the numerical code we developed to solve these equations and present some comparisons between the numerically generated solutions and the perturbative results established in the previous section. Since most of this material has already been discussed in Paper 1, we limit the discussion of our code to a brief outline.

Fig. (1) displays the numerical grid on which all variables are discretized. The z-axis is covered with two sets of coordinates: ZA_k at zone faces and ZB_k at zone centers, with k = 1, 2, ..., km defining the different zones. Periodic boundary conditions are imposed by identifying zones 1 and km-1, and zones 2 and km. We choose z = 0 to lie at ZB_2 and z = L at ZB_{km}. All variables are defined at zone centers; when converting a differential equation to a difference equation, we average all variables if necessary to center the equation correctly on the ZB grid.

Figure 1. The numerical grid. Coordinates ZA_k define zone faces and ZB_k label the zone centers.

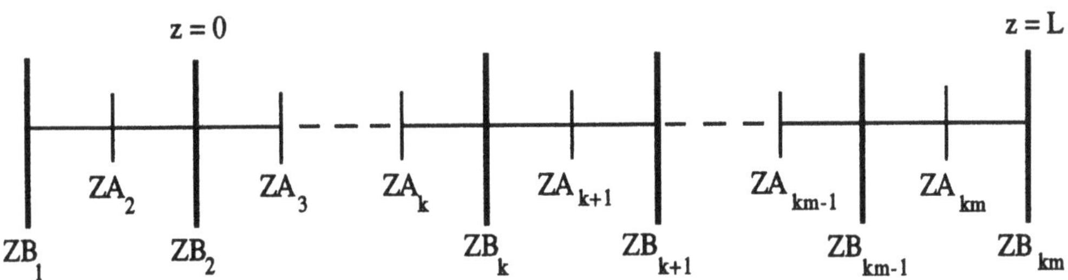

The momentum constraint (14) is solved quite easily and adequately with a simple trapezoidal integration scheme producing a truncation error ~ $(\Delta z)^2$. The nonlinear Hamiltonian constraint is more difficult to solve. The York decomposition transforms this equation into a form (13) which guarantees the existence of a unique positive bounded solution (O'Murchadha & York 1973). In finding this solution, we follow a strategy based on that developed by Evans. We first linearize Eq. (13) about some solution ϕ_o and discretize the differential equation to yield a symmetric matrix of linear equations. We then solve these equations with an optimized Incomplete Cholesky Conjugate Gradient (ICCG) matrix solver (Kershaw 1978, Evans 1986). This solution is then used as the new value of ϕ_o and this cycle is repeated until successive iterations converge.

We have tested this procedure extensively for a number of different situations, including a case in which we have an exact analytic solution for ϕ. Convergence of the initial guesses ϕ_o was obtained in all cases to one part in 10^{10} in just a few iterations (typically 3 or 4). Errors were ~ $(\Delta z)^2$ or better, as expected by the central difference approximation to our differential equations. We present some results in Figs. (2) and (3) which are representative of the accuracy with which our code solves the initial value equations. All calculations were made with 128 nodes (127 zones) to cover the z-axis ranging from zero to one.

Figs. (2a), (2b) and (2c) represent standing wave perturbations of the axisymmetric Kasner cosmology. We present solutions to both the momentum and Hamiltonian constraints for the free data given by Eqs. (21a,b). The labels (a), (b), and (c) on top of

Figure 2. Numerical and analytic solutions to standing wave perturbations of the axisymmetric Kasner cosmology.

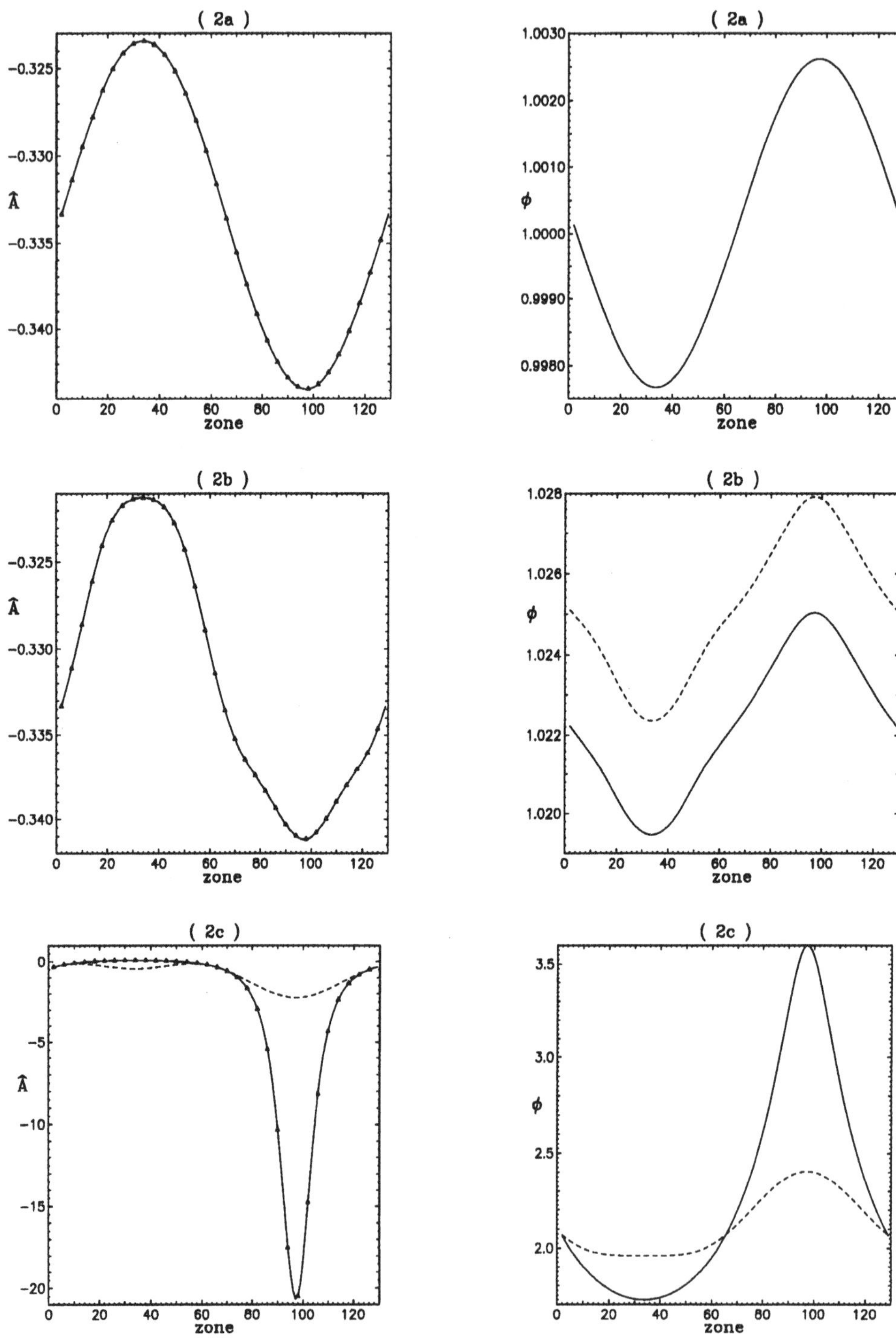

Figure 3. Numerical and analytic solutions to pulse-like perturbations of flat space.

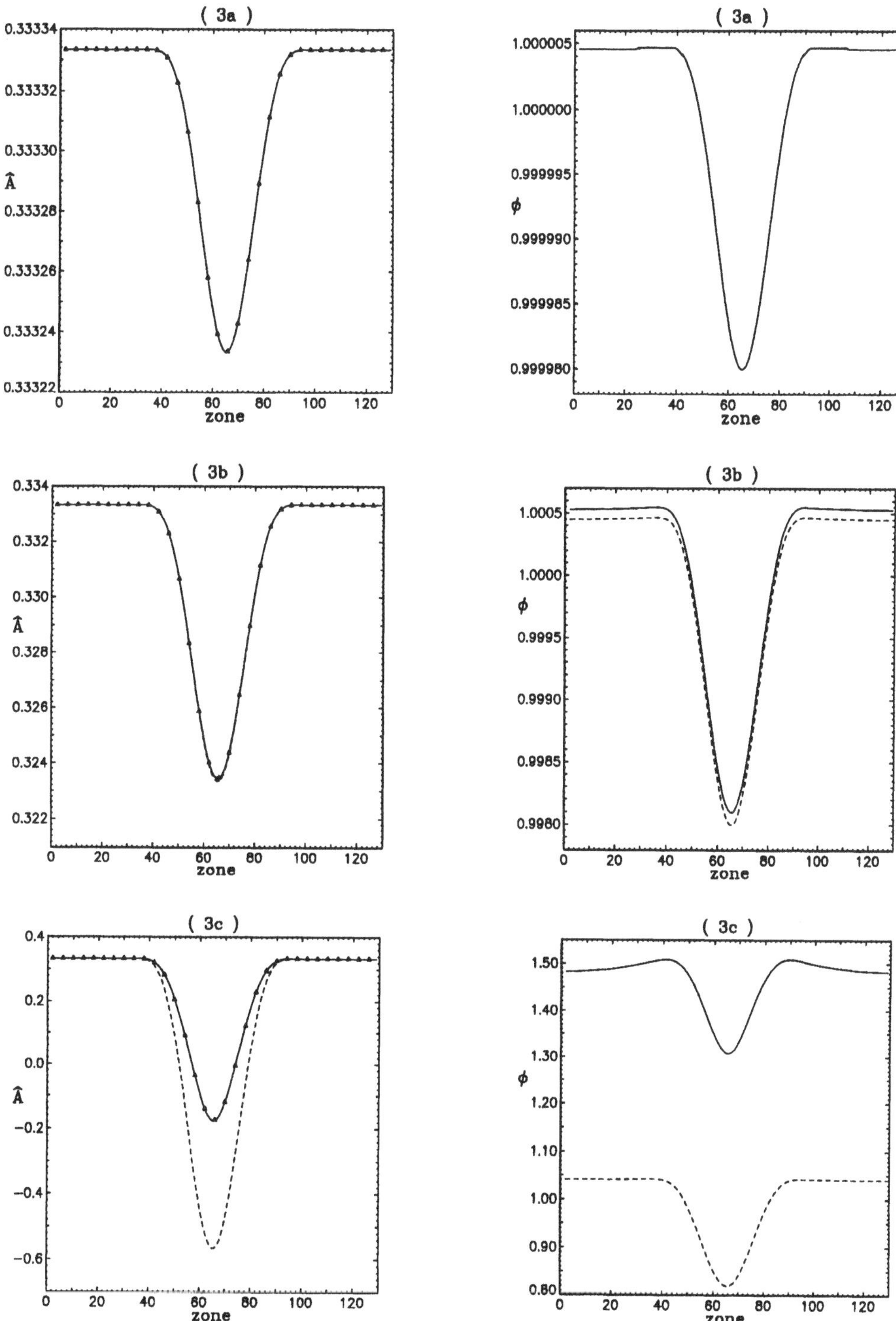

each figure indicate the perturbation amplitude, that is : (a) $a = b = 0.01$, (b) $a = 0.01$ and $b = 0.9$, and (c) $a = 0.9$ and $b = 0.01$. The line template for these figures is as follows: solid lines are the code results, dashed lines represent the second order perturbative results presented in Paper 1, and triangles are the exact analytic solutions obtained only for the momentum variables. Table 1 displays the errors between the exact and the numerically computed momentum as well as the difference between the second order perturbative and numerically computed Hamiltonian constraint.

Fig. (3) shows the pulse-like perturbations of flat space discussed above. The labels on top of each figure indicate (a) $a = b = 0.0001$, (b) $a = b = 0.01$, and (c) $a = b = 0.9$. The line templates are the same as in Fig. (2) except that now the dashed lines represent first order perturbation solutions. Table 2 shows the errors between the exact analytical and numerically computed momentum as well as the errors between the first order perturbative and computed solutions to the Hamiltonian constraint.

Table 1. Absolute and cumulative errors for the runs shown in Fig. (2). See Paper 1 for definitions of these errors.

	MOMENTUM		HAMILTONIAN	
amplitudes	maximum absolute error in single zone	cumulative relative error over entire grid	maximum absolute error in single zone	cumulative relative error over entire grid
a = b = .01	3×10^{-6}	6×10^{-6}	5×10^{-7}	3×10^{-7}
a = .01, b = .9	8×10^{-6}	1×10^{-5}	3×10^{-3} *	3×10^{-3} *
a = .9, b = .01	4×10^{-3}	2×10^{-4}	——— *	——— *

Table 2. Absolute and cumulative errors for the runs shown in Fig. (3).

	MOMENTUM		HAMILTONIAN	
amplitudes	maximum absolute error in single zone	cumulative relative error over entire grid	maximum absolute error in single zone	cumulative relative error over entire grid
a = b = .0001	1×10^{-7}	1×10^{-7}	1×10^{-8}	9×10^{-9}
a = b = .01	1×10^{-5}	1×10^{-5}	1×10^{-4} *	8×10^{-5} *
a = b = .9	2×10^{-3}	2×10^{-3}	——— *	——— *

The asterisks in Tables 1 and 2 indicate that the corresponding entries are not true error indicators because the analytic result is only a perturbative solution. In fact, no entries are made for the larger amplitude cases.

FUTURE DIRECTIONS

We have demonstrated that we can successfully solve the initial value problem numerically for vacuum cosmologies with plane symmetry. The next stage in the development of this project is to evolve the spacelike slices containing initial data forward in time. This necessitates solving first order coupled hyberbolic partial differential equations for the evolution as well as second order elliptic equations enforcing the constraints. This will be further complicated by the shift and lapse functions which will be chosen to evolve our coordinate system in a predetermined gauge. Although we have addressed and solved some of these problems, we will not present our results here but refer the reader to future papers.

Once we have a fully developed and properly tested evolution code, we may begin to address some of the questions raised in the introduction. Our approach will be that of computational synergetics (Zabusky 1984). We will use the code to search for new physical behaviour and in the process determine the effect of all relevant parameters. This knowledge should then prove advantageous to any analytical treatment of this behaviour.

ACKNOWLEDGEMENTS

This research was supported by NSF grants PHY-8404931, PHY-8417918, PHY-8451732, PHY-8706315, and PHY-8806567, and by Cray Research, Inc. The computations were carried out at the National Center for Supercomputing Applications at the University of Illinois.

REFERENCES

Anninos, P., Centrella, J. M., Matzner, R. (1988). Phys. Rev. D, submitted.

Belinski, V., Zakharov, V. (1978). Sov. Phys. JETP, 48, 985.

________________ . (1980). Sov. Phys. JETP, 50, 1.

Berger, B. K. (1974). Ann. Phys., 83, 458.

Brill, D. (1982). In Spacetime and Geometry, eds. R. A. Matzner and L. C. Shepley (Austin: University of Texas Press), p. 59.

Carr, B., Verdaguer, E. (1983). Phys. Rev. D, 28, 2995.

Centrella, J. M., Wilson, J. (1983). Ap. J., 273, 428.

________________ . (1984). Ap. J. Suppl., 54, 229.

Centrella, J. M. (1986). In Dynamical Spacetimes and Numerical Relativity, ed. J. M. Centrella (Cambridge: Cambridge University Press), p. 123.

Centrella, J. M., Matzner, R., Rothman, T., and Wilson, J. (1986). Nuclear Physics B, 266, 171.

Evans, C. R. (1986). In Dynamical Spacetimes and Numerical Relativity, ed. J. M.Centrella.(Cambridge: Cambridge University Press), p. 3.

Gowdy, R. H. (1971). Phys. Rev. Letters, 27, 826.

Ibanez, J., Verdaguer, E. (1983), Phys. Rev. Lett., 51, 1313.

____________________ . (1985), Phys. Rev. D, 31, 251.

Kershaw, D. S. (1978). J. Comp. Phys., 26, 43.

Kurki-Suonio, H., Centrella, J. M., Matzner, R., and Wilson, J. (1987). Phys. Rev. D, 35, 435.

O' Murchadha, N., York, J. W. (1973). J. Math. Phys., 14, 1551.

York, J. W. (1979). In Sources of Gravitational Radiation, ed. L. L. Smarr (Cambridge: Cambridge University Press), p. 83.

Zabusky, N. (1984). Physics Today, 37, no. 7, 36.

An Algorithmic Overview of an Einstein Solver

Roger Ove

National Center for Supercomputing Applications

University of Illinois at Urbana-Champaign

INTRODUCTION

An Einstein space-time generator is discussed, with an emphasis on tools for the analysis of the resulting geometries. The general ideas behind the algorithm are laid out, particularly those that contribute to its effectiveness in dealing with large scale problems on supercomputers. An example of the use of such a system to provide evidence in support of a conjecture is briefly discussed.

MOTIVATION AND PHILOSOPHY

The purpose of this work is to probe the nature of solutions to the Einstein equations, in situations not amenable to analytic treatment. At present much of our understanding of these equations is based upon solutions with a great deal of symmetry or cases that are algrebraically special. It is not always clear that such examples illustrate general characteristics of the field equations, or merely artifacts of symmetry. By resorting to numerical generation of solutions, these symmetries can be relaxed and the nature of the "generic" solution exposed.

It is expected that the general properties of spacetimes are in many cases independent of matter content (assuming well-behaved matter). The general theorems regarding gravitational collapse and singularity formation support this view, and go through regardless of the matter content. For this reason we restrict ourselves to the vacuum. In addition, because we are interested in the nature of the field equations and not peculiarities of coordinate systems, the simplest case (spatial hypersurfaces diffeomorphic to three-tori) is chosen. Although these spacetimes are "cosmological", in that the hypersurfaces are closed, we are not studying cosmology (which to this author is concerned with the origins of the physical universe). Nor is the the intent to uncover "properties of the torus", although certainly any insight into the effects of the spatial topology in the global evolution is of interest. Much of what is done here is readily adaptable to other manifolds, although in other cases one could not so readily decouple the behavior of the field equations from the complexities of the coordinates.

Limitations of present computers enforce the imposition of a single spatial Killing vector, so that we are left with two dimensional fields evolving in time. Because of this, and the fact that the model is cosmological (there is no axis of symmetry), any local singularity will be a one-dimensional object extending all the way around the spacetime. Any results relating to the formation of singularities in such spacetimes are relevant only to such spatially closed spacetimes, and in this sense the work is specific to the cosmological case.

SUMMARY OF THE FORMALISM

Only a brief overview of the formalism used will be given here. A more detailed description can be found elsewhere.[1] The general idea is to borrow from the five dimensional Kaluza-Klein theory, allowing the Killing vector to play the role of the "extra" dimension. Doing so casts the four dimensional Einstein equations in a form equivalent to three dimensional gravitation (which has no degrees of freedom) coupled to a scalar and vector field. In the five dimensional theory this vector field would obey the Maxwell equations and hence have two degrees of freedom, but in the lower dimensional case this is reduced to one.

The metric takes the form

$$ds^2 = e^{-2\phi} d\sigma^2 + e^{2\phi} \left\{ dx^3 + \beta_a dx^a + \beta_0 dt \right\}^2$$

where

$$d\sigma^2 = -\tilde{N}^2 dt^2 + g_{ab} \left(dx^a + \tilde{N}^a dt \right) \left(dx^b + \tilde{N}^b dt \right)$$

All variables are functions of t, x^1, and x^2 only, as x^3 has been singled out as the direction of symmetry. Terms denoted by a tilde are two dimensional quantities, and the letters $(a, b, c, ...)$ will be used for their indices. Four dimensional quantities will have their indices denoted by lower case Greek letters, while $(i, j, k, ...)$ will be used for three dimensional quantities.

Parametrized in this way, the Einstein action is

$$S = \int dt\, d^2x \left\{ \pi^{ab} g_{ab,t} + e^a \beta_{a,t} + p\phi_{,t} - \tilde{N}\tilde{H} - \tilde{N}^a \tilde{H}_a - \beta_0 e^a_{\ ,a} \right\}$$

where

$$\begin{aligned}
\tilde{H} =& \frac{1}{\sqrt{g}}\left\{\pi^{ab}\pi_{ab} - (\pi_a^{\ a})^2 - \frac{1}{8}p^2 + \frac{1}{2}e^{-4\phi}g_{ab}e^a e^b\right\} \\
& \sqrt{g}\left\{-{}^{(2)}R + 2g^{ab}\phi_{,a}\phi_{,b} + \frac{1}{4}e^{4\phi}g^{ac}g^{bd}\left(\beta_{a,b} - \beta_{b,a}\right)\left(\beta_{c,d} - \beta_{d,c}\right)\right\} \\
\tilde{H}_a =& -2\pi_{a\ ;b}^{\ b} + p\phi_{,a} + e^b\left(\beta_{b,a} - \beta_{a,b}\right)
\end{aligned}$$

The dynamics of the equations are represented by the scalar field ϕ and the vector field β, and these and their momenta are the only fields that are explicitly evolved. Additional quantities are determined by imposing coordinate conditions and solving the Hamiltonian and momentum constraints, all of which can be cast as elliptic equations (and except for the Hamiltonian constraint, linear elliptic equations).

The formalism used here expresses the dynamics in terms of the norm of the Killing vector and its twist, and in practice a scalar potential related to the twist potential (a true four dimensional scalar) replaces the β degree of freedom. Originally it was believed that this pseudo-twist potential was also a four dimensional scalar, although analytic investigation did not make this clear. Numerical evidence conclusively indicates that it is not a true scalar, since cases have been observed in which its perturbative waves propagate at speeds in excess of the local speed of light. Unfortunately, some time was wasted attempting to construct an analytic demonstration before this accidental observation was made.

OVERVIEW OF THE METHODS

Predictor Corrector

The equations comprise a set of nonlinear hyperbolic equations satisfying elliptic constraints and coordinate conditions. In the continuum limit, the constraints are preserved exactly by the evolution equations. However, as this feature is due to a continuous invariance of the theory (by Noether's theorem), a discrete conserved quantity can not be expected to exist in the differenced case. It should be emphasized that it is the discreteness in time alone that impacts the failure of the constraints to be preserved. The system which is discrete spatially but continuous in time will preserve differenced versions of the constraints. This suggests using a method that is higher order in time than in space, and hence we turn to predictor-corrector methods.

While quite effective and popular for solving ODE's, true predictor-corrector methods are rarely used for hyperbolic partial differential equations. We exclude two-step methods such as McCormick's or improved Euler from the discussion, and use the term predictor-corrector to refer to higher order schemes (as is also the practice in discussions of ODE methods).

The lack of popularity of the method seems to be due to problems with instabilities, and only unfavorable remarks can be found in the literature. However, the basic ideas and quadrature formulations are as applicable to PDE's as they are to ODE's, and the linearized stability behavior can easily be shown to be equivalent to that of simple leapfrog differencing. Why then the bad reputation? It would appear that many early attempts did not succeed because of failure to iterate the corrector.

The basic idea behind predictor-corrector schemes is to cheaply predict the value of the derivative (usually by an inaccurate interior quadrature formula), and then correct this approximation by applying a more accurate quadrature formula. In the case of ODE's, analysis typically assumes that the corrector eliminates any error induced by the predictor. However, it could become necessary even in this case to iterate the corrector (for instance, if the time step were too large). In the case of PDE's, time steps are usually chosen to satisfy stability requirements rather than improve accuracy, so the need to iterate becomes much more important.

The Sectioned Approach

These potential difficulties led in the present case to a sectioned predictor-corrector method. A section of spacetime is generated (typically eight time steps) by some method, in this case a pure extrapolation of the time derivatives (predictor) followed by a 4th order accurate application of quadrature (corrector). After a section has been generated, it is repeatedly improved with a more accurate 4th order quadrature formula. This iterative improvement is most important in the first section, since initially no past data is present and the generated data contains errors of the first order.

The first generation of data in a section can be produced with little regard to the accuracy of the elliptic equations (constraints and coordinate conditions). There is little sense in expending a great deal of time solving them, only to repeat the calculation when their dependent data changes. The accuracy required of the elliptic solvers is increased as a section is iteratively improved. Obviously, such an approach requires a solver with negligible startup time.

The Einstein equations are time reversible and it is desirable that the discrete equations also have this feature, which they do not. To partially correct for this failure, the initial extrapolation gives the value of the time derivative at $t + dt/2$. This is followed by a quadrature formula contrived to make use of this value and past values to update the field at $t + dt$. Thus the scheme mimics the simple leapfrog method (which is time symmetric), and time reversal errors occur only in high order terms. The quadrature formula used to iteratively improve the sections is completely time symmetric. The result of this attention to time symmetry is very low numerical dissipation.

The quadrature formulas used here were generated as needed with the symbolic manipulation program maple. Although they could quite easily be computed by hand, doing so in an automatic manner facilitates rapid comparison of different schemes, while reducing the likelyhood of algebraic error. Maple was used extensively to aid all calculations involved in this work.

Generating a spacetime in sections has obvious advantages when one wishes to analyze the geometry. A drawback is the memory required to store the recent history. The work discussed here was done on a Cray X-MP48 with an SSD (Solid State Device, a RAM disk), and required 4 million words of memory and 17 million words of SSD storage.

The Elliptic Equations

Alternating direction implicit (ADI) methods were once quite popular, but have in recent times been supplanted by more widely applicable conjugate gradient methods. However, one of the results of ADI research was the discovery that for certain classes of "special" problems, extraordinarily fast convergence could be achieved. It was because of the lack of such rapid convergence in the general case that the method fell out of favor.

Here we make use of the freedom to choose coordinates to transform in such a way that the linear elliptic operators are of the preferred ADI type. The two metric is made conformally flat by a suitable choice of the shift vector. All of the elliptic equations as a consequence involve an operator of the form

$$\Delta = f^{ab}\partial_a\partial_b - A(x^a)$$

where

$$f^{ab} = f^{ab}(t), \quad det(f) = 1$$

The derivative part of this operator has a known spectrum, and entire families of error modes can be eliminated (to machine accuracy) with a single ADI sweep. This statement is true regardless of the form of the (positive) term $A(x^a)$, since that part of the operator is involved in the inversion for both even and odd sweeps. In addition, linearity of the equations and method imply that once a mode has been eliminated it will not be excited by subsequent sweeps. This applies to the Hamiltonian constraint as well, for although it is nonlinear it is solved by iteratively solving a linear equation of this form.

The existence of such a conformally flat global coordinate system is of course peculiar to the 2 torus. However, in other interesting cases the situation is even simpler. For example, metrics on the 2 sphere are always conformal to the standard spherical metric, and the case of a patch with boundary can be mapped to the unit diagonal metric. In either of these cases there is no need to allow the flat metric f^{ab} to evolve (so that even if the calculating the spectrum involved some computation, it need only be done once).

The "Variational Trick"

Fast convergence or not, there is still a heavy price to be paid for having to solve elliptic equations at every time step. To alleviate this, a variational scheme is used to "propagate" the solutions of elliptic equations. As far as the final result is concerned, this is equivalent to optimal extrapolation, without the need to find the optimal basis.

The basic idea is to expand ϕ (representing a solution to one of the elliptic equations) as a linear combination of iterates in a section, and then find the best coefficients. All of the linear elliptic equations that arise in this problem can be derived from a discrete action of the form

$$S = \frac{1}{2} A_{ij} \phi^i \phi^j - K_i \phi^i$$

where i, j are spatial indices (where each index runs over every node of a hypersurface), and A_{ij} is of definite sign. Express ϕ as an expansion of basis functions

$$\phi^i = a^\alpha \psi^i_\alpha$$

and define

$$\tilde{A}_{\alpha\beta} \equiv A_{ij} \psi^i_\alpha \psi^j_\beta \qquad \tilde{K}_\alpha \equiv K_i \psi^i_\alpha$$

The idea is to use as basis functions the values of the field at nearby time steps. These "past" iterates (they could be future iterates as well) are denoted by the index α.

If we now extremize the action with respect to the parameters a_α, we can obtain and solve (assuming the basis is well conditioned) the equation

$$\tilde{A}_{\alpha\beta} \equiv A_{ij} \psi^i_\alpha \psi^j_\beta \qquad \tilde{K}_\alpha \equiv K_i \psi^i_\alpha$$

for the vector a_α. If this matrix is not singular, then $a^\alpha \psi^i_\alpha$ minimizes the error, in the norm defined by the original operator A_{ij}:

$$\|\delta\| \equiv \left|A_{ij}\delta^i\delta^j\right|$$

If the matrix is singular, or numerically singular, then ordinary extrapolation with a chosen basis is an alternative. The domain in which the variational method fails (such as when all past iterates are equal) is one where simple extrapolation tends to work reasonably well.

To see the relationship with extrapolation, choose a set of basis functions (such as $1, t, t^2, ...$), equal or less then to the number of iterates in a section. For each spatial point (x, y), extrapolate in time by a least squares fit. The result will be of the form

$$\phi_{new}(x, y, t) = a_\alpha(t)\,\psi^\alpha(x, y, t)$$

where a_α is a function of t only. This is the same form that results from the variation, except for the essential difference that in that case the result was extremized over all possible choices of basis functions. Therefore the variational scheme is equivalent to this linear extrapolation with an optimal basis. The great advantage is that it eliminates the need to find the optimal basis (optimal with respect to the original elliptic equation).

This technique is used as a cheap alternative to solving elliptic equations, and as a means of "evolving" their solutions. As such it provides a means of integrating the elliptic and hyperbolic parts of the overall problem. It is particularly useful in relaxing to the solution in a section that has been crudely computed. Currently the effectiveness is limited by the fact that SSD is used to store the past iterates, which involves the overhead of reading and writing to that device. This should not be a problem on machines with larger directly addressable memories.

An important property of this variational procedure is its favorable performance scaling as the size of the problem is increased. For very small problems such as may be run on a microcomputer, the methods used here are inferior to simply solving each elliptic equation with successive over-relaxation. However, for large problems with 10^5 nodes per hypersurface, we are able to advance a time step in under 10 cpu seconds (Cray X-MP48) while maintaining the solutions of 7 elliptic equations to a high degree of accuracy. For comparison, the popular conjugate gradient linear elliptic solvers typically achieve 1 order of magnitude error reduction per cpu second for problems of this size on such a machine.

ANALYSIS OF SPACETIMES

As mentioned earlier, an advantage of generating a spacetime in sections is that it facilitates analysis of the geometry during the evolution. The current program incorporates several geometric probes, which are invoked after a section has been generated.

Null geodesics

The paths of photons can be used to aid in the detection of singularities, assuming the photons are sufficiently soft that they do not influence the gravitational dynamics. In this limit the photons follow the null geodesics of the spacetime. If the expansion of null geodesics is negative for all families of null geodesics normal to the surface of a compact region, then it can be shown that in a finite proper time a singularity will form. Such a surface is called a trapped surface. For example, in the Schwarzschild solution (in coordinates such that the radial coordinate remains spacelike, such as a maximally sliced gauge) the proper distance between initially outgoing photons will tend to zero in the vicinity of the singularity. In the numerically evolved spacetime we monitor the null expansion normal to strategically located 2-surfaces. For concreteness, the paths of "actual" photons are evolved as well.

The null geodesic equation for this parametrization can most easily be generated by extremizing the length of the line element, and then imposing the symmetry and the null condition. It is interesting to note that in this case the symmetry can not be imposed in the action before doing the variation.

We start with the action

$$
\begin{aligned}
S =& \frac{1}{2} \int d\lambda \; g_{\mu\nu} \dot{x}^\mu \dot{x}^\nu \\
=& \frac{1}{2} \int d\lambda \left\{ e^{-2\phi} \left[-\tilde{N}^2 \dot{t}^2 + g_{ab} \left(\dot{x}^a + \tilde{N}^a \dot{t} \right) \left(\dot{x}^b + \tilde{N}^b \dot{t} \right) \right] + e^{2\phi} \left(\dot{z} + \tilde{\beta}_a \dot{x}^a \right)^2 \right\}
\end{aligned}
$$

where a "dot" denotes a derivative with respect to the affine parameter λ. The variation in the symmetry direction δz results in

$$
\frac{d}{d\lambda} \left[e^{2\phi} \left(\dot{z} + \tilde{\beta}_a \dot{x}^a \right) \right] = 0
$$

which identifies the conserved "angular momentum" associated with the symmetry. Making use of this analogy, we express the conditions of symmetry and the null condition in the form

$$L = e^{2\phi}\left(\dot{z} + \tilde{\beta}_a \dot{x}^a\right)$$
$$L^2 = -\tilde{N}^2 \dot{t}^2 + g_{ab}\left(\dot{x}^a + \tilde{N}^a \dot{t}\right)\left(\dot{x}^b + \tilde{N}^b \dot{t}\right)$$

where L is a constant for a given photon. These conditions are imposed exactly during the evolution of a photon.

In a general coordinate system constructed on a family of spacelike hypersurfaces, it may be impossible to impose a condition such as "that photon is moving left" . This is because the coordinates may be shifting at greater than the speed of light, a possibility even though the hypersurface is spacelike. For example, the (1,1) inverse metric component can be negative and its associated unit vector timelike, while the normal to the hypersurface remains timelike (and hence the surface spacelike). As a consequence, it is not possible to solve the null condition for the derivative of t for arbitrary values of the spatial derivates. We bypass this difficulty by defining and evolving shifted momentum quantities P_a

$$P_a \equiv g_{ab}\left(\dot{x}^a + \tilde{N}^a \dot{t}\right)$$

The quantities that are evolved are x^a and P_a, and the null condition is imposed with

$$\dot{t} = \frac{1}{\tilde{N}}\sqrt{L^2 + g_{ab} P^a P^b}$$

which gives a real result for any P_a. Alternatively, if the derivatives of the spatial coordinates were directly evolved, then differencing errors could build up until it is no longer possible for the tangent vector to be null. The choice of sign in the above equation is arbitrary, and amounts to a choice of whether the affine parameter is increasing or decreasing.

In choosing initial photon data it is still desirable to specify actual directions rather than shifted momenta. This is done, but the values are subject to change if necessary. The final direction is chosen to be as close to the requested direction as possible.

Null expansion

Obviously it is not possible to exhaustively search all regions of all spacetimes for the existence of a trapped surface. The point of view taken here is that we are looking for physical affects that would be obvious if analytic dynamic solutions to the Einstein equations were readily available. Therefore the approach is to construct initial data such that, if a singularity were to form, it would form in a predictable location on the spatial hypersurfaces. If nature is in any way cunning and forms singularities only for a small set of initial data, then a positive result would be a stroke of luck. In defense of this attitude, it is quite easy to produce local trapped surfaces in a collapsing T^3xR universe with very similar initial data.

More detailed discussions of null expansion and trapped surfaces can be found elsewhere.[2] Here only a brief sketch will be given, to indicate any differences in convention or implementation details.

The 2–surfaces will be defined by

$$f(x,y) = 0 \qquad t - t_0 = 0$$

and this is the first simplification, restricting the surfaces to lie within the spatial hypersurfaces. The null vectors normal to the surface are defined by

$$\xi^{\pm}_{\beta} = \pm\alpha\left(\mu^{\pm}\delta^0_{\beta} + f_{,\beta}\right)$$

where α and μ are defined such that

$$\begin{aligned} g^{\mu\nu}\xi^{\pm}_{\mu}\xi^{\pm}_{\nu} &= 0 \\ g^{\mu\nu}\xi^{+}_{\mu}\xi^{-}_{\nu} &= -1 \end{aligned}$$

The latter condition is somewhat irrelevant, being of convenience if one wishes to construct a projection operator from the null vectors, which is not being done here. It is imposed here for consistency with other references.

The tangent space is spanned by the vectors

$$\begin{aligned} &Z^{\mu} = e^{-\phi}\delta^{\mu}_{z} \\ &V^{\mu} \quad \left(determined\ by\ V \cdot Z = V \cdot \xi^{\pm} = 0\ and\ V \cdot V = 1\right) \end{aligned}$$

Using this basis to define the projection operator onto the surface

$$P^{\nu}_{\mu} = V_{\mu}V^{\nu} + Z_{\mu}Z^{\nu}$$

the null expansion is defined to be

$$-g^{\mu\nu}\xi^{\pm}_{\alpha;\beta}P^{\alpha}_{\mu}P^{\beta}_{\nu}$$

For the numerical experiments, an additional simplification is imposed to avoid having to pass arbitrary parametrized curves to a subroutine. The surfaces are required to be rectangular cylinders

$$f(x,y) = \begin{cases} x - x_0 \\ y - y_0 \end{cases}$$

This does not seriously effect the generality of what can be done, as one would not expect the existence of trapped surface to depend critically on their shape. Also, if the initial data is "symmetric" (assuming the boundary is sufficiently far away) around the center of the hypersurface then the expansion need only be checked along surfaces orthogonal to a line radially outwards.

CMC Hypersurfaces

The time slicing prescription used here is not the usual constant mean curvature (CMC) condition normally used for cosmological spacetimes. Rather, the symmetry is made use of to define a more natural 2 dimensional variant

$$\frac{\tilde{g}_{ab}\tilde{\pi}^{ab}}{\sqrt{g}} = \tau(t)$$

where τ is an specified function of time (related to the 2 dimensional expansion of the spatial 2–surface).

It is of interest to be able to find the true CMC hypersurfaces, in order to establish whether unusual behavior is physically meaningful or merely a breakdown of the coordinate system due to nonstandard slicing. Given a surface defined by $\omega(x,y) = \omega_0$, the trace of the extrinsic curvature (divergence of the unit normal) is

$$Tr(K) \equiv n^a_{;a} = \frac{1}{\sqrt{-g}}\left(\sqrt{-g}\frac{g^{\mu\nu}\omega_{,\nu}}{\sqrt{-g^{\alpha\beta}\omega_{,\alpha}\omega_{,\beta}}}\right)_{,\mu}$$

If this quantity is a constant then the surface is a CMC hypersurface.

Finding such surfaces in an arbitrary spacetime generated with coordinates not suited to them is a nontrivial task. In situations with more symmetry where the metric is given analytically, the problem can be reduced to that of integrating an ODE.[3] This scheme requires that it be possible to locally "invert" the metric components, that is, find the spacetime point at which a component that has a certain value. Also, the approach of finding a surface with a prescribed curvature is problematical in a numerically generated spacetime, even in this case where a section of the spacetime is available for inspection (since in general there may be no surface that lies entirely in a given section).

The approach described here is more suited to numerically generated spacetimes, since instead of finding the surface we "propagate" its generating function. It is restricted to situations where one expects that (at least locally in time) there is a unique CMC hypersurface for value of the extrinsic curvature in some range. The 3–torus is one situation where this is believed to be true. Clearly none of this is applicable to the asymptotically flat case

If we assume that a family of such surfaces exists, then it is feasible to allow them to define their own generating functions. That is, we impose the condition

$$\omega = \frac{1}{\sqrt{-g}} \left(\sqrt{-g} \frac{g^{\mu\nu}\omega_{,\nu}}{\sqrt{-g^{\alpha\beta}\omega_{,\alpha}\omega_{,\beta}}} \right)_{,\mu}$$

Of course we could have chosen any locally monotonic function of ω, but this could adversely affect the stability of what follows. For example, choosing $-\omega$ would reverse the stability properties. Which sign to choose is dependent upon the sign of the leading linear operator on the right hand side of this equation. Rather than finding a surface with a specified constant curvature, we will generate ω throughout the spacetime by iterating this equation and then the desired surfaces will simply be those where this function is a constant. The existence of time derivatives on the right hand side (which of course cancel if the coordinates were tuned to the surfaces) is one of the features that clouds the issues of existence and uniqueness. It is interesting to note how the nonlinearity of this elliptic problem alters these issues. If the problem were linear, then in addition to requiring periodic spatial boundary conditions, ω would have to be specified on initial and final hypersurfaces. This is quite unlike the case for quasilinear hyperbolic equations, where existence and uniqueness (based upon characteristics) is unchanged from the linear theory.

This equation can be generated from an action principle, by extremizing the area of the surface. However, the problem is somewhat ill posed, as the action is not only unbounded below but also not necessarily real. Physically what this means is that as we vary the surface it may not remain spacelike everywhere, in which case the measure is undefined. This is why determination of maximal and CMC surfaces is a vastly more difficult problem in Lorentzian geometries than in the Riemannian case (the classic soap bubble problem). In practice the surfaces do exhibit a strong tendency to kink into the imaginary space (if only we could visualize this!). Numerical tests indicate that it is better to update ω for an entire section before making the replacements, a practice which tends to maintain smoother approximations to the time derivatives.

Iterating this equation in the most obvious way would involve

1. Guessing ω in a section.

2. Evaluating the rhs.
3. Replacing ω.

However, this is clearly unstable, which can be seen by linearizing and making the analogy with linear elliptic solvers. It could be stablized most easily by adding linear terms to both sides of the equation

$$\omega' + A\omega' = \frac{1}{\sqrt{-g}}\left(\sqrt{-g}\frac{g^{\mu\nu}\omega_{,\nu}}{\sqrt{-g^{\alpha\beta}\omega_{,\alpha}\omega_{,\beta}}}\right)_{,\mu} + A\omega$$

The analogy with linear iterative solvers indicates that this is equivalent to the Jacobi method, and hence we would expect the convergence to be extremely slow. Performance can be improved by moving parts of the rhs operator to the lhs and inverting with fast linear solver, obtaining the solution to the nonlinear equation by iteration. The previously mentioned Jacobi related method is of course also the application of linear iteration to solve a nonlinear equation (this difference is one of graininess, the Jacobi approach being fine grained). Extracting the dominant term leads to the most coarse grained iteration

$$\left(h^{ij}\partial_i\partial_j + \frac{1}{\alpha}\right)\omega' = -\frac{\omega_{,\nu}}{\sqrt{-g}}\left(\sqrt{-g}g^{\mu\nu}\right)_{,\mu} - \frac{\alpha^2}{2}g^{\mu\nu}g^{\alpha\beta}_{\ ,\mu}\omega_{,\nu}\omega_{,\alpha}\omega_{,\beta}$$
$$- \left(h^{00}\partial_0\partial_0 + 2h^{0i}\partial_0\partial_i\right)\omega$$

where

$$h^{\mu\nu} = g^{\mu\nu} + n^{\mu}n^{\nu} \qquad \alpha = \left(-g^{\alpha\beta}\omega_{,\alpha}\omega_{,\beta}\right)^{-\frac{1}{2}}$$

Thus with the dominant term extracted the iteration would converge in a single step if the problem were linear and the time derivative terms on the right dropped out. Because the coordinates are not suited to CMC surfaces, the operator that must be inverted is not of the preferred ADI type. A compromise is to move only the desired operator to the lhs and add the necessary stabilizing terms. In practice it is unnecessary to evaluate the rhs after the operator has been split out, as it can simplify be added to both sides (evaluating unprimed ω on the rhs). Taking this route makes it easier to evaluate alternative schemes.

In the above discussion the term "stable" is loosely used. In all cases what is meant is linearized stability around a solution that is known to exist (such as in the vicinity of the Kasner solution). A test of stability for perturbed surface in a simple Kasner spacetime

$$g_{\mu\nu} = diag\left(-1,\ 1,\ t^2,\ 1\right) \qquad \omega_0\left(x, y, t\right) = -\frac{1}{t} + \epsilon \sin\left(nx\right)$$

leads to

$$\omega_{new} = -\frac{1}{t} + f(x) \qquad f_{,xx} - \frac{1}{t^2} f = -\frac{5}{2} t \left(n\epsilon \cos(nx) \right)^2$$

which implies quadratic convergence. Numerical experiments for more complex cases indicate rapid convergence, but a very narrow radius of convergence. This necessitates that solutions for spacetimes far from known results must be found by iteratively constructing the initial data in conjunction with the CMC hypersurface data.

An Application

The primary purpose of this project was to provide evidence for or against various conjectures of classical relativity, in particular Cosmic Censorship. An important related issue is the effect of symmetry on singularity formation in expanding spacetimes, important because of the existence of a powerful result in the two Killing vector case. Also, this issue is of importance in understanding the limitations of what can be inferred from out numerical framework, as the entire formulation relies heavily on the presence of symmetry.

First consider the the case of a vacuum T^3xR spacetime with two spatial Killing vectors (the case of three Killing vectors is obviously trivial). The dynamics of such Gowdy spacetimes[4][5] can be expressed in a form similar to the nonlinear sigma model, where the norms of the Killing vectors play the roles of two interacting scalar fields. It has been proven for this system that a foliation exists in which the trace of the extrinsic curvature is a constant on each hypersurface, and that the foliation covers the entire spacetime.[6] The theorem implies that a black hole cannot form as the spacetime evolves away from the cosmological singularity, since the hypersurfaces tend towards $K = 0$ and the black hole (assuming that it is a crushing singularity) could not be approached by such surfaces. Apparently the fact that the singularity would cover the entire surface of a 2-torus, due to the symmetry, prevents it from forming.

If the symmetry is relaxed then we have a somewhat different picture. There is nothing to stop the formation of the black hole in this case, since locally the spatial hypersurfaces do not "know" they are tori. Given a spatial hypersurface with a sufficiently localized disturbance (otherwise smooth), we can approximate the region with an asymptotically flat model. In such a scenario it has been established beyond reasonable doubt that black holes can form. If the singularity is crushing then the foliation can not be complete, so the above theorem for the Gowdy metrics is not extendible to spacetimes with no symmetry without being modified in some way.

For the case of one Killing vector, the problem can be attacked numerically. As described elsewhere [7], the evidence seems to suggest that the one Killing vector case is similar to the Gowdy case, and that local singularities do not form. Also, while attempting to contrive initial data likely to form a local singularity, some analytic progress was made and it can be shown that for a wide class of initial data there exists no trapped surface of the type the experimental "apparatus" is designed to detect.

The author would like to thank the organizing committee for accepting this manuscript well past the deadline, and also many thanks to R. Ove for painstakingly typesetting it at the very last moment

REFERENCES

[1] R. Ove, *Dynamical Spacetimes and Numerical Relativity.* Cambridge: Cambridge University Press, 1986.

[2] S. Hawking and G. Ellis, *The Large-Scale Structure of Spacetime.* Cambridge: Cambridge University Press, 1973.

[3] D. Eardley and L. Smarr *Phys. Rev. D*, vol. 19, p. 887, 1979.

[4] R. Gowdy *Phys. Rev. Lett.*, vol. 27, p. 826 and 1102, 1971.

[5] R. Gowdy *Ann. Phys.*, vol. 83, p. 203, 1974.

[6] J. Isenberg and V. Moncrief *Comm. Math. Phys.*, vol. 86, p. 485, 1982.

[7] R. Ove, *Symmetry and Singularity Formation in Expanding Spacetimes.* 1988. in preparation.

A PDE Compiler for Full-Metric Numerical Relativity

Jonathan Thornburg
Dept of Geophysics & Astronomy
The University of British Columbia
Vancouver BC V6T 1W5
Canada

Abstract

We are interested in doing $3+1$ numerical relativity with a full (all components nonzero) metric, such as arises in the Smarr/York minimal distortion gauge. Here we present our technique for overcoming the resulting complexity of the Einstein equations, by using a "PDE compiler" to automatically generate finite differencing code from a high-level description of the tensor differential operators involved.

We have constructed a prototype PDE compiler, and a full-metric numerical relativity code based on it. The code is fully 3-covariant and uses minimal distortion spatial coordinates to evolve a vacuum axisymmetric spacetime containing a single black hole present in the initial data. It uses 4th order centered finite differencing on a non-staggered uniform grid.

We have found that the combination of relativity code written at the level of tensor differential operators and finite differencing code generated automatically by a PDE compiler works well. It greatly simplifies the programming of the relativity code, allowing us to concentrate on the physics rather than the details of the finite differencing. This raising of the level of abstraction at which we think about the code has turned out to be perhaps the most valuable benefit of using a PDE compiler.

1 Introduction

In this paper we discuss some of the techniques we have developed for doing $3+1$ numerical relativity with a full (all components nonzero) metric. This work is developmental in nature, aiming to develop tools and techniques for the future study of interesting physical systems. As a long term goal, we are particularily interested in the study of black hole collisions, for example the decay and merger of black hole binary systems.

Thus we focus on ideas and techniques which generalize easily to fully 3-dimensional systems (although our present numerical code is restricted to the axisymmetric case) and can adapt to complicated, changing topologies such as occur in multiple black hole systems. We restrict our spacetimes to be vacuum (at least outside the black holes) and asymptotically flat.

Here we briefly review (part of) our motivation for adopting a full metric, then discuss some of our implementation techniques for dealing with the resulting complexity of the $3+1$ equations. A more detailed discussion of these topics, as well as of our slicing condition, time evolution scheme, manner of dealing with black holes, and other details of our code, will appear elsewhere (Thornburg (1988)).

2 Minimal Distortion Coordinates

Minimal distortion coordinates were first proposed by Smarr and York (1978a,b); York (1979) gives a readable introductory discussion of their definition and properties. As suggested by the name, they are (can be) defined to minimise a natural measure of the distortion of the coordinates from one time step to the next.

They have a number of desirable theoretical properties:

- They are fully 3-covariant and can adapt to any spatial topology, including one with black holes present. 3-covariance implies that the effects of changes in the (arbitrary) coordinate choice on the initial slice are limited to inducing the same recoordinatization on later slices – no additional geometric changes occur. This helps to disentangle the physics from the coordinate effects.

- As the name implies, they should reduce the coordinate distortion that tends to build up as the evolution progresses. This distortion has been a major problem in past attempts to construct codes that can run for long periods of time, both because of the numerical problems it causes and because it makes it difficult to interpret the codes' output.

- They are a natural strong-field generalization of the Dirac, ADM, and de Donder (4-harmonic) gauges. In particular, they share the excellent weak-field radiation propagation properties of these gauges.

Isenberg (1979) discusses these properties in more detail, as well as several others related to the canonical formulation of gravity.

However, minimal distortion coordinates are difficult to implement, and do not seem to have been used in any multidimensional codes so far. The main problem is that they leave all components of the metric (and other field tensors) nonzero, which greatly complicates the $3+1$ equations. As well, a complicated (3-)vector elliptic equation must be solved at each time step to compute the minimal distortion shift vector.

We now discuss some of the techniques we have developed to deal with the complexity of full-metric numerical relativity.

3 A PDE Compiler

The main technique we use to deal with the complexity of minimal distortion coordinates and the resulting full metric, is the use of a "PDE Compiler" to automatically generate finite differencing code from a high-level description of the tensor differential operators involved. Nakamura et al. (1987) have used a somewhat similar technique based on the "Reduce" symbolic algebra system.

It's important to realise that, *written at the tensor differential operator level*, neither the minimal distortion equations nor indeed any of the $3+1$ equations for a full metric are particularily complicated – the complexity only appears when they're expanded to scalar form and finite differenced. Thus the PDE compiler's (human-prepared) input, the specification of the tensor differential operators, can remain tractable even for a full metric.

The PDE compiler processes this input into finite differencing code in a conventional algebraic language (Fortran, C, Ada, etc.). This finite differencing code is then combined with other (non-finite-differencing computation, utility, etc.) code to form the overall numerical relativity program.

We have found that almost all the operations needed in a $3+1$ code can be cast into the PDE compiler framework, so the PDE compiler's input makes up the bulk of the code. In light of this we regard this input *as* the code, written in a specialised "PDE language", and the PDE compiler's output (the finite differencing code) as merely another computer-generated intermediate file, like the object modules generated by a Fortran, C, Ada, etc. compiler for input to a linker/loader.

This view of the PDE compiler technique, reformulating the problem in a specialised language (the PDE compiler input) which is implemented by translating it into an existing one (Fortran, C, Ada, etc.), has proven to be of great conceptual value. It's a common paradigm in computer science, particularily in artificial intelligence (AI); the AI context has in fact provided many of our implementation ideas and techniques.

PDE compilers are not yet off-the-shelf software components in the way algebraic language compilers are – they must still be constructed from scratch. Indeed, LeBlanc (1985) predicted their use would only start to become important "toward the end of the next ten years". There is little literature available on them, and much of our research has been directed towards gaining a better understanding of the issues involved their construction and use in a numerical relativity setting.

To do this, we have constructed a prototype PDE compiler, and a numerical relativity code, `Mk.2`, based on it. All of `Mk.2`'s spatial finite differencing is generated by the PDE compiler. (Our exclusion of the temporal finite differencing is for implementation convenience only, and is not intrinsic to the PDE compiler technique.) `Mk.2` evolves a vacuum axisymmetric spacetime containing a single black hole present in the initial data. `Mk.2` is fully 3-covariant, using minimal distortion spatial coordinates and a full metric. It uses (PDE-compiler-generated) centered 4th order finite differencing on a non-staggered uniform grid.

Internally, `Mk.2` has a layered structure, with relativity code (almost all written in our "PDE language") on the top and the PDE compiler and other utility routines on the bottom. All of `Mk.2`, including the PDE compiler and the finite differencing code it generates, is written in C. This partitioning of the code, with the relativity and the finite differencing kept separate, has proven very valuable. The relativity code is greatly simplified by not having to deal with finite differencing, which helps to keep the physics of the continuum equations from being lost in the complexity of their numerical solution. The PDE compiler, dealing only with finite differencing, can be debugged on simple model problems such as those discussed below.

As suggested by the name, the internal structure of our PDE compiler itself is in many ways similar to that of an algebraic language compiler. It's composed of two phases – a "front end" and a "back end". The front end does mainly symbolic processing (tensor to scalar conversion, application of spacetime symmetries or differential identities, etc.) to reduce the input to a canonical intermediate form. The back end selects the difference schemes and generates the finite differencing code. We discuss this internal structure in more detail below.

3.1 An Example

To clarify just what our PDE compiler does and how it does it, we consider as an example its treatment of the scalar Laplacian operator in an axisymmetric 3-dimensional space with coordinates (x^1, x^2, x^3) and a full metric. (This problem, in the context of the scalar Poisson equation with the further restriction that the metric be diagonal, has in fact been the main test case for debugging our PDE compiler.)

The first stage in solving this problem is to rewrite the Laplacian operator in tensor form and make all the partial derivatives explicit,

$$\begin{aligned} \nabla^2 &= g^{ij} \nabla_i \nabla_j && (1) \\ &= g^{ij} \partial_{ij} - g^{ij} \Gamma^k_{ij} \partial_k && (2) \end{aligned}$$

where we assume that g^{ij} and $g^{ij}\Gamma^k_{ij}$ have already been computed. (`Mk.2` computes g^{ij} by numerically inverting the 3×3 matrix g_{ij} at each grid point; all other grid functions are then computed via our PDE compiler in the manner discussed hereinafter.) Our present PDE compiler requires all differential operators to be written in terms of coordinate partial derivatives (and Christoffel symbols if necessary), as in (2). However, input at the semantic level of (1) is certainly possible with a more sophisticated PDE compiler.

Given (some encoding of) this input, the front end of our PDE compiler first does the tensor to scalar conversion by expanding the dummy index summations,

$$\begin{aligned}\nabla^2 &= g^{11}\partial_{11} + g^{12}\partial_{12} + g^{13}\partial_{13} \\ &\quad + g^{21}\partial_{21} + g^{22}\partial_{22} + g^{23}\partial_{23} \\ &\quad + g^{31}\partial_{31} + g^{32}\partial_{32} + g^{33}\partial_{33} \\ &\quad - (g^{ij}\Gamma^1_{ij})\partial_1 - (g^{ij}\Gamma^2_{ij})\partial_2 - (g^{ij}\Gamma^3_{ij})\partial_3 \end{aligned} \tag{3}$$

then applies the axisymmetry assumption $\partial_3 = 0$ and the identity $\partial_{21} = \partial_{12}$ to obtain

$$\nabla^2 = g^{11}\partial_{11} + 2g^{12}\partial_{12} + g^{22}\partial_{22} - (g^{ij}\Gamma^1_{ij})\partial_1 - (g^{ij}\Gamma^2_{ij})\partial_2 \tag{4}$$

In this form, the Laplacian operator is ready to be finite differenced by our PDE compiler's back end. The general problem of difference scheme selection is often described as more an art than a science, and an expert system (in the AI sense) might well be appropriate here in the future. For the present, our PDE compiler incorporates heuristics to make a default choice, which can be overruled if need be. Our present heuristics are very simple, choosing the usual centered 4th or 2nd order differencing schemes for each term independently, except near the grid boundaries where the differencing is off-centered if needed. These heuristics have proved adequate for our present work, but will probably need extending to deal with more general physical systems (hydrodynamics, shocks, etc.) and more sophisticated differencing schemes (monotonic, compact, implicit, etc.).

Our present PDE compiler represents operators by their explicit molecule (we use this word in the sense of "a little matrix of finite differencing coefficients") components at each grid point. Returning to our example, and considering the 2nd order case in the grid interior for simplicity, it selects the finite difference approximation

$$\begin{aligned}\nabla^2 &= g^{11}\begin{bmatrix} +1 \\ -2 \\ +1 \end{bmatrix} + 2g^{12}\begin{bmatrix} +1/4 & 0 & -1/4 \\ 0 & 0 & 0 \\ -1/4 & 0 & +1/4 \end{bmatrix} + g^{22}\begin{bmatrix} +1 & -2 & +1 \end{bmatrix} \\ &\quad - (g^{ij}\Gamma^1_{ij})\begin{bmatrix} -1/2 \\ 0 \\ +1/2 \end{bmatrix} - (g^{ij}\Gamma^2_{ij})\begin{bmatrix} -1/2 & 0 & +1/2 \end{bmatrix} \end{aligned} \tag{5}$$

to (4) and generates the (schematic) code shown in figure 1 to compute the corresponding Laplacian molecule.

3.2 Operators

We now discuss how a molecule L discretely approximating a given (continuum) operator $\mathcal{L}$ is *used.* There are two cases: It may be used to solve for a discrete approximation U to the solution U of $\mathcal{L}U = f$, by solving (the large sparse linear

Figure 1: Schematic code produced by our PDE compiler to compute the molecule components for the scalar Laplacian operator on the grid interior. Each block of finite differencing code is labelled with the term in (5) it computes.

The actual code generated by our PDE compiler is stylistically somewhat different, and is written in C for implementation convenience. There are also significant differences in how the data is organised in memory, as discussed in the text.

```
c
c Schematic Fortran code to compute scalar Laplacian molecule components
c on the grid interior.  The grid is of size (N_i,N_j), with coordinate
c spacing (delta_1,delta_2).  Centered 2nd order finite differencing is
c used, and boundary handling is omitted.
c
c The molecule itself is represented by the 9 arrays Mxy(i,j), where each
c of "x" and "y" may be either "m", "0", or "p" to denote an offset of -1,
c 0, or +1 respectively.  The the actual molecule components at the grid
c point (i,j) are thus:
c       [ Mmm(i,j)  Mm0(i,j)  Mmp(i,j) ]
c       [ M0m(i,j)  M00(i,j)  M0p(i,j) ]
c       [ Mpm(i,j)  Mp0(i,j)  Mpp(i,j) ]
c
c The inverse metric components are assumed to be stored in the arrays
c g_uu_11(i,j), g_uu_12(i,j), and g_uu_22(i,j), and the contracted
c Christoffel symbols in the arrays Gamma_u_1(i,j) and Gamma_u_2(i,j).
c
        factor_1 = 1.0 / delta_1
        factor_2 = 1.0 / delta_2
        factor_11 = 1.0 / (delta_1 * delta_1)
        factor_12 = 1.0 / (delta_1 * delta_2)
        factor_22 = 1.0 / (delta_2 * delta_2)

                do 20 i = 2, N_i-1
                do 10 j = 2, N_j-1

                Mmm(i,j) = 0.0
                Mm0(i,j) = 0.0
                Mmp(i,j) = 0.0
                M0m(i,j) = 0.0
                M00(i,j) = 0.0
                M0p(i,j) = 0.0
                Mpm(i,j) = 0.0
                Mp0(i,j) = 0.0
                Mpp(i,j) = 0.0

                temp = g_uu_11 * factor_11
                Mm0(i,j) = Mm0(i,j) + +1.0 * temp
                M00(i,j) = M00(i,j) + -2.0 * temp
                Mp0(i,j) = Mp0(i,j) + +1.0 * temp
```

$$\Big\} g^{11} \begin{bmatrix} +1 \\ -2 \\ +1 \end{bmatrix}$$

```
                temp = 2.0 * g_uu_12 * factor_12
                Mmm(i,j) = Mmm(i,j) + +0.25 * temp
                Mmp(i,j) = Mmp(i,j) + -0.25 * temp
                Mpm(i,j) = Mpm(i,j) + -0.25 * temp
                Mpp(i,j) = Mpp(i,j) + +0.25 * temp
```

$$\Big\} 2g^{12} \begin{bmatrix} +1/4 & 0 & -1/4 \\ 0 & 0 & 0 \\ -1/4 & 0 & +1/4 \end{bmatrix}$$

```
                temp = g_uu_22 * factor_22
                M0m(i,j) = M0m(i,j) + +1.0 * temp
                M00(i,j) = M00(i,j) + -2.0 * temp
                M0p(i,j) = M0p(i,j) + +1.0 * temp
```

$$\Big\} g^{22} \begin{bmatrix} +1 & -2 & +1 \end{bmatrix}$$

```
                temp = - Gamma_u_1 * factor_1
                Mm0(i,j) = Mm0(i,j) + -0.5 * temp
                Mp0(i,j) = Mp0(i,j) + +0.5 * temp
```

$$\Big\} -(g^{ij}\Gamma^1_{ij}) \begin{bmatrix} -1/2 \\ 0 \\ +1/2 \end{bmatrix}$$

```
                temp = - Gamma_u_2 * factor_2
                M0m(i,j) = M0m(i,j) + -0.5 * temp
                M0p(i,j) = M0p(i,j) + +0.5 * temp
```

$$\Big\} -(g^{ij}\Gamma^2_{ij}) \begin{bmatrix} -1/2 & 0 & +1/2 \end{bmatrix}$$

```
 10             continue
 20             continue
```

system) $\mathtt{LU} = \mathtt{f}$, where $\mathtt{f}$ is a discrete approximation to f. Or, it may be used to compute a discrete approximation $\mathtt{f}$ to $f = \mathcal{L}U$ where U is known, by convolving it with a discrete approximation $\mathtt{U}$ to U. We refer to $\mathcal{L}$ in these two cases as an "LHS" and "RHS" operator respectively, since it appears on the left and right hand sides respectively of the equations $\mathcal{L}U = f$ and $f = \mathcal{L}U$.

Along with our PDE compiler, we have constructed routines to support both of these ways to use molecules. Our generic "molecule solver" routine numerically solves $\mathtt{LU} = \mathtt{f}$ as defined above, where $\mathtt{L}$ may be any molecule and $\mathtt{f}$ any grid function. This routine is "generic" in the sense that it can solve *any* system of this form, independent of the precise nature of the molecule $\mathtt{L}$. It currently uses a direct solver based on the SPARSPAK routines of George and Liu (1981) (modified to handle nonsymmetric matrices) for the actual linear equation solution. Such direct solvers are highly robust, even when extended to nonlinear operators as described below. In our developmental setting, this robustness and absence of "tuning parameters" is very valuable. However, in other environments more efficient iterative solvers may be appropriate; these aren't precluded by the PDE compiler technique.

If an LHS operator is nonlinear, we use the usual multivariable Newton iteration. This gives excellent (quadratic) convergence, much better than the functional interation often used. However, it requires the Jacobian, which can be somewhat inconvenient to compute. The most elegant solution would be to compute it by *symbolically* differentiating the operator, then numerically evaluating (presumably with the help of the PDE compiler) this symbolic result. Our present PDE compiler can't do this type of symbolic operation, but this capability would be useful in a future PDE compiler.

We have also constrcted a generic "molecule evaluator" to evaluate $\mathtt{f} = \mathtt{LU}$ given the molecule $\mathtt{L}$ and the "operand" grid function $\mathtt{U}$, by convolving $\mathtt{L}$ with $\mathtt{U}$. (This routine actually works in a manner analogous to our PDE compiler – by generating C code which, when compiled and run, will do the convolution.)

We now consider the treatment of more general *tensor* operators. For example, we can define a "Ricci operator" which operates on the metric tensor g_{ij} to give the Ricci tensor

$$R_{ij} = -\tfrac{1}{2}g^{kl}(\partial_{ij}g_{kl} + \partial_{kl}g_{ij} - \partial_{il}g_{jk} - \partial_{jk}g_{il}) + (\Gamma\Gamma \text{ terms}) \qquad (6)$$

(we'll ignore the $\Gamma\Gamma$ terms in this discussion; we also assume that g^{ij} has already been computed). It's then possible to (and `Mk.2` does) calculate the Ricci tensor by constructing a molecule discretely approximating the Ricci operator, then convolving this molecule with the metric tensor.

We must thus generalise our notion of "molecule" to accommodate such tensor operators. In general each component of the (tensor) result depends on each component of the (tensor) operand, so we actually need a *matrix* of molecules at each grid point with matrix subscripts corresponding to tensor components, in the same way we need a matrix of numbers to define a linear transformation between vectors. (As discussed below, we only store the algebraicly independent components of tensors that have symmetries; dealing with this in a PDE compiler is basically just a matter of careful bookkeeping.) For example, we represent the Ricci operator mentioned above by a 6×6 matrix of molecules at each grid point, since g_{ij} and R_{ij} each have 6 algebraicly independent components.

For an RHS operator such as this, the explicit computation of molecule components is unnecessary – the operand (g_{ij}) could be finite differenced directly, saving the storage occupied by the molecule components. This can be very large – eg. the Ricci operator requires a total of 900(324) floating point numbers at each grid point for 4th(2nd) order finite differencing.

We have in fact found that RHS operators are about an order of magnitude more common than LHS operators in our formulation of the $3 + 1$ equations, and a properly optimized treatment of RHS operators is a major priority for our next generation PDE compiler. As a temporary expedient, our present code stores only a single row of the molecule matrix, and dynamically aliases it to each row in succession of the "virtual" matrix. This is still highly inefficient, with 150(54) numbers stored at each grid point for the Ricci operator. We tolerate this only because of the prototypical nature of our code.

Operators (both LHS and RHS) could be represented much more compactly by simply storing the numeric coefficients of each coordinate partial derivative operator at each grid point. For a scalar operator, this takes only 6 numbers at each grid point in 2-D or 10 in 3-D, compared with 25(9) in 2-D or 125(27) in 3-D for the explicit molecule components. Given this representation, the explicit molecule components could be regenerated on-the-fly when they're needed by the molecule solver for an LHS operator. For an RHS operator applied to an operand grid function $\mathtt{f}$, it would be natural to precompute each coordinate partial derivative of $\mathtt{f}$, so any such operator could then be evaluated by a simple dot product of its coefficient vector with the precomputed derivatives. We plan to explore this representation in our next generation PDE compiler.

3.3 The z Axis

In numerical relativity we often encounter coordinate singularities. For example, we often use (r, θ, ϕ) coordinates with the usual polar spherical topology, which have a singularity on the z axis. (Our discussion throughout will be in terms of this example, although our PDE compiler actually handles a slightly more general type of coordinate singularity. It might also be possible to handle the singularity at the origin by the techniques we describe here, but we suspect not.) In these coordinates, tensor components will in general blow up (or vanish) on the z axis.

The standard remedy for this problem is to factor out appropriate powers of $\sin\theta$ from all grid functions, so the "scaled" values stored in the numerical grid remain finite and nonzero on the z axis. We use a minor variation of this, in which powers of θ itself are factored out. (This simplifies the corresponding changes which need to be made to operators, as discussed below.) Thus we write, for example,

$$(g_{ij})_{\text{true}} = (g_{ij})_{\text{stored}} \cdot \begin{bmatrix} 1 & 1 & \theta \\ 1 & 1 & \theta \\ \theta & \theta & \theta^2 \end{bmatrix} \tag{7}$$

where the multiplication is element-by-element (like APL).

When the definitions of grid functions are modified in this manner, compensating changes must also be made to the operators applied to them:

$$\partial_\theta(\theta^n f) = \theta^n(\partial_\theta + \frac{n}{\theta}I)f \tag{8}$$

$$\partial_{\theta\theta}(\theta^n f) = \theta^n(\partial_{\theta\theta} + \frac{2n}{\theta}\partial_\theta + \frac{n(n-1)}{\theta^2}I)f \tag{9}$$

where f is the scaled grid function stored in the grid and n is the "scaling factor" (so $\theta^n f$ is the true value).

As well, axisymmetry places a further constraint on the behavior of grid functions near and on the z axis. In order to avoid a conical singularity, all theta derivatives must vanish there, so L'Hospital's rule gives the substitution

$$\partial_\theta \to \theta\,\partial_{\theta\theta} \qquad \text{(on the } z \text{ axis only)} \tag{10}$$

The resulting scaled grid functions remain well-behaved on the z axis, as do the operators and hence the molecules relating them. Thus no changes are needed to our molecule solver or evaluator. However, our PDE compiler must apply (8), (9), and (10) as part of the front end's (symbolic) processing of a differential operator. It does this *transparently*, i.e. its input (the relativity code) is written as if the coordinates were nonsingular. This transparency is important in order to retain the tractability of the PDE compiler's input for the full metric $3+1$ equations.

This scaling system works well on simple problems (we have had no difficulty solving the scalar Poisson equation with a $1/\theta$-like solution), but has proved insufficient to handle more complicated tensor problems such as those of numerical relativity. A much more sophisticated treatment of the finite differencing, incorporating the "numerical regularization" technique of Evans (1984), will probably be needed to handle such systems.

A major part of the problem also lies with our (straightforward) tensor-level formulation of the full metric $3+1$ equations – these exhibit $\infty - \infty$ cancellations on the z axis when used with topologically polar spherical coordinates. For example,

$R_{\theta\theta}$ is $O(1)$ on the z axis, but our expression (6) for it contains the $1/\theta^2$ term $g^{\phi\phi}\partial_{\theta\theta}g_{\phi\phi}$. We have also found similar cancellations in our formulation of the vector Laplacian operator used in computing the minimal distortion shift vector. We are still investigating how to deal with these $\infty - \infty$ problems; we suspect that the solution will lie in more sophisticated symbolic processing to analytically cancel the offending terms before they're finite differenced.

As a temporary expedient to allow testing of other parts of the code, we currently run `Mk.2` on a domain which excludes the (spatial) region $\theta < \theta_{\min}$ for some positive constant $\theta_{\min}$ (currently 30°), with analytical-solution Dirichlet boundary conditions on the artificial $\theta = \theta_{\min}$ boundary.

3.4 Implementation Considerations

Our PDE compiler handles boundary conditions as modifications of the operators to be compiled on the boundary. These modification must be specified by the higher level (relativity) code, as they're logically part of the continuum specification of the problem. (The modified operators receive no special treatment by our PDE compiler; indeed it's not even "aware" that they're boundary operators.) For example, the Poisson equation $\nabla^2 U = s$ with Dirichlet boundary conditions $U = 0$ on the boundary would be specified as the solution of $\mathcal{L}U = f$ on the entire problem domain (including the boundary), where

$$\mathcal{L} = \begin{cases} \nabla^2 & \text{in the grid interior} \\ I & \text{on the grid boundary} \end{cases} \tag{11}$$

$$f = \begin{cases} s & \text{in the grid interior} \\ 0 & \text{on the grid boundary} \end{cases} \tag{12}$$

We implement off-centered differencing near the grid boundaries in a slightly unusual way. The usual technique is to introduce fictitious points outside the boundary, extrapolate values for them, then finite difference with the usual (centered) molecules. Instead, our PDE compiler generates code to actually compute the off-centered molecule components, which are then used directly. We hoped that this technique, which is mathematically equivalent to the usual one, would be more convenient to program. It wasn't, and we plan to return to the usual technique in our next generation PDE compiler and numerical relativity code.

Our PDE compiler and numerical relativity code store all grid functions and molecule components for a given grid point contiguously in memory to improve cache and virtual memory performance. Although we haven't tested it, this data layout should be quite important – Streit (1981) reports a factor of two speedup gained by modifying an existing code in this manner. Unfortunately, because of our inefficient treatment of RHS operators, our code stores so much data at each grid point (about 850 floating point numbers, 80% of them molecule components) that its actual memory locality is very poor. We hope to greatly reduce this memory requirement in our next generation numerical relativity code.

We have also constructed a specialized "numeric database" system, which serves to insulate code using it (including almost all of the relativity code and much of the PDE compiler) from the exact layout of the different grid functions in (contiguous) memory at a given grid point. This system provides a layer of indirection between the code's reference to a grid function and the actual memory location at a given grid point. Our main use of this is to store only the algebraicly independent components of symmetric tensors, while still alowing code using the system to be written as if all components were present. We also use this system for aliasing molecules to save storage, as mentioned above.

Our overall experience with the "numeric database" system has been favorable, in that it has allowed us to write the code using it, in particular almost all of the relativity code, as if all components of all tensors are stored in the grid. However, this system has also provided us with a painful demonstration of the dangers of incomplete data abstraction. Certain stereotyped code sequences (eg. to "locate a tensor component") are commonplace in our PDE compiler and even occur occasionally in the relativity code, preventing us from easily making changes to the "numeric database" system or indeed to our data layout itself. We plan to provide a more flexible and fully abstracted descendant of this system in our next generation PDE compiler and numerical relativity code.

3.5 Symbolic Manipulation

We have several times referred to various tasks our PDE compiler performs as *symbolic manipulation* tasks. Indeed, one of the main suprises we encountered in constructing our PDE compiler was the (large) amount of symbolic manipulation in it, and we now consider a PDE compiler to be *mainly* a symbolic manipulation program. As such, we feel strongly that it should be written in a suitable (symbolic programming) environment. There are in fact several options for the language/programming environment the PDE compiler should be written in:

1. The PDE compiler (and presumably its input as well) could be written in the language of a symbolic algebra system. This is an elegant approach, but it implies the preselection of the representations, conventions, etc. of the symbolic algebra system, whether appropriate or not to the PDE compiler application. Also, such systems are often somewhat behind the state of the art as *programming* environments (for example, although many are Lisp-based, few are based on *modern* Lisps).

2. The PDE compiler could be written directly in (a modern) Lisp, possibly with a symbolic algebra system employed to do preprocessing of the sort

discussed above for linearizing nonlinear operators or handling the z axis. The main advantages of this option are the flexibility it offers and the superb programming environment of a modern Lisp system (but see Trickey (1988) for a somewhat contrary view).

3. As above, but with the number crunching done in Lisp itself. This somewhat radical option allows full use of the rich Lisp programming and debugging environment, but places great demands on the number crunching efficiency of the Lisp system. These *can* be met, as witness the Livermore S-1 Lisp compiler described by Brooks et al. (1982), but usually aren't.

4. The PDE compiler could be written directly in an algebraic language (Fortran, C, Ada, etc.). Given this choice, there are two natural forms the PDE compiler's input language can take: it can be a custom-built language (parsed in the PDE compiler by a custom-built parser), or it can be a sequence of subroutine calls in the algebraic language (parsed by the algebraic language compiler itself). In either case this option places fairly minimal demands on computer facilities and human expertise, but it provides a rather inhospitable environment for symbolic manipulation.

It's important to note that these options refer to the language/programming environment the PDE compiler itself is written in, not (except for option 3) that of the PDE-compiler-generated finite differencing code. Thus (again excepting option 3) any inefficiencies in the symbolic algebra or Lisp system the PDE compiler is based on do *not* effect the efficiency of the numerical calculations, only the efficiency of the PDE compilation(s) done as part of the code construction process.

Because of its simplicity and minimal machine requirements, we chose option 4 (with the PDE compiler input a sequence of C subroutine calls) for our present system. In hindsight, this choice of option 4 was a serious mistake, and we plan to use option 2 in our next generation PDE compiler.

4 Conclusions

By using a PDE compiler, we have been able to overcome the complexity of the general (full metric) coordinate $3+1$ equations. This has allowed us to use minimal distortion spatial coordinates, apparently for the first time in a multidimensional code. Except for severe z axis problems, our code is fully functional; we are presently engaged in testing it to assess its accuracy.

Our present PDE compiler is prototypical in nature, and has (in hindsight) several major design flaws which prevent its use in a realistic code. It's highly inefficient, with overheads of an order of magnitude or so in both space and time. It also generates very bulky output (finite differencing) code – about 120,000 lines for our relativity code using 4th order finite differencing. These flaws are irritating, but not fundamental to the PDE compiler technique, and our design for a next generation Lisp-based PDE compiler should resolve them.

We have found the use of 4th order finite differencing to be highly beneficial, with dramatic gains in accuracy far outweighing the modest costs in increased work per grid point. Because all our spatial finite differencing is generated automatically by the PDE compiler, there was little extra work involved in using 4th order – the original conversion from 2nd to 4th order took about one programmer-day (mostly spent fixing unrelated bugs), required *no* changes to the relativity code, and worked perfectly on the first test run. We estimate that a (future) trial of 6th order would take only a programmer-hour or so. With our next generation PDE compiler we expect to be able to try more sophisticated differencing schemes (monotonic, compact, implicit, etc.) with similarly little effort.

Our code is partitioned into two (disjoint) layers: relativity code on the top, the PDE compiler on the bottom. This "separation of knowledge" has proved very valuable. The relativity code is quite close to a literal transcription of our (straightforward) formulation of the $3+1$ equations, which makes it comparatively easy to write and debug. The PDE compiler is generic enough that we have been able to do one most of its debugging on simple model problems.

By automating the finite differencing process, the use of a PDE compiler greatly reduces the effort needed to try new ideas in the (continuum) formulation of the problem. For example, the minimal distortion shift vector is defined by a 3-vector elliptic equation (to be solved at each time step), assuming the lapse is known. Fairly late in the development of our code, we converted it to a slicing condition requiring a simultaneous solution (at each time step) of a 4-vector elliptic equation for the lapse *and* the shift vector. This required some rearrangement of the terms in the original equation, as well as the addition of additional lapse terms. We feel that the PDE compiler technique was invaluable in enabling us to complete the conversion in about 3 programmer-days.

But we have found the most important benefit of using a PDE compiler to be an unexpected one: it has changed the way we *think* about our code. By freeing us from the details of finite differencing, it has allowed us to think in terms of continuum (tensor) differential operators. This raising of the level at which we think is reminiscent of moving from assembler to (say) Fortran, where one soon becomes accustomed to thinking in terms of (say) variables, do loops, and common blocks, rather than registers, addressing modes, and stack pointers. (Moving from a scalar algebraic language (Fortran, C, Ada, etc.) to APL also comes to mind, where one (hopefully!) learns after a time to think in terms of array operations.)

Particularily in a developmental setting, the ease of experimentation that comes with the PDE compiler technique is very valuable. We believe that, with further research, PDE compilers can be developed to the point where they can also be used in efficient "production" codes.

5 Acknowledgments

My supervisor, W. G. Unruh, has provided many stimulating conversations, as well as several key ideas and suggestions. The idea of using a full metric was originally suggested to me by T. Piran. My conception of a PDE compiler originated and was refined in the course of an extended electronic mail discussion with M. Choptuik, who has also assisted me in many other ways too numerous to mention.

I thank the University of British Columbia, the Natural Sciences and Engineering Research Council, and the Isaac Walton Killam foundation for financial support; NSERC also supported the computer facility used to do this research. This research was done as part of my UBC Ph.D thesis.

6 References

Bardeen, J. M. and Piran, T. (1983): "General Relativistic Axisymmetric Rotating Systems: Coordinates and Equations", Physics Reports **96**(4), 205–250.

Brooks, R. A., Gabriel, R. P., and Steele, G. L. Jr. (1982): "An Optimizing Compiler for Lexically Scoped LISP", Proceedings of the ACM Sigplan '82 Symposium on Compiler Construction, published as ACM Sigplan Notices **17**(6), 261–275.

Centrella, J. M., LeBlanc, J. M., and Bowers, R. L. (1985): "Numerical Astrophysics", Jones and Bartlett, Boston.

Choptuik, M. and Unruh, W. G. (1986): "An Introduction to the Multigrid Method for Numerical Relativists", General Relativity and Gravitation **18**(8), 813–843.

Evans, C. R. (1984): "A Method For Numerical Relativity: Simulation of Axisymmetric Gravitational Collapse and Gravitational Radiation Generation" University of Texas (Austin) Ph.D thesis.

George, A. and Liu, J. W. (1986): "Computer Solution of Large Sparse Positive Definite Systems", Prentice-Hall, Englewood Cliffs.

Isenberg, J. A. (1979): "The Construction of Spacetimes from Initial Data", University of Maryland Ph.D thesis.

Leblanc, J. M. (1985): "The Future of Numerical Astrophysics", pp. 534–548 in Centrella, LeBlanc, and Bowers (1985).

Nakamura, T., Oohara, K., and Kojima, Y. (1987): "General Relativistic Collapse to Black Holes and Gravitational Waves from Black Holes", Progress of Theoretical Physics Supplement **90**, 1–218.

Smarr, L. L., Ed. (1979): "Sources of Gravitational Radiation", Cambridge University Press, Cambridge.

Smarr, L. and York, J. W. Jr. (1978a): "Radiation Gauge in General Relativity", Physical Review D **17**(8), 1945–1956.

Smarr, L. and York, J. W. Jr. (1978b): "Kinematical Conditions in the Construction of Spacetime", Physical Review D **17**(10), 2529–2551.

Streit, R. L. (1981): "Solution of Large Hermitian Eigenproblems on Virtual and Cache Memory Computers", ACM SIGNUM Newsletter **16**(2), 6–7.

Thornburg, J. (1988): University of British Columbia Ph.D thesis (expected December 1988).

Trickey, H. (1988): "C++ versus Lisp: A Case Study" ACM Sigplan Notices **23**(2), 9–18.

York, J. W. (1979): "Kinematics and Dynamics of General Relativity", pp. 83–126 in Smarr (1979).

NUMERICAL EVOLUTION ON NULL CONES

R. Gómez and J. Winicour

Department of Physics and Astronomy
University of Pittsburgh
Pittsburgh, PA 15260, USA

Abstract. A numerical algorithm designed to evolve the null cone characteristic initial value problem for general relativity is discussed in the context of nonlinear scalar waves in Minkowski spacetime.

Introduction

In the early1960's, Bondi's (Bondi *et al* 1962) use of null coordinates to study the asymptotic properties of the gravitational field cleared up previous ambiguities and inconsistencies and opened up a golden age of gravitational radiation theory. But this formalism was not capable of tackling the global problem of relating radiation fields to interior sources. Here we discuss a numerical version of that approach that can be implemented globally. (Isaacson *et al* 1983) We first present the underlying physical scheme, then the properties of the numerical code and conclude with some runs for nonlinear scalar fields which demonstrate its effectiveness.

The Null Cone Initial Value Problem

The approach is based upon a family of outgoing null cones emanating from some central geodesic world line. Null coordinates $x^\alpha = (u, r, x^A)$ with $x^A = (\theta, \phi)$ are introduced, where the retarded time u labels the null cones in terms of proper time along the central geodesic, r measures the luminosity distance along the null cones and the x^A are angular coordinate for the null rays, as shown in Fig. 1. In order to emphasize the underlying ideas we illustrate it in terms of electromagnetic fields in Minkowski space.

First, gauge freedom is used to eliminate the outward null component of the vector potential, i.e. to set $A_r = 0$. In this gauge, Maxwell's equations reduce to three equations of the following form.

$$(r^2 A_{u,r})_{,r} = H_u(A_B)$$
$$A_{B,ur} = H_B(A_B, A_u),$$

where the H's are determined from their arguments by local operations intrinsic to a single $u = constant$ null cone.

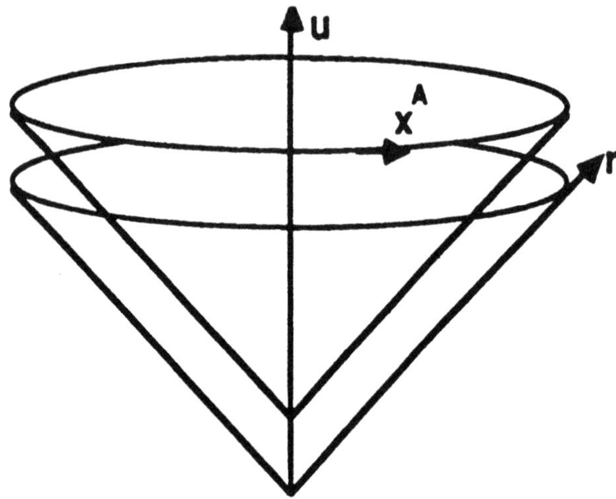

Figure1. Null cone coordinates.

This leads to the following simple scheme for the null cone initial value problem. The angular components A_B comprise constraint free initial null data. (It is a feature of the characteristic initial value problem that the u-derivative of A_B is not part of the initial data). Then, by an explicit hierachy of radial integrations, the above equations give the time derivative $A_{B,u}$ of the null data. The integration constants are determined by integrating from the origin where, by a combination of gauge and smoothness conditions, $A_u = A_B = 0$. The strategy behind this beautifully simple scheme carries over, virtually unchanged, to general relativity coupled to hydrodynamics and to all gauge theories that we are aware of. There appear additional intermediate variables but there still remains an explicit hierachy of radial integrations which lead from constraint free null data to their u-derivative.

In addition to the absence of constraints on the initial data and the lack of any elliptic equations, this approach has other important advantages for studying radiation fields. The only gauge dependence arises from the freedom in choice of the initial position and velocity of the central geodesic which determines the retarded time coordinate of the null cones. The radiation appears immediately at null infinity on the initial hypersurface. This has facilitated a clear study of the relationship between radiation and initial data sets and has been an important factor in the achievements described below. Furthermore, null infinity may be compactified,using the Penrose (1963) conformal method, by the simple transformation $r = x/(1-x)$, with each null cone covered by the range $0 \leq x \leq 1$. By conformal rescalings all fields are smooth in this compactified formalism. This is in contrast to the situation at spatial infinity where analogous techniques are not compatible with smoothness and can not be adapted to a finite difference scheme. In this way, the null approach avoids the awkwardness of extracting waveforms from the behavior of fields at finite space-time points. Numerically, this is an efficient scheme since only a small percentage of grid points are necessary to describe the asymptotic region. In the general relativistic case, the introduction of a compactified grid which includes null infinity allows the study of energy in terms of the Bondi mass and news function in strict accordance with their geometrical formulation as representations of the Bondi-Metzner-Sachs group.. Numerical studies of this type have already been succesfully carried out (Isaacson *et al* 1983; Gomez *et al* 1986). Further numerical investigations based upon this approach should open the way to study how the general relativistic dynamics of the interior affects other geometrically defined asymptotic quantities such as angular momentum, supermomentum and the Newman-Penrose quantities.

An important problem raised by this approach is how to prescribe initial null data in a way which does not introduce extraneous incoming radiation. Without some method for doing this, ambiguities would arise in distiguishing between interior matter sources and incoming fields as the origin of outgoing radiation. The condition that the gravitational null data be shearfree, while correct in the vacuum case, leads to large amounts of incoming radiation in the presence of matter. This problem has been satisfactorily solved by determining the null data in terms of a Newtonian limit based upon the Cartan formulation of Newtonian gravitational theory (Winicour 1984). In Cartan's formulation, the absolute time hypersurfaces of Newtonian theory are null hypersurfaces. Identification of these hypersurfaces as the limit of the general relativistic *outgoing* null cones leads to the elimination of *incoming* radiation. An outgrowth of this approach has been the establishment of the Einstein quadrupole radiation formula as an initial value theorem on the quasi-Newtonian null data in the form (Winicour 1987)

$$NEWS^{(0)} = Q_{,uuu}$$

where the left side is the general relativistic news function to leading post-Newtonian order and the right side is the third time derivative of the background Newtonian quadrupole moment, with both sides evaluated on the same null hypersurface. The motivation of this theorem by a numerical approach is another example of the synergysm which results from mixing theoretical and numerical considerations.

The Evolution Algorithm

Here, we will not delve further into the fascinating relationship between initial null data and the properties of their resulting radiation fields. The numerical codes which lead from null data to time derivative of null data, on a single cone, are in excellent shape. Some recent exact solutions (Bicak *et al* 1988) are being used to further improve their accuracy. Instead we will focus here on the more difficult problem of numerical evolution through a sequence of null cones. Early attempts at this for general relativity broke down in the near field region, for reasons that we now understand through studies of scalar fields. (See the discussion concerning Eq. (3)). The remaining discussion will concentrate on scalar evolution but the algorithm is constructed so as to carry over to gravitational and gauge fields.

In order to evolve the system, the radial integration for the u-derivative of the null data at u_0 must be converted into a finite difference equation which gives the null data at $u_0 + \Delta u$. To begin, consider solutions of the homogeneous scalar wave equation

$$\nabla^\alpha \nabla_\alpha g = 0.$$

in a 2-dimensional (possibly curved) space-time. For any null parallelogram with vertices at the points P, Q, R and S, as depicted in Fig. 2, the identity

$$g_Q = g_P + g_S - g_R.$$

can be established by integrating the wave equation over the interior Σ of the parallelogram. This identity can be carried over to spherically symmetric, source free scalar waves Φ in 4-dimensional space-time by identifying $g = r\Phi$. It generalizes to nonspherical scalar waves with nonlinear source terms, satisfying

$$\nabla^{\alpha}\nabla_{\alpha}\Phi = F(\Phi), \tag{1}$$

in the form

$$g_Q = g_P + g_S - g_R + \int_{\Sigma} dA(-\frac{L^2 g}{r^2} + rF(\frac{g}{r})). \tag{2}$$

Here, the contributions to the area integral over Σ arise from the angular part of the wave operator, represented by the angular momentum operator L^2 with eigenvalues $l(l+1)$ and the source term F, which combine to form the effective source for the spherical part of the wave operator.

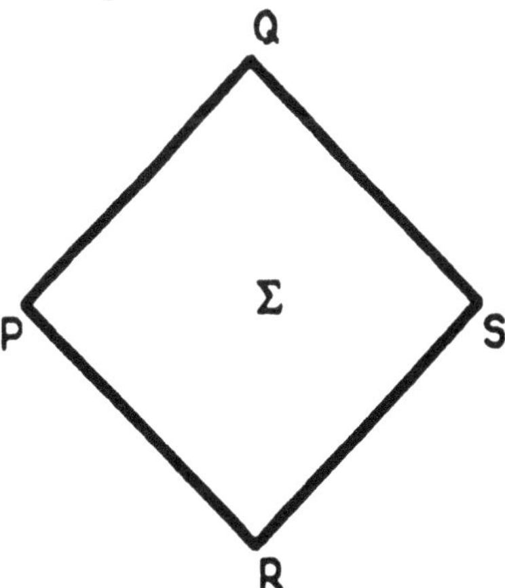

Figure 2. Points for the null parallelogram identity.

In null cone coordinates, (2) gives rise to the following explicit marching algorithm for evolution of a scalar field. Let the null parallelogram span two adjacent null cones at u_0 and at $u_0 + \Delta u$, as depicted in Fig. 3, for a fixed choice of angles. Suppose for now that the points P, Q, R and S lie on the numerical grid and that g has been determined on the entire u_0 cone and on the $u_0 + \Delta u$ cone radially outward from the origin to the point P. Then (2) determines g at the next radial point Q provided the integral over Σ can be calculated. But this integral can be approximated, to second order accuracy in grid size, by replacing the integrand by its average between the known points P and S, at which g has already been determined. After carrying out this procedure for the grid points at all angles with radius corresponding to Q, the procedure can be repeated to determine the field at the next radially outward grid point, the point T in Fig. 3. After completing this radial march to null infinity, the field g is then evaluated on the next null cone at $u_0 + 2\Delta u$. In practice, the points P, Q, R and S cannot be chosen to lie exactly on the grid because the velocity of light in terms of the compactified radial grid coordinate x is not constant. As a consequence, the field g at each of the points P, Q, R and S is approximated to second order accuracy by an interpolation between grid points. However, cancellations arise between these four interpolations so that the combination appearing in (2) is fourth order accurate. The net result, including the approximation to the integral, is that the numerical version of the exact equation (2) steps g radially outward one zone with an error of fourth order in grid size. The resulting global error in g, after evolving a finite retarded time, would then be expected to be second order accurate.

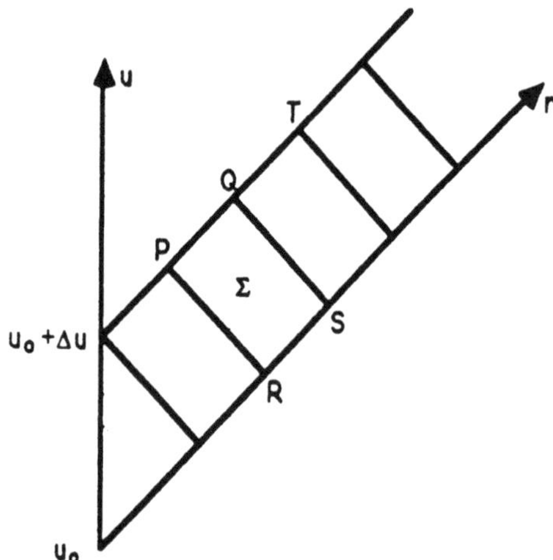

Figure 3. Scheme for the marching algorithm.

These considerations are confirmed by convergence tests of the code, Convergence measurements for data with various angular mutipole dependences are shown in Fig. 4 which plots global error versus grid size Δx (while Δy and Δu are kept at fixed ratios to Δx). The log-log plots approach a slope of 2 for small Δx, in agreement with second order accuracy.

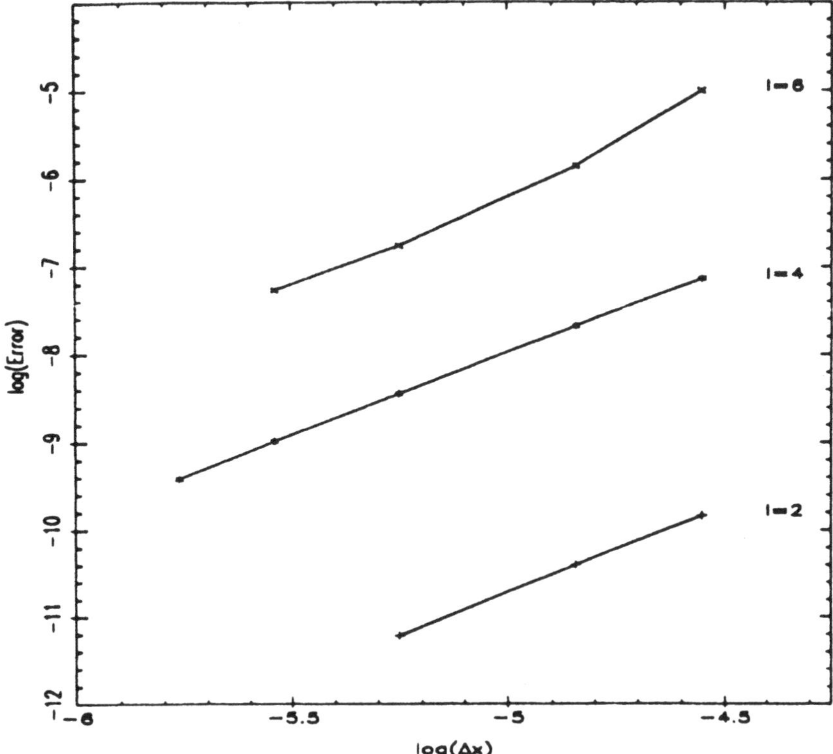

Figure 4. Convergence measurements.

Because of the explicit nature of this algorithm, its stability might be expected to depend upon an analogue of the Courant-Friedrichs-Levy condition that the physical domain of dependence be contained in the numerical domain of dependence. This requirement has a tricky geometric aspect to it in the null cone case. Let Q (see Fig. 3) be the new point in the evolution process. As $\Delta u \to 0$, the intersection of the past light cone of Q with the previously evolved outgoing null cone at u_0 approaches the outgoing null ray from the vertex to S. Consequently, for small Δu the past light cone of Q will lie inside the grid points at neighboring angles to S, *i.e.* the physical domain of dependence of Q will lie inside the numerical domain of dependence. However, as Δu increases, the intersection of Q's past light cone with u_0 spreads rapidly in the angular directions. This is analogous to the aberration effect in which the visual field is sharply peaked in the forward direction for velocities near the velocity of light but which opens up rapidly as the velocity becomes nonrelativistic. As a result, if the grid displacement Δu is increased, while holding Δr, $\Delta\theta$ and $\Delta\phi$ fixed, the physical domain of dependence rapidly spreads outside the numerical domain of dependence. Analytic studies show that this effect is most severe for points Q at small radii and that the domain of dependence requirement leads to the inequality

$$\Delta u \leq k\Delta r(\Delta y)^2 \tag{3}$$

where k is a factor of order one whose exact value depends on the way boundary conditions at the origin are treated. (For the present code, $k = 4$). The dependence of Δu on the square of the angular step size, as opposed to the first power as in the Cauchy evolution case, seems to be the price paid for the simplicity of the null cone algorithm.

Interpretive Features

Graphs of g at consecutive retarded times u require a different interpretation than the more familiar graphs with respect to $t = u + r$. We present here some simple examples for the source free linear wave equation to illustrate how to interpret null cone pictures. First, consider the spherically symmetric wave equation, for which $g = r\Phi$ is equivalent to the amplitude of a one dimensional wave on a string clamped at $r = 0$. Figure 8 illustrates the linear (numerical) evolution of an incoming spherical wave consisting initially of a single pulse. In (u, r) coordinates, the pulse travels without dispersion toward the origin with speed $\Delta r/\Delta u = 1/2$ but since Fig. 8 is expressed in terms of the compactified grid coordinate $x = r/(1+r)$ the shape of the pulse distorts slightly. As the pulse hits the origin it is "reflected" instantaneously to future null infinity ($x = 1$) because the speed of an outgoing wave is infinite in retarded coordinates. This reflection simultaneously lowers the remaining incoming section of the pulse, as is manifest in Fig. 8. The value of g at null infinity is the amplitude of the outgoing radiation and the radiation power is proportional to $(g_{,u})^2$. Note that g has a constant (horizontal) slope at null infinity.

This corresponds to the conservation of a Newman-Penrose (1968) quantity for the scalar field, which in this case consists of the monopole term in the $O(1/r^2)$ part of Φ. Examples to be discussed later show that this Newman-Penrose quantity is not conserved for certain nonlinear scalar fields.

The superposition of an incoming pulse with the reflected outgoing pulse (of opposite sign) produce a node in g at the origin, as apparent in Fig. 8. Now consider two identical, nonoverlaping incoming pulses, each having the shape of the initial pulse in Fig. 8, in (t, r) coordinates. The incoming pulse and the refected outgoing pulse would interfere to produce nodes throughout a region of constant t. In null cone coordinates, the nodes appear instead on the timelike line along which corresponding points of the incoming and reflected pulses cross. This is depicted in Fig. 5 by superimposing many retarded time snapshots during the crossing interval. The sharpness of the line of nodes in this figure clearly demonstrates that the code accurately transports the pulses with the correct velocity.

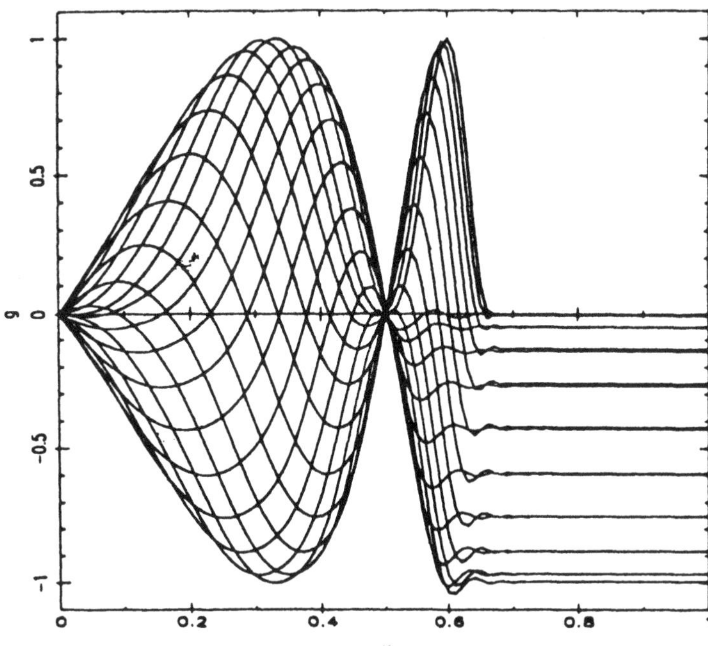

Figure 5. A timelike line of nodes.

Figure 6 illustrates the numerical evolution of a wave with a mixture of monopole and quadrupole angular dependence. The angular coordinate used in

the code is $y = -\cos\theta$. Of course, the corresponding exact solution can be easily constructed. (In fact, solutions of this type were used to make the convergence measurements displayed in Fig. 4.) The initial data in Fig. 6 has been designed to represent a purely incoming wave. The monopole component of the wave travels to the origin as in the previous example but backscattering of the the quadrupole component by the angular momentum barrier immediately produces quadrupole radiation at null infinity. This becomes evident in Fig. 6 as the longer lived monopole part (characterized by its flat angular dependence) separates from the quadrupole part. The code accurately tracks the angular dependence of the wave, with numerical noise arising chiefly due to the steep radial slopes in the choice of initial data.

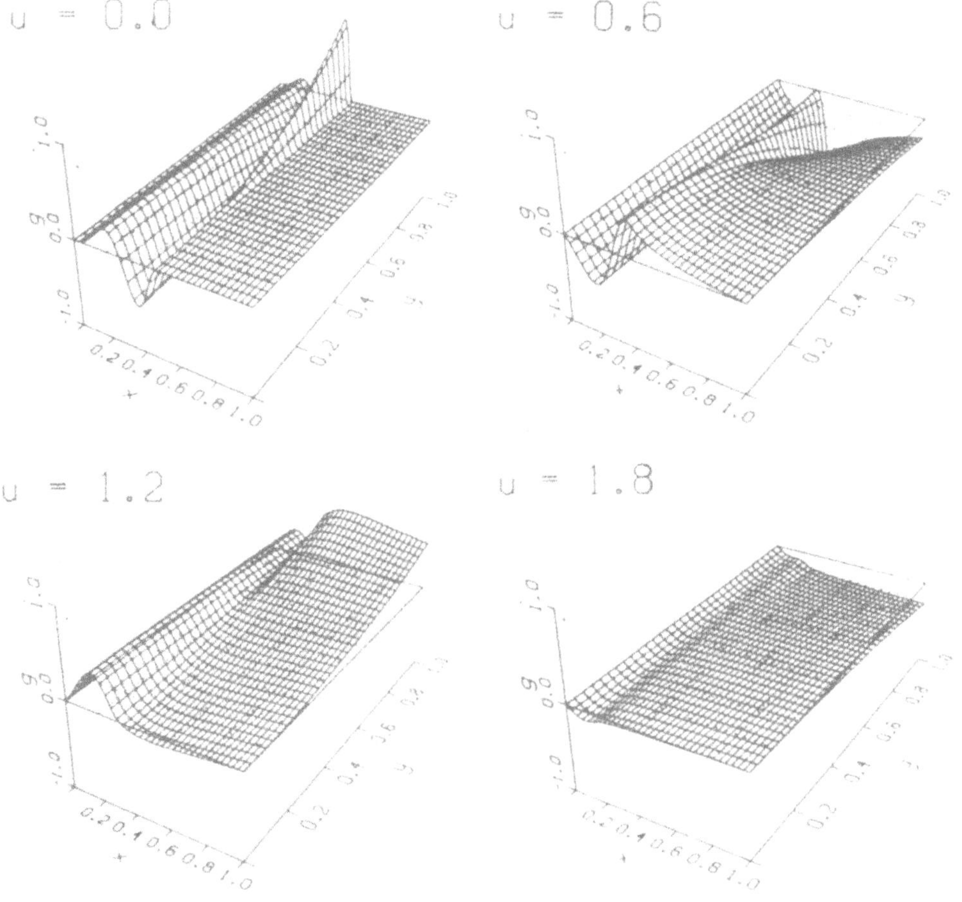

Figure 6. Evolution of a monopole-quadrupole wave.

Nonlinear Results

Nonlinear, self interacting wave equations can be obtained from a Lagrangian with potential $V(\Phi)$,

$$L = \frac{1}{2}(\nabla^\alpha \Phi)\nabla_\alpha \Phi + V(\Phi).$$

This gives rise to the field equation (1) with $F = -\partial V/\partial \Phi$.

Φ^4 theory. The choice $V(\Phi) = k\Phi^4$ has attracted considerable attention in model quantum field theories and for purely mathematical reasons because of its conformal invariance. It is also of special interest because Φ^4 is the lowest order monomial potential consistent with $O(1/r)$ radiation fields. Comparison with the evolution of the spherically symmetric linear pulse is displayed in Fig. 8. Several nonlinear effects of Φ^4 theory are evident. Notably, there is oscillation in the Φ^4 potential well as the wave progresses. This distorts the pulse shape and creates nodes in the waveform as these oscillations take place at faster rates in regions of greater Φ. Additionally, there is backscattering so that radiation from the pulse, which was purely incoming in the linear case, shows up at null infinity almost instantaneously. Note that the slope of g at null infinity does not remain horizontal, as in the linear case. This corresponds to another unique feature of Φ^4 theory. The Newman-Penrose quantity which measures this slope, as discussed earlier, is not conserved as opposed to the linear case or the case of a potential of higher order. As the system radiates more and more of its energy it behaves more linear and this Newman-Penrose quantity changes more slowly. But typically the Newman-Penrose quantity will not be small when this quasilinear regime is reached. Consequently, the system develops a long time scale tail which, from analytic considerations, decays asymptotically as $1/u$. This behavior is evident in Fig. 8, in which the linear field completely disappears at retarded time $u = 2$ when the back edge of the initial pulse reaches the origin but in which the Φ^4 field persists in the slowly decaying tail mode.

Using a calculus based upon conformal invariance, Hughston and Hurd (1983) have developed a method to obtain some exact solutions of Φ^4 theory.

An example with axial symmetry, and with t and y reflection symmetries, is

$$\Phi = \left(\frac{8a^2}{8a^2kr^2y^2 + (u(u+2r) + 2a^2k)^2} \right)^{1/2} . \tag{4}$$

This solution has singularities in the equatorial plane ($y = 0$) on the hyperbolae $t^2 - r^2 = -2a^2k$. For negative k, the singular hyperbolae consist of two spacelike sheets, as depicted in Fig. 7. We have taken advantage of this solution to calibrate our nonlinear code as well as to check its ability to track the formation of a singularity. This is achieved by beginning the numerical evolution on an initial null cone lying in the smooth region between the hyperbolae, the null cone u_0 in Fig. 8. Convergence tests again confirm global second order accuracy of the code. In addition, energy conservation tests show that the sum of the energy $E(u)$ and the integrated radiation power have the constant value $E(u_0)$ to good accuracy, typically 1% on a grid of 65 radial by 33 angular points. Energy conservation serves as an important check on numerical validity when no exact solutions are available, as in the next model theory.

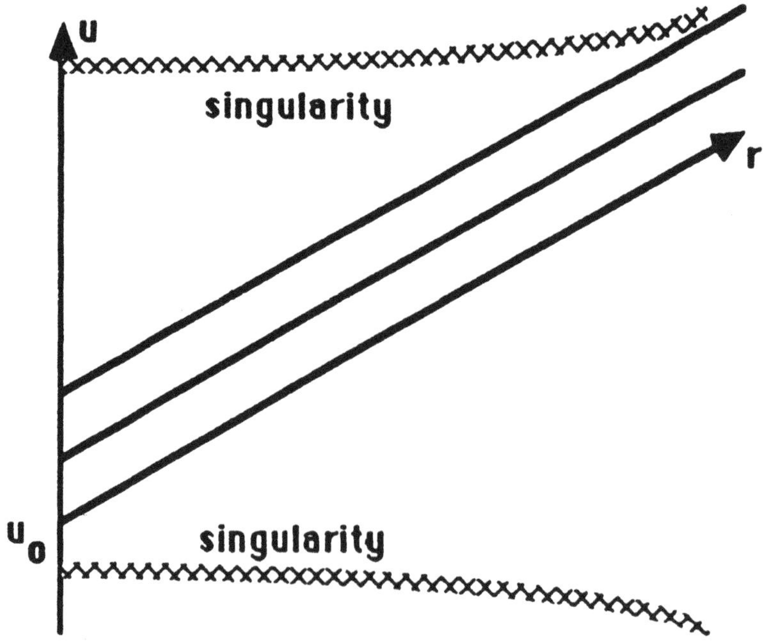

Figure 7. The equatorial plane of the solution (4) with $k < 0$.

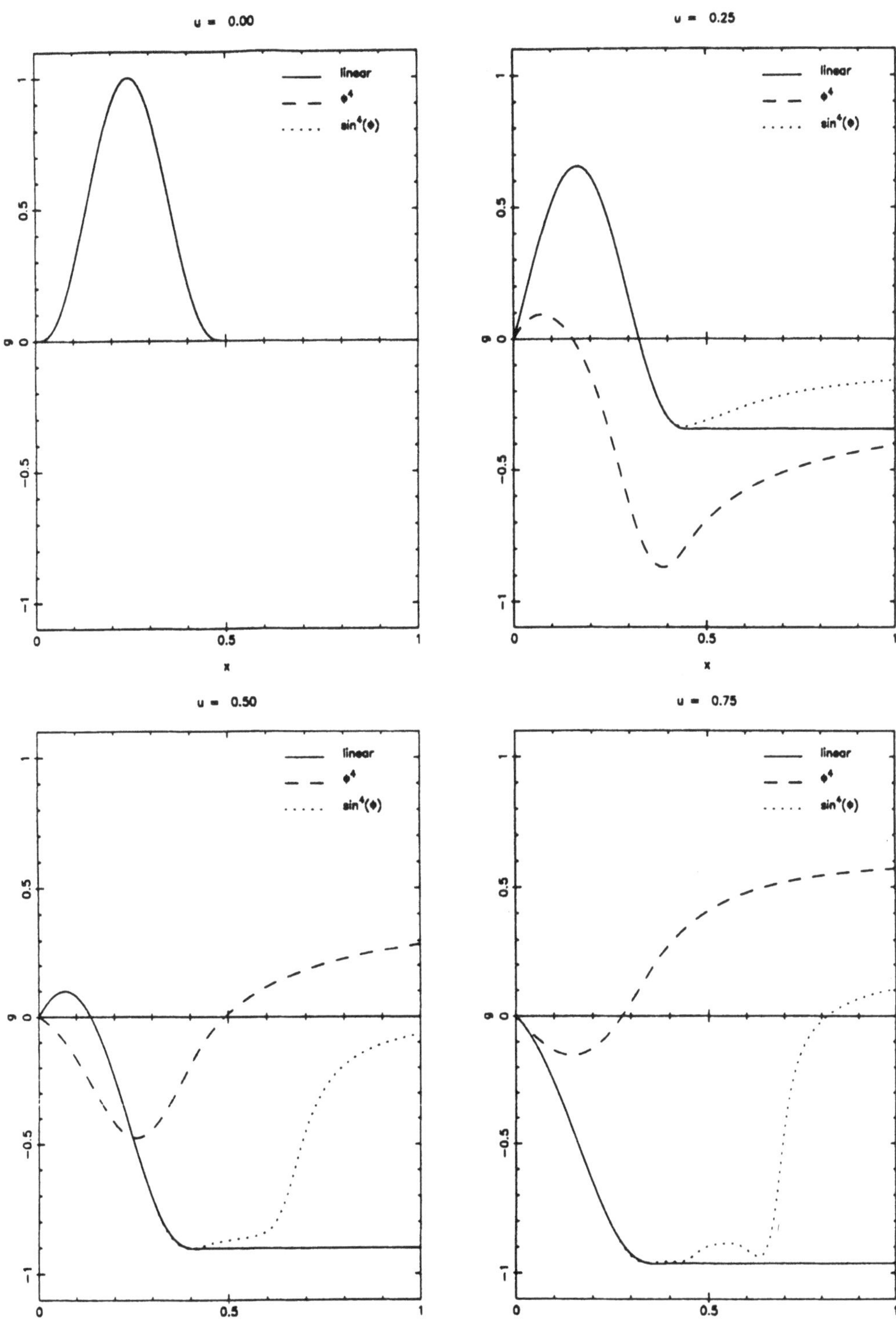

Figure 8. Evolution sequence of a spherical pulse.

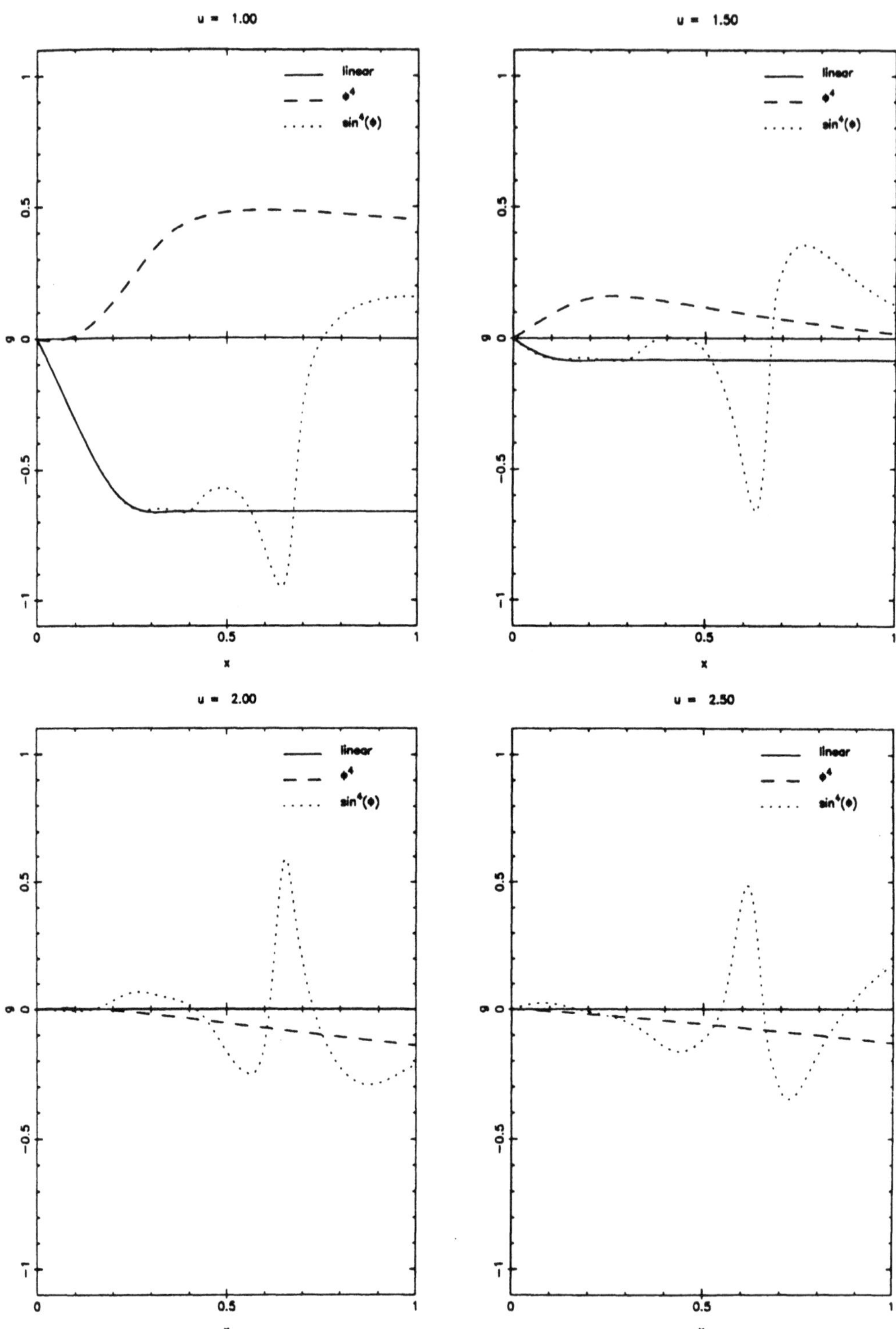

Figure 8. Evolution sequence of a spherical pulse (continued).

$\sin^4\Phi$ theory. The choice $V = k\sin^4\Phi$ behaves like the Φ^4 model when $F << \pi$, i.e. in the quasilinear regime. This implies that these models have the same asymptotic radiative behavior at null infinity. In the nonlinear regime, the $\sin^4\Phi$ model should exhibit the phenomena which we have discussed above: oscillation, dispersion, backscattering, nonconservation of the Newman-Penrose quantity and a $1/u$ tail decay. But, in addition, the theory might be expected to display some aspect of the solitary wave behavior familiar for the one dimensional sine-Gordon equation. There are no exact three dimensional versions of those one dimensional features which arise from topological conservation laws, such as the stability of kink solutions. That is because the connectedness of points at infinity is different in the three dimensional case. However, the kink stability may alternatively be established by energy considerations whose analogue can also be applied in three dimensions, as follows. Consider a region on a null cone in which $\Phi = n\pi$, with n an integer so that the field is situated at a zero of the potential V. Then the energy density of the field vanishes throughout this region because energy density on a null hypersurface does not involve the u derivative of the field. Now suppose two such regions on the same null cone, but with different integer n, were surrounded by some interpolating field which is reasonably smooth and goes to zero at null infinity. In order to break up this quasikink configuration, energy must flow into the constant regions to readjust the values of Φ in the two regions so as to match. This requires at least enough energy to lift Φ in one of the regions above the maximum of the potential. But if these regions are large this will be a slow process and the quasikink will be long lived. Thus, in the spherically symmetric case, one would expect to find long lived phenomena in the region far from the vertex of the cone. On the other hand, as a spherically symmetric quasikink collapses toward the origin the effect of three dimensions is to shrink its volume so that it requires relatively little ambient energy flux to equalize the two potential minima. Near the origin, quasikinks will be very fragile, at least in the spherically symmetric case.

A more detailed discussion of quasisolitary wave behavior in three dimensions will appear elsewhere. Here we conclude with a numerical example which illustrates the effectiveness of the code in revealing the above features. We choose the same initial null data for a spherically symmetric pulse which we have previously evolved in the linear and Φ^4 models. In Fig. 8, these evolutions are superimposed with that of the $\sin^4\Phi$ model. At early times, it is evident that the $\sin^4\Phi$ evolution closely approximates the linear evolution near the origin and behaves similarly to the Φ^4 evolution near null infinity. For intermediate values of r, a transition region develops with kinklike structure (although this is not apparent in Fig. 8 which plots g rather than Φ). This gives rise to quasistable behavior so that at $u = 2$, when the linear solution has radiated away and the Φ^4 solution has begun its decay in a $1/u$ tail, kinklike structure persists in the $\sin^4\Phi$ solution. At approximately $u = 30$, corresponding to 15 lifetimes of the linear solution, the $\sin^4\Phi$ evolution finally settles into a $1/u$ tail. A typical kinklike structure which develops at intermediate times is displayed in Fig. 9, which plots Φ versus r.

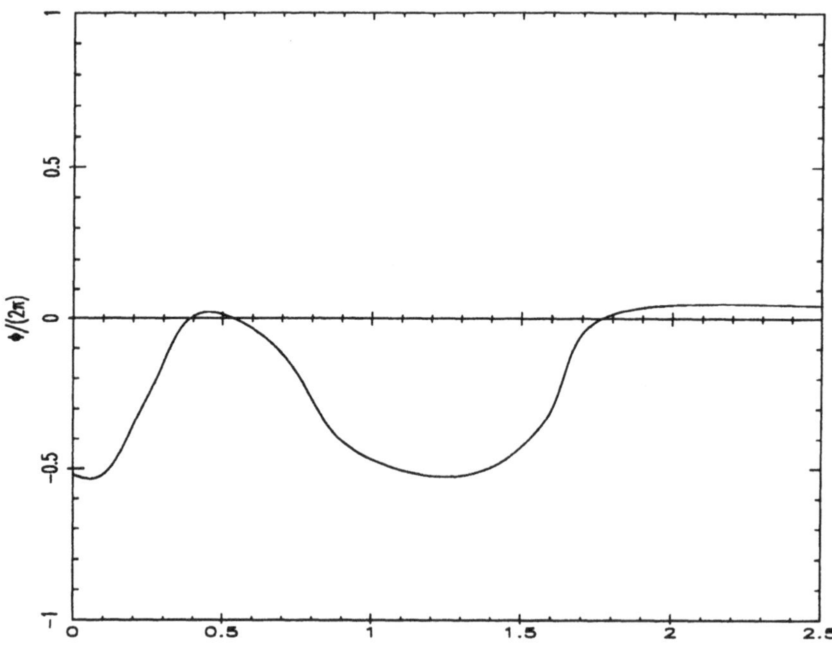

Figure 9. Kinklike structure formed during $\sin^4\Phi$ evolution.

References

Bondi, H., van der Burg, M.G. & Metzner, A.W.K. (1962). Proc. R. Soc. A **270**, 103.

Bičák, J., Reilly, P. & Winicour J. (1988). Gen. Rel. Grav. **20**, 171.

Gómez, R., Isaacson, R. A., Welling, J. S. & Winicour, J. (1986). In *Dynamical Spacetimes and Numerical Relativity*, J.M. Centrella, ed. (Cambridge University Press, Cambridge).

Hughston, L.P. & Hurd, T.R. (1983). Phys. Rep. **100**, 273.

Isaacson, R.A., Welling, J.S. and Winicour, J. (1985). J. Math. Phys. **26**, 2859.

Newman, E.T. and Penrose, R. (1968). Proc. R. Soc. London **A 305**, 175.

Penrose, R. (1963). Phys. Rev. Lett. **10**, 66.

Winicour, J. (1984). J. Math. Phys. **25**, 2506.

Winicour, J. (1987). J. Math. Phys. **28**, 668.

Normal Modes Coupled to Gravitational Waves in a Relativistic Star

Yasufumi Kojima

Department of Physics, Kyoto University, Kyoto 606, Japan

ABSTRACT

Motivated by the suggestion of a model system consisting of a star and gravitational waves, normal modes of a relativistic star are re-examined by relativistic perturbation equations. It is found that there are two classes of normal modes in the realistic radiating system. The differences between the two classes are clear in the decay rates of the complex eigenfrequencies, in the dependence of the rates on the coupling to gravitational waves and in the eigenfunctions. One class is a well-known class of weakly damped modes, which are usual modes of stellar oscillations such as the f- and p-modes. The weakly damped modes represent the slow decays of the stellar oscillations due to gravitational radiation. The other is a class of strongly damped modes, which are closely related to the freedom of gravitational waves itself. Hence, there are no modes corresponding to these modes in the Newtonian limit. The strongly damped modes represent rapid decays of the gravitational waves penetrated into the star. The energy is almost deposited in the gravitational waves so that the oscillations are quickly damped.

1. INTRODUCTION

In recent years, there have been remarkable developments in the numerical relativity. In the frontier of the numerical relativity, fully three-dimensional problems are dealt as shown elsewhere in this book. We also never fail to grasp physical properties. Schutz(1986) emphasized the healthy interaction between analytical and numerical relativity. That is, numerical relativity should provide some hints or approximations to analytical relativity. On the other hand, analytic relativity is useful for checks of numerical relativity. A model solvable by an analytic way is often useful to understand physical mechanisms. As one example, there is a problem concerning normal oscillations of a relativistic star (Schutz(1986)). From a simple toy model (Schutz(1986), Kokotas and Schutz (1986)), we expect undiscovered class of normal modes. Therefore we first review the toy model briefly.

2. A TOY MODEL

The model system consists of two strings. One is of finite length $2l$ with fastened ends and the other is semi-infinite with a fastened end. We impose the outgoing wave condition on the semi-infinite string at infinity. A massless spring connects two strings at the distance l from the fastened end of each string. In this model, the finite and semi-infinite strings represent a star and the space-time, respectively. Small vibrations of each string correspond to the stellar oscillations and gravitational waves. We assume the form $e^{-i\omega t}$ for the time variation of the vibrations of this system. Then the eigenfrequency ω can be obtained by solving

$$z(e^{-z} + e^{z}) = k(e^{-z} - e^{z})(2 + e^{2z}), \tag{2.1}$$

where $z = i\omega l/v$. v is the wave-propagation speed in the strings and $k(>0)$ denotes the dimensionless coupling constant of the spring connecting two strings. For small k , solutions of Eq.(2.1) fall into two families of modes;

family 1 of normal modes

$$\omega_n = \left[\pi(n+\frac{1}{2}) + \frac{k}{\pi(n+\frac{1}{2})}\right]\frac{v}{l} - 2i\left[\frac{k}{\pi(n+\frac{1}{2})}\right]^2\frac{v}{l} \tag{2.2}$$

for $k \ll 1$, where n is an integer.

family 2 of normal modes

$$\omega_n = \pi(n+\frac{1}{2})(1+\frac{1}{2a})\frac{v}{l} - ia\frac{v}{l} \tag{2.3}$$

for $k \ll 1$ and $n\pi \ll a$, where a is the larger of the two solutions of the transcendental equation;

$$a = ke^{2a}. \tag{2.4}$$

Frequencies of modes in both families are complex and the imaginary parts are negative. This fact means damped oscillations. The larger absolute value of the imaginary part means the greater damping rate. For small k , modes in the family 1 represents weakly damped mode, while those in the family 2 represents strongly damped modes.

We here summarize the properties of these normal modes.

Properties of a weakly damped mode

(1) As k approaches 0, the imaginary part of a normal mode approaches 0. The frequency in the limit $k = 0$ is the vibration frequency of the finite string.

(2) The decay rate increases as the coupling becomes stronger, that is, k increases.

(3) The oscillation frequency increases and the decay rate decreases, as the number of nodes n increases.

(4) In the eigenfunctions, the amplitude of the vibration of the finite string is larger than that of the semi-infinite string.

Thus the weakly damped mode represents normal oscillation of the finite string. The oscillation is damped slowly due to a little leakage of energy through the semi-infinite string. In the realistic problem, this mode corresponds to the decay mode of the stellar pulsation due to gravitational radiation, which can be successfully described by quadrupole formula.

Properties of a strongly damped mode

(1) As k approaches 0, a in eq.(2.4) goes to infinity, so that the decay rate goes to infinity. In the limit $k = 0$, this mode can not exist.

(2) As k increases, a decreases, that is, the decay rate decreases.

(3) The oscillation frequency increases with the number of nodes n. However, the decay rate does not depend on n.

(4) In the eigenfunction, the amplitude of the vibration of the semi-infinite string is larger than that of the finite string.

Thus the strongly damped mode represents normal oscillation of the semi-infinite string. The decay of the oscillation is very rapid because the energy is almost deposited in the semi-infinite string. The coupling of two strings is important for the existence of the mode in the family 2, because the boundary conditions would not be satisfied only by the semi-infinite string. The corresponding mode to the strongly damped mode has never been seen in numerical calculations for relativistic stars, but it seems certain it should be there (Schutz (1986)). Motivated by the suggestion of this toy model, we re-examine the classical problem to obtain normal modes of a relativistic star. Much attention is paid to undiscovered strongly damped modes.

3. BASIC EQUATION

3.1 perturbations inside a star

There is a long history in the relativistic pulsations of a spherical star. Thorne and his coworkers (Thorne and Campolattro (1967), Price and Thorne (1969), Thorne (1969), Ipser and Thorne (1973)) began to study this problem in 1967. More recently Lindblom and Detweiler (1983, 1985) examined quadrupole oscillation for various neutron stars. We therefore summarize the method of the linear perturbation briefly. Details are also written elsewhere (Kojima (1987), Nakamura, Oohara and Kojima (1987)). We use the units of $c = G = 1$ only in this section. The basic equations are a fourth-order system of equations inside the star. The metric perturbations in the Regge-Wheeler gauge are given by

$$ds^2 = -e^{\nu}(1 - r^l H_0 Y_{lm} e^{-i\omega t})dt^2 - 2i\omega r^{l+1} H_1 Y_{lm} e^{-i\omega t} dt dr$$

$$+e^{\lambda}(1 + r^l H_0 Y_{lm} e^{-i\omega t})dr^2 + r^2(1 + r^l K Y_{lm} e^{-i\omega t})(d\theta^2 + \sin^2\theta d\phi^2). \tag{3.1}$$

The fluid perturbations are given by the Lagrangian displacements

$$[\delta r, \delta\theta, \delta\phi] = [r^{l-1} W Y_{lm}, r^{l-2} V \partial_\theta Y_{lm}, r^{l-2} V (\sin\theta)^{-2} \partial_\phi Y_{lm}] e^{-i\omega t}. \tag{3.2}$$

Here $e^{\nu(r)}$ and $e^{\lambda(r)}$ are the metric functions that characterize the equilibrium stellar model and $H_0(r), H_1(r), K(r), W(r)$ and $V(r)$ characterize the metric and fluid perturbations and $Y_{lm}(\theta, \phi)$ is the spherical harmonic.

From the perturbed Einstein equations, we have a fourth-order system of equations among K, H_1, H_0 and W. The function V is written by a linear combination for K, H_1, H_0 and W. As the boundary conditions of the fourth-order system of equations, there are two conditions at the center of the star and one at the stellar surface from the regularity conditions. Thus we have one solution inside the star for any ω.

3.2 perturbations outside a star

Outside the star, the system of perturbation equations is reduced to a second-order system of equations. The perturbation equations are the same as those for a black hole. They can be described by either the Regge-Wheeler equation (Regge and Wheeler (1957)) or the Zerilli equation (Zerilli (1970)). They are related to each other by a certain transformation (Chandrasekhar (1983)). We use the Regge- Wheeler equation because the potential term is simpler. The equation is given by

$$\left[\left(1-\frac{2M}{r}\right)\frac{d}{dr}\left(1-\frac{2M}{r}\right)\frac{d}{dr}+\omega^2-\left(1-\frac{2M}{r}\right)\left(\frac{l(l+1)r-6M}{r^3}\right)\right]X=0, \quad (3.3)$$

where M is the total mass of the star.

At the stellar surface ($r = R$), the function X and the first derivative of the function X' are related to the interior perturbation functions. In addition to these boundary conditions, we impose the purely outgoing wave condition at infinity;

$$X \to X_0 \exp[i\omega(r + 2M \ln(r - 2M))] = X_0 \exp(i\omega r)(r - 2M)^{2i\omega M}. \quad (3.4)$$

where X_0 is a constant. Thus we have an eigenvalue problem. Eigenvalues are the values for which the function X in Eq.(3.3) satisfies the boundary conditions at the stellar surface and at infinity. We will have eigenvalues by solving the eigenvalue problem numerically.

3.3 a method of solving normal modes

The problem to obtain a normal mode with a large imaginary part is difficult in the point of numerical calculations, because the large imaginary part of the complex frequency causes numerical instability in solving the differential equations. One example is to obtain quasi-normal modes of a black hole. In order to avoid the numerical instability, the integration should be limited to smaller region. In order to do so, it is necessary to expand the function X by r and replace the boundary condition at infinity by the condition approximated at a smaller point. This procedure is taken by Chandrasekhar and Detweiler (1975). Recently, Leaver (1985, 1986) showed a nice technique to obtain the quasi-normal modes of a black hole. He expanded the function X into a certain power series and reduced the differential equation to a recurrence form among the expansion coefficients. Eigenvalues and eigenfunctions are obtained from the condition that the expansion series should be convergent. According to him, we expand X as

$$X = (r - 2M)^{2i\omega M} e^{i\omega r} \sum_{n=0}^{\infty} a_n (1 - \frac{R}{r})^n, \quad (3.5)$$

where a_n's are expansion coefficients which are determined by the four term recurrence relation,

$$\alpha_n a_{n+2} + \beta_n a_{n+1} + \gamma_n a_n + \delta_n a_{n-1} = 0 (n = 0, 1, 2...) \tag{3.6}$$

and one condition, $a_{-1} = 0$. Here $\alpha_n, \beta_n, \gamma_n$ and $\delta_n (n \geq 0)$ depend on ω, R, M and l. The explicit forms of them are omitted here. If the expansion series is convergent, the function X given by Eq.(3.5) satisfies the boundary condition (3.4) at infinity. The general solution of Eq.(3.6) can be spanned by three linearly independent solutions. They are classified in terms of the asymptotic form of a_n for large n;

$$a_n \approx \left(\frac{-2M}{R-2M}\right)^n n^{(4i\omega M - 1)}, \tag{3.7}$$

$$a_n \approx n^{-(i\omega(2M+R)+\frac{3}{4})} \exp(-2\sqrt{-2i\omega Rn}) \tag{3.8}$$

and

$$a_n \approx n^{-(i\omega(2M+R)+\frac{3}{4})} \exp(+2\sqrt{-2i\omega Rn}). \tag{3.9}$$

The first solution, Eq.(3.7) is $|a_{n+1}/a_n| < 1$ for large n if $R > 4M$, so that the series in Eq.(3.5) converges. The solution with a minus sign in the exponent, Eq.(3.8) will be also converge while the solution with a plus sign in the exponent, Eq.(3.9) diverges. Thus a desirable solution is a linear combination of Eq.(3.7) and Eq.(3.8) for large n. We solve Eq.(3.6) from large n and determine the ratio of the magnitudes of solutions (3.7) and (3.8) by the one condition $a_{-1} = 0$. By recurring in the direction of decreasing n , we exclude the undesirable solution, Eq.(3.9), even if it is included as numerical errors, because the solution given by Eq.(3.9) decreases in this direction. Thus for a given ω , there is a unique convergent solution outside the star. Eigenvalues are the values for which the functions X' and X inside the star match to the functions X' and X outside the star smoothly at $r = R$. They form a set of discrete values.

4. NUMERICAL RESULTS

We restrict our examination to the quadrupole mode ($l = 2$). Moreover we may restrict our calculations to Re $(\omega) > 0$ because if $\omega = \omega_1 + i\omega_2$ is an eigenvalue, then $\omega = -\omega_1 + i\omega_2$ is also an eigenvalue, where ω_1 and ω_2 are real. We use simple stellar models with polytropic equations of state because we expect the existence of these modes does not depend on the detailed stellar structure. We adopt

$$Gp/c^4 = 100(G\rho/c^2)^2. \tag{4.1}$$

Normal frequencies of the f-mode for the stellar models with this equation of state are well-known (Balbinski, Detweiler, Lindblom, and Schutz (1985)). Therefore we can test our numerical codes for the normal modes of the f-mode. After a check of numerical calculations for such a case, we have searched for complex eigenfrequencies with large imaginary parts. Since we consider homentropic case, frequencies of g-modes are zero. In addition to well-known class of weakly damped modes, another class of normal modes are found. Thus we have indeed two classes of normal modes.

Table 1 Normal frequencies

GM/Rc^2	Modes	$\mathrm{Re}(\omega(R^3/GM)^{1/2})$	$\mathrm{Im}(\omega(R^3/GM)^{1/2})$
	f	1.21	-2.53×10^{-4}
	p_1	3.09	-3.82×10^{-5}
0.109		3.31	-7.56
		9.95	-8.05
		19.08	-10.93
	f	1.16	-6.06×10^{-4}
	p_1	2.65	-1.06×10^{-4}
0.211		3.64	-4.12
		8.60	-5.71
		10.24	-3.79

Table 1. Frequencies of normal modes are listed for several models of neutron stars for the equations of state Eq.(4.1). Five modes are shown for each stellar model. The f- and p_1 -modes belong to the class of weakly damped modes and the others belong to the class of strongly damped modes.

The imaginary parts of the complex eigenvalues are negative for all modes obtained by our numerical calculations. This fact means the normal oscillations are all the damped oscillations. The larger absolute values of the imaginary parts mean stronger decaying rates. Thus normal modes with the larger absolute values of the imaginary parts are strongly damped modes while normal modes with the smaller absolute values of the imaginary parts are weakly damped modes. As examples, we tabulate eigenvalues for various stellar models in Table 1. We present the two lowest frequencies of weakly damped modes and the three lowest frequencies of strongly damped modes. There are, of course, higher modes besides these modes. It is apparent that there exist two classes which are completely different in the magnitude of the imaginary parts of the normal frequencies.

The imaginary parts are much smaller than the real parts for the weakly damped modes. They correspond to the f- and p_1- modes according to the usual classification. The values of eigenfrequencies for the stellar oscillations nearly coincide with those in the Newtonian case. These modes of the relativistic stellar oscillations are examined by many authors. This class of the normal modes represents the slowly decaying oscillations of the star due to gravitational radiation. In addition to these weakly damped modes, there are eigenfrequencies with large damping rates. The imaginary parts of the complex frequencies are comparable to the real parts in magnitude.

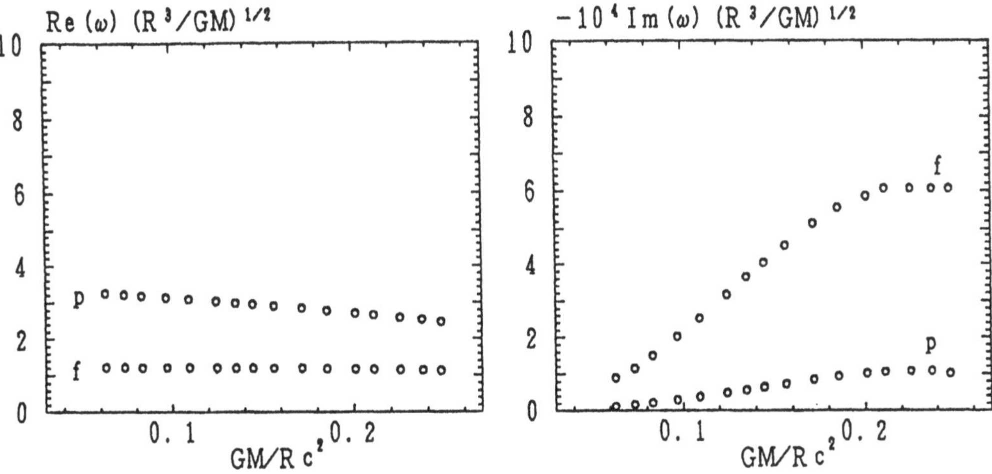

Fig.1. Normal frequencies of the stellar model with the equation of state (4.1) are shown as a function of the relativistic factor GM/Rc^2. Real and imaginary parts of the complex frequencies for weakly damped modes are shown. The imaginary parts are normalized by $10^{-4}(GM/R^3)^{1/2}$ and the real part by $(GM/R^3)^{1/2}$. Circles denote the calculated values.

In Figs.1and 2, we show four normal frequencies of a stellar model with the equation of state (4.1) as a function of the relativistic factor GM/Rc^2 , where M and R are the total mass and the radius of the star. Coupling to gravitational waves will become stronger with the increase of this factor. We normalize the frequencies by $(GM/R^3)^{1/2}$ except for the imaginary parts of the weakly damped modes. They are normalized by $10^{-4}(GM/R^3)^{1/2}$.

The absolute values of the imaginary parts in weakly damped modes increase with the factor GM/Rc^2 , while the real parts are almost independent of it as shown in Fig.1. This fact means that the emission rates of the gravitational waves increase and the stellar oscillations are damped quickly as the system becomes more relativistic. Moreover this figure shows that the absolute values of the imaginary parts for the p_1- mode are always less than those for the f-mode. Gravitational emission is less effective in higher modes because the oscillations are more confined to a relatively outer region of the star and produce smaller change of the quadrupole moment.

On the other hand, the behavior of the imaginary parts for the strongly damped modes is completely different from that for the weakly damped modes as shown in Fig.2. The imaginary part is roughly same value for different two strongly damped modes. The absolute values of the imaginary parts for the strongly damped modes roughly decrease with the increase of the factor GM/Rc^2, although there are some irregularities. This fact means that more slowly damped modes are allowed to exist in the system more strongly coupled to gravitational waves.

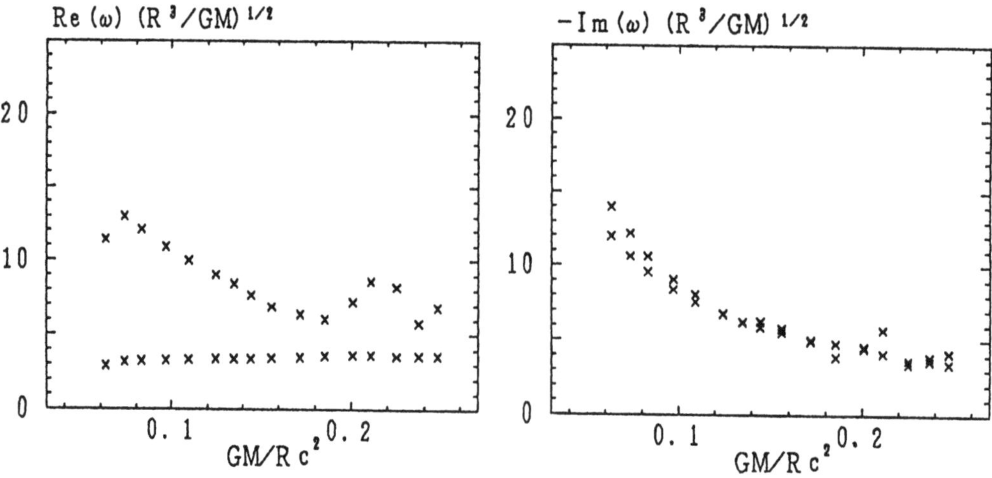

Fig.2. The same for Fig.1, but for strongly damped modes. The frequencies are normalized by by $(GM/R^3)^{1/2}$. Crosses denote the calculated values.

Finally we compare two classes of normal modes in terms of eigenfunctions. We illustrate the metric perturbation K and the relative displacement of matter in the radial direction $\delta r/r$ inside the star in Fig.3. We may set $K = 1$ at the center of the star,because only the relative magnitudes of the eigenfunctions are meaningful. The number of nodes increases as the frequency becomes higher for both classes of modes as expected from general theory of oscillations. A remarkable fact is that $\delta r/r$ for strongly damped modes is always very small compared with that for weakly damped modes. The strongly damped modes are related to the rapid decay of gravitational waves penetrated into the star without almost disturbing the matter. On the other hand, the displacement of the matter is essential for the weakly damped modes. Thus the difference in the properties of the weakly damped modes and the strongly damped modes is shown clearly in the behavior of eigenfunctions.

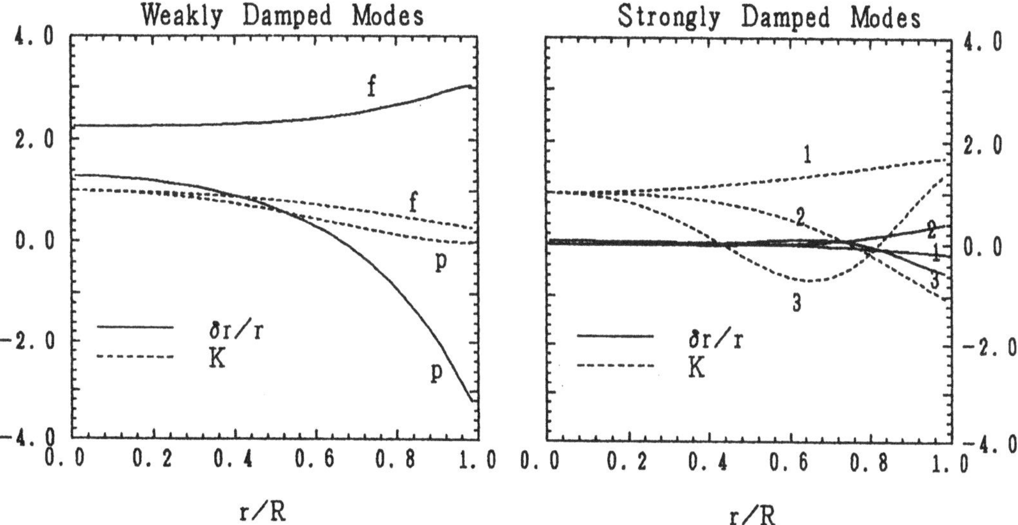

Fig.3. Eigenfunctions inside the star are shown for weakly damped modes and strongly damped modes of the stellar model with the equation of state (4.1) and $GM/Rc^2 = 0.109$. K and $\delta r/r$ represent the metric perturbation and the radial displacement of the matter in an equatorial plane, respectively. We always set $K = 1$ at the center of the star.

5. SUMMARY OF THE RESULTS

Numerical results show the actual system of the star and gravitational waves is very analogous to the toy model. It is found that there are two families of normal modes in the physical system just like the toy model. We confirm the properties of these normal modes suggested by the toy model, if we replace GM/Rc^2 by k. This factor GM/Rc^2 indicates the strength of the coupling between stellar oscillations and gravitational waves like k in the toy model.

The weakly damped modes, which have been discussed by many authors, correspond to the f- and p-modes in the usual classification of stellar oscillations. Imaginary parts of the normal modes are more than a factor 10^{-3} smaller than real parts of them. In the Newtonian limit, these eigenfrequencies coincide with the values calculated by the Newtonian theory of the gravity. The decay rates increase with a relativistic factor GM/Rc^2. When the system becomes more relativistic and stellar oscillations couple to gravitational waves more tightly, the emission rates of gravitational waves increase, that is, the stellar oscillations are damped more quickly. Thus this family of the weakly damped modes represents the slow decay of stellar oscillations.

In addition to this family of normal modes, there is another family of normal modes. There are no modes corresponding to them in the Newtonian limit. The imaginary parts of normal modes are comparable to the real parts in magnitude for these modes. As the system becomes relativistic, the decay rates decrease, because it becomes difficult for the waves to escape rapidly from such a strongly coupled system. The decay rates are almost same in different modes which belong to the family of strongly damped modes. Displacements of the matter inside the star are very small in these modes. In this sense, these modes are related to the decaying modes of initially excited gravitational waves.

6. CONCLUDING REMARKS

We need further healthy contacts between numerical and analytical relativity. No one tries to search for the strongly damped modes without the suggestions of the simple model. The properties of the toy model is passed through the test of the linear perturbation. Next they will be subject to the test by fully relativistic calculations. This is a future problem.

In the Dyson's toy model as cited in Schutz (1986), some modes of the family of strongly damped modes become unstable for a strong coupling case. We have no evidence for such an unstable mode in our numerical calculations. Such a mode may set in for the extremely relativistic case of $GM/Rc^2 > 1/4$ because our numerical calculations are limited to $GM/Rc^2 < 1/4$. (See §3.) It is necessary to explore another method for such a case. This is also a future problem.

Details of this paper are written in Kojima (1988).

ACKNOWLEDGMENTS

I would like to thank the organizers of this workshop for making it possible for me to attend the workshop. I am indebted to the Japan Society for the Promotions of the Science for financial aids.

References

Balbinski, E., Detweiler, S., Lindblom, L. and Schutz, B.F. (1985) Mon. Not. R. Astron. Soc. **213,** 553.

Chandrasekhar, S. and Detweiler, S. (1975) Proc. R. Soc. Lond. **A344,** 441

Chandrasekhar, S. (1983) *Mathematical Theory of Black Holes,* (Clarendon press, Oxford), Chap. 4.

Detweiler, S. and Lindblom, L.(1985). Astrophys.J. **292,** 12.

Ipser, J.R. and Thorne, K.S. (1973). Astrophys.J. **181,** 181.

Kojima, Y. (1987). Prog. Theor. Phys. **77,** 297.

Kojima, Y. (1988). Prog. Theor. Phys. **79,** 665.

Kokotas, K.D. and Schutz, B.F. (1986). Gen. Rel. Grav. **18,** 913.

Leaver, E.W. (1985). Proc. R. Soc. Lond. **A402,** 285;
(1986). J. Math. Phys. **27,** 1238;
(1986). Phys. Rev. **D34,** 384.

Lindblom, L. and Detweiler,S. (1983). Astrophys.J. Suppl. **53,** 73.

Nakamura, T., Oohara, K. and Kojima, Y. (1987). Prog. Theor. Phys. Suppl. **90,** 1.

Price, R. and Thorne, K.S. (1969). Astrophys.J. **155,** 163.

Regge, T. and Wheeler, J.A. (1957). Phys. Rev. **108,** 1063.

Schutz, B.F. (1986). in *Dynamical spacetimes and Numerical Relativity* , ed. by J.M. Centrella. (Cambrige University Press).

Thorne, K.S and Campolattro, A. (1967). Astrophys.J. **149,** 591.

Thorne, K.S. (1969). Astrophys.J. **158,** 1; 997.

Zerilli, F.J. (1970). Phys. Rev. **D2,** 2141.

COSMIC CENSORSHIP AND NUMERICAL RELATIVITY

Dalia S. Goldwirth†, Amos Ori and Tsvi Piran†*
Racah Institute for Physics
The Hebrew University, Jerusalem, Israel

I. INTRODUCTION.

The frontiers of numerical relativity, have followed, so far the development of computers. With construction of faster and larger computers numerical relativists have built bigger and larger codes in a hope that we can learn more physics from a large code then from a smaller one. The first general relativistic code [1] was one dimensional. The second generation codes, that were developed in the seventies and the early eighties [2-5] were two dimensional. The development of supercomputer directed most of the current efforts towards three dimensional codes. We have heard in this meeting about several three dimensional codes that are being constructed and tested now. The development of hardware is not always a blessing. The development of color terminals diverted a large fraction of the efforts of the numerical relativity community from physics to computer graphics.

The problems that are addressed by numerical relativity became more and more elaborated. The first code studied the essence of spherical collapse. The second generation codes, the two dimensional codes, explored the emission of gravitational radiation from collisions [3] or from rotating collapse [5]. With the third generation three dimensional codes we hope to be able to calculate eventually everything, but currents goals are calculations of emission of gravitational radiation from non head on collisions, from binary mergers and from rotating collapse in which the core breaks to a few objects - the collapse plunge and pursuit scenario. The newer larger codes also try to make "more realistic" calculations by using more sophisticated equations of state and larger grids, improving physical and numerical accuracies.

One may wonder is this the only way to advance in numerical relativity? Our answer to this question is negative. We should not be satisfied just with larger computations of the same kind. We should try to expand the boundaries of numerical relativity and enlarge the domain of questions that can be addressed using these techniques. One new possible frontier is the application of numerical techniques to questions in quantum gravity [e.g. 6,7]. Another possibility is to remain within classical general relativity but to address questions that *were not*

---

† also at the Institute for Theoretical Physics, University of California, Santa Barbara, CA 93106, USA

* talk given by Tsvi Piran

allowed to be addressed by numerical methods before. One such question is the Cosmic Censorship conjecture [8] and the appearance of naked singularities. This question is probably the most important open question in classical general relativity [9]. However it was considered a tabu for numerical relativity since it involves singularities - and one immediately questions the validity of a numerical solution when singularities are involved. If a singularity is observed one suspects that it is a numerical artifact. If the spacetime is regular one suspects that the observed regularity is a result of the numerical smoothing. We believe that one can overcome these problems and that we should address questions like cosmic censorship by a proper synergism between analytic methods and numerical techniques. In fact one can built, for special cases, extremely accurate codes that can explore singular or almost singular spacetimes. We present in this paper two such codes. We hope that following these examples the trend in numerical relativity will change towards construction of codes that will open new frontiers.

II. GRAVITATIONAL COLLAPSE OF A MASSLESS SCALAR FILED

Christodoulou [10], has shown, recently, that for weak enough initial spherical massless scalar field, there exists a regular solution for arbitrary long time for the coupled general relativistic and scalar field equations. The weak scalar field converges towards the origin, bounces and disperses to infinity. One expects a gravitational collapse to a black hole when the initial field is stronger. However Christodoulou [11] has described a special initial data that might lead to a naked singularity (this initial data represents a generic approach towards a singularity of a collapsing massless scalar field, if such a singularity exists).

We have constructed a characteristic code for studying numerically, the collapse of massless scalar field [12]. We formulate the problem in a characteristic way so that we need to solve only ODEs and we obtain very accurate solutions. Our numerical results show that it is unlikely that Christodoulou's special initial data will produce a naked singularity.

II.1 EQUATIONS AND FORMALISM.

To achieve an accurate numerical solution we formulate the problem in a way that will lead to characteristic equations. Surprisingly the formulation that is most useful for analytic studies [10] is also the best for the numerical calculations. We express the metric of the spherically symmetric spacetime in the form

$$ds^2 = -g\bar{g}du^2 - 2\frac{g}{\bar{g}}dudr + r^2 d\Omega^2 \quad , \tag{2.1.1}$$

and we define the integrated scalar filed:

$$\phi = \bar{h} = \frac{1}{r}\int_0^r h dr \quad . \tag{2.1.2}$$

The nontrivial Einstein equations are:

$$G_{rr} = \frac{2}{r}\frac{g_{,r}}{g} = 8\pi \bar{h}_{,r}^2 \tag{2.1.3}$$

and

$$G_{ru} = \frac{2}{r^2}\bar{g}[\frac{g}{\bar{g}} + r(\frac{g_{,r}}{g} - \frac{\bar{g}_{,r}}{\bar{g}}) - 1] = 8\pi\bar{g}\bar{h}^2_{,r} \quad . \tag{2.1.4}$$

Regularity at the origin leads to $g(u,0) = \bar{g}(u,0)$. The boundary conditions $h(u,0) = \bar{h}(u,0)$ forces us to integrate the equations outwards and to impose the normalization $g(u,0) = \bar{g}(u,0) = 1$. This corresponds to selecting the time coordinate as the proper time of an observer at the origin. The more common normalization $g(u,\infty) = \bar{g}(u,\infty) = 1$, which corresponds to selecting the time as the proper time of an observer at infinity cannot be used here. With these boundary conditions the solution at r depends only on the solution at $r' < r$ and we can integrate Eqs. 2.1.3 and 2.1.4:

$$g = \exp\left[4\pi\int_0^r \frac{(h-\bar{h})^2}{r}dr\right] \tag{2.1.5}$$

and

$$\bar{g} = \frac{1}{r}\int_0^r g dr \quad . \tag{2.1.6}$$

The incoming null geodesics are the characteristics of the problem. We write the scalar field evolution equation:

$$h_{,u} - \frac{1}{2}\bar{g}h_{,r} = \frac{1}{2}(g-\bar{g})(h-\bar{h}) \quad . \tag{2.1.7}$$

as an ODE:

$$\frac{dh}{du} = \frac{1}{2r}(g-\bar{g})(h-\bar{h}) \quad , \tag{2.1.8a}$$

along the incoming null geodesics:

$$\frac{dr}{du} = -\frac{1}{2}\bar{g} \quad . \tag{2.1.8b}$$

The mass contained within the sphere of radius r (at a retarded time u), $M(u,r)$ equals:

$$M(u,r) \equiv \frac{r}{2}(1 - \frac{\bar{g}}{g}) \quad . \tag{2.1.9}$$

Using $\partial M/\partial r = 2\pi\frac{\bar{g}}{g}(h-\bar{h})^2$, we express M as:

$$M = 2\pi\int_0^r \frac{\bar{g}}{g}(h-\bar{h})^2 dr \quad . \tag{2.1.10}$$

$2M/r$ provides a measure on the strength of the gravitational field. A black hole forms when $2M/r_h = 1$. It follows from Eq. 2.1.9 that $2M/r_h = 1$ requires that $\bar{g}/g = 0$. With the normalization $g(u,0) = \bar{g}(u,0) = 1$, g and $\bar{g}$ diverges on r_h in such a way that $\bar{g}/g \to 0$. The divergence of the metric function forces us to stop the evolution before the horizon forms. However, we can infer on black hole formation from the divergence of the metric functions and from the fact that $2M/r \to 1$. In actual computations we can reach $2M/r$ values as high as 0.99 . (Note that with the normalization $g(\infty) = 1$, g and $\bar{g}$ are nonzero and finite everywhere. Again the horizon cannot be crossed but now g and $\bar{g}$ will approach zero as the interior approaches a black hole. However, we have explained earlier that we cannot use this normalization here.)

II.2 THE NUMERICAL SOLUTION

We define a radial grid r_n where $n = 1....N$. Each grid point represents a radial null geodesic. We write Eqs. 2.1.8a-b as a set of $2N$ coupled differential equations:

$$\frac{dh_n}{du} = \frac{1}{2r_n}(g_n - \bar{g}_n)(h_n - \bar{h}_n) \quad , \tag{2.2.1a}$$

$$\frac{dr_n}{du} = -\frac{1}{2}\bar{g}_n \quad . \tag{2.2.1b}$$

h and g satisfy the boundary conditions:

$$\bar{h}_1 = h_1 \qquad g_1 = \bar{g}_1 = 1 \quad . \tag{2.2.2}$$

Eqs. 2.2.1a-b include an implicit integration along r, for a fixed u, to obtain $\bar{h}$, g and $\bar{g}$. We write these integrals as:

$$\bar{h}_n = \frac{1}{r_n}\sum_{i=1}^{n}\omega_i h_i \tag{2.2.3a}$$

$$g_n = \exp[4\pi\sum_{i=1}^{n}\omega_i\frac{(h_i - \bar{h}_i)^2}{r_i}] \tag{2.2.3b}$$

$$\bar{g}_n = \frac{1}{r_n}\sum_{i=1}^{n}\omega_i g_i \tag{2.2.3c}$$

Where the coefficients ω_i are determined by a 3 point Simpson method for unequally spaced abscissas (even if the grid is evenly spaced initially, it will not remain so during the evolution, see Fig. 1). The integration is based on overlapping parabolas. We solve the equation:

$$f(r) \simeq a_n r^2 + b_n r + c_n \tag{2.2.4a}$$

for r_{n-2}, r_{n-1} and r_n to obtain a_n, b_n and c_n and then we evaluate the integral I_n at r_n as:

$$I_n = I_{n-2} + \int_{r_{n-2}}^{r_n} f(r)dr$$

$$= I_{n-2} + \frac{a_{n-2} + a_n}{2}(\frac{r_n^3 - r_{n-2}^3}{3}) + \frac{b_{n-2} + b_n}{2}(\frac{r_n^2 - r_{n-2}^2}{2}) + \frac{c_{n-2} + c_n}{2}(r_n - r_{n-2}). \tag{2.2.4b}$$

Specifically:

$$\bar{h}_n = \frac{1}{r_n}[\bar{h}_{n-2}r_{n-2} + \sum_{i=n-2}^{n} \omega_i h_i] \quad , \tag{2.2.5a}$$

$$g_n = \exp[\ln g_{n-2} + 4\pi \sum_{i=n-2}^{n} \omega_i \frac{(h_i - \bar{h}_i)^2}{r_i}] \quad , \tag{2.2.5b}$$

$$\bar{g}_n = \frac{1}{r_n}[\bar{g}_{n-2}r_{n-2} + \sum_{i=n-2}^{n} \omega_i g_i] \quad . \tag{2.2.5c}$$

For $n = 2$ we use the Trapezium rule:

$$I_2 = \frac{1}{2}(f(r_1) + f(r_2))(r_2 - r_1) \quad . \tag{2.2.6}$$

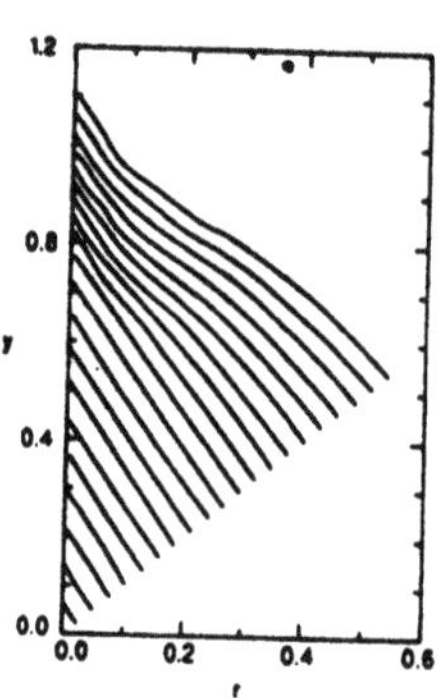

Fig. 1: *Null geodesics (for A=1.8), in the r and $r + \sqrt{2}u$ plane.*

We solve the 2N ordinary differential equations using three different standard methods: Sixth order Runge-Kutta, twelve order implicit Adams method and fifth order Gear's method [13].

The time step, du, is determined so that in each step the change in the null trajectory r_n is less than half the distance between it and the null trajectory r_{n-1}: I.e.

$$du < \frac{r_n - r_{n-1}}{\bar{g}_n} \quad . \tag{2.2.7}$$

Once a null trajectory reaches the origin $r = 0$ it bounces and disperses "instantaneously" along the outgoing null geodesic $u = Const$ to infinity (see Fig. 1). At this stage we exclude this grid point, from the solution. The calculation comes to its end when all the matter arrives at the origin and disperses to infinity, i.e. when $r_N = 0$. When the initial field is weak, $\bar{g}_n \simeq 1$ and the calculation ends at the retarded time $u_{final} \simeq 2r_N$. u_{final} decreases when the initial field increases.

As we approach the stage when a black hole forms $\bar{g}_n \to \infty$, condition 2.2.7 yields $du \to 0$ and this halts the calculation before the horizon appears. We identify the horizon's location from the maximal value of $2M/r$, a typical value for which is 0.99 .

We use the errors in $Max[2M/r]$ as a measure on the combined accumulated numerical errors. Using $A = 1.8$ (see next section), for which the gravitation field is quite strong, we have checked the integration along r by comparing calculations with two grids with 400 and 800 points. The largest absolute error (which appeared towards the end of the calculation), was less then 10^{-4} (Fig. 2a). The largest relative error was 3%. We have checked the integration along u by comparing three different integration methods. Comparison of the solutions revealed differences of about 10^{-6} (Fig. 2b).

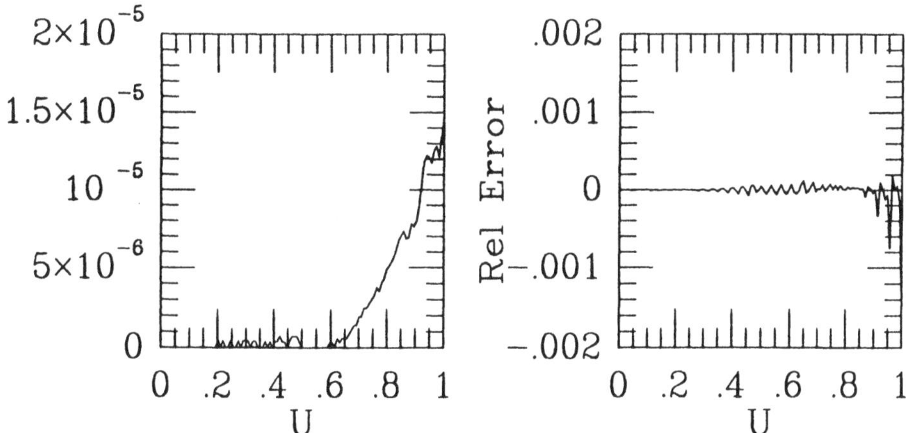

Fig. 2: *relative error in $Max[2M/r]$ vs. u (for A=1.8): a. (left) : Comparison between two different numerical methods for solving ODE; b. (right) : Comparison between two different grid sizes.*

Choptuik [14] has constructed recently a 3+1 finite differencing code that studies the same problem. He employs a different coordinate system. In spite of that we were able to compare our codes and the agreement is very good. A detailed description of this comparison will be given elsewhere.

II.3 EVOLUTION OF A GAUSSIAN WAVE PACKET

We have chosen an initial Gaussian ϕ:

$$\phi(u=0,r) = \bar{h}(u=0,r) = Ar^2 \exp[-(\frac{r-x_0}{D})^2] \quad , \tag{2.3.1}$$

with $D = .1$, $x_0 = .2$ and $r_N = .54$ on a grid of 400 points in the radial direction. We have performed a series of calculations for different values of A. For a small amplitude, e.g. $A = .01$, the gravitational effects are negligible and the results are similar to the Newtonian ones.

For $A < .1$ the gravitation field is very weak, and the scalar field simply bounces and disperses to infinity. An increase of the strength of the initial field leads to an increase in $2M/r$ and causes the null geodesics to converge. A second peak appears in $2M/r$ when the amplitude of A increases. A black hole appears when $A \geq 1.8745$ (see Fig. 3).

Our calculation stops, due to the divergence of the metric functions, just before the apparent horizon forms. As A increases a black hole forms earlier and with a larger final mass M_f. The graph of M_f/M_i (M_i being the initial mass) as function of A displays a jump at $A = 1.8745$: below that value $M_f = 0$, above it M_f has a finite value (see Fig. 4).

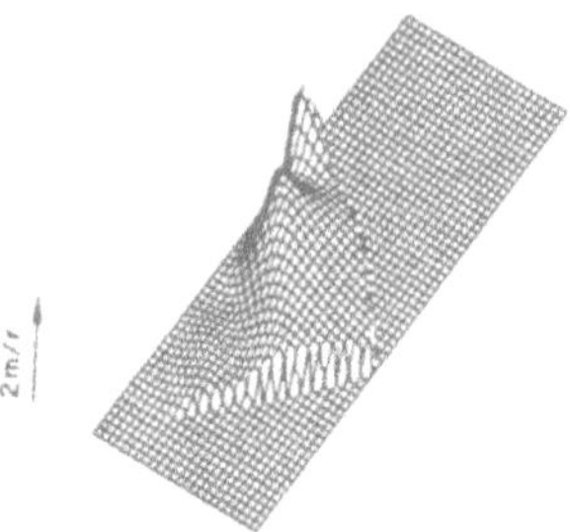

Fig. 3: *$2M/r$ for $A = 1.8745$, (vertical scale of 0-1). The formation of black hole can be seen when $2M/r$ approaches unity.*

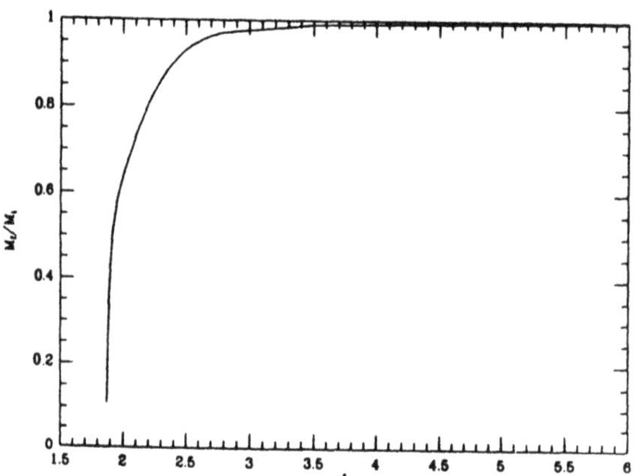

Fig. 4: *M_f/M_i vs. A for $A \geq 1.8745$.*

II.4 CHRISTODOULOU'S SPECIAL SOLUTION

We turn now the question whether it is possible to choose initial data which will evolve into a naked singularity. In an unpublished work Christodoulou [15] has suggested to consider as initial data at $u = -1$ a scalar field whose future evolution satisfies

$$\phi(u,r) \equiv \bar{h}(u,r) = \bar{h}(-1, \frac{r}{|u|}) - \kappa ln(|u|) \quad . \tag{2.4.1}$$

Defining

$$\bar{h}(-1,x) = -\int_x^\infty \theta(\tilde{x})d\tilde{x} \quad , \tag{2.4.2}$$

where $x = r/|u|$ leads to:

$$h(u,r) = h(-1,\frac{r}{|u|}) - \kappa ln(|u|) - x\theta(x) \quad , \tag{2.4.3a}$$

$$g(u,r) = \exp[4\pi \int_0^{x|u|} |u|x\theta^2(x)dx] = g(-1,\frac{r}{|u|}) \tag{2.4.3b}$$

and

$$\bar{g}(u,r) = \frac{1}{r}\int_0^{\frac{r}{|u|}} g(-1,\frac{r}{|u|})|u|d(\frac{r}{|u|}) = \bar{g}(-1,\frac{r}{|u|}) \quad . \tag{2.4.3c}$$

The requirement that ϕ satisfies the evolution equation, Eq. 2.1.7, yields an ordinary differential equation for θ:

$$x(\frac{1}{2}\bar{g} - x)\frac{d\theta}{dx} = \kappa - \theta(x)[\frac{1}{2}(g+\bar{g}) - 2x] \tag{2.4.4}$$

Eq. 2.4.4 together with Eq. 2.4.3b-c, specify the initial data for $0 \le r \le x_1$. The lhs of Eq. 2.4.4 vanishing at x_1, where $x_1 = \frac{1}{2}\bar{g}$ and a solution satisfying Eq. 2.4.1 does not exist for $r > x_1$. With initial data for ϕ given at $u = -1$, ϕ becomes singular at $(r = 0\ ,\ u = 0)$ and is regular elsewhere for $-1 \le u \le 0$. Therefore it is a potential counter example to cosmic censorship.

The solution of Eq. 2.4.4, which we call the interior solution, determines the spacetime only in the domain of dependence of the initial region $\{u = -1, 0 \le r \le x_1\}$. To obtain a global solution one has to specify initial data, $\theta(x)$, on $r > x_1$ and worry whether a black hole forms before $u = 0$ in the exterior region. With Christodoulou's construction it is impossible to specify vacuum as exterior initial data. Eqs. 2.4.4 and 2.4.3a yield $h(x_1) \ne 0$ and the choice $h = 0$ for $r > x_1$ means a discontinuous initial data. One must supplement the solution of Eq. 2.4.1, 2.4.3a-c and 2.4.4 with a non-vanishing causal exterior initial data at $r > x_1$ whose evolution might not be trivial. The global structure of the solution depends on the region exterior to the null geodesic, $r_1(u)$, that passes through $(u = -1, x_1)$. In particular if a black hole forms at the exterior before $u = 0$, the singularity at $(r = 0,\ u = 0)$ is not naked. The role of our numerical solution is to check whether it is possible to have an exterior extension that will not lead to a black hole before $u = 0$.

We have solved Eq. 2.4.4 numerically using standard methods of integrations for ordinary differential equations. We achieve easily high accuracy in spite of the fact that there is a regular singular point at x_1 where: $\frac{1}{2}\bar{g}(x_1) = x_1$. By gradually reducing the integration step we approach x_1 and obtain an accurate estimate for $\theta(x_1)$ and x_1. We evaluate κ, using Eq. 2.4.4, from the calculated values of x_1 and $\theta(x_1)$ and compare it with the original κ to obtain an estimate of the numerical accuracy. For small κ values ($\kappa < .4/\sqrt{4\pi}$) the difference between the two is less then 10^{-4} and it is less then 10^{-5} for the larger values ($\kappa > .4/\sqrt{4\pi}$).

Different values of κ yield different initial conditions on $[0, x_1]$ where $x_1 = x_1(\kappa)$. $\theta(x_1)$ decreases with κ, while $2M/r|_{x_1}$ and x_1 increase with κ. The larger

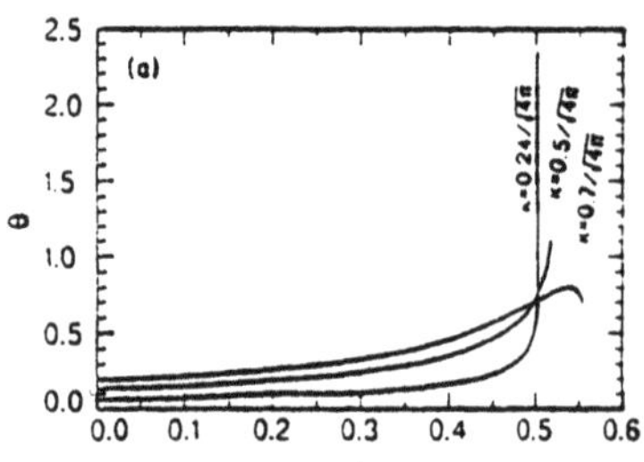

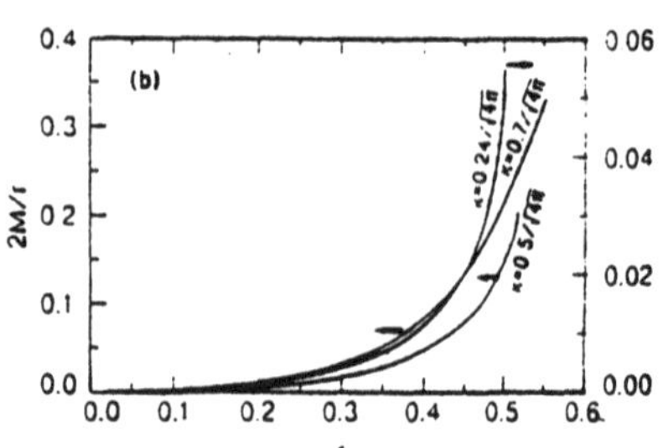

Fig. 5: *Initial data at* $u = -1$, *and* $0 \leq r \leq x_1$. *With* $\kappa = .24/\sqrt{4\pi}$, $\kappa = .5/\sqrt{4\pi}$ *and* $\kappa = .7/\sqrt{4\pi}$: *a.(left)* θ *vs.* r, *b.(right)* $2M/r$ *vs.* r.

$2M/r|_{x_1}$ the more likely it is to form a black hole. On the other hand as κ increases $\theta(x)$ becomes less steep and the numerical errors decrease (see Fig. 5).

The inner solution provides inner boundary conditions for the exterior solution on $r_1(u)$. Using $r_{1,u} = x_1$ and the initial condition $r_1(u = -1) = x_1$ we obtain:

$$h(u, r_1) = h(-1, x_1) - \kappa ln(|u|) \tag{2.4.5a}$$

$$\bar{h}(u, r_1) = \bar{h}(-1, x_1) - \kappa ln(|u|) \tag{2.4.5b}$$

$$g(u, r_1) = g(-1, x_1) \tag{2.4.5c}$$

$$\bar{g}(u, r_1) = 2x_1 \tag{2.4.5d}$$

$$\frac{2M(u, r_1)}{r_1} = \frac{2M(-1, x_1)}{x_1} \tag{2.4.5e}$$

We choose an external initial data and we calculate the exterior solution using the inner boundary conditions given by Eqs. 2.4.5a-e. Fig. 6 describes the evolution of $2M/r$ for different κ values and for two kinds of external continuation functions.

We describe only the exterior region, i.e. the domain of dependence of the region $\{u = -1, x_1 < r < r_{max}\}$. We have found that (see also Fig. 7) a black hole always forms before $u = 0$.

II.5 CONCLUSIONS

We have studied numerically spherical collapse of a massless scalar field. For a weak field, $2M/r$ is much smaller than unity and the scalar field converges inwards, bounces and disperses to infinity. When the initial field is stronger, $2M/r$ approaches unity and a black hole forms. In our coordinate system, g and $\bar{g}$ diverges and we can not follow a black hole appearance in this calculation. We can, still, estimate where and when a black hole horizon appears from graphs of $2M/r$, or g. A black hole appearance is also marked clearly on the trajectories of null geodesics which converge inwards in the (r,t) plane.

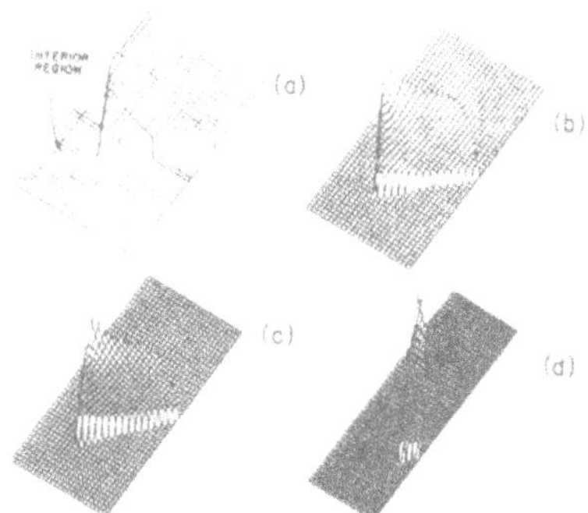

Fig. 6: *$2M/r$ in the exterior region for a.* $\kappa = .24\sqrt{4\pi}$ $\theta(x > x_1) = \theta(x_1)2^{25}/(1+\frac{x}{x_1})^{25}$. *b.* $\kappa = .5/\sqrt{4\pi}$ *and* $\theta(x > x_1) = \theta(x_1)2^{25}/(1+\frac{x}{x_1})^{25}$. *c.* $\kappa = .7/\sqrt{4\pi}$ *and* $\theta(x > x_1) = \theta(x_1)2^{25}/(1+\frac{x}{x_1})^{25}$. *d.* $\kappa = .5/\sqrt{4\pi}$ *and* $\theta(x > x_1) = \theta(x_1)(1-x)^{50}/(1-x_1)^{50}$

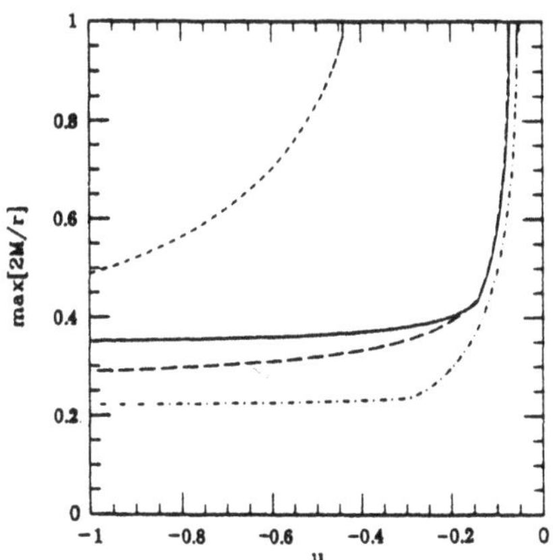

Fig. 7: *$Max[2M/r]$ vs. u, for different values of κ and continuation functions.*
$\kappa = .24/\sqrt{4\pi}$ *and* $\theta(x > x_1) = 2^{25}\theta(x_1)/(1+\frac{x}{x_1})^{25}$ - *short dashed line,*
$\kappa = .5/\sqrt{4\pi}$ *and* $\theta(x > x_1) = 2^{25}\theta(x_1)/(1+\frac{x}{x_1})^{25}$ - *long dashed line,*
$\kappa = .7/\sqrt{4\pi}$ *and* $\theta(x > x_1) = 2^{25}\theta(x_1)/(1+\frac{x}{x_1})^{25}$ - *solid line,*
$\kappa = .5/\sqrt{4\pi}$ *and* $\theta(x > x_1) = \theta(x_1)(1-x)^{50}/(1-x_1)^{50}$ - *dashed dotted line.*
A black hole always forms before $u = 0$.

We have evolved various extensions of Christodoulou's special initial conditions. In all the cases a black hole formed before $u = 0$. We believe that it is unlikely that one can find an extension that will lead to a naked singularity. Recent results of Christodoulou [16] indicate that if a massless spherical scalar field forms a naked singularity, the scalar field must approach the form given by Eq. 2.4.1.

Our results show, however, that such a singularity is most likely hidden inside a black hole. We conclude, therefore, that it is very unlikely spherical gravitational collapse of a massless scalar field will lead to a naked singularity.

III. SELF SIMILAR SPHERICAL COLLAPSE

By now, several authors have described counter examples [9,15-20] to the cosmic censorship hypothesis. Some examples result from aphysical initial data [18-20]. The other examples deal mostly with pressureless matter (either dust of null fluid) and include either shell crossing (which can also appear with nonzero but bound pressure [15]) or shell focusing singularities. Shell crossing singularities can be disregarded if we allow δ function distribution. Shell focusing singularities are more difficult to get rid off, however it is generally believed that these singularities will not appear if we treat dust self-consistently (using the collisionless Boltzman equation [9]) or if pressure is introduced.

We show [21-22] that a self-similar general relativistic spherical collapse of an adiabatic perfect fluid with an equation of state $p = (\gamma - 1)\rho \equiv k\rho$ and low enough γ values, results in a naked singularity. The singularity is tangent to an event horizon which surrounds a massive singularity and the redshift along a null geodesic from the singularity to any observer is infinite.

To obtain the numerical solution we solve two sets of coupled ODEs, corresponding to different coordinate systems. Each coordinate system has its own coordinate singularities. To bypass these singularities we transform numerically from one coordinate system to another. This method is a classical manifestation of usage of different coordinate patches, which is taught in introductory differential geometry but rarely used in practice.

III.1 EQUATIONS AND FORMALISM

We consider here a spherical space-time with a perfect fluid. The only perfect fluid barotropic equation of state which is consistent with self similarity is:

$$p = (\gamma - 1)\rho \equiv k\rho \tag{3.1.1}$$

where ρ in the total energy density, p is the pressure and k a constant. We use "Schwarzschild like" coordinates (r, t):

$$ds^2 = -e^{\nu}dt^2 + e^{\lambda}dr^2 + r^2 d\Omega^2 \tag{3.1.2}$$

where $d\Omega^2 = d\theta^2 + sin^2\theta d^2\phi$. Eq. 3.1.1 and the condition $-e^{\nu}u^{t2} + e^{\lambda}u^{r2} = -1$, leave us with four independent functions: ν, λ, ρ and u^r. The self similarity condition yields:

$$\rho \equiv \frac{D(x)}{4\pi r^2} \ ; \ \lambda = \lambda(x) \ ; \ \nu = \nu(x) \ ; \ u^r = u^r(x) \ ; \ u^t = u^t(x) \tag{3.1.3}$$

where $x \equiv r/t$ is the self-similarity variable and the lines $x = const$, are the self-similarity lines.

The Einstein spherical self-similar equations in these coordinates become:

$$T_0^0: \quad 2D(u_t u^t - k u_r u^r) = (1 - e^{-\lambda}) + x e^{-\lambda}\lambda' \qquad (3.1.4a)$$

$$T_1^1: \quad 2D(u_r u^r - k u_t u^t) = (1 - e^{-\lambda}) - x e^{-\lambda}\nu' \qquad (3.1.4b)$$

$$T_1^0: \quad 2(k+1)D u^r u_t = x^2 e^{-\lambda}\lambda' \qquad (3.1.4c)$$

$$u^\alpha T_{\alpha\ ;\beta}^{\ \beta}: \quad -k[u^r(D' - \frac{2D}{x}) - x u^t D'] + (k+1)e^{-(\lambda+\nu)/2}[(De^{\nu+\lambda)/2}u^r)' - x(De^{(\lambda+\nu)/2}u^t)'] = 0 \qquad (3.1.4d)$$

where $'$ means a derivative w.r.t. x. These equations are supplemented by boundary conditions which follow from the requirement of regularity at the origin for $t < 0$: $u^r(0) = 0$ and $\lambda = 0$, and an arbitrary choice of $\nu = 0$.

The mass $m(r,t)$ is defined by $e^{-\lambda} \equiv 1 - 2m/r$. We define a related self-similarity variable $\mathcal{M}$, as:

$$\mathcal{M}(x) \equiv \frac{m}{r} = \frac{1}{2}(1 - e^{-\lambda}) \quad . \qquad (3.1.5)$$

Using M we can rewrite 3.1.4a-c as:

$$x\mathcal{M}' = D[1 - (k+1)u_r u^r] - \mathcal{M} \qquad (3.1.4a')$$

$$-\frac{1}{2}x\nu' e^{-\nu} = D[(1 - (k+1)u_t u^t] - \mathcal{M} \qquad (3.1.4b')$$

$$x\mathcal{M}' = (k+1)\frac{D}{x}e^{\lambda}u^r u_r \qquad (3.1.4c')$$

We will also need another set of coordinates, the comoving coordinates (T, R), which we define by requiring: $U^\mu = U^T \delta_T^\mu$. The comoving line element is:

$$ds^2 = -e^{\Psi}dT^2 + e^{\Lambda}dR^2 + r^2 d\Omega^2 \quad . \qquad (3.1.6)$$

The normalization condition $u_\mu u^\mu = -1$ yields: $u^T = e^{-\Psi}$. We define a comoving similarity variable, $y = R/T$, and write the hydrodynamic and metric functions as:

$$\rho(R,T) = \frac{D(y)}{4\pi r^2} \ ; \ \Psi(R,T) = \Psi(y) \ ; \ \Lambda(R,T) = \Lambda(y) \ ; \ r(R,T) = \tilde{r}(y)T \qquad (3.1.7)$$

Substitution of the self-similarity condition (Eq. 3.1.7) into the comoving spherical Einstein equations yields:

$$T_0^1: \ -2y\tilde{r}'' - \Psi'(\tilde{r} - y\tilde{r}') + y\Lambda'\tilde{r}' = 0 \qquad (3.1.8a)$$

$$T_0^0: \ 1 + e^{-\Psi}(\tilde{r} - y\tilde{r}')(\tilde{r} - y\tilde{r}' - y\tilde{r}\Lambda') - e^{\Lambda}(2\tilde{r}\tilde{r}'' + \tilde{r}'^2 - \tilde{r}\tilde{r}'\Lambda') = 0 \qquad (3.1.8b)$$

$$T_{1\ ;\mu}^{\mu}: \ \Psi' = -\frac{2k}{k+1}\frac{\tilde{D}'}{\tilde{D}} \qquad (3.1.8c)$$

$$T_{0\ ;\mu}^{\mu}: \ -\Lambda' = \frac{2}{k+1}(\frac{\tilde{D}'}{\tilde{D}} + \frac{2}{y}) + 4(\frac{\tilde{r}'}{\tilde{r}} - \frac{1}{y}) \qquad (3.1.8d)$$

where we have defined $\tilde{D} \equiv D/\tilde{r}^2 = 4\pi\rho T^2$, and $'$ means now a derivative w.r.t. y.

Self similarity is an intrinsic geometrical property of the space-time. Hence if we can express a solution in comoving coordinates coordinates there exist corresponding "Schwarzschild like " self similar coordinates. To derive the transformation formulae we use the fact that both metrics are diagonal:

$$0 = g^{tr} = e^{-\Lambda}(\frac{\partial t}{\partial R})(\frac{\partial r}{\partial R}) - e^{\Psi}(\frac{\partial t}{\partial T})(\frac{\partial r}{\partial T}) \tag{3.1.9}$$

We define $\tilde{t} = t/T$ and we look for a solution of Eq. 3.1.9 of the form

$$\tilde{t}(R,T) = \tilde{t}(y) \tag{3.1.10}$$

We obtain a first order differential equation:

$$e^{-\Lambda}\tilde{t}'\tilde{r}' - e^{-\Psi}(\tilde{t} - y\tilde{t}')(\tilde{r} - y\tilde{r}') = 0 \tag{3.1.11}$$

that defines $\tilde{t}(y)$ up to an integration constant. The latter is fixed by the requirement that at the regular center $T = t$, hence $\tilde{t}(0) = 1$. We use now the definition of both similarity variables to write:

$$x = r/t = \tilde{r}/\tilde{t} = x(y) \quad . \tag{3.1.12}$$

The worldlines $x = const$, correspond to $y = const'$. Both are the invariant curves of the homothetic killing vector that exists in the space-time. Generally the inverse function, $y = y(x)$ exists, and we can transfer all functions of y to functions of x. The rest of the metric functions are:

$$e^{-\lambda} = e^{-\Lambda}\tilde{r}'^2 e^{-\Psi}(\tilde{r} - y\tilde{r}')^2 \tag{3.1.13a}$$

and

$$e^{-\nu} = e^{-\Psi}(\tilde{t} - y\tilde{r}')^2 - e^{-\Lambda}\tilde{t}'^2 \tag{3.1.13b}$$

The velocities are:

$$u^r = e^{-\Psi/2}(\tilde{r} - y\tilde{r}') \quad , \tag{3.1.14a}$$

$$u^t = e^{-\Psi/2}(\tilde{t} - y\tilde{t}') \quad . \tag{3.1.14b}$$

D is the same in both coordinate systems. The r.h.s. of Eqs. 3.1.13a,b and Eqs. 3.1.14a,b is a function of y only. Using Eq. 3.1.12 one finds that the metric functions, the velocities and D are all functions of x.

III.2 THE CENTRAL SINGULARITY

The point (0,0) is singular in every self similar solution. To see this we consider the density at the regular center:

$$\rho(T,0) = \frac{\tilde{D}(0)}{4\pi T^2} \tag{3.2.1}$$

and the density diverges. The density diverges also when we approach the origin (0,0) along any self similarity line, of the form $y = y_0$:

$$\rho(T, y_0 T) = \frac{\tilde{D}(y_0)}{4\pi T^2} \tag{3.2.2}$$

The density at $T = 0$ for $R \neq 0$ equals:

$$\lim_{T\to 0} \rho(T, R \neq 0) = \lim_{T\to 0} \frac{D_\infty}{4\pi r(R,T)^2} \quad , \tag{3.2.3}$$

where $D_\infty \equiv \lim_{y\to\infty} D$. Generally D_∞ is finite (in all solutions apart from a one parameter family of solutions which we do not discuss here [see 28]) and $\lim_{T\to 0} r(R,T) \neq 0$ for $R \neq 0$, hence $\lim_{T\to 0} \rho(T, R \neq 0)$ is finite.

The singularity at the origin follows directly from the self similar structure of the solution, (Eqs. 3.1.3 and 3.1.7) and it does not reflect any singularity of the solutions of Eqs. 3.1.4a,d or 3.1.8a,d. *This singularity is the one that can be seen from infinity in some solutions.* The divergence of the density in this singularity results also in a divergence of the curvature scalars there. In general in a perfect fluid the divergence of the density leads, necessarily, to divergence of the scalar $R^\beta_\alpha R^\alpha_\beta$ which satisfies $R^\beta_\alpha R^\alpha_\beta = \rho^2 + 3p^2$. The Ricci Scalar, $R = \rho - 3p$, diverges also if $k \neq 1/3$.

When the singularity at (0,0) is naked there is a Cauchy horizon emerging from it. In these cases we continue the solution analytically across the Cauchy horizon. The following discussion of the structure of the space-time beyond the Cauchy horizon depends on this continuation. When the singularity is not naked the evolution that follows the formation of the singularity at the origin follows directly from the initial data. In both cases the singularity that forms at (0,0) acquires mass at a rate proportional to T. At $T > 0$ there is a massive singularity at the center ($r = 0$). But the center doesn't remain at $R = 0$. The singularity is at $R = y_{ms} T$.

Another family of solutions, which we call "Asymptotically Friedmann Solutions" exists. In these solutions the whole $T = 0$ spacelike surface is singular. This singularity resembles the collapse to $r = 0$ of a closed Friedmann solution. We will not consider these solutions here (see [28] for a detailed discussion of classification of general relativistic self similar solutions.)

III.3 THE NUMERICAL SOLUTION.

The solution is characterized by two parameters, k and D_0. For a given choice of these parameters we integrate Eqs. Eqs. 3.1.4a-d from $x = 0$ towards $x = \infty$ and Eqs. 3.1.8a-d. from $y = 0$ towards $y = \infty$ numerically. These equations are ordinary non linear differential equations that we solve using standard numerical techniques. These equations contain, however several singular points (e.g. the sonic point, the regular center, the massive singularity and coordinate singularities) and the essential difficulty in the solution is to pass these points. The singularities are of different types. The central singularity which we have discussed in the previous section is a true singular point but it is not a singular point of the equations. There are a few points in which one coordinate system or another is singular. At these points we use the transformations that we have derived earlier to pass from a solution obtained in the coordinate system that becomes singular to the other one which is regular and in this way we bypass the singularity. Finally there are the "sonic point" and the massive singularity which are singular points in all the solutions. In these points (as well as in the other singular points) we expand the solution analytically, and we use the expansion to overcome the numerical problems and to study analytically the behavior of the solution there. In fact the numerical solution serves mostly as a bridge between these asymptotic expansions.

To check the accuracy of our solutions we have performed the following tests:

1. Comparison of the solution with different grid sizes.
2. Comparison of the solutions in the two different coordinate systems.
3. Comparison with analytic solutions like Friedmann or dust.
4. Comparison with the Newtonian limit (when it is valid).
5. Comparison with the analytic asymptotic expansions.

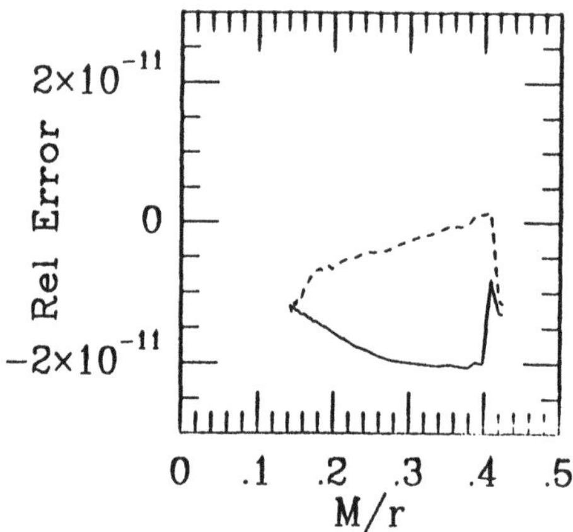

Fig. 8: *Relative errors in D (solid line) and in u_r (dashed line) versus M/r for calculations in two different coordinate systems.*

The most interesting test, from a relativistic point of view is the comparison of the solution in two different coordinate systems (see Fig. 8). We compare scalar quantities that are obtained from solutions of Eqs. 3.1.4a-d vs. those that are obtained from a solution of Eqs. 3.1.8a-d. Another test of the numerical solution that we present is the comparison between an analytic self similar dust solution and our numerical solution (Fig. 9)

The results that we present here are significant to at least four digits. There is no numerical problem in improving this accuracy. However, none of the properties of the solution that we describe depend critically on the accuracy of the solution.

III.4 THE SONIC POINT

The self-similar equations become degenerate at the "sonic point" (denoted $x = x_{sp}$ or $y = y_{sp}$). At this point the fluid moves at the speed of sound relative to the self-similarity line $x = x_{sp}$. The self-similar equations (Eqs. 3.1.4a-d, 3.1.8a-d) are at $x = x_{sp}$. The behavior of the solutions at the sonic point resembles the behavior of Newtonian self similar solutions [23-24].

In some solutions D' and $u^{r\prime}$ diverge at the sonic point and the solution cannot be extended beyond it. We will consider, in the sequel, only solutions for which all

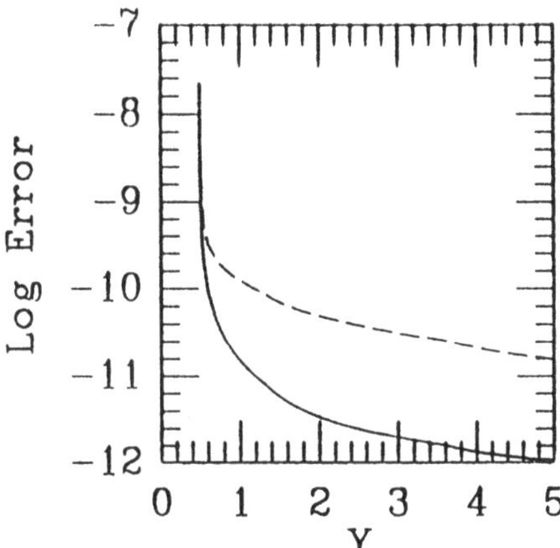

Fig. 9: *Relative errors in D (solid line) and in M/r (dashed line) between calculated and analytic self-similar dust solutions*

(first order) derivatives are finite at x_s and the solution can be continued beyond it. This requirement forms a "band structure" in the space of solutions: The D_0 line is divided into an infinite set of "permitted" segments (for which the solutions can be continued) and "forbidden" segments (for which the solutions necessarily terminate at the sonic point). In every "permitted" band, only few solutions are analytic at x_{sp}. All the other solutions are smooth, but not analytic there.

The sonic point is, generally, a branching point in the solution. Therefore, an additional parameter is required to specify the solution beyond x_{sp} and the space of solutions is a two-parameter set for $|x|>|x_{sp}|$. On the other hand, if one focuses on solutions which are analytic at x_{sp}, the space of solutions is reduced to a discrete set (there is a discrete set of D_o values for which x_{sp} is approached analytically, and for every such solution there is only one analytic continuation beyond x_{sp}).

We will focus in the following on the GRPL solutions which are the relativistic analogues of the Penston-Larson [25-26] Newtonian self similar solutions.

III.5 GENERAL FEATURES OF THE "BLACK HOLE TYPE" SELF SIMILAR SOLUTION.

We limit the discussion to what we call "black hole" type solutions [28]. Fig. 10 is a schematic space-time diagram of such solution in Comoving coordinates. Figs. 11a-b describe a specific (k=0.01) GRPL solution (Fig. 11a for $t < 0$, and Fig. 11b for $T > 0$). We see that at $t < 0$ (before the singularity forms) the central region is almost Newtonian. The density profile develops a central divergence but the gravitational field remains rather weak. Hence, a black hole does not form. Fig. 11b displays a singularity at $y = y_{ms}$ where D , u^r and M diverge and $\tilde{r} = 0$. The matter shells collapse into the central singularity a along $y = y_{ms}$ (see Fig. 10).

The asymptotic behavior near y_s is given by:

$$D \approx D_{ms}\delta y^{1/3-k} \quad ; \quad \tilde{r} \approx \tilde{r}_{ms}\delta y^{2/3} \quad ; \quad \mathcal{M} \approx \mathcal{M}_{ms}\delta y^{-2/3}$$

$$u^r \approx u_{ms}\delta y^{-1/3} \quad ; \quad e^{\psi} \propto \delta y^{2k} \quad ; \quad e^{\lambda} \propto \delta y^{-2/3} \tag{3.5.1}$$

where $\delta y \equiv (y - y_{ms})^{1/(1-k)}$. It follows directly for the form of the metric functions that the singularity at y_{ms} is spacelike. The divergence of $\mathcal{M}$ at the singularity

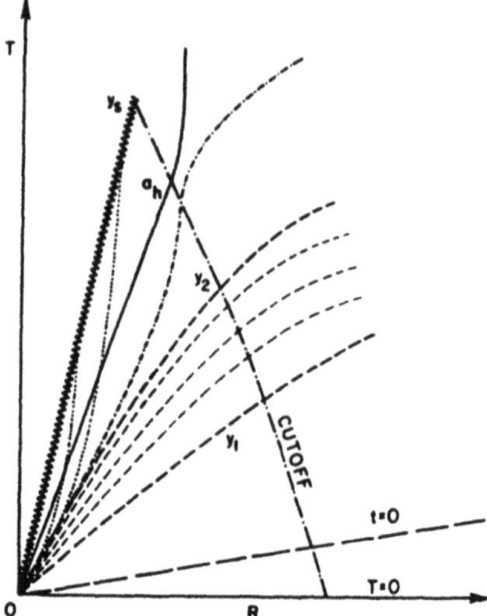

Fig. 10: *A schematic spacetime diagram of the collapse in comoving coordinates. The singularity at y_s is represented by a sawtooth-like line. The apparent horizon is denoted by 'ah'. The cutoff is denoted by a long dashed dotted line. Dashed lines denote null geodesics that are between y_1 and y_2 and escape to infinity. Dotted lines denote null geodesics that are between y_2 and y_s and fall back into the singularity. The short dashed and dotted line denotes a geodesics that is between y_2 and y_s and would have fallen into the singularity, but it escapes to infinity because of the cutoff.*

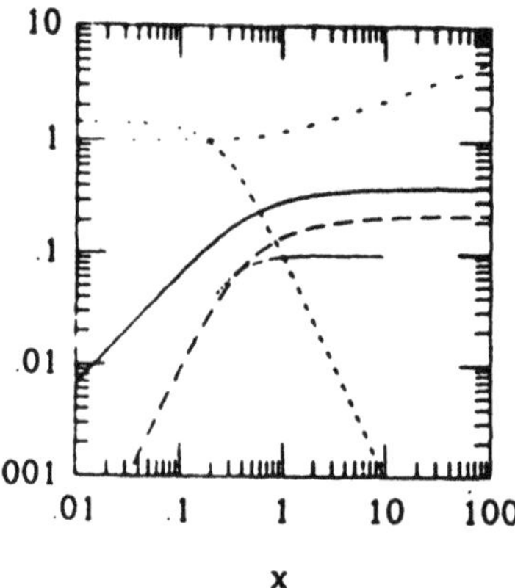

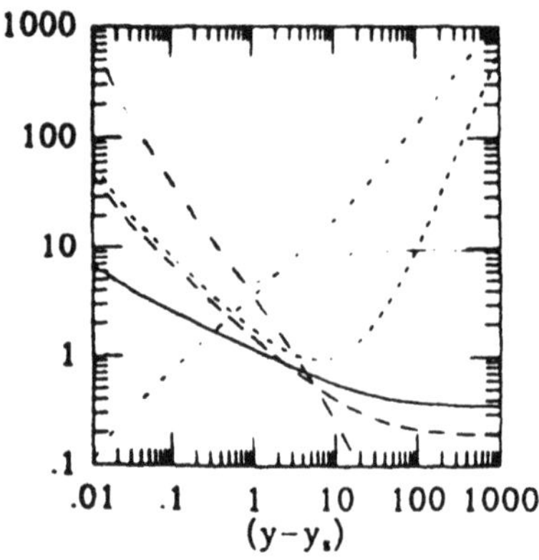

Fig. 11a: *Self similar collapse expressed in "Schwarzschild like" coordinates for $t < 0$ ($\gamma = 1.01$ and $d_0 = 1.4377$): $|u^r|$ (solid line), $x^2 d = 4\pi\rho r^2$ (dotted line), d (short dashed line), $2m/r$ (long dashed line) and $|g_{tt}|$ (dashed dotted line). Note that $2m/r < 1$ for all x values.* **Fig. 11b:** *Self similar collapse in comoving coordinates for $T > 0$ ($\gamma = 1.01$ and $d_0 = 1.4377$): $|u^r|$ (solid line), $100 \times 4\pi\rho r^2 = 100\tilde{r}^2 \times D$, (dotted line), $1000 \times D$ (long dashed dotted line), $2m/r$ (long dashed line), $\tilde{r} = r/T$ (short dashed dotted line) and $F(y) = y^2 g_{RR}(y)/|g_{TT}(y)|$ (short dashed line). Note that $2m/r = 1$ at y slightly below y_2, where $F(y_2) = 1$.*

is related to the fact that the singularity is massive. The mass of the singularity grows linearly with T :

$$\frac{m_{singularity}}{T} = \lim_{y \to y_{ms}} \mathcal{M}\tilde{r} = \mathcal{M}_{ms} r_{ms} = const \quad . \tag{3.5.2}$$

The divergence of $\mathcal{M}$ also indicates that there are trapped surfaces near the massive singularity. Hence there is a black hole in these solutions.

The Comoving coordinates become singular at two other self similar lines:

1. $T = 0$ where g_{TT} diverges.
2. $R = 0$ (the regular center) where g_{RR} diverges.

These coordinate singularities are bypassed by transforming to "Schwarzschild like" coordinates or by transformation to other comoving coordinates which are not self similar one. Similar singularities occur at the "Schwarzschild like" coordinates.

III.6 NULL GEODESICS AND CAUSAL STRUCTURE

The radial null geodesics (RNG) of the self similar solutions satisfy:

$$\frac{dR}{dT} = \pm\sqrt{\frac{g_{TT}}{g_{RR}}} = F(R/T) \quad , \tag{3.6.1}$$

where + corresponds to an outgoing geodesic and - to an ingoing one.

We call a solution of Eq. 3.6.1 which is of the form: $R = y_0 T$ (where y_0 is a constant), a "simple RNG". The negative solutions of the algebraic equation:

$$F(y) \equiv -\frac{y^2 g_{RR}(y)}{g_{TT}(y)} = 1 \quad , \tag{3.6.2}$$

describe ingoing simple RNGs that reach the singularity at (0,0). The positive solutions describe outgoing RNGs that leave the singularity at (0,0).

We denote by $y_1^+, ...y_n^+$ all the positive solutions of Eq. 3.6.2 with $y_1^+, > y_2^+ > ... > y_n^+ > y_{ms}$. We rewrite Eq. 3.6.1 as:

$$\frac{dy}{dl} = \frac{dR}{dT} - \frac{R}{T} = \pm\sqrt{-\frac{g_{TT}}{g_{RR}}} - y \equiv g^{\pm}(y) \tag{3.6.3}$$

where $l \equiv \ln T$ and the $\pm$ sign correspond to outgoing and ingoing geodesics. The simple outgoing RNGs satisfy: $g^+(y) = 0$ the simple ingoing ones satisfy: $g^-(y) = 0$. Consider now the RNGs in the range $y_2^+ < y < y_1^+$. Outgoing RNGs cannot cross each other (apart from in a singular point). Hence these geodesics do not cross y_1^+ or y_2^+. g^+ has a constant sign since it does not vanish in this range. According to Eq. 3.6.3 an RNG, $y(l)$, in this range will be monotonous functions, with $y_{\pm\infty} \equiv \lim_{l\to\pm\infty} y(l)$. Clearly $y_2^+ \leq y_{\pm\infty} \leq y_1^+$ and $g^+(y_\infty) = 0$. Hence if $g^+ > 0$ then $y_{+\infty} = y_1^+$ and $y_\infty = y_1^+$ otherwise. Each RNG spans the whole range (y_2^+, y_1^+). For $g^+ > 0$ it is tangent to y_2^+ at (T=0, R = 0) (which corresponds to $l = -\infty$) and it is tangent to y_1^+ at $T \to \infty$. The role of y_1^+ and y_2^+ is reversed if $g^+ < 0$.

We can divide the self-similar solutions to solutions containing a naked singularity and solutions that do not contain one. $\lim F = \infty$ for both $y \to y_{ms}$ and

$y \to \infty$ F has a minimal value F_{min} between this two values. When $F_{min} > 1$ there are no simple outgoing RNG and the singularity at (0,0) is not naked. When $F_{min} < 1$ there are two values y_1^+ and y_2^+ for which $F = 1$. In this cases the singularity at (0,0) is naked. The outgoing simple RNGs at y_1^+ and y_2^+ and all the curved RNGs between them reach infinity. y_1^+ is a Cauchy horizon and y_2^+ is the event horizon. All the outgoing geodesics that leave (0,0) in the range $y_{ms} < y < y_2^+$ are initially tangent to y_2^+. These geodesics expand (r increases with T) until they reach the apparent horizon y_{ah}. They recollapse later into the massive singularity at y_{ms} and reach it within a finite T value. All the GRPL solutions with $k < k_l$ are naked. We have calculated numerically the limiting value : $k_l \approx 0.0105$.

III.7 MATCHING TO AN ASYMPTOTICALLY FLAT REGIONS AND THE CUTOFF

It follows directly from the asymptotic expansion of all the self-similar solutions that we have described so far that $\mathcal{M} = m/r$ is constant at $T = 0$. Consequently, the total mass diverges and $\lim_{r\to\infty} g_{rr} \neq 1$ and $\lim_{r\to\infty} g_{tt} \neq -1$. The solutions are not asymptotically flat. To obtain an asymptotically flat space-time we must introduce a cutoff to the self-similar solution and match it to an asymptotically flat solution. We have seen earlier that some of the self-similar solutions contain naked singularities. We can introduce a cutoff in such a way that while the global causal solution is changed, the local solution near the singularity is not influenced by the cutoff. The new solution is asymptotically flat and it contains a naked singularity.

We introduce the cutoff to the initial values (on some space-like hypersurface) in such a way that the density is the self-similar density for $0 \leq r \leq r_i^o$. The density decreases smoothly to zero in the range $r_i^o \leq r \leq r_e^o$ and it vanishes identically for $r > r_e^o$. The total mass with this profile is finite. The cutoff perturbs, however, the self-similar solution. The internal front of the perturbation moves inwards along the worldline $r_i(t)$. The solution is not self-similar for $r > r_i(t)$. The external perturbation front is $r_e(t)$, and the space-time is empty for $r > r_e(t)$. Both $r_e(t)$ and $r_i(t)$ are time-like. The cutoff that we introduce is spherical. Hence there are no gravitational perturbations. The profile that we introduce is less dense than the self-similar one, hence the internal perturbation front propagates inwards as a rarefaction wave. This wave moves inwards at the speed of sound.

An incoming radial curve that moves at the speed of sound,a_s, satisfies:

$$\frac{dR}{dT} = -a_s\sqrt{\frac{-g_{TT}}{g_{RR}}} \tag{3.7.1}$$

This equation resembles Eq. 3.6.1 and the space of solutions is also quite similar. There is one simple solution satisfying $R = Y_{sp}T$ at the sonic point. In addition there are two classes of curved solutions. One interior to Y_{sp} includes curves that reach the origin at $T < 0$. The other exterior to Y_{sp} includes curves that reach the origin at $T > 0$. Clearly, if the initial cutoff point r_i, satisfies $r_i^0 > r_{sp}$ (where r_{sp} is the location of the sonic point on the initial slice) then $r_i(t)$ will reach the origin only after the singularity forms at the origin. In such a case the cutoff will not influence the structure of the singularity and its nearby region.

Common hydrodynamic considerations, suggest that the rarefaction wave moves inwards at the speed of sound. Clearly, the fastest that any perturbation can move is the speed of light. If we choose $r_i^0 > r^- = x^- t$, the ingoing null rays that leave r_i^0 will reach the $r = 0$ at $T > 0$. In this case it is guaranteed that the region near the singularity will not be influenced by the cutoff, even if for some reason the rarefraction wave becomes transonic.

To examine the influence of the cutoff on the global causal structure we must distinguish between space-time with and without a naked singularity. We consider here only the naked singularity case. In his case the overall behavior of the incoming RNGs is not influenced by the cutoff (apart of a change in the details of the curves $r(t)$ in the cutoff region. All the outgoing RNGs that emerge before y_2 reach the cutoff line and continue from these to null infinity as regular radial geodesics in the Schwarzschild metric. y_1 remains the Cauchy horizon of the space-time. In the complete self-similar space-time all the geodesics that emerge after y_2 fall back into the massive singularity. With the cutoff some of those RNGs will escape to null infinity. Hence, with the cutoff, y_2 is no longer the event horizon. The event horizon is the RNG that meets the apparent horizon at the cutoff line. In the vacuum both horizons coincide at $r = 2m_b$.

III.8 CONCLUSIONS

We have shown that there exist a family of general relativistic solutions, describing self-similar spherical collapse of an adiabatic perfect fluid, that include naked singularities and provide a counter example to the Cosmic Censorship hypothesis. These naked singularities resemble the shell focusing naked singularities that are observed in pressureless collapse, in spite of the fact that our matter field has non vanishing and unbound pressure (the pressure is, however, always small relative to the total energy density). Unlike the dust singularities [34,35,28], these singularities are strong (in the sense defined by Tipler [36]). Clearly this solution is not sufficient to abandon the cosmic censorship hypothesis. One can think about a few caveats before doing so. First, the redshift along any null geodesic emerging from the singularity towards an external observer is infinite, hence energy cannot escape from this singularity (unless the singularity has an infinite luminosity [26,28]). Furthermore such naked singularity occurs only for a relatively low (and possibly aphysical) γ. We might find physical reasons to rule out such matter sources. Finally, it is not clear if this causal structure is stable under perturbations and it is possible that these solutions are only of 'measure zero' and might be ignored.

This numerical solution that we have obtained here is very accurate. It employs basic concepts of differential geometry like covering the space-time with different coordinate patches to overcome coordinate singularities. As far as we know such techniques have not been used in numerical relativity before.

It is an open question whether the solutions that we have found here are stable and whether the naked singularity is an artifact of the self-similarity. Both questions could, in principle, be studied using standard methods of numerical relativity by evolving, using the standard ADM formalism, our self similar initial data with or without small perturbations. At least the stability question can be answered in this way.

ACKNOWLEDGEMENT

We would like to thanks D. Christodoulou for helpful conversations. This research was supported by a BSF grant to the Hebrew University.

REFERENCES

[1] May, M. M. and White, R. H. Phys. Rev. **141**, 1232, 1966.
[2] Wilson, J. R. Astrophys. J., **173** , 431, 1972; Wilson, J. R. in *Sources of Gravitational Radiation*, ed. Smarr, L., Cambridge University Press, England, 1979.
[3] Smarr. L. L. Ann. N.Y. Acad. Sci. **302**, 569, 1977
[4] Nakamura, T. and Sato, H. Prog. Theor. Phys. **66**, 2038, 1981; Nakamura, T. Ann. N.Y. Acad. Sci, **442**, 56, 1984; Nakamura, T. Oohara, K. and Kojima, Y. Prog. Theor. Phys. Supp, **90**, 1, 1987
[5] Stark, R. F. and Piran, T., Phys. Rev. Lett., 55, 891, 1985; Stark, R. F. and Piran, T. in *Proceeding of the IV Marcel Grossmmann meeting*, ed. R. Ruffini, North Holland, 1985; Stark, R. F. and Piran, T. Computer Physics Reports, **5**, 221-264, 1987; Stark, R. F. and Piran, T. in *Proceeding of the Yamada conference*, eds. Sato, H. and Nakamura, T. World Scientific, 1986; Piran, T. and Stark, R. F., in *Dynamical Space Time and Numerical Relativity*, ed. Centrella J. Cambridge University Press, 1977; Piran, T. and Stark, R., F., in *The 12th. Texas Symposium*, eds. Livio, M. and Shaviv, G. Annals of the N.Y. Acad. Sci., **470**, 247, 1987.
[6] Hartle, J. N. preprint, 1988.
[7] Hamber, H. W. and Williams, R. W. Nucl. Phys. **B269**, 712, 1986.
[8] Penrose, R., Rivista del Nuovo Cimento. Serie I, 1 Numero Special, 252, 1969.
[9] Eardley, D. M. in *Gravitation and Astrophysics" Cargese 1986* , Carter, B. and Hartle J. Editors, Plenum Press, New York, 1987.
[10] Christodoulou, D., *12th Texas Symposium* Eds. Livio, M. and Shaviv, G., Annals. N. Y. Acad. Sci., **470**, 147, 1987; Christodoulou, D., 1986, Comm. Math. Phys. **105**, 337; Christodoulou, D., 1986, Comm. Math. Phys. **106**, 587; Christodoulou, D., 1987, Comm. Math. Phys. **109**, 591; Christodoulou, D., 1987, Comm. Math. Phys. **109**, 613.
[11] Christodoulou, D., Privet communication, 1984.
[12] Goldwirth, D. S., Msc. Thesis, Hebrew University, Jerusalem, 1986; Goldwirth, D. S., and Piran, T. Phys. Rev. **D36** 3575-3581, 1987
[13] Lambert, L.D., Computational Methods in Ordinary Differential Equations, John Wiley & Sons, New York.
[14] Choptuik, M., Phd. Thesis, University of British Columbia, 1986.; Choptuik, M., this meeting.
[15] Yodzis P., Seifert H. J. and Muller zoom Hagen H., Comm. Math. Phys. **34**, 135, 1973.
[16] Eardley, D. M. and Smarr, L. Phys. Rev. **D19**, 2239, 1979; Eardley, D. M. Gen. Rel. and Gravitation **10**, 1033, 1979; Christodoulou D., Comm. Math. Phys. **93**, 171, 1984.
[17] Hiscock, W. A., Williams, L. G. and Eardley, D. M., Phys. Rev. **D26**, 751, 1982.
[18] Lake K. and Hellaby C., Phys. Rev. D **24**, 3019, 1981 ; Lake K., Phys. Rev. D **26**, 518, 1982; Steinmuller B., King A. R. and Lasota P. J., Phys Lett. **51A**, 191, 1975.
[19] Yodzis P., Seifert H. J. and Muller zoom Hagen H., Common. Math. Phys. **37**, 29, 1974.
[20] Kuroda Y., Pro. Theor. Phys. **72**, 63, 1984.
[21] Ori A. and Piran T ; Phys. Rev. Lett. **59** , 2137-2140, 1987; Ori A. and Piran,

T. Gen. Rel. and Gravitation **20**, 7-13, 1988.
[22] Ori A. and Piran T. preprint, 1988.
[23] Hunter, C. Astrophys. J., **218**, 834, 1977; Hunter, C. Mon. Not. R. Astr. Soc., **223**, 391, 1986.
[24] Whitworth, A. and Summers, D., Mon. Not. R. Astr. Soc., **214**, 1, 1985.
[25] Penston, M. V., Mon. Not. R. Astr. Soc. **144**, 449, 1969.
[26] Larson, R.B., Mon. Not. R. Astr. Soc., **145**, 271, 1969.
[27] Newman, R. P. A. C. Class. and Quantum Grav. **3**, 527, 1986.
[28] Lake, K. Phys. Rev. Lett. **60**, 241, 1988.
[29] Tipler, F. J. Clark, C. J. S. and Ellis, G. F. R. in *General Relativity and Gravitation* ed. Held, A., Plenum Press New York, 1980.

For EU product safety concerns, contact us at Calle de José Abascal, 56–1°,
28003 Madrid, Spain or eugpsr@cambridge.org.

www.ingramcontent.com/pod-product-compliance
Ingram Content Group UK Ltd.
Pitfield, Milton Keynes, MK11 3LW, UK
UKHW050728090726
473066UK00013B/1057

* 9 7 8 0 5 2 1 1 1 5 9 5 7 *